51st Annual Conference of the Environmental Design Research Association 2020

AA005401

Tempe, Arizona, USA
4 – 7 April 2020

Editors:

Chingwen Cheng
Olivier Vallerand
Linda Nubani
Deni Ruggeri
Claudia Bernasconi

ISBN: 978-1-7138-1577-8

Printed from e-media with permission by:

Curran Associates, Inc.
57 Morehouse Lane
Red Hook, NY 12571

Some format issues inherent in the e-media version may also appear in this print version.

The Environmental Research Design Association (EDRA)

2010 Pear St.
#1060 Pinole, CA
94564-9998

www.edra.org

Additional copies of this publication are available from:

Curran Associates, Inc.
57 Morehouse Lane
Red Hook, NY 12571 USA
Phone: 845-758-0400
Fax: 845-758-2633
Email: curran@proceedings.com
Web: www.proceedings.com

Hosted by The Design School at
Arizona State University
Organized by EDRA 51
Conference Committee
in collaboration with
EDRA Board of Directors

TEMPE
ARIZONA
APRIL 4-7
2020

Proceedings of the 51st Annual Conference of the Environmental Design Research Association

Editors:
Chingwen Cheng, Ph.D.
Olivier Vallerand, Ph.D.
Linda Nubani, Ph.D.
Deni Ruggeri, Ph.D.
Claudia Bernasconi

Socially Embedded Collaboration

TRANSFORM

edra51

Environmental Design Research Association

EDRA51 Program Committee
Chingwen Cheng, Ph.D., Chair
Karen Franck, Ph.D., Member
Renae Mantooth, Member

EDRA51 Conference Committee
Chingwen Cheng, Ph. D., Chair
Olivier Vallerand, Ph.D., Co-Chair, Program Coordinator
Milagros Zingoni, Co-Chair, Logistics Coordinator
Mauricio Mejia, Ph.D., Co-Chair, Posters Coordinator
Magnus Feil, Co-Chair, General Coordinator
Brie Smith, Co-Chair, General Coordinator

EDRA51 Virtual Conference Committee
Deni Ruggeri, Ph.D.
Claudia Bernasconi
Renae Mantooth
Olivier Vallerand, Ph.D.
Chingwen Cheng, Ph.D.

EDRA51 Bronze Sponsorship
The Design School, Arizona State University

The Design School
ASU Herberger Institute for Design and the Arts
Arizona State University

EDRA Board of Directors
Linda Nubani, Ph.D., Chair
Deni Ruggeri, Ph.D., Past Chair, Executive Director
Chingwen Cheng, Ph.D., Chair-Elect
Claudia Bernasconi, Secretary
David de la Peña, Ph.D., Treasurer
Robert Nichols, Board Member
Robert Gifford, Ph.D., Board Member
Lesa Lorusso, Ph.D., Board Member
Sharmin Kader, Ph.D., Board Member
Renae Mantooth, Student Representative Board Member
Karen Franck, Ph.D., Emerita Board Member

EDRA Organizational Members
Arizona State University
Michigan State University
North Dakota State University
Pennsylvania State University
Rutgers Center for Green Building
University of California Davis
University of Colorado Boulder
University of Idaho
University of Manitoba
University of Tennessee Knoxville

Corporate Members
DLR Group
HGA Architects and Engineers

EDRA Placemakers
Gowri Betrabet Gulwadi
Keith Diaz Moore
Peter Hourihan
Lynn Paxson
Gary Evans
Peter Hourihan

EDRA Wayfinders
Sherry Ahrentzen
Cherif Amor
David Boeck
Paula Horrigan
Bonnie Sanborn
Dan Stokols

2010 Pear St. #1060 Pinole, CA 94564-9998
www.edra.org

Table of Contents

Introduction

2020 is a peculiar year. At the time I am writing this introduction to the EDRA 51 proceedings, the world is three months into the COVID-19 pandemic global health crisis, one month into lifting the stay-at-home order in Arizona, and two weeks into world-wide protests for Black Lives Matter, an anti-racism movement initiated in the United States. The persistence of systemic racism inherited from slavery and the colonial past is perpetuated in the built environment and many communities around the world today. The recent pandemic crisis and prolonged climate change impacts have revealed the inequitable burdens and exacerbated injustice in minority and vulnerable communities. This is the time to deeply reflect and criticize the current structures and systems in place that are relevant to our work in environmental design research for building just, resilient, and sustainable futures.

This year, the Environmental Design Research Association (EDRA) invited practitioners, researchers, and educators to join the 51st annual conference to collectively rethink how socially embedded, inclusive, and collaborative environmental design may serve as a foundation for transforming society and addressing the major challenges we are facing in the world. Over five hundred and fifty people answered our call for proposals in the early stage. Unfortunately, due to the COVID-19 outbreak, the in-person conference could not take place as planned by the local hosts—the EDRA 51 Conference Committee at The Design School, Arizona State University, in Tempe, Arizona, joined by other ASU faculty, student, and local practitioner volunteers. The participants would have enjoyed the warm sunshine in beautiful spring days of Arizona and various mobile sessions and activities we planned for you: the Taliesin West Frank Lloyd Wright UNESCO world heritage site, Paolo Soleri's arcology laboratory of Arcosanti, the original space research in desert environment of Biosphere 2, the many wonders of the Sonoran Desert region at Tucson's Desert Museum, Scottsdale's McDowell Sonoran Preserves, Phoenix's Sonoran and South Mountain Parks and Preserves, the river social-ecological systems along the Rio Salado, the cultural heritage of historical and midcentury Phoenix's architectural sites, the sustainability practices and campus design at Arizona State University, the biomimicry research and practice, and learning about soundscape through a unique sound walk experience. We would have welcomed you and congratulated the EDRA, CORE, and GREAT PLACES award winners at an outdoor plaza luncheon at ASU The Design School with our local delights, and celebrated with old and new friends at the evening banquet in the world renowned Phoenix Botanical Garden accompanied with unique desert landscapes, drinks, food, music, and dance under the stars. Arizona missed you. I sincerely hope you will have a chance to come visit in the future.

When COVID-19 hit the United States in March, many conferences faced the most strenuous situation of deciding whether to cancel, postpone, or shift to a virtual platform and how to handle registrations. The EDRA51 Conference Committee worked closely with the EDRA Board and EDRA Executive Director. Each decision we made together was difficult and taken seriously. With the tremendous help from a newly formed EDRA51 Virtual Committee less than a month before the start of the conference, and with dozens of volunteers who helped to host and manage each session, I am grateful that the EDRA 51 was able to take place virtually with over 350 participants coming from over 14 countries through 270 individual and group presentations and 90 visual presentations, and three plenary sessions—an opening keynote, a closing keynote, and a panel discussion. We addressed how socially embedded collaboration with communities transform environmental design research at all scales to move communities beyond status-quo toward more sustainable and resilient models with adaptive and transformative capacities. In lieu of mobile sessions, EDRA51 responded to the emerging challenges of pandemic urbanism through an impromptu virtual workshop. In addition, EDRA51 hosted a virtual happy hour, serving as a nectar to quench the thirst of connecting with EDRAers in the time when shelter-in-place and social/physical distancing have become a new norm for many. Thanks to everyone who participated and made our first virtual EDRA conference a success.

Thanks to all reviewers who helped in the peer-review process for the conference papers and abstracts, and thanks to the editorial board for overseeing the quality of this document. The EDRA51 Proceedings is a collection presenting the multiple facets of how environmental design research collaborates with academia, not-for-profit organizations, public sectors, and private sectors to lead the charge in transforming people's relationships with each other, their communities, and their environments toward net positive futures. The EDRA51 Proceedings is organized by presentation types—referred full papers, abstracts-individual presentations, abstracts-group presesntations, abstracts-pre-conference intensives, abstracts-video shorts, abstracts-posters—in addition to the following conference tracks:

A - Transform Design for Resilience & Sustainability
B - Transform Education for Our Common Future
C - Transform in Transitions
D - Transform Places with People and the Environment
E - Transform Processes and Outcomes for Equity

Today, our world is more connected and more complex with persistent and prolonged issues, as well as new challenges that exist in every community, including but not limited to: systemic racism, social and environmental justice, climate change and climate justice, environmental quality, biodiversity, health, technology, communication, and education. Our work must continue to reveal and address those challenges, produce new knowledge in environmental research, develop innovative and integrated environmental design solutions and practices, and contribute to policy-making for actionable change. EDRA is fully committed to this work with you.

We look forward to seeing you in EDRA 52!

Sincerely,
Chingwen Cheng, PhD, PLA, LEED AP
EDRA Chair-Elect & Program Committee Chair
EDRA 51 Conference Committee Chair

Special Thanks to EDRA 51 Volunteers:

Rebecca Bachman
Taraneh Darvish
Hope Freije
Marzieh Rezaei Ghaleh
Alysha Helmrich
Ana-Karinna Hidalgo
Md Islam
Smriti Jain
Mariana Junqueira
Karen Kim
Rosa McDonald
Rhiannon Neuville
Peter Price
David Richardson
Nikolas Smilovsky
Uchita Vaid

EDRA 51 Logo Designer:
Gabrielle Conrad-Amlicke

Cover Designer:
Marzieh Rezaei Ghaleh

edra 51

Tempe, Arizona

Keynote Address & Plenary Sessions

Keynote Sessions

The Hip Hop Architecture

Michael Ford

Sunday, April 5, 6:00pm-7:30pm

During this keynote, Michael Ford discussed how he created The Hip Hop Architecture Camp, an international intuitive using hip hop culture as a catalyst to introduce youth to architecture, design, and urban planning. Ford shared details of projects which consulted with The Hip Hop Architecture Camp as a tool for community engagement, including The Universal Hip Hop Museum.

Michael Ford is known as The Hip Hop Architect as he explores architecture and urban planning through the lens of hip hop culture. Ford, is the creator of The Hip Hop Architecture Camp®, an international, award winning youth camp which uses Hip Hop Culture as a catalyst to help youth understand, critique, and generate architecture. Ford's work has been featured in a variety of places including The Oprah Winfrey Network, The TODAY Show, Rolling Stone Magazine, ESPN's The Undefeated, and Vibe Magazine. Currently, Ford is leading the design of The Universal Hip Hop Museum in The Bronx.

Connection Versus Correction

Pono Shim

Tuesday, April 7, 4:45pm-6:00pm

Connect first and connection isn't negotiations or reasoning! Our Political, Economic, and Educational systems have primarily focused on correction when solving problems and entering communities. This system's heritage is based on Colonization and the effects are the same today as they were centuries ago, traumatizing. In exercising our correction muscles our connection muscles have atrophied. How can we connect deeply?

Pono Shim is a storyteller and he sees life unfurl in story and then shares stories to connect. Pono Shim, Pono Shim has earned a local and national reputation for being an influential thought leader as well as a respected executive of a multi-million dollar nonprofit organization. Pono has a deep understanding of cultural, economic and historical issues facing Native Hawaiians and other under-served and marginalized communities and his work includes advancing cultural knowledge as a means of solving the problems of today. Pragmatic yet visionary, his expertise in storytelling and facilitating difficult conversations led him to establish the Higher Skills Academy that shifts traditional thinking and thought processes to help re-frame challenges and more effectively address problems.

Opening Ceremony with Land Recognition & Special Remarks

Jacob Moore

Sunday, April 5, 8:00am-8:30am

Jacob Moore is responsible for the intergovernmental affairs between ASU and tribal nations and communities. Previously, Jacob Moore was managing partner for Generation Seven Strategic Partners, LLC, and also worked as an Economic Development Analyst and Special Assistant on Congressional and Legislative for the Salt River Pima-Maricopa Indian Community.

Jacob previously served as a member of the Arizona State Board of Education. Mr. Moore is currently on the board of directors for the Arizona Community Foundation, the ASU Morrison Institute, WestEd and the Arizona Minority Education Policy Analysis Center (AMEPAC). Mr. Moore earned a Bachelor of Science degree in Finance and an Executive MBA from the Arizona State University's W.P. Carey College of Business and is a citizen of the Tohono O'odham Nation.

Plenary Session

Post-Colonial Decolonization Design with Communities: Transform our design thinking, research, and practice moving forward

Organizer: Chingwen Cheng
Moderator: Mark Roseland
Panelists: Wanda Dalla Costa, Irasema Coronado, Gyan Nyaupane

Tuesday, April 7, 10:30am-12:00pm

Redistribution of power dynamic in planning process for achieving full citizen participation has been considered the ultimate goal for community design in theory. Nevertheless, the current practice of community design methods are often limited to preconception of the products to be delivered to the public. Moreover, those products are mostly derived from the mainstream of the worldview that is now globalized. In addition, the participatory method predominately taught and practiced in the mainstream of planning and design education and professional practices can be considered as another form of colonization as the processes of colonialism and imperialism. This panel aims to challenge current education and practice engaging with indigenous and minority communities in Arizona, the region, and the world and to explore new insights in ways for genuine, meaningful, and effective community engagement in the design process. As scientists and designers seek to design products and implement solutions to achieve sustainable development goals globally, this panel is particularly important for understanding the underlying ethics and values for the critical mindsets of decolonizing design in the participatory process that retains, rediscovers, and reinstalls the power to communities.

Chingwen Cheng is Assistant Professor of Landscape Architecture at The Design School and Senior Sustainability Scientist at Arizona State University. Dr. Cheng is dedicated to engaging transdisciplinary research for the advancement of green infrastructure planning and design in building resilient communities. Her work on Climate Justicescape investigated spatial patterns of vulnerable communities exposed to climate change associated hazards with a lack of adaptive capacity in the distributive justice of green infrastructure planning and design outcomes. Her recent inquiry investigates procedural justice in the community design process in building resilience capacity for climate change actions. Dr. Cheng was a visiting scholar at the Risk Society and Policy Research Center and Global Change Research Center of National Taiwan University. She is a registered Professional Landscape Architect and LEED Accredited Professional with extensive practices in watershed planning, stormwater management, Low-Impact Development, community engagement, and urban design in the US.

Mark Roseland

is professor and past director of Sustainable Community Resources and Development and Senior Sustainability Scientist at Arizona State University. Before coming to ASU, he was at Simon Fraser University (SFU) in Vancouver, Canada, where he was professor of planning in the School of Resource and Environmental Management and director of the Centre for Sustainable Development. He is a Registered Professional Planner and full member of the Canadian Institute of Planners, and he has worked as Chief City Planner for a municipality in the Metro Vancouver area. He has been cited by The Vancouver Sun as "one of Vancouver's top 50 living public intellectuals" and has received both the SFU Sustainability Network Award for Excellence in Research on Sustainability and the SFU President's Award on Leadership in Sustainability. His best-selling book "Toward Sustainable Communities: Solutions for Citizens and Their Governments" is in its 4th edition. He was a founding director of Catalyst Community Developments, which leverages real estate assets to create vibrant, affordable and inspiring places for people to live and work in. He is the founder of Pando|Sustainable Communities, a multilingual online network to promote collaboration between sustainable communities researchers and practitioners.

Wanda Dalla Costa,

AIA, LEED A.P. is a member of the Saddle Lake Cree Nation. She is a practicing architect, a professor and a YBCA 100 2019 honoree. At Arizona State University, she is the director and founder of the Indigenous Design Collaborative, a community-driven design and construction program, which brings together tribal community members, industry and a multidisciplinary team of ASU students and faculty to co-design and co-develop solutions for tribal communities. Dalla Costa holds a Master of Design Research in City Design from SCI-Arc, and a Master of Architecture from the U of C.

Transforming with Indigenous Design

Dalla Costa will highlight two aspects of her work. First, Indigenous Futurities, a term coined by Dr. Grace Dillon, which challenges assumptions consigning Native American peoples and lifeways to the past. It proposes using creative thinking as a pathway toward Indigenous revival, using past-future visions, or self-determined representations and alternative narratives about our identities and futures to reimagine our futures. Second, the Indigenous Placekeeping Framework, created by Dalla Costa in 2015, to work with tribal partners on applied work at the Indigenous Design Collaborative at ASU.

Irasema Coronado

is professor and director of the School of Transborder Studies. Her research focuses on the politics of the U.S. Mexico border region; including environmental policy and binational cooperation. She has conducted participatory action-based research in colonias in the border region focusing on access to water. She served on several EPA citizen advisory boards: The Good Neighbor Environmental Board, National Environmental Justice Advisory Council and the National Advisory Committee Council for Environmental Policy and Technology and the National Advisory Environmental Health Sciences Council of NIEHS. She was the executive director of the Commission of Environmental Cooperation of North America.

Transforming the borders

Coronado focuses on the importance of engaging communities in a respectful and appropriate manner when conducting research in diverse communities. She highlights the importance of bidirectional communication when developing research projects and shares some lessons learned from her experience working in colonias populares in Nogales, Sonora, colonias in the border region of Texas and New Mexico. Her involvement in the environmental justice movement in the United States is showcased in a case study collaborated with Familias Unidas Del Chamizal in El Paso, Texas. Coronado aims to challenge researchers and practitioners to be proactive and preemptive rather than reactive to address challenges and opportunities in working with community under the framework of decolonizing design process.

Gyan Nyaupane is professor of School of Community Resources and Development at Arizona State University. He has expertise in natural resource management, community development, and sustainable tourism. His research primarily focuses on understanding the complex relationships between humans and the natural environment. His research has been instrumental to envisioning and planning policies for federal, state, tribal, and local land management agencies in Asia, Africa, and the USA. Recently, he led a research project with Indigenous communities in Arizona, Nepal, and Australia.

Transforming Relationships with the Natural Environment

Indigenous communities across the globe have traditionally maintained deep relationships with the natural environments for survival since the beginning of human civilization. Colonials established a new form of utilitarian relationship with the natural environment at the expense of Indigenous people's moral relationship and livelihood. When Indigenous peoples' connection with the environment was weakened or being eliminated, they lost their identities, culture, and livelihoods. Indigenous knowledge developed over generations for managing lands and natural resources for their livelihoods has also been lost or being impacted by the hegemonic Western knowledge. Although there have been increasing calls on the urgency of engaging with the indigenous knowledge systems to reconnect them and resolve environmental challenges and improve the livelihoods of Indigenous communities, this cannot be achieved without bringing indigenous communities and their knowledge systems into planning and policies. Drawing from the recent studies conducted with the indigenous communities in Arizona, Australia, and Nepal, Dr. Nyaupane highlights the challenges as well as the initiatives taken by the communities to overcome them.

edraid

Tempe, Arizona

Awards & Recognition

EDRA Awards and Recognition Committee

Sharmin Kader, PhD, Committee Chair

Robert Nichols, EDRA Career & Achievement Awards Chair

Sue Weidemann, PhD, Michael Brill Award Chair

Amin Mojtahedi, Certificate of Research Excellence (CORE) Chair

Claudia Bernasconi, Great Places Awards Chair

Carla Brisotto, PhD, Great Places Awards Co-chair

Tadd Heidgerken, Great Places Awards

Rosa MacDonald, student volunteer, Great Places Awards

2020 Michael Brill Grant Recipient

The Michael Brill Award for Urban Communication, Design, and Human Behavior seeks to encourage research projects that connect design and communication theories to human responses in the designed environment. The intent of the Brill Award is to help support innovative work addressing human needs at the intersection of design, human behavior, and communication. The environments under consideration in this research may range in scale from specific interior environments to large-scale urban environments and may also include virtual environments.

Kyoungmee Kate Byun

Assistant Professor, Northern Arizona University

Exploration of human perception towards physical settings of an exhibition as a narrative environment: A study using Virtual Reality (VR)

The purpose of this study is to understand the importance of narrative environments for exhibition design and examine how physical factors of narrative environments affect human perception to facilitate communication. Spatial narrative theory has developed in the last 25 years to underpin the analysis and the design of narrative environments through application in landscape and narratives (Potteiger, 1998), architectural and urban design narratives (Psarra, 2009; Ryan, 2016), and psychological spatial theories (Merleau-Ponty, 2005), and construction space (Lefebvre, 1991). Based on four narrative theories, distinct factors of narrative environments; form, scale, and light, will be the variables for the study. 3D virtual gallery space is intentionally designed to examine difference physical factors (e.g., form, scale, and light). A defined group of 30 participants (total 6 groups) will be exposed to 3D virtual gallery space to explore their perceptions on the effectiveness of interactions between physical settings and narratives. All participants will complete a Likert scale survey and have semi-structed interviews for data collection. The content of the exhibition will be narratives by qualitative photovoice analysis from a project titled, "Understanding HIV Risk and Resilience among Adolescents who have been Orphaned by HIV/AIDS in Hai Phong, Vietnam Using photovoice and Social Network Analysis".

EDRA 2020 Career Award: Ulrich Schramm, Ph.D.

Ulrich Schramm was born in Stuttgart, Germany. He has been a Professor in the Department of Facility Management since 2002. He is a trained architect and experienced facility programmer. He received his doctorate in Architecture from the University of Stuttgart and a post-doctoral fellowship from the German Research Foundation (Deutsch Forschungsgemeinschaft, DFG) for his stay at the University of Cincinnati as a Visiting Professor of Architecture. Results of his studies within the International Building Performance Evaluation (IBPE) consortium and the campus-based research project Intelligent Building Technologies (InteG-F) have been presented at conferences of the Enviornmental Design Research Association (EDRA) since 1995 and published in several articles and book chapters. Some recent chapters he has written were published in Enhancing Building Performance (Wiley-Blackwell, 2012) and Architecture beyond criticism: Expert Judgement and Performance Evaluation (Routledge, 2015). He also co-authored the first book on Facility Programming in Germany Nutzerorientierte Bedarfsplanung: Prozessqualität für nachhaltige Gebäude (Springer, 2011). Currently he is serving at the German Institute for Standardization (Deutsches Institut fur Normung, DIN) in Berlin as expert and co-chair of the revised edition of the German building standard DIN 18205 Brief for Building Design.

EDRA 2020 Career Award: Ann Sloan Devlin, Ph.D.

Over the course of four decades in academia, Ann Sloan Devlin has acquired a mastery in the area of environmental psychology, specifically related to research in wayfinding (spatial cognition) and evidence-based health care design, as well as in research methods. She teaches at Connecticut College in New London, Connecticut, where she is the May Buckley Sadowski '19 Professor of Psychology.

She has published extensively on healthcare environments, in particular the role of the facility's appearance in people's judgments of expected care. She has published 6 books and over 40 articles in peer-reviewed journals as well as numerous book chapters and conference proceedings.

From 2016-2018, she was editor-in-chief of the journal Environment and Behavior and is past edra board member and secretary. At Connecticut College, she has received three separate faculty awards: the John S. King Faculty Teaching Award, the Helen Brooks Regan Faculty Leadership award, and the Nancy Batson Nisbet Rash Faculty Award for Excellence in Research. She is a Fellow of Division 34 of the American Psychological Association.

2020 Certificate of Research Excellence (CORE) Recipients & Jurors

The Environmental Design Research Association (EDRA) is pleased to announce the recipients of the Certificate of Research Excellence (CORE) for 2020. CORE recognizes rigorous, valuable, and impactful practice-based research that sparks innovation and promotes best practice in environmental design. The credibility of academic research is typically judged on the basis of well-established protocols and peer-review processes – but what constitutes research quality in the context of the practice of environmental design? The tools used to assess academic research do not easily translate to practice-based research, a challenge for researchers and practitioners alike. In response, EDRA developed CORE to serve as a measuring stick to assess the value of research that goes beyond rigor alone. EDRA CORE's unique evaluation framework identifies practice-based environmental design research that is not only rigorous, but also offers measurable value, meets industry challenges, and advances design thinking; recognizes the importance of research in practice; provides a compass to guide designers, researchers, organizations, and manufacturers; and builds on EDRA's tradition of inquiry, reflection, and collaboration and its commitment to innovation. The following practice-based design research projects met the CORE criteria and were selected for recognition, based on the review of a cross-disciplinary panel of reviewers consisting of academic researchers, practice-based researchers, designers, and industry thought leaders. The CORE recipients have been recognized during EDRA51, April 4-7, 2020

2020 EDRA CORE Reviewers

BRIAN K SCHERMER, PhD, AIA
Associate Professor, University of Wisconsin-Milwaukee; Principal, Workshop Architects

DIANA NICHOLAS, NCIDQ, AIA, LEED GA
Assistant Professor and Director MS Design Research, Drexel University

EMILY CHMIELEWSKI, EDAC
Senior Design Researcher, Perkins Eastman

FIONA DE VOS, PhD
Founder and Owner, Studio dVO

GIYOUNG PARK, PhD, AIA, LEED AP BD+C, WELL AP
Senior Design Researcher, Sam's Club Technology

GOWRI BETRABET GULWADI, PhD, LEED AP
Professor, Interior Design, University of Northern Iowa

HAROON SATTAR, NCIDQ, LEED AP
Associate Professor University of Central Missouri

KIMARI PHILLIPS, MA, EDAC
Senior Research Analyst, LPA, Inc.

NAM-KYU PARK
Associate Professor, University of Florida

RENAE RICH
Research Analyst, HDR

SHARMIN KADER, PhD
Design Research Associate MTS

SHEILA F. CAHNMAN, FAIA, FACHA, LEED AP
President, JumpGarden Consulting LLC

SHIREEN KANAKRI, PhD
Assistant Professor, Ball State University

TERRI ZBOROWSKY, PhD
Design Researcher, HGA Architects

ZHE WANG, PhD
Henan University, Center for Senior Healthcare

Resilience through Regeneration: The Economics of Repurposing Vacant Land with Green Infrastructure

MERIT CERTIFICATION

Texas A&M University, University of Delaware, Charity Productions

Study Team: Galen Newman, Rui Zhu, Dingding Ren, Dongying Li, Jennifer Horney, Katie Kirsch, Garett Sansom, Charles X White

Nearly 17% of each large U.S. cities' land area is vacant; this has increased by 1.3% since 1998. When floods strikes, vacancies catapult. The Texas coast is one of themost frequently flood impacted areas globally. Houston, TX has one of the highest numbers of flood-related fatalities in the past 50 years. From 2001 to 2017, Harris County incurred 20 billion dollars in flood-related damages. Houston reports nearly 11% of its land as vacant, many concentrated in marginalized communities, which are highly vulnerable to flooding. Many underserved urban areas affected by flood disasters are also becoming increasingly ecologically and socially fragmented due to the accumulation of vacant properties. These unused lots can potentially provide land for ecological/hydrological land uses. Despite overall population in-migrations in flood prone regions, many marginalized neighborhoods are characterized by excessive amounts of vacancies. Rather than chasing development-based incentives for regenerating vacant lots in these areas, a balance should be sought between new developmental land uses and green infrastructure (GI) to help counteract stormwater runoff and flood effects, or resilience through regeneration. This research examines the costs and benefits of retrofitting green infrastructure into vacant lands in underserved, flood prone communities using landscape performance models across three master plans in Houston, Texas. The research asks, what are the economic costs and benefits of retrofitting GI into underserved communities as a strategy for vacant land regeneration? It uses landscape performance measures across three master plans for lower-income, minority dominant, flood-prone neighborhoods in Houston, Texas, USA to evaluate the economic and hydrologic impacts of GI regeneration projects. Results suggest that, when using this approach, 1) flood risk significantly decreases, 2) short term, upfront economic costs increase, and 3) the long-term economic return on investment is much higher.

Integrating a Resilience Scorecard and Landscape Performance into a Geodesign Process

MERIT CERTIFICATION

Texas A&M University, University of Delaware

Study Team: Galen Newman, Zixu Qiao, Matthew Malecha, Siyu Yu, Susan Puumala, Jennifer Horney, Jaekyung Lee, Youjung Kim, RyunJung Lee, Phil Berke

Geodesign has become more integrated in a number of applications (Wilson, 2014). For example, Geodesign can be applied as a mechanism to inventory, analyze and project a future state of affairs for geographic space (Goodchild, 2012). The framework for a Geodesign process (Stenitz, 2012) specifies six key models to be produced, including representation, process, evaluation, change, impact and decision. Simultaneously, as part of an initiative proposed by the Landscape Architecture Foundation, academic institutions have been collaborating to quantitatively assess the environmental, economic and social benefits of urban design projects (Yang and Binger, 2016). This effort, known as landscape performance, encourages evidence-based designs that are grounded in quantitative performance measures. Unfortunately, many of these performance measures have not yet been fully incorporated into the Geodesign process. There are other analytical-planning methods, such as the resilience scorecard (Berke et al., 2015), which use quantitative performance measures to reduce losses from hazard events through conditional analyses and policy review, rather than projections or postimplementation evaluation metrics. However, while these types of analyses can provide a sound foundation for evaluation models, they are still quite separate from Geodesign approaches.

In this research, the approach allows the integration of the concepts of flood resilience and community design with the fields of landscape architecture, regional planning, land use management, and hydrology in a community of approximately 100,000 residents located on the Texas Gulf Coast. The effects of climate change, such as sea level rise, have had observable ecological, social, and economic impacts on the built environment in this region. Sea level rise has already had a significant impact on Gulf Coast communities, resulting in wetland loss, increased coastal erosion/inundation, and increases in the duration and frequency of flooding from storm surge (Horton et al., 2014). The National Oceanic and Atmospheric Administration (NOAA, 2015) predicts that (in a med-high scenario) the mean sea level will rise at least 0.82 inches per year in the U.S. Gulf Coast, reaching 6.29 feet by 2100. In 2008, Hurricane Ike caused extensive damage to the Texas Gulf Coast, causing 113 deaths and $29.5 billion in damage; approximately 200 of the damaged homes were located in League City, TX (Rego & Li, 2010).

League City, due to its location on the Texas Gulf Coast, is highly vulnerable to flood events and other issues related to sea level rise. Working in partnership with city officials, the authors developed and executed a Geodesign process which integrated the resilience scorecard (Berke et al., 2015) as the evaluation model, a vertical buffer tool (as a process model) to project sea level rise, and landscape performance (as an impact model). The project included an assessment of flood vulnerability and projection of the 100 year floodplain in the year 2100, accounting for increases due to sea level rise. This information is used as a guide for the location of future development, as well as to inform the development of a master plan for a resilient community within a 97-acre site surrounded by urban development in League City. In light of the community's vulnerabilities and the sea level rise projections revealed through the Geodesign process, a series of adaptive flood attenuation mechanisms for protecting the newly designed community from flood events and the eventual impacts of sea level rise are suggested. Finally, landscape performance projections are conducted to measure potential impacts of the proposed master plan.

Developing a Social Media Data Based Post-Occupancy Evaluation Method: New Understandings of Social and Physiological Attributes of Designed Waterscapes

Oklahoma State University, North Dakota State University, Shaoxing University

Study Team: Bo Zhang, Yang Song, Yiwen Zheng

Social media supplies of huge, ecient, and ever-growing data which can be taken advantage of by researchers to better understand landscape design. This project uses Instagram, by far the most popular free online service, as the major data source to explore the usage and perception of Fort worth water gardens.

Evaluation of Ambulatory Clinic Designs: Core Modules with Shared Vs. Patient-Only Corridors

HDR, Advocate Aurora Health

Study Team: Jeri Brittin, Mike Doiel, Peter Messina

The goal of this evaluation was to understand how recently-implemented ambulatory clinic module designs perform from the perspective of providers and staff. Many clinics included in this evaluation were based on a new core module concept with central team space to allow all providers and staff the ability to work in close proximity together on patient care.

A major difference between clinic designs was use of a separate patient corridor so that patients would not come into the team space, versus a shared corridor with patients entering exam rooms from the central core area. 11 clinics were included in the study and data were collected from site visits, interviews, and surveys.

Supporting Staff and Student Outcomes through Design of a School for Special Needs Students: Spero Academy

HDR, Spero Academy

Study Team: Jeri Brittin, Francesqca Jimenez, Brian Giebink, Curtis (Chipp) Windham

As part of a larger study, HDR conducted a staff survey at Spero Academy to measure outcomes associated with a new school design. In order to understand the differences between the old and new environments, the survey was administered at two points in time, once during occupancy of the previous school facility (pre) and once again eight months later following the transition to the new school (post).

The survey focused on the functionality of the educational spaces for current user needs and a variety of important staff outcomes, including collaboration and engagement.

Quantitative data were obtained through standardized survey questions and qualitative data were obtained through open-ended questions.

Learning from the Past: Two Infusion Centers Inform a New Design

<u>MERIT CERTIFICATION</u>

HGA, University of Wisconsin-Milwaukee, St. Elizabeth Healthcare

Study Team: Kara Freihoefer, Nancy Doyle, Mahshid Jalalianhosseini, Amanda Simpson, Lou Ann Bunker-Hellmich

Each year, approximately 650,000 cancer patients receive chemotherapy in outpatient oncology clinics (i.e., infusion centers) throughout the United States (CDC, 2018). Infusion centers have become important settings in which patients receive life-saving cancer treatment.

Literature on outpatient infusion settings is scarce (Kamimura, et al., 2012) and few studies have considered both staff and patients' points of view. Considering there is limited studies on these settings, this research serves as baseline benchmarking data for comparison with other studies on cancer infusion centers, in addition to informing future infusion center designs.

This study compared staff workflow, communication and collaboration, and staff and patient satisfaction with multiple aspects of two cancer infusion centers—an open bay design and a semi-private bay design—both located on a regional medical center campus in the eastern United States. An additional goalof the study was to gather baseline data for comparison in a future post-occupancy evaluation of a new infusion center that will replace the existing infusion centers studied.

EDRA Great Places Awards 2020

2020 GREAT PLACES AWARDS

The EDRA Great Places Awards are unique among programs that honor professional and scholarly excellence in environmental design. They seek to recognize work that combines expertise in design, research, and practice, and contributes to the creation of dynamic, humane places that engage our attention and imagination. Award-winning projects reflect an interdisciplinary approach that is enduring, human-centered, sustainable, and concerned with the experiential relationship between people and their environment (built and natural) over time. We invite participation from a range of design and research disciplines — particularly projects whose significance extends beyond any one profession or field. All submissions should show how research and/or public participation is linked to--or part of--an environmental design practice, and vice versa. Submissions should also demonstrate how an understanding of the experience of place may be used to generate insightful design.

The Great Places Awards Jurors

Antje Steinmuller
Place Design Juror

Alpa Nawre, PLA, LEED AP
Place Planning Juror

Kevin Conger, ASLA, LEED AP
Place Design Juror

Russell Baltimore, AIA
Place Planning Juror

Keith Diaz Moore, Ph.D, AIA
Place Research Juror

Sanjoy Mazumdar, Ph.D.
Place Research Juror

Stephen Vogel, FAIA
Book Award Juror

Federica Goffi, Ph.D.
Book Award Juror

The Great Places Awards Winners

2020 Great Places Award Winner - Place Design: "City Thread"

Molly Hunker and Greg Corso

The jury selected this project as a memorable contribution to the mission of the Environmental Design Research Organization. The project excels in supporting a wide range of activities with a simple, flexible, and recognizable structure, resulting in an infrastructure for urban life in an alleyway. Produced with stakeholder input, the installation unifies different voices by enabling a broad range of possible programs without prescribing any. Jurors appreciated the strong, iconic design language, and congratulate the design team on achieving a project that presents a skillful balance of responding to specific current needs and remaining open to future change.

2020 Great Places Award Honorable Mention - Place Design: "Gateway to Growing: Maring-Hunt Community Built Market Pavilion, Outdoor Kitchen, and Nature Play Pockets"

Pamela Harwood

The jurors commend this project's exemplary process of engagement with neighbors and students that led to a lively destination for daily neighborhood use and seasonal festivals alike. This careful work with neighborhood needs is complemented by a rigorous design and construction process.

2020 Great Places Award Winner - Place Planning: "The Integrated Development of Hatirjheel Area"

Ishtiaque Zahir, Ehsan Khan, and Iqbal Habib

This is truly a wonderful water restoration and public realm reclamation project in the context of Dhaka, the capital of Bangladesh. The masterplan is implemented and represents an outstanding achievement in the way it sets a precedent for multifunctional public realm development. The interventions are well designed and will have a lasting positive impact on this city. Initially conceived as a wetland restoration project, it eventually became a comprehensive urban renewal project that promotes east-west vehicular and pedestrian connectivity, reduces vehicular traffic congestion, reintroduces waterways as a means of transportation, retains water during monsoon, manages waste water and drainage, protects the wetland from further encroachment, and provides much needed recreational and civic facilities for citizens.

2020 Great Places Award Honorable Mention - Place Planning: "Brownie Neighborhood Park"

Kristine M Stiphany, Jason Sowell, and Nathan Brigmon

Clearly articulated goals of providing safe, public greens by re-using urban fragments are achieved in this proposal for a low-income neighborhood in Austin, Texas. The design and graphics are concise and well developed. This proposal represents a very attainable development with a clear impact on the residents of this area. Conceived during a series of public engagement workshops that encouraged residents and users to participate in the design of the park, this design provides a vision for the future development of this park.

2020 Great Places Award Winner - Place Research: "Double Duty Parks: Two Studies on the Performance of Social Spaces as Resilient Flood Infrastructure"

Anya Domlesky

The timeliness and intersectionality of the need to reduce the impacts of climate change together with social responsiveness make "Double Duty Parks" standout. The two case studies presented approaches to making flood-resilient infrastructure into attractive, accessible public parks/spaces. The research provided a good understanding of both the projected performance of the design to reduce environmental damage due to disasters, and what design features led to these projects becoming utilized, social and recreational spaces. The case study information made available at landscapeperformance.org is quite exceptional and provides an exemplar for how to present practice-based research. Moreover, the projects resulted in increasing public health and well-being along with an educational and interpretive component of enabling local citizens to learn about the dangers and protections. Overall, these excellent projects lead the way in showing how multiple complex goals can be incorporated in the creation of highly successful environments.

2020 Great Places Award Honorable Mention - Place Research: "New Parkland Hospital Evaluation Research "

Jeri Brittin

This is very strong applied research, presented in a compelling fashion. The project began with the idea that the physical environment of hospitals can contribute to patient outcomes, patient and staff experience, and quality of care in health care settings and that Post Occupancy Evaluation (POE) research can help in this. The hypotheses explored these using multiple data sources including archival data (patient & safety records), surveys, and focus groups. The research raised questions about the universal efficacy of decentralized nursing. Although the research results were not all unequivocal, the new hospital facilities resulted in increased patient satisfaction, seemed to be liked by the staff, and from the pictures appear spacious, welcoming and color coordinated. The study has been well disseminated and its impact well documented. The project is a great example of how evidence from POE research can lead to good design.

2020 Great Places Award Winner - Book: "Reciprocal Landscapes. Stories of Material Movements"

Jane Hutton

Inspired by the work of Robert Smithson 'non-site', this memorable and timely book stands out as an novel perspective on 'everyday landscape construction materials' in the context of well-known public landscapes in the city of New York– revealing otherwise invisible stories of the sites from which the materials originate. The book is articulated through a clear structure, and a chapter is devoted to each material with the stories of the original sites -considering social, political and ecological entanglements of material practices. The book is well written, accessible, and a must read.

2020 Great Places Award Honorable Mention - Book: "Streets Reconsidered: Inclusive Design for the Public Realm"

Daniel Iacofano and Mukul Malhotra

This is a compelling book that calls for a fundamental rethinking of America's streets by designing them for living in addition to driving. Case studies from around the world illustrate significant strategies such as accommodating people of all ages and abilities, varied travel modes, social interaction and economically viable business transactions. The design aspects are illustrated through wonderful photographs and beautiful drawings and renderings. Easily read and clearly explained and illustrated, the book should be in the library of every municipal planning department, urban design firm, landscape architecture offices and community organizations promoting healthy living and great human spaces.

edra51

Tempe, Arizona

Refereed Full Papers

A6 - Transform Design for Resilience & Sustainability: Planning, Design & Behavior VI

Methodology for Measuring Arable Land in Urban and Suburban Morphology

Arash Alborzi (University of Florida); Martin Gold (University of Florida)

1. ABSTRACT

This study proposes a method by which municipalities can measure arable land in urban and suburban morphology. The proposed method quantifies and qualifies potential arable areas within different land uses and urban morphology. Because of the academic, geographic, and agricultural conditions of North Central Florida, Gainesville, Florida has been selected as the pilot case study context to propose, test, and evaluate a method of measuring arable land in urban/suburban contexts. Immanent threats such as sea level rise and population growth are two acute factors that combined will put more people on a shrinking land mass during the 21st century in Florida. Recently, progressive municipalities have implemented urban agriculture as a secondary land use within existing urban uses in order to launch a viable, nutritious, resilient, and affordable food system in support of sustaining the health and viability of communities. The proposed methodology includes the following five steps:

1) Identify varied morphology throughout Gainesville based on land use and density.

2) Select five-acre representative samples from each morphology.

3) Analyze the cultivable area within each sample utilizing Google Map, Google Earth, and AutoCAD software.

4) Qualify and quantify the areas of cultivability and ownership status – privately, publicly, or institutionally owned.

5) Propose methods of extrapolating the data to calculate arability in the community

The result of this study estimate land inventories by which municipalities could make strategic decisions toward implementing urban agriculture as a new land use within the existing land use policies that contribute to the development of urban form.

1.1. Keywords

Urban Agriculture, Arable, Suburban, Vegetable Garden.

2. INTRODUCTION

The 20th Century saw the collapse of the family farm, the rise of big agriculture, and the corporate commoditization of food production and distribution. Although the corporate model provided a robust food industry that delivers a stable variety of foods at affordable prices throughout the US, the USDA reports that nutritional quality has decreased and the consumption of fats, oils, sugars, and salts is at an alarmingly high rate based on per capita consumption. With local food, farmers markets, slow food, organic farming, food hubs, and farm-to-table movements gaining in popularity, the farm and food distribution network is diversifying and improving.

One strategic model for food production and perhaps a strategy for global food shortages that utilizes the three dimensions of sustainability in Urban Agriculture (Longo, 2016). Urban agriculture utilizes urban components such as parks, medians, rooftops, parking lots, and yards for the cultivation of food in or around a city. Based on the Brundtland Report, "sustainable agriculture" is defined as an agriculture which is formed on an economically viable and socially accepted development process that sustains environmental resources for the present and the future generations (Aubry et al. 2012). In an urban agriculture city,, the city dwellers produce their food locally on their urban fields. Newly defined cultivation components across the cities could be counted as social spaces that advance social interactions and utilize sustainable cultivation methods such as reducing chemical pesticides and decreases unintended environmental problems. Figure 1.1 above shows urban agriculture as a system within in the three dimensions of sustainability.

Since the 1990s, researchers and advocates of urban agriculture have discussed the systems and benefits of urban agriculture. Based on the reviews, urban agriculture creates and revitalizes healthy food availability, social interactions, urban recreational activities, urban economic well-being, and low-income communities. Recent literature reviews indicate that urban agriculture benefits as alternatives to the threats of industrial and conventional agriculture have motivated

Figure 1. Urban Agriculture Systems.

municipalities, policymakers, and food movement initiatives to apply urban agriculture in cities (Angotti, 2015). Although there are reviews discussing the benefits of urban agriculture, only a limited number have offered ways by which municipalities could practically implement urban agriculture as a new land use or at what scales it might be applied.

This study proposes urban agriculture as a secondary land use that could underpin a viable, nutritious, responsive, and affordable food system. The study proposes a new method which carefully quantifies potential arable area within existing urban and suburban morphology. The result of this study is presented as graphic maps of land inventories that illustrate urban farming as an overlay zoning land-use designation through Geographic Information systems (GIS). The results will help municipalities, decision makers, and planners to estimate the area of arable land and analyze how urban agriculture might be feasible and profitable within cities. Based on the academic, geographic, and agricultural conditions of North Central Florida, Gainesville has been selected as the pilot case study context for this dissertation. The State of Florida faces major challenges for providing food in the future, including the projected population growth, sea level rise, and new construction.

2.1. Pilot Study Context – Gainesville, Florida

Gainesville has a rich background of agriculture with a naturally sunny, humid, subtropical climate that receives 51" of rain annually, around 222 sunny days, and is placed in one of the US strategic land areas which are suitable for growing crops. The primary factors for productive cultivation, enough sunlight and water plus a year-round growing season, exist in Gainesville. Moreover, the University of Florida Institute for Food and Agricultural Sciences (IFAS) is headquartered in Gainesville with 87 extension offices – one in each of the Florida counties. Gainesville naturally protected from the direct effects of sea level rise and has limited flood events at an elevation of 100' above sea level (Gold and Padua, 2018). Based on a background study of Gainesville's geographical and climate characteristics along with it's urban and suburban morphology, a variety of scales of cultivation are possible. This research will be an effort empower citizens and civic leaders to allow Gainesville to become a model community for agri-urbanism.

2.2. Threats and Opportunities

According to the U. S. Census Bureau (2018), the state of Florida is the third most populous state in the US as of 2017. Population growth in Florida is mainly associated with both domestic and international immigration on a continuing basis (Zwick and Carr, 2006). Florida's economy, society, ecology and climate attract people on a global scale. Industries such as tourism, agriculture, international trade, aerospace, and financial services have formed Florida's economy since the early to middle of the 19th Century (Smith, 2005). Socially, the State population has been formed by different ethnic groups and Florida is counted as the eighth most diverse state (McCann, 2018). Florida could be considered socially sustainable because of its diverse population — one of the fundamental factors of social sustainability. Geographically, most of Florida occupies a peninsula between the Gulf of Mexico, the Atlantic Ocean and the Straits of Florida. This geographic condition makes the climate and environment suitable for year-round outdoor living, agriculture, and tourism.

Florida is encountering challenges in the early 21st century that are projected to increase through the end of the Century. The Bureau of the Census identifies Florida as the ninth fastest-growing population state in the US. While this projected population growth might be beneficial for Florida's economy, responding to the fundamental needs of people such as housing, food, water, and shelter will be crucial in the future. The Bureau of the Census reported in 2008 that Florida is the third most coastline populated state in the US. with nearly 75% of Floridians living in the coastline counties (Wilson and Fischetti, 2010). Coastal areas are the most vulnerable to sea level rise. Furthermore, Miami-Dade, Broward, and Palm Beach counties are the top three populous counties of Florida which include 30 percent of the state's total population. A common sea level rise forecast of 90-cm seawater is predicted to inundate areas including 5% of Miami-Dade County (Kopp, 2014).

The prevalent land-use development pattern in Florida has been to transform farmland into suburban residential single family developments. At the time of this writing, this development trend continues to proliferate fueled by population growth and migration from disaster prone areas on the coasts. Cities, therefore, will be facing urban transformation in peripheries including land-use changes, agricultural losses to residential or commercial development, and higher densities of occupation as major sustainability challenge. Food shortages, environmental impacts, and social issues are likely to be exacerbated unless strategies are adopted by decision makers, urban planners, designers, and architects to improve the potential outcomes.

3. RESEARCH OBJECTIVES

Urban agriculture, as a potential new land use within traditional existing land uses, will require municipalities to have access to land inventories and measures of arability by which they could for healthier, economically resilient, and more sustainable communities. In fact, the method should provide both the distribution of arable land and project potential food production for a variety of existing land uses. The proposed method in this study seeks to provide useful data for decision makers, urban planners, and urban agriculture consultants to find and incentivize potential arable areas.

4. METHODOLOGY: Morphological Distribution of Arable Land

The method estimates the area of urban cultivable lands and provides land area data necessary for estimating urban agriculture viability and the level of support or incentives toward implementation. Knowing how much land is available in each land use makes it possible for municipalities to estimate future agricultural yields.

Front yards, back yards, flat roofs, urban parks, sidewalk groves, boulevards, and squares are included as potential arable areas. This proposed method identifies varied morphology throughout Gainesville based land use and density. Five-acre representative samples are selected from each identified morphology. The arable area of each sample is analyzed by utilizing Google Map, Google Earth, and AutoCAD software. Finally, the criteria used to qualify for arability is the land ownership status – private, public, or institution owned. Four consecutive steps are utilized to quantify potential arable areas within different land uses and urban morphology in Gainesville, FL. On-line map databases, Google Map and Google Earth, are utilized to visually identify arable lands and that information is imported into AutoCAD for measurement and area calculations. Mathematical comparisons of the data are done with spreadsheets. The steps of the method are as follows:

1. Morphology Identification
2. Samples Selection
3. Cultivable Area Measurement and Analysis
4. Qualifying and Quantifying Arability

4.1. Morphology Identification

The physical form of a city, including building layout, road patterns, land uses, and green space, creates the morphology of the city (Lai et al. 2018). Based on similar morphologies, urban/suburban regions with similar physical forms could have similar arability potential. In this step, similar urban forms in Gainesville are identified and sample areas selected to evaluate the diversity within a given land use. The sample data is evaluated to identify the arable potential within that particular morphology. Additionally, The range of potential arability is evaluated across the land uses measured. This information allows a more general assessment of the arable potential for the global land use and provide estimates of community viability for production, incentivization, and the types of agriculture that might be optimized withing a land use category. The land use designations are listed as follows:

- Commercial
- Mixed Use (Downtown)

Pre 1970s
- Single Family
- Residential Low
- Residential Medium
- Residential High

1970s to 2000
- Single Family
- Residential Low
- Residential Medium
- Residential High

Post 2000
- Single Family

By a chronologic analysis of the aerial images of Gainesville in the decades of the 19th and 20th century's, there has been a substantial change in urban morphology, including urban grids and major roads, that happened between 1961 and 1968 and then again in 2000. The result of this analysis suggests that existing Gainesville land uses can be divided into two main categories, including pre-1970, 1970-2000, and then post 2000. In order to identify arability withing those morphologies the studied regions are divided into the three major categories of "Pre-1970", "1970-2000" and "post 2000". Figure 2 shows the aerial images in 1961 and 1968.

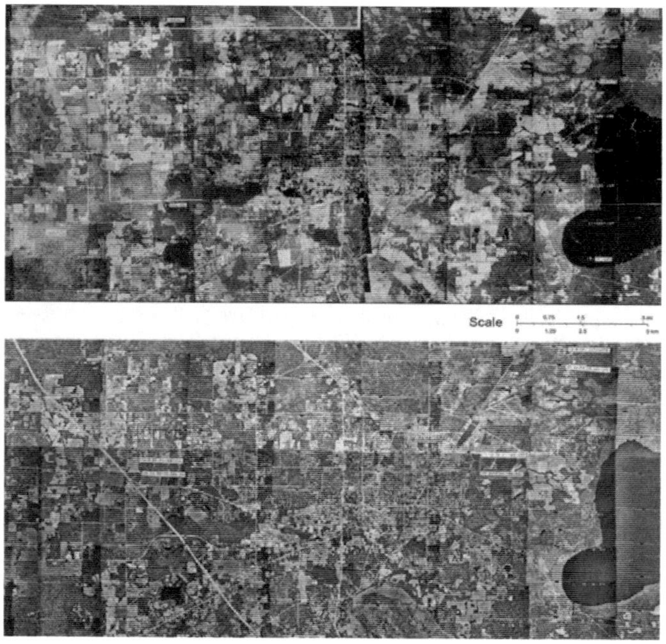

Figure 2. Gainesville aerial images, upper: 1961, lower: 1968, Alachua County Property Appraiser

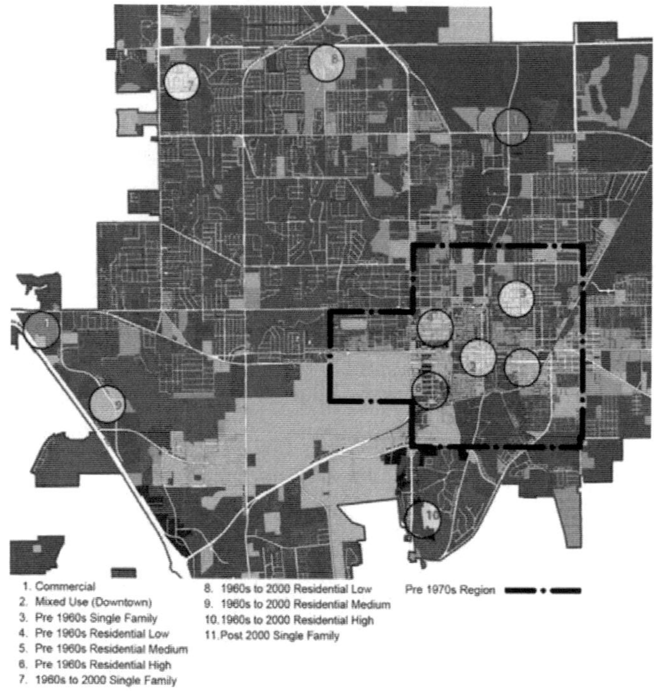

Figure 3. Representative Samples in Gainesville Annexation History.

4.2. Sample Selection

In this step, five-acre representative samples from each morphology are selected by using Google Map or Google Earth. Figure 3 shows several representative samples selected from each specified morphology. The number of representative samples in the proposed methodology is effective. Generally, in big data research which is done by municipalities or governmental institutions, a huge number of samples can be selected in order to estimate more accurate potential arable areas.

4.3. Cultivable area Analysis

In order to analyze the area of arable lands in each clustered region, five-acre samples in each region are selected by using either Google Map or Google Earth; and then their areas are calculated in AutoCAD software. In order to analyze further, a selected sample from Google is imported to AutoCAD. The sample areas and designated regions are then drawn in scale (Figure 4). In a sample analysis, the regions of each property, its built buildings and access, sidewalks and groves, streets, boulevards, public squares, and ROW (right of way) of

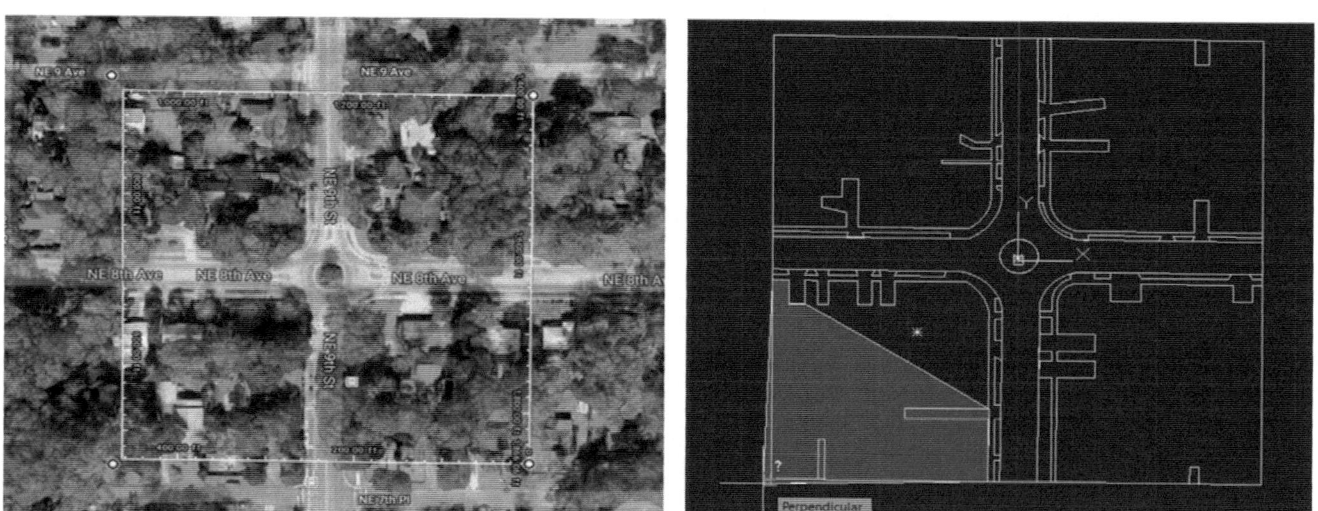

Figure 4. Sample Analysis in Google Map and AutoCAD.

each road are drawn in the AutoCAD file. ROW drawings specify the private and public ownership. Finally, the area of each drawn regions are calculated in the AutoCAD file and the results are classified in an Excel file based on their category.

4.4. Qualifying and Quantifying Arability

The result of this method are categorized in an excel file in terms of their ownership status to qualify and quantify the areas of cultivability and ownership status – privately, publicly, or institutionally owned. The ownership status defines the scopes of responsibility of the citizens and the city. This classification is significantly useful in the future land use decision making. The ownership status includes the following categories:

Private Ownership:

- Arable Land
- Arable Roofs
- Built

Public Ownership:

- Arable ROW
- Squares and Boulevards
- Streets and Sidewalks

Civic Domain:

- Cultivable Parks
- Non-Cultivable Parks

5. RESULTS

The results of this study show that 66% of the studied samples area is arable, and 34% is non-arable. Based on the ownership status analysis, public ownership includes 21%, and private ownership includes 79% of the total sample areas. The results also indicate that the arable areas of the private ownership have 60%, the non-arable areas of the private ownership have 19%, the arable areas of the public ownership have 6%, and the non-arable areas of the public ownership have 15% of the total sample areas. In average, 76% of the private ownership lands are arable, and 24% are built environment. In the public ownership, 31.9% of the lands are arable ROW, 67.7% are Streets and sidewalks, and 0.4% are Arable Squares and boulevards. Figure 5 shows the results in detail. Since the analyzed samples have been selected from the regions where land use, morphologic, and physical characteristics are similar, it can be concluded that the average of the results achieved in each region can be used as the region's arable area.

6. CONCLUSIONS

This study demonstrated a measuring potential for arable areas in urban and suburban morphology, using Gainesville, Florida as a pilot case study. The results indicate that the potential arable areas are substantial and could be leveraged as a viable food industry in the City of Gainesville. This methodology will continue to be

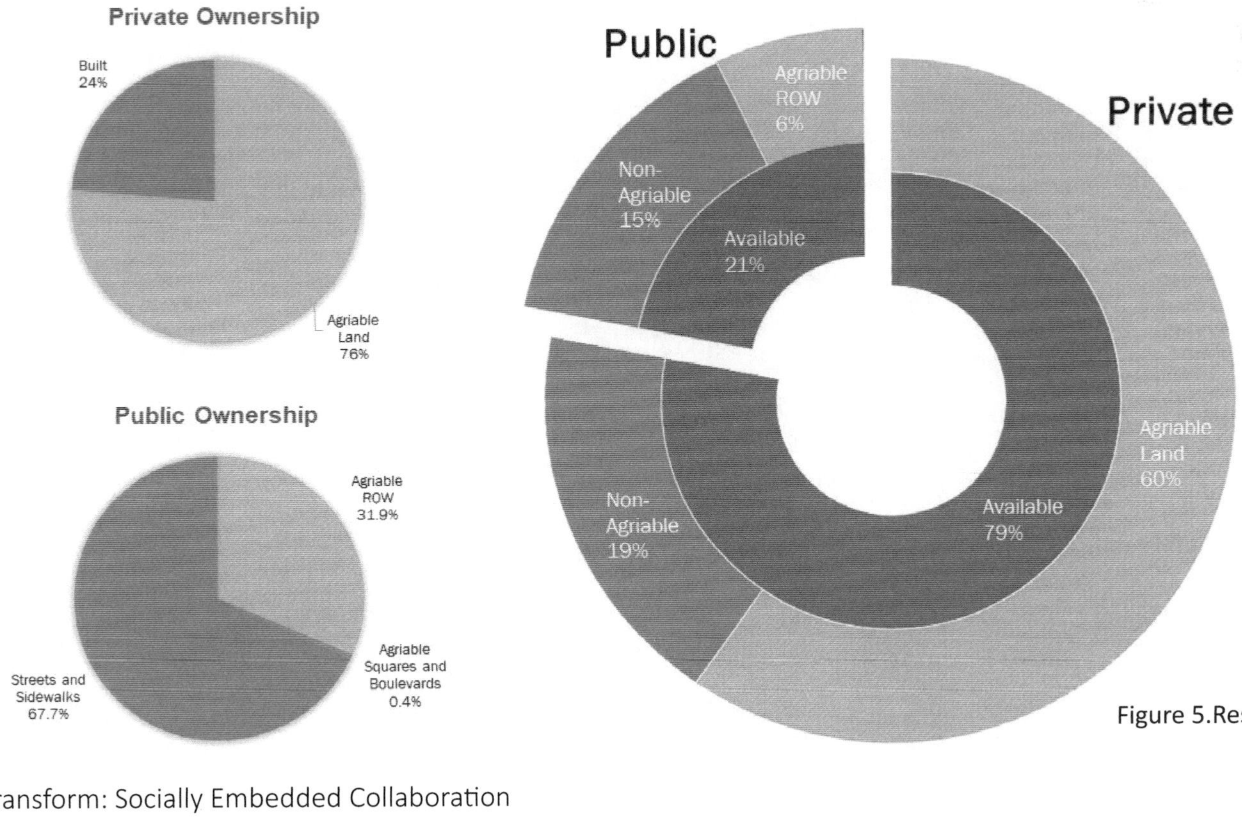

Figure 5.Results.

developed and refined over the next year with the goal of translating the findings into user accessible data for the implementation of urban agriculture at residential and commercial scales. Subsequently, the municipality may evaluate the viability of land use designations or overlays as a policy that could catalyze implementation as part of a comprehensive strategic plan by the city to support citizens and stakeholders with appropriate rules, regulations, and codes.

REFERENCES

Angotti, T. (2015). Urban agriculture: long-term strategy or impossible dream? Public Health, 129(4), 336–341.

Aubry, C., Ramamonjisoa, J., Dabat, M.-H., Rakotoarisoa, J., Rakotondraibe, J., & Rabeharisoa, L. (2012). Urban agriculture and land use in cities: An approach with the multi-functionality and sustainability concepts in the case of Antananarivo (Madagascar). Land Use Policy, 29(2), 429–439.

Gold, M., & Padua, M. (2018). Agri-Urbanism (A. Alborzi, Ed). Gainesville, Florida: University of Florida, School of Architecture.

Longo, P. (2016). Food Justice and Sustainability: A New Revolution. Agriculture and Agricultural Science Procedia, 8, 31–36.

McCann, A. Most & Least Diverse States in America, Retrieved September 7, 2019 from: https://wallethub.com/edu/most-least-diverse-states-in-america/38262/.

Smith, S. K. (2005). Florida Population Growth: Past, Present and Future. Gainesville, Florida - University of Florida. Retrieved from: https://www.bebr.ufl.edu/sites/default/files/Research%20Reports/FloridaPop2005_0.pdf

US Census Bureau. (2018, March 29). Idaho is Nation's Fastest-Growing State, Census Bureau Reports. Retrieved from https://www.census.gov/newsroom/press-releases/2017/estimates-idaho.html.

Zwick, P. D., & Carr, M. H. (2006). Florida 2060 A Population Distribution Scenario for the State of Florida. Gainesville, Florida – GeoPlan Center, University of Florida. Retrieved from: https://api.ning.com/files/f3Cpy2Teci0gMoNEKe4WAzNzxyGZt1W1qZagIMR6H7E-tfG4gMLByYXtWwR20BrbORJ2pf79-9sczCaJJQfN0KXuhCwefGiJ/Florida2060.pdf.

A8 - Transform Design for Resilience & Sustainability

An Environmental-Economic Friendly Construction, Robot-Aided Fabrication with Clay: An Experimental Study

Shermineh Afsary (University of Kansas); Paola Sanguinetti (University of Kansas)

1. ABSTRACT

To achieve environmental and economic benefits, and facilitating the production of architectural elements, this research in 3D printing technologies for architecture, engineering, and construction focuses on the use of building materials that can be integrated into an additive manufacturing (AM) framework. This research will investigate the design, modeling, simulation of 3D-printing and production process of a self-supported/ stackable clay system by robotic fabrication. The robot-aided fabrication with different cementitious mixtures, also described as AM, as an automated production process, is an emerging technology that can be used in the construction industry to achieve economic, environmental, and other benefits such as solving complexity and providing design freedom, increasing the speed and etc. Most of the research in AM have been done on concrete, and less on Clay. Using Clay in this industry is important as an environmental-economic friendly material. Some of the previous research have been done on Robot Aided Fabrication with clay, however this research contribute to finding a new material mixture and tiling system which support a stable and flexible shape. The Methodology of this research is experimental study. After reviewing different printing and material parameters in 3D printing, with focusing on the application of the clay, this research conducted experiments on the rheology of the printing material and printing parameters to create a modular three-dimensional tiling system with clay. A case study is a self-supported, stackable clay system using a robotic arm and presenting a workflow involving modeling, simulation, and production. Material and printing parameters are evaluated. The results point to further studies to optimize the printing process for appropriate fluidity and stability.

1.1 Keywords

Fabrication and Robotic crafts, Digital Fabrication, Automated Construction, Additive Construction, Clay 3D printing.

2. INTRODUCTION

Robot-aided fabrication with cementitious materials as an additive manufacturing (AM) has the potential to transform the architecture, engineering and construction industry by automating highly customized production processes. AM processes have been in development for the past 20 years (Pegna, 1997). Research in the application of AM in architecture, engineering and construction industry (AEC) has evolved from trying to scale-up 3D printing desktop applications to the development of automated systems that extrude and layer building materials. Material optimization and geometric complexity are key issues when dealing with AM for building components. This paper reviews AM research of cement and clay mixtures and identifies parameters to be considered to improve design to fabrication workflows. A case study using clay presents a workflow involving modeling, simulation and production.

2.1. Developments in 3D Printing as a Material Layering Process for Construction

There are various types of 3D printing processes with dry or wet application of material layers. D-shape was developed by Enrico Dini in 2013 (Bos, et al, 2016; Colla, & Dini, 2013). This 3D printing layering technique is a dry powder-bed based printing to deliver sand and a layer of magnesium binder material at a constant flow rate. Contour crafting was developed to automate the layering of wet cementitious grout and troweling (Khoshnevis, 2004; Khoshnevis, et al., 2001; Khoshnevis, et al., 2006). Both of these systems use a gantry to deposit horizontal layers of material. The contour crafting system has a gantry and a nozzle with a computer-controlled trowel to smooth the material being extruded (Wu, Wang & Wang, 2016). 3D printed building structures or components have been produced reach 10 m in length, and 3 m in height (Bos, et al., 2016).

2.2. Material Parameters

Focusing on the wet process, the challenge in layering extruded wet material without compromising the overall structural strength of the 3D print is commonly understood. With clay extrusion, material solidification happens fast, whereas with concrete and cement-based materials, curing time affects the status of the layers adhering to each other (Bos, et al., 2016).

2.2.1. Rheology of Cementitious Material for 3D Printing

Extruded concrete needs to set fast with low level of slump, because the material is unsupported after leaving the extruder. A high-performance cement-based mortar has been developed for 3D concrete printing (Lim et al, 2012). Le et al. (2011) mention that the binder material is a mix of cement, fly ash, and silica fume. A retarder or accelerator can be added. The mix contains polypropylene microfibers to reduce shrinkage and deformation. An approach more appropriate to concrete 3D printing could be to develop a fiber-reinforced printable concrete with enough ductility and tensile strength. An extensive review by Yoo and Yoon (2016) on Ultra-High-Performance Fiber Reinforced Concrete (UHPFRC) reinforced with steel fibers shows optimum strength, strain hardening, stiffness, fracture energy, and crack opening behavior under various loading conditions.

2.2.2. Rheology of Clay Material for 3D Printing

Houben and Guillaud (1994) have introduced two types of clay-based mixtures that can be adapted for 3D printing. A mixture with cement stabilization has 15% gravel, 50% sand, 15% silt and 20% clay. A mixture with lime stabilization has 15 % gravel, 30 % sand, 20% silt and 35% clay. While different types of clay such as porcelain and stoneware, behave differently during 3D printing, it is possible to control the rheology of clay by changing the amount of water and additives. For example, changes in the viscosity of the clay may cause the 3D printed object to collapse; or, it may be difficult to extrude the clay-based mixture (Gürsoy, 2018). Rosenwasser, et al. (2017) found that fine porcelain has exceptionally low plasticity and fine particles, which lead to an exceptional hardness once fired and can lead to cracking. They also found that materials containing grog can be more appropriate for preventing cracking and reducing shrinkage. They developed a clay mixture containing 1.5 gallons of water added to each 50-pound bag of clay with a power drill in a covered 5-gallon bucket to prevent evaporation. They also applied glaze by spray gun for coating to increase strength.

2.3. Printing Parameters

A comparison between three layering printing processes is presented in Table 1, showing layer thickness and other printing factors. Most 3D printers for cementitious material found in the literature deposit 6–25 mm thick layers (Bos et al., 2016; Lim, et al, 2012). The Contour Crafting reports a layer thickness of 13 mm resulting from a nozzle with a 15 mm diameter (Hwang, et al., 2004). Kontovourkisthe and Tryfonos (2018) propose that a 1mm layer deposition of clay is ideal for 3D printing. Researchers at The Institute for Advanced Architecture of Catalonia has focused on large-scale 3D printing. They have developed a large-scale 3D printer to produce for a modular system (IAAC, 2017). The World's Advanced Saving Project (WASP) has developed a 40-foot tall Delta 3D printer with the building dimensions of 19 x 20 x 21 ft (5.8 x 6.0 x 6.3 m). (Kontovourkis, & Tryfonos, 2018). Gursey (2018) has tested the varying the nozzle type during the printing process: from a 3mm circular nozzle in the first four rows, and they switching to an elliptical nozzle to complete the print. The layer heights and nozzle heights were also systematically changed in these experiments, increasing the distance between the nozzle tip and the base plate from 1cm to 3cm. Rosenwasser, et al. (2017) at Cornell University used the clay non-woven approach to apply traditional crafting methods and materials in conjunction with robotic fabrication methods.

3. Case Study- 3D Printed Tiling System

This case study documents the workflow involving design, modeling, simulation, and the 3D-printing process, using a robotic arm to 3D print a stackable tiling system using a clay mixture. The methodology is presented in three sections, 1) the design of the tiling system to achieve multiple stackable options and maintain structural forms; 2) modeling strategy to represent a single tile and simulate to verify by the 3D printing; 3) printing process in which material and printing parameters are varied to optimize the structural integrity of the print.

3.1. Tiling System Design

A modular three-dimensional tile (Figure 1) is designed with five diamonds with cantilevers to create a dynamic tiling system with aesthetic. The diamond shape has a stable and flexible shape, which can create multi-facet surfaces. The self-supported clay system is divided to small modular tiles to be able to pre-fabricated. A variety of tiling configurations can be created by this modular tile by putting and stacking them together in x, y and z direction (Figure 2). Possible tiling systems by installation of this modular tile in different directions is modeled in Rhino Software (Figure 3). This tiling system is applicable in pre-fabricated architecture for facade design or interior design. Design of the shape of the design is according to the design of traditional clay tiles.

Table 1. Comparison of Extrusion- based (wet-based) 3D printing between concrete and clay

	CONCRETE	CLAY
Material	6— 25 mm (Lim, et al, 2012) 40 × 10 mm by the Eindhoven University of technology (Bos, et al, 2016)	>=4 mm 1 to -25 mm
Nozzle diameter or layer thickness	9-20 mm	>=1 mm up to 20 mm 1. The world's largest delta-style 3D printer (WASP) founded by Massimo Moretti can 3D print with higher range of layer thickness with clay (Moretti, 2016
Reinforcement	>1 m dimension (Lim, et al, 2012)	>1 m dimension. Up to 12 m tall and 20 m diameter by the word largest delta 3D printer (Moretti, 2016).
Print size	A mix of cement, fly ash, and unidentified silica fume. A retarder or accelerator can be added (Le, et al, 2011) In-house Printable Concrete (Lim, et al, 2012). A high-performance cement-based mortar for 3D Concrete Printing with 54% sand, 36% reactive cementitious compounds and 10% water (Lim, et al, 2012).	Less fine materials containing of grog decreases breakability (Rosenwasser, Mantell and Jenny, 2017). 1.5 gallons of water with 50 pounds of clay with a power drill. Glazing as a coat increases strength (Rosenwasser, Mantell and Jenny, 2017). A mix of gravel, sand, silt and clay (Houben and Guillaud, Hu, 1994) With glaze/ without glaze
Material mixture	High strength, Limited printing dimension by the printing frame, 5.4 m (L)×4.4 m (W)×5.4 m (H) (Lim, et al, 2012).	Low environmental footprint, local availability, cost-effective modular fabrication for a customized design in architecture.

3.2. Modeling and Simulation strategies

The initial computational modeling approach involved creating three-dimensional surfaces through a set of designed polylines and curves using an algorithm developed in Grasshopper. The plug-in KUKA PRC is used to generate the tool path for the KUKA robotic arm and export an input file using the KUKA Robot Language (KRL) (Figure 6). The KUKA|prc add-on of robots in architecture (Braumann & Brell-Cokcan, 2011) create an inverse kinematic (IK) simulation within Grasshopper to predict and find the issues before 3D printing by Robots (Erdine et al, 2017).

The two polylines are shifted in the x-y plane and lofted to create a surface (Figure 4). The surface is input as a Boundary Representation and deconstructed into a series of profiles contours and the layer's spacing is adjusted based on the material mixture (Figure 5). After converting curves to polylines, control points are extracted using the grasshopper script to convert the model to a set of points. The points are used to generate planes that are reoriented to adjust the toolpath, and maintain a point-to-point linear movement (Figure 7). In the next step, the 3D printing process using the robotic arm is simulated for calibration and verification (Figure 8). The script developed with Grasshopper and KUKA|PRC is directly transferred to the robot controller (Figure 9).

Figure 1. Single modular clay tile.

Figure 2. Various possible tiling system.

Figure 3. One of the possible Tiling system by stacking the modular tiles on z direction with their cantilevers.

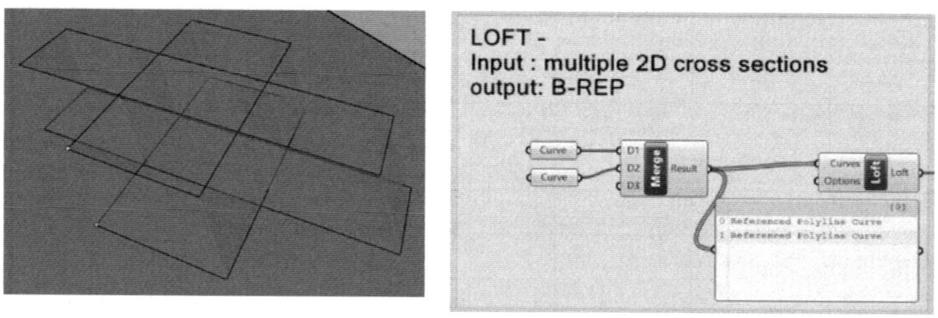

Figure 4. LOFT – The scripting of the tile with grasshopper. Input is multiple 2D polylines to merge them together for the Loft command to fit a surface through the selected polylines.

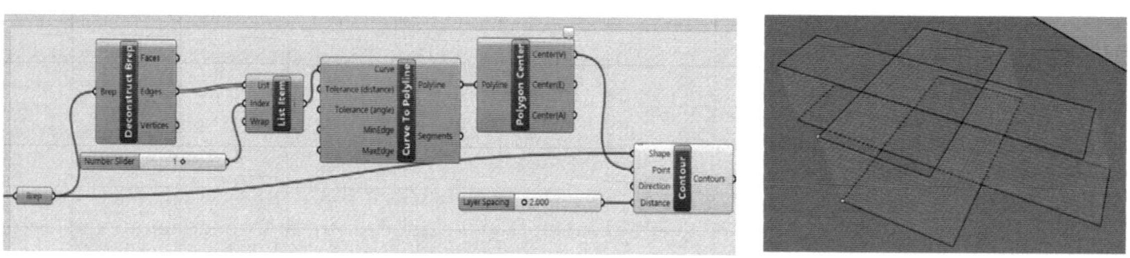

Figure 5. The process of preparing the modeling through Grasshopper Scripting for the tool path by slicing the model to the layers.

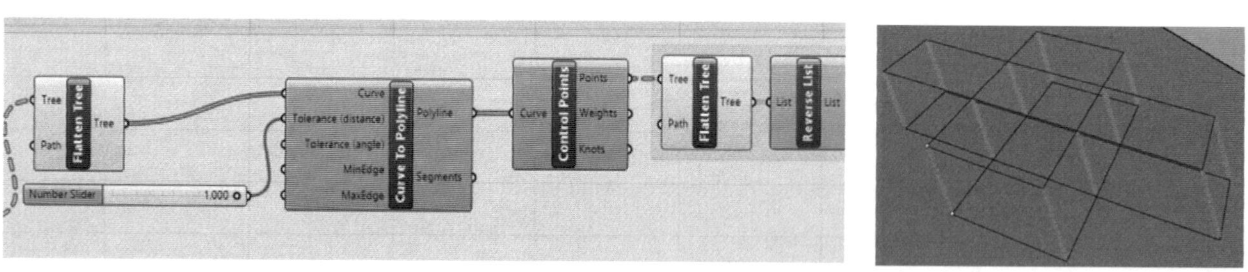

Figure 6. Reverse List to reverse toolpath for the KUKA Robot Language (KRL)

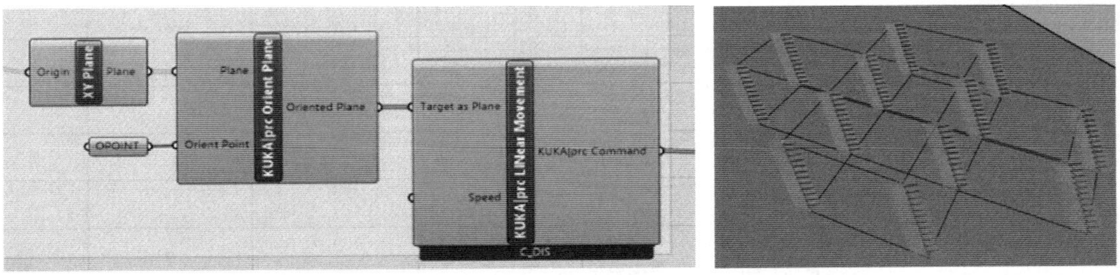

Figure 7. Adjusting the model for robotic tool path of KUKA linear movement

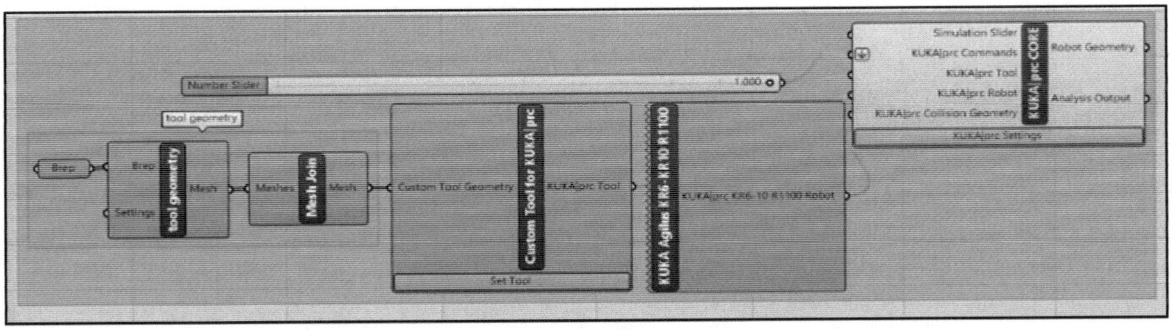

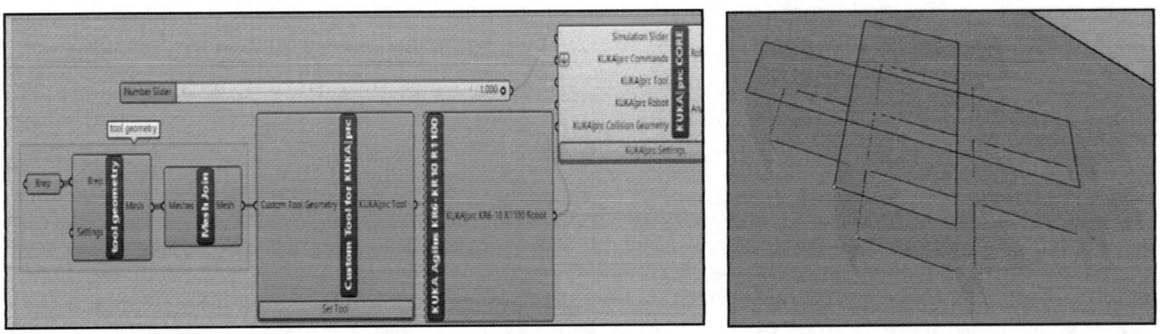

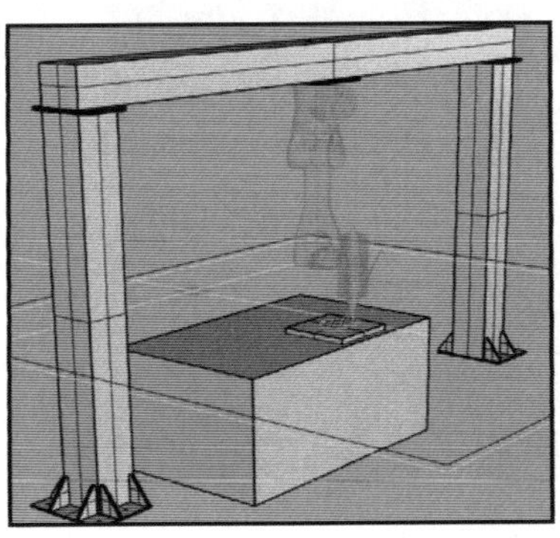

(Left, above) Figure 8. Simulating and observing real-time results of the robot by KUKA|prc

(Right) Figure 9. The process of the KUKA robot and pre-setting before saving the file as a SRC file to 3D print with the robot.

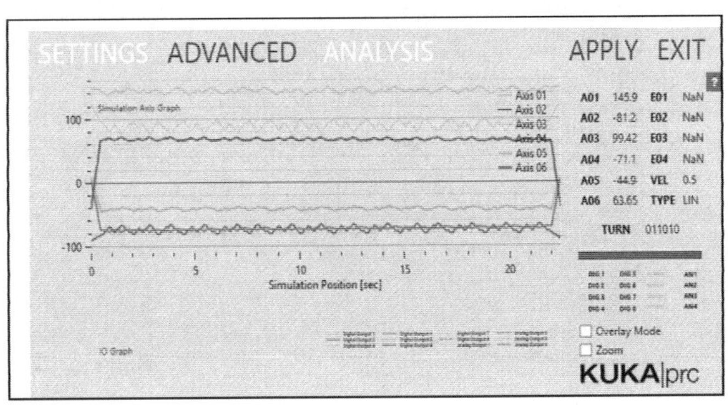

3.3. Printing Process

In the experiments of 3D printing, material mixture parameters of the clay and printing parameters including speed, air pressure, size of nozzle and optimum height of the tile is tested.

3.3.1. Material Parameters

The mixture of the material is developed in three attempts in our experiments to attain desirable pomp ability, fluidity and printability. The three mixtures in the four attempts is presented in the tables (Table 2, 3 and 4). In the mixture 1, 800 grams clay and 360 grams water is used. The cantilevers of the three- dimensional tile collapsed before drying (Figure 15). Then, some additives is added to improve the quality of the mixture to make it buildable. These additives are talc, soda ash, and Sodium Silicate. Talc is a mineral that eliminates fluidity, soda ash is a water-soluble material, and Sodium Silicate also known as liquid glass is used as a concrete sealer. The weight of each of the minerals in the mixture 2 is presented in the table (Table2).This mixture is used for experiment 2 and 3 with different size of the nozzle. It is realized that it is needed to decrease the amount of water and increase the amount of Om4 Clay of the mixture to decrease its fluidity and increase its stickiness. Finally, the mixture 3 used in experiment 4 had the best fluidity, pomp ability, printability and stability. To attain proper consistency of the material, a mixer is used. The material is poured in the cartridges with imposing pressure on it to prevent having air bubbles in the mixture and attain consistency. The cartridge is fixed on the arm of the robot to posit the material (Figure 11 & 12).

Table 2. The mixture 1 used in Experiment 1

Ingredients	Weight (gram)
Om4 Clay	800
Water	360

Table 3. The mixture 2 Used in the Experiment 2 and 3.

Ingredients	Weight (gram)
Om4 Clay	400
Talc	400
Soda Ash	8
Sodium Silicate	2.4
Water	360

Table 4. The mixture 3 used in the Experiment 4- Final Mixture

Ingredients	Weight (gram)
Om4 Clay	450
Talc	400
Soda Ash	8
Sodium Silicate	2.4
Water	330

3.3.2 3D Printing Parameters- Speed and Air pressure, Size of Nozzle and Optimum Height of the Tile

After the tube is filled and put in the robot's arm, the SRC file is transferred to KUKA robot. Then, the robot is adjusted with the robot's controller to 3Dprint (Figure 13). To find the proper speed and air pressure for our

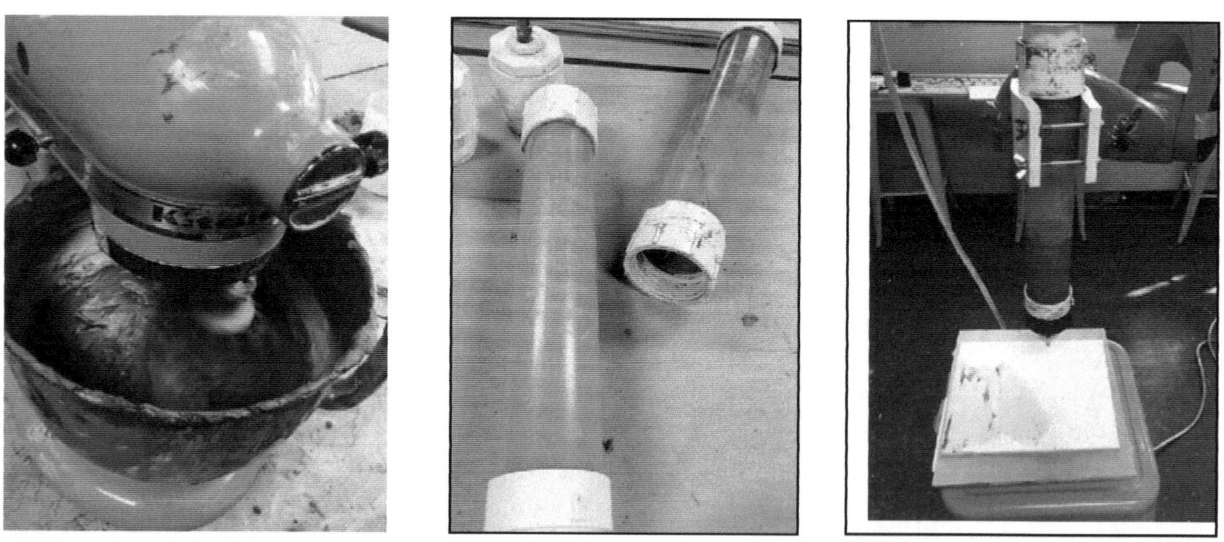

Figure 10 (a, b, c) Mixing process to attain consistency in material, and cartridges to be be fixed to the robotic extruder system.

model our material and model, different attempts are done. The speed of the program override is 51% and for the speed of the jog, override is 10% in the experiment 4 (Figure 15). Pneumatic systems involved forcing air into a hose connected to the cartridge filled with the clay, which pushes clay through a nozzle after the pressure is applied. This system is less controlled since there is a delay between stopping the flow of the air and starting the flow of clay. In this system, air pressure is important, since it can control the flow rates of the clay. In our case study, the proper pressure in the experiment 4 is approximately 30 PSI (Figure 14). However, this pressure and speed sometimes vary a little during the 3D printing because the consistency of the clay can change a little and needs to change these measures to bring out the material from the cartridge. For the size of Nozzle, the nozzle diameter size of Ø 4mm and Ø 6mm is tested.

4. RESULTS

The tile created with the nozzle Ø 4mm deformed and the layers of the cantilevers collapsed before drying (Figure 16). The tile created by the nozzle Ø 6mm is stable before drying (Figure 16 & 17). About the consistency of the material, in experiment 1 and 3, there is a lot of bubbles in the material, because it didn't been mixed well with the mixture and any strong pressure is not imposed in the cartridge while pouring them (Figure 18 & 19). The Optimum Height of one Tile to support the cantilevers is 2.5 cm. First a tile is created more than 2.5 cm of height, but the cantilevers collapsed easily at the beginning, before drying. It could be found out that tiles with 2.5 cm height are constructible. Therefore, instead of having tiles with more than 2.5cm height, tiles with 2.5 cm height can be created and dried, then attach together. Final output of 3D printing of the tile with various tiling system is presented in the following pictures (Figure 20, 21 & 22). In the experiments 1 to 4 printing parameters can be explored of Robot- aided 3D printing for AM with a cementitious- based material. The 3D printing output, the prototype itself, provides useful information to be observed and compared to find out the proper printing parameters in this case study. A key challenge during the 3D printing process is Rheology of the material to have a printable material with precise level of fluidity, stickiness and consistency to balance the enough stability of the tile with proper strength to support cantilevers as well as to have pomp ability during the concrete pouring pressure by the Robot arm and its nozzle. This challenge is met by testing the different percentage of minerals and water in the mixture with additive materials simultaneously as to attain a proper printing parameters including size of minerals and water in the mixture with additive materials simultaneously as to attain a proper printing parameters including size of nozzle, air pressure, speed of 3D printing and optimum height of tile. The comparison between material and printing parameters of experiment 1 to 4 and their results is summarized in Table 5.

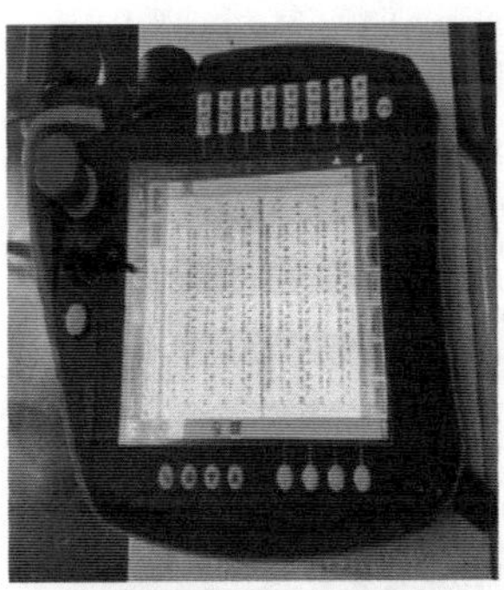

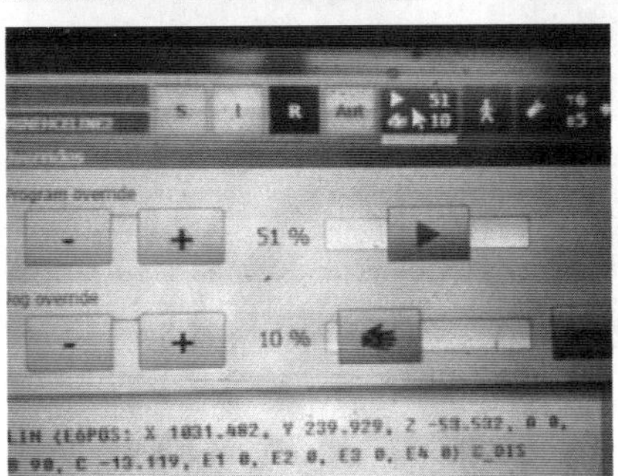

Figure 11. KUKA manipulator interface to adjust the printing speed and (c) right. Air pressure controller

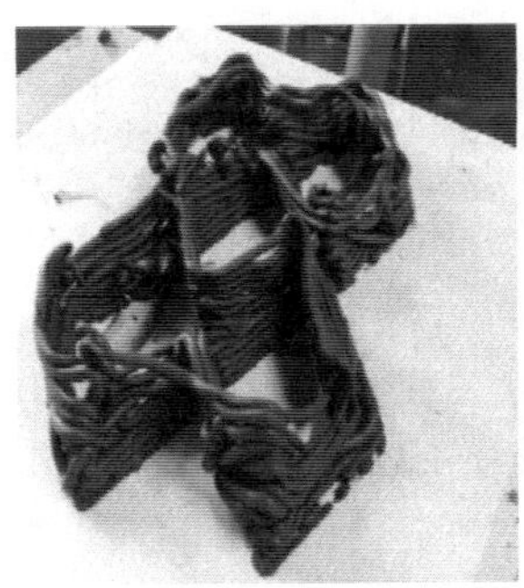

Figure 12 (Left). Experiment 1 with mixture 1.
Figure 13 (Right). Experiment 2 with mixture 2 and nozzle Ø 15mm- The layers of the cantilevers collapsed before drying.

Figure 14 (Left). Experiment 3 with mixture 2 and nozzle Ø 6 mm.

Figure 15 (Right). Experiment 4 with mixture 3 and nozzle Ø 6 mm.

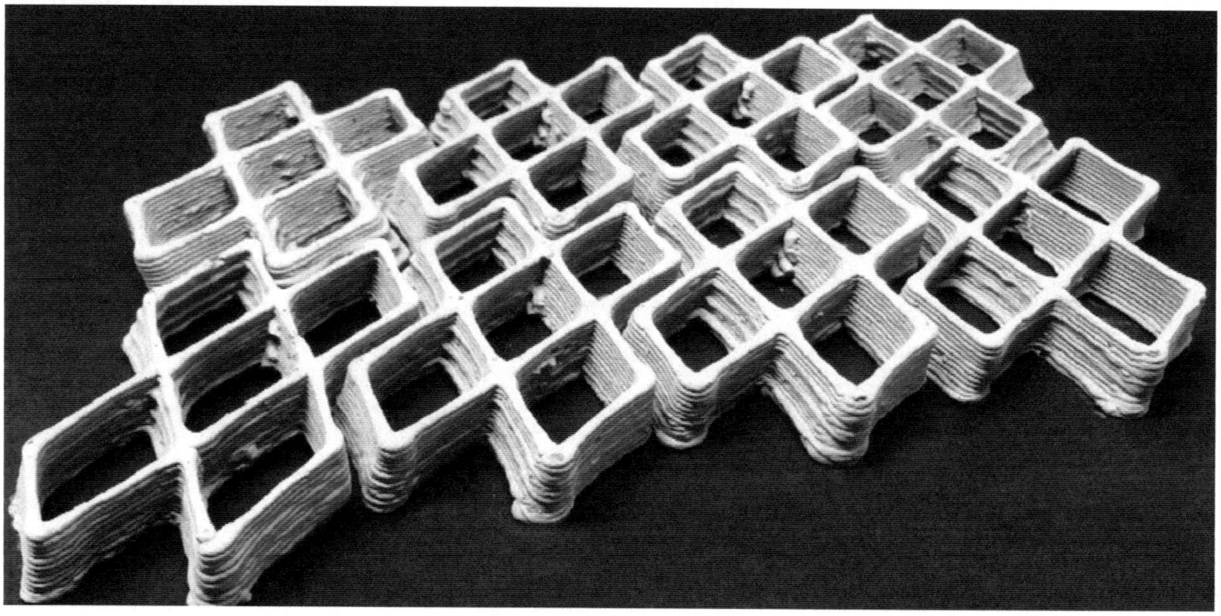

Figure 16 (a). Tiling system of stacking tiles in one surface of x-y

Figure 16 (b) left. Tiling system of stacking three tiles in z direction then putting them together in x-y surface
Figure 16 (c) right. Tiling system of stacking all the tiles on z direction in the way that each tiles has the extrusion of opposite direction than the tile of its underneath

Table 5. Comparison between material and printing parameters of four experiments and their results.

<table>
<tr><td colspan="4">Experiment 1</td></tr>
<tr><td colspan="2">Material Parameters</td><td>Printing Parameters</td><td>Results</td></tr>
<tr><td>Om4 Clay</td><td>800</td><td rowspan="2">nozzle Ø 6 mm
Speed of the program override 60%
Speed of the jog override 20%</td><td rowspan="2">The three- dimensional tile collapsed before drying.</td></tr>
<tr><td>Water</td><td>360</td></tr>
<tr><td colspan="4">Experiment 2</td></tr>
<tr><td colspan="2">Material Parameters</td><td>Printing Parameters</td><td>Results</td></tr>
<tr><td>Om4 Clay</td><td>400</td><td rowspan="5">nozzle Ø 4mm
Speed of the program override 51%
Speed of the jog override 10%</td><td rowspan="5">The layers of the cantilevers collapsed before drying due to the small size of the nozzle.
High fluidity of material
Low stickiness of material</td></tr>
<tr><td>Talc</td><td>400</td></tr>
<tr><td>Soda Ash</td><td>8</td></tr>
<tr><td>Sodium Silicate</td><td>2.4</td></tr>
<tr><td>Water</td><td>360</td></tr>
<tr><td colspan="4">Experiment 3</td></tr>
<tr><td colspan="2">Material Parameters</td><td>Printing Parameters</td><td>Results</td></tr>
<tr><td colspan="2">Same as material mixture in experiment 2</td><td>nozzle Ø 6 mm
Speed of the program override 51%
Speed of the jog override 10%</td><td>Low level of consistency of material causes a lot of bubbles in the material, because it is not mixed well with the mixture and any strong pressure is not imposed in the cartridge
High fluidity of material
Low stickiness of material</td></tr>
<tr><td colspan="4">Experiment 4</td></tr>
<tr><td colspan="2">Material Parameters</td><td>Printing Parameters</td><td>Results</td></tr>
<tr><td>Om4 Clay</td><td>450</td><td rowspan="5">Speed of the program override 51%
Speed of the jog override 10%
Nozzle Ø 6 mm
Using the mixture more and imposing stronger pressure while pouring the material in the cartridges more than experiments 1 to 3 to attain consistency of the material
Two height of the tile is tested, more than 2.5 cm and 2.5 cm.</td><td rowspan="5">Proper fluidity and stickiness: With decreasing the percentage of water and increasing the percentage of the Om4 clay, the fluidity decreased and stickiness increased to the proper level

Proper consistency

The optimum height of the tile is 2.5 cm. The tiles with height of more than 2.5 cm collapsed, while the tiles with 2.5 cm height were stable.</td></tr>
<tr><td>Talc</td><td>400</td></tr>
<tr><td>Soda Ash</td><td>8</td></tr>
<tr><td>Sodium Silicate</td><td>2.4</td></tr>
<tr><td>Water</td><td>330</td></tr>
</table>

5. DISCUSSION

In this experimental research on creating a three-dimensional cementitious tiling system, printability, buildability, transportability, stability and less breakability are the important factors to be attained. One of the complexity of the tiles is building its cantilevers. To prevent from collapsing these cantilevers, material mixture and consistency, Nozzle size, Robot speed and air pressure play a key role. In this regard, there is three attempts to improve our material mixture and printing parameters. In Comparison of experiment 1, 2, 3 and 4 it can be found out that the mixture 3 in experiment 4 has the best level of fluidity and sickness to become pump able in 3D printing. Moreover, it can be realized that the optimum size of nozzle Ø 6 mm with the optimum size of the tile, which is 2.5 cm. The importance of consistency of material is also very important, as in experiment 4, a consistent material is created with a proper mixture, proper size of nozzle and speed and air pressure in 3D printing. A more integrated automated approach is needed to 3D print rheological materials in the construction industry. Data from three different aspects needs to be linked to improve the AM process in Architectural design and material optimization. The architectural design process should be improved by AM as the building component geometry needs to be analyzed to optimize the horizontal deposition of material and its structural performance In terms of material optimization process, material characteristics, setting and hardening properties need to be analyzed to optimize the tool path, printing speed, material layer, and flow rate.

6. CONCLUSIONS

The study of the followability, formability, and hardening time of the printing material is critical to improve the AM process. Expected significance of the field of AMoC is reduction of material waste in construction Formwork and construction material; and Optimizing design for construction workflow. This case study explored on modeling, simulation, printing and material parameters of Robot-aided 3D printing for AM. This allows an ease and effective development of robotic automated system for 3D printing with clay. Nevertheless, future work can incorporate AMoC in Robotic 3D printing on various forms of tiling systems in order to test the efficiency of concrete and cementitious- based material flow.

REFERENCES

Bos, F., Wolfs, R., Ahmed, Z., & Salet, T. (2016). Additive manufacturing of concrete in construction: potentials and challenges of 3D concrete printing. Virtual and physical prototyping, 11(3), 209–225. Received from https://doi.org/10.1080/17452759.2016.1209867.

Braumann, J., Brell-Cokcan, S. (2011). Parametric robot control: integrated CAD/CAM for architectural design, Association for Computer Aided Design in Architecture (ACADIA), Lincoln, Nebraska, USA, 2011: Joshua M. Taron ed. Proc., 241- 252.

Colla, V. and Dini, E. (2013). Large-scale 3D printing: from deep sea to the Moon. In: E. Canessa, C. Fonda, and M. Zennaro, eds, Low-cost 3D printing, for science. Education & sustainable development, 7(2), 127–132.

Erdine, E., Kallegias, A., Moreira, A., Fernando, L., Devadass, P.,& Sungur, A. (2017). Robot-aided fabrication of interwoven reinforced concrete structures. SimAUD, Toronto. Canada, May 2017, Society for Modeling & Simulation International (SCS).

Gürsoy, B. (2018). From control to uncertainty in 3D Printing with Clay fabrication. Virtual & physical prototyping, 2(36), 21–30.

Houben, Hugo and Guillaud, Hubert. (1994). Earth construction, a comprehensive guide, London: Intermediate Technology Publications.

Hwang D, Khoshnevis, B. (2004), Concrete wall fabrication by contour crafting, Paper Presented at the 21st International Symposium on Automation and Robotics in Construction ISARC, South Korea, 2004.

IAAC, Institute for Advanced Architecture of Catalonia (2017), Terra Performa., from https://iaac.net/research-projects/

Khoshnevis, B. (2004). Automated construction by contour crafting- related robotics and information technologies, Automation in Construction, 13(1), 5-19.

Khoshnevis, B., R., Russell, H. Kwon, S. Bukkapatnam (2001), Contour crafting– a layered fabrication technique. IEEE Robotics and Automation Magazine, 8 (3), 33–42.

Khoshnevis, B., Huang, D, Ke-Thia,, Y, Zhenghao,Y (2006). Mega-scale fabrication by contour crafting. International journal of Industrial and Systems Engineering, 1(3), 301–320.

Kontovourkis, O and Tryfonos, G (2018), Integrating parametric design with robotic additive manufacturing for 3D clay printing: An experimental study. 35 th International Symposium on Automation and Robotics in Construction ISARC, Berlin, Germany, July 2018.

Le, T., et al. (2011). Hardened properties of high performance printing concrete, Cement and Concrete Research, 42, 558–566.

Lim, S., Buswell, R. A., Le, T. T., Austin, S. A., Gibb, A. G. F., & Thorpe, T (2012). Developments in construction-scale additive manufacturing processes, Automation in Construction, 21, 262–268.

Moretti, Massimo (2016). The world's largest delta-style 3D printer. The World's advanced Saving Project (WASP), from https://inhabitat.com/worlds-largest-delta-style-3d-printer-can-print-nearly-zero-cost-housing-out-of-mud/

Pegna, J. (1997). Exploratory investigation of solid freeform construction, Automation in Construction 5 (5), 427-437.

Rosenwasser, D., Mantell, S., & Sabin, J. (2017). Clay non-woven, 37 th Annual Conference of the Association for Computer Aided Design in Architecture (ACADIA), 502- 511.

Wu, P., Wang, J., & Wang, X. (2016). A critical review of the use of 3-D printing in the construction industry. Automation in Construction, 68, 21–31.

Yoo, D.-Y and Yoon, Y.-S. (2016). A review on structural behavior, design, and application of ultra-high-performance fiber reinforced concrete. International Journal of Concrete Structures and Materials, 10 (2), 125–142.

A10 - Transform Design for Resilience & Sustainability

Building Performance Evaluation in Hospital Ward Planning By Nursing Travel Distance Simulation Using BIM

Midori Takagi (Mie University); Kyohei Yamaguchi (Mie University); Takahiro Takeuchi (Yasui Architects); Akikazu Kato (Mie University);

1. ABSTRACT

The studies of nursing travel distances of the hospital ward have been carried out by many researchers from 1955 (Nuffield Provincial Hospitals Trust 1955). In these researches, the prediction of nursing travel distances by changing patient rooms and/or placements for hospital ward planning using data of nursing travel distances survey was carried out (Taniguchi, et.al. 1984), and the prediction of number of nursing movement by changing the locations of nursing bases (Toriyama et.al. 2008). However, such simulations were carried out manually or using dedicated simulation software, so that it is difficult for designers and facility managers to use them in practice. Therefore, this paper aims to reveal the usefulness of nursing travel distance simulation using Building Information Modeling (BIM), and also to show the effectiveness for the prediction of nursing travel distance by changing the locations of nursing bases and patient rooms in the Building Performance Evaluation (BPE) life cycle of planning, programming, design, and management of hospital wards. A survey of nursing travel distance was carried out in two hospital wards in Japan to grasp the actual nursing activities, and it was conducted to extract the elements that affect nursing travel distance. Following the analysis, the simulation was carried out to make predictions by changing the position of the staff base and changing the placements of patients. Through this simulation, it became clear that 1) by improving the use method of nursing base, it was possible to reduce 27% of travel distance, 2) by changing patient placements, the difference in increase of travel distance totaled 30%.

1.1. Keywords

Building Information Modeling, Simulation, Hospital Ward Planning, Nursing Activities, Nursing Travel Distance

2. BACKGROUND AND PURPOSE OF RESEARCH

The studies of nursing travel distances and nursing activities of the hospital ward have been carried out by many researchers such as the Nuffield Foundation in 1955 (Nuffield Provincial Hospitals Trust 1955) and Yale Traffic Index in 1975 (Thompson et.al. 1975) using various simulations and analysis.

The first observational survey on nursing activity in Japan was by Yoshitake published in 1964 (Yoshitake, et.al 1964). The study pointed out that the two main factors influencing the length of nursing activity were the locations of seriously ill patients and the locations of nurse stations. Furthermore, after investigating the movement of physicians as well as nurses, the study pointed out the significance to consider the movements of nurses rather than those of physicians when designing hospital wards.

As a part of relevant research, the prediction of nursing travel distances was carried out by changing patient rooms and/or placements for hospital ward planning using data from nursing travel distances survey (Taniguchi, et.al. 1984). According to this survey, it was argued that the reduction of nursing travel could be expected by changing the patient placements, and that a reduction rate of about 5% was possible for the nursing efficiency. The study also pointed out the impact of patient placement changes in relevance to the ward floor plan typology. The prediction of nursing movement by changing the locations of nursing bases was also carried out (Toriyama et.al. 2008). In this study, the location and number of staff bases was used at parameters of digital information, and they were able to be located near patient's rooms as Personal Nursing Bases (PNB). The study argues that the functions of PNB reduced the flow of nursing and change the places of work. However, such simulations were carried out manually or using dedicated simulation software, and it is difficult for designers and facility managers to use them in practice. It should be noted that there are some interesting developments in the field in relations to space syntax analysis (Hendrich 2009), and it will be reviewed using the survey results in this study.

Recently, the use of Building Information

Modeling (BIM) has become a global trend to improve the entire life cycle of building. Various simulations can be carried out on BIM to merit the practice of planning and design, construction management, and facility management. Moreover, BIM software has begun to provide an environment called visual programming, in which anyone may develop and/or customize simulations at ease.

Building Performance Evaluation (BPE) grew out of Post-Occupancy Evaluation (POE). The process model for BPE takes into consideration performance evaluation over the entire life cycle of buildings in six phases and feedback loops: 1. Strategic Planning – Effectiveness Review, 2. Programming - Program Review, 3. Design – Design Review, 4. Construction – Commissioning, 5. Occupancy – Post-Occupancy Evaluation, 6. Re-Use/ Recycling – Market/Needs Analysis. (Preiser 2018) In conjunction to use of BIM, the introduction of BPE should provide a drastic improvement in the building life cycle.

In this study, a survey of nursing travel distance was carried out in two hospital wards in Japan to grasp the actual nursing activities, which is a good example of POE itself. Moreover, one of the authors was an architect who designed the hospital, built in 1986, and he has been involved in its strategic planning phase throughout the present occupancy phase. Because this survey contributed to its recent renovation planning, the process is a good case study of BPE itself. Finally, the factors affecting the nursing travel distance were extracted, so that they can be used in a nursing travel simulation system developed based on BIM.

The study aims to show the effectiveness of such simulation system for the prediction of nursing travel distance and to show the evaluation of hospital ward planning and management by changing the locations of nursing bases and patient rooms in the BPE life cycle of planning, programming, design, and management of hospital wards.

3. DEVELOPMENT OF SIMULATION SYSTEM USING BIM

3.1. Simulation Development Environment

In this study, a visual programming tool named Dynamo was used to extend the functions of AutoDesk Revit, a BIM software. In addition, Python 2.7 is used as a text programming language to write command to Dynamo. By performing simulation using BIM software, which has become a major design tool in recent years, the following can be considered as a seamless simulation, starting from stages of a planning, programming, design, and construction management, and finally in the occupancy/use stage utilizing the BIM model.

3.2. Simulation System of Nursing Travel Distances

This simulation program visually expresses the frequency of nursing travels on assumed optimum routes mainly in ward's corridors and calculates the totals of travel distances and time. The assumed optimum routes are given using A-star search algorithm, a computer algorithm widely used in pathfinding. By drawing nursing travel lines using these routes, most commonly used travel routes are easily clarified. As the system flow, probable destinations are shown using Excel chart and an undirected graph is created. Then, the degree of congestion and distance for each route are calculated by using a nursing travel flow line search program that applied above search algorithm on Dynamo. Finally, the analysis results are shown in Revit drawing and Excel chart. Except for the two process stages of "Draw nursing travel line" and "Determine destination" on Revit, the most of process is automated. This simulation was designed to reduce the amount labor for processing before the simulation, so that it may be used on various conditions and by various users.

4. OUTLINE OF NURSING TRAVEL DISTANCES SURVEY

In this study, the simulation is performed based on using surveyed data of actual travels by nurses on duty. The surveyed hospital ward uses Partnership Nursing System (PNS) that was developed by the nursing department at University of Fukui Hospital. In PNS, two nurses form a team partnership to take care of the patients using the high-tech medical equipment including the electronic medical record system (University of Fukui Hospital 2009). Because the nursing requires the deep devotion, the profound experience, the wide knowledge, and the heavy physical labor, each partner nurse in PNS is advantageous to the other to complement and cooperate with each other. The PNS partner works throughout one year from committee activities to actual tasks in the hospital ward and shares the results and responsibilities. Thus, PNS provides the safe and high-quality services for patients,

4.1. Overview of Surveyed Hospital Wards

The strategic planning for the hospital was carried out in 1970's based on the analysis using Huff Model, a method to predict consumers' choice of a facility using a gravity model equation developed by economist David Huff in 1960's (Yanagisawa 1978). The programming, design and construction was carried out in 1980's, and the hospital was newly opened in 1986. In the recent renovation planning, the transition of wards management methods was expected. As shown in Figure 1, at the nursing base in the original design principle was that the entire ward was under one nursing management unit of 90 beds and consisting of three nursing units of 30 beds with distributed nursing bases directly supporting nursing activities. The ward plan-type was named as a ward with distributed nursing bases.

However, due to differences in the amount of nursing work at each wards and changes in the nursing system, the operation was currently divided into two nursing units, namely E-Ward with 36 beds and W-Ward with 53 beds. Accordingly, the W-Ward used the staff station (SS) in the center of the ward as the main nursing base, and the E-Ward used the nurse corner (NC) in the center of the nursing unit as the main nursing base. The W-Wards also had an NC at the center of nursing units, but it was currently not used as the main nursing base. In both wards, the high-tech medical equipment was used including the electronic medical record system. This current survey was conducted in two hospital wards (E- and W-ward) both of the internal medicine discipline. The nursing system was in the introduction stage of Partnership Nursing System (PNS) in E-Ward, which adopted a combined method of PNS and conventional Team Nursing. In E-Ward, in the morning session PNS was used and in the afternoon session Team Nursing, so that the number of available nurses was increased in the afternoon session. The W-Ward used conventional Team Nursing. On the other hand, late shift nurses and early shift nurses in both wards worked according to fulfill certain functions including task charges of goods management, drug preparation, dinner distribution, evening care, etc.

4.2. Survey Method

1. Nursing Travel Distances Survey:

This survey was conducted using a time study method that one investigator tracked and manually recorded activities of one nurse. The investigator recorded the time, action, location, and portable items whenever the surveyed nurse made action and/or changed location. The surveyed nurses were two day-shift nurses, two day-long-shift nurses, one late-shift nurse, and two night-shift nurses. The shift system taken by the wards was a variable two shift system, that some day shift nurses worked on a longer 13 hours period, day-long-shift, longer than 8 hours period in conventional three shift system. On the day of the survey, W-ward had overtime work, so survey was conducted including overtime hours (this study took into account the content of work and added data for overtime work). In this paper, the surveyed nurses were displayed using the symbol (ward name_shift_number) in Table 1.

Table 1. Outline of H-Hospital Wards

Subject of Investigation	E_Ward		W_Ward	
Nursing Staff System	7:1 (Patient to Nurse Ratio)		7:1 (Patient to Nurse Ratio)	
Ave. Length on Stay (day)	16.4(Dec.2018)		14.8(Dec.2018)	
Investigation Date	Dec. 05-06, 2018		Dec. 12-13, 2018	
Number of Patients	23		43	
Number pf Beds	36		50	
Nursing System	A.M. : PNS P.M. : Team Nursing		Team Nursing	
Outline of Nuse's Duties	○Head Nurse<1>		○Head Nurse<1>	
	A Pair	●Day_Shift(E_D_1)<1> ●Day Long_Shift(E_DL_1)<1>	A team	●Leader_Day_Shift(W_D_1)<1> ◐Day Long_Shift(W_DL_1)<2> ○Day_Shift<1> ○Early_Shift<1> ○Afternoon_Shift<1>
●:Subject of Investigation ◐:Subject of Investigation (one of two nurses) <x>: x is the Number of Nurse	B Pair	●Day_Shift(E_D_2)<1> ●Day Long_Shift(E_DL_2)<1>	B team	●Leader_Day_Shift(W_D_2)<1> ●Day Long_Shift(W_DL_2)<1> ○Day_Shift<1> ○Early_Shift<1> ○Afternoon_Shift<1>
	C Pair	○Day_Shift<1> ○Early_Shift<1>		
	Late	◐Member(E_L)<2>	Late	●Leader (W_L)<1> ○Member<2>
	Night	●Leader (E_N_1)<1> ●Member(E_N_2)<1>	Night	●Leader (W_N_1)<1> ◐Member(W_N_2)<2>
Investigation Time	Day_Shift 8:30~17:15 Day Long_Shift 8:30~21:30 Late_Shift 12:45~21:30 Night_Shift 20:30~8:30		Day_Shift 8:30~18:30 Day Long_Shift 8:30~21:30 Late_Shift 12:45~21:30 Night_Shift 20:30~8:30	

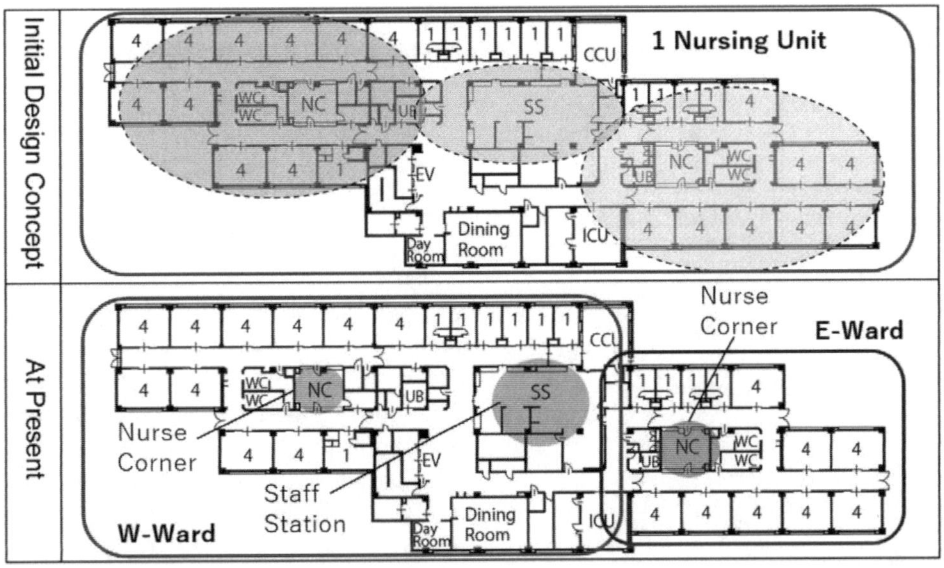

Figure 1. Method of Management in Hospital Ward

2. Patient's Nursing Necessity Survey:

The patient's nursing necessity index, which was introduced recently in Japanese hospital management to show the degree of necessary nursing, and that of bed rooms locations were recorded for 2 weeks before and after the survey date.

5. ANALYSIS OF FACTORS AFFECTING NURSING TRAVEL DISTANCES

Analyses was conducted regarding the contents of nursing activities, work location, and the frequency of visits to bed rooms with an assumption that these were the factors mostly affecting the travel distances of nurses. Unfortunately, on the survey day, there were no patients in the ICU and no seriously ill patients in the CCU, and these rooms were excluded from the analysis.

5.1. Nursing Activities and Work Locations

The analysis of nursing activities revealed that nurses in W-ward spent more time for movements than those in E-ward (Figure 2). On the other hand, the duration percentage of direct nursing care in W-ward was found to be larger than that of E-ward. The cause-effect relations were not such, although it was assumed that longer walking time may hamper the time provision for direct nursing. Remembering that overtime duties occurred in W-ward on the day, it should be pointed out that the conditions of patients called for more amount of direct nursing and the longer walking time aggravated the overtime situations.

Secondly, regarding the impact of nursing travel distances in the PNS, the nursing activities analysis was conducted in E-ward which was at the introductory stage of PNS. In comparison to movement time among nurses who were the partners (A-pair: E_D_1, E_DL_1, B-pair: E_D_2, E_DL_2 (see Table 1)), the figures showed a similar trend. By principle, PNS partners complement and cooperate with each other in an equal position. The partnership would not realize if one were to prevail, so that the other will travel to fetch or dispose some material and other deeds. Thus, the similar travel time can be evaluated as a normal operation of PNS in E-ward. On the other hand, when focusing on the difference in movement time of A-pair and B-pair, that of the B-pair was found generally higher. Therefore, it is assumed that there is a high possibility that the nursing travel distances depend on the locations of bed rooms in charge by the partners, meaning that the distance between bed rooms and nursing base is significant.

Moreover, the staying time rate in the corridor was longer for W-ward (Figure 3). Together with the above movement time ratio, nurses whose main nursing base is SS, at the end of the nursing unit, have a longer staying time in corridor and travel time. Therefore, it can be inferred that NC as the main nursing base will reduce the nursing travel distances.

5.2. Analysis of Number of Bed Rooms Visits

(1) Relationship between Number of Bed Rooms Visits and Nursing Necessity Index

Nursing Necessity Index indicates the degree of nursing care needed for certain patient, which forms the accumulated base for needed number of nurses. It is presently used in many Japanese hospitals because it is also used as an important factor in the application for hospital charge reimbursements. This index mainly

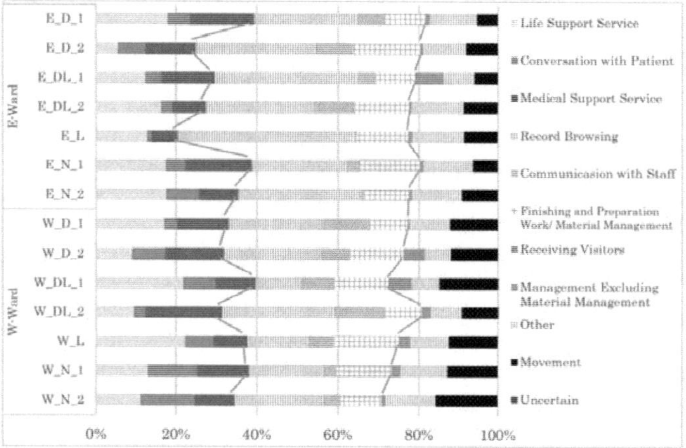

Figure 2. Analysis on Nursing Activities

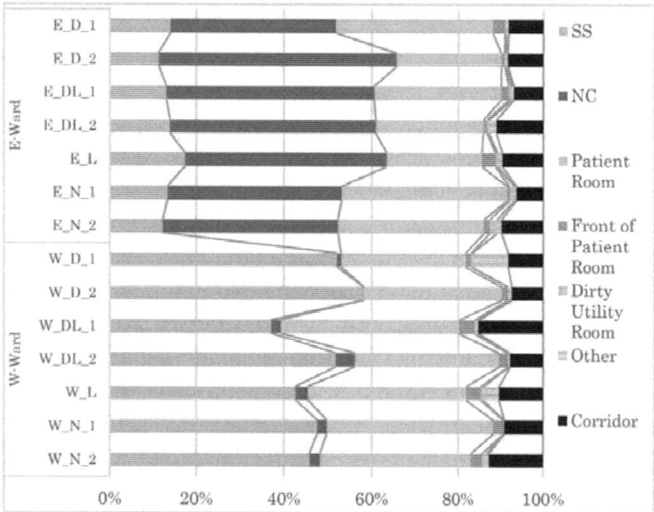

Figure 3. Analysis on Place of Nursing Activities

consists of three items: A-item for measuring the degree of monitoring and treatment, B-item for assessing the degree of patient's physical conditions such as the possibility to roll over by oneself etc., and C-item for assessing the degree of provided medical procedures such as surgery. From the regression analysis between Nursing Necessity Index and the number of bed rooms visits, the factors related to the nursing travel distances were clarified. As the result, above B-item was found to show a strong relationship with the number of bed rooms visits, meaning that the degree of patient's physical conditions is the major cause of nurses' visits to bed rooms.

(2) Relationship between Number of Bed Rooms Visits and Distance from Nursing Base to Bed Rooms

The regression analysis on the relationship between the distance to bed rooms from main nursing base and the number of bed rooms visits was carried out. The distance to bed rooms from nursing base is defined as the distances to the door of bed rooms from the main staying location at nursing base. The nursing base is NC in E-ward and SS in W-ward. However, no strong correlation was found, meaning the closeness to bed rooms does not necessarily result in more visits.

(3) Relationship between Bed Rooms Placement and Nursing Necessity Index

The daily Nursing Necessity Index of patients and their locations in surveyed wards for the period of two weeks, a week before and after the survey date was secured, and the regression analysis of the relationship between the Nursing Necessity Index and the bed room placement that is the distance between patient room's door and the nursing base rooms was carried out. It was found that the patient with smaller nursing necessity index were placed in more remote bed rooms from the nursing base. This tendency was particularly noticeable in W-ward. The interviews with the head nurse revealed that patients with higher B-item Index were placed closer to the nursing base when possible. From these, the location of nursing base and the bed rooms placements were considered as major factors affecting the nursing travel distance.

(4) Relationship between the Time and Motion

The space syntax theory offers a method to describe configured space as a network of discrete spatial units in which purely relational attributes of the units. Initial results with one variable of space syntax for the purpose of illustrating the utility of the technique and its potential implications for the nurse work environment was analyzed. Figure 4 shows that the RFID movement patterns of 53 nurses, covering 143 nursing shifts from 5 of the 36 medical-surgical units in the overall study (Hendrich 2009).

These correlations suggested that nurses employed distinct movement strategies. Analysis of these data demonstrated a negative correlation between number of entries to patient rooms and average time spent per visit, but a positive correlation between entries and total time spent in patient rooms. It showed that creating an environment that makes a greater number of entries should result in more frequent trips to patient rooms and more time spent in patient rooms, even though the average trip is of relatively shorter duration.

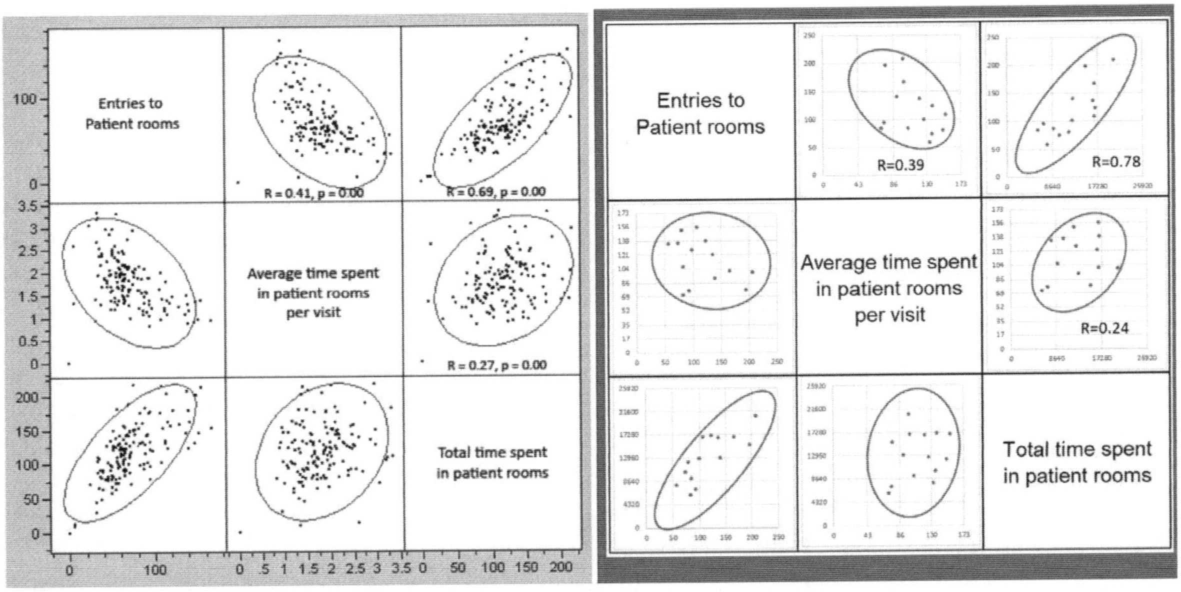

Figure 4. Multivariate scatter plot of entries to patient rooms, average time spent in patient rooms per visit, and total time spent in patient rooms. Left: Unit-Related Factors That Affect Nursing Time with Patients: Spatial Analysis of the Time and Motion Study; Right: Result in this Study

6. Analysis of Ward Planning and Management by Nursing Movement Simulation Using BIM

The developed system of simulation using the BIM was used to perform the analysis of nursing travel distances.

6.1. Understanding the Current Nursing Travel Distance of Each Nurse

First, to test the suitability of assumed optimum routes in the simulation system, the current nursing travel distances on the survey date was confirmed through the simulation. Since this was based on the route actually tracked by nurses, considerably accurate values were achieved as an output.

Comparing the nursing travel distance in E-ward and W-ward (A-line in Table 2), the movement distance of W-ward was found to be larger for all nurses in than those of E-ward. In particular, W_DL_1 day-long shift nurse visited so many bed rooms that the travel distance during the working day exceeded 9 km. The longer length of nursing travel distance in W-ward was considered to be resulting from the fact that the main nursing base SS was placed at the end of W-ward, instead of NC in the center. This was supported by the fact that the nursing travel distance of nurses who were in charge of bed rooms closer to the nursing base SS was smaller among W-ward nurses and the figures were similar to those of nurses in E-ward (Table 2).

(1) Average Walking Speed of Nurses Based on Calculated Travel Distance and Actual Travel Time

In this study, the assumed walking time was calculated using the distance calculated by the nursing travel distance simulation and using the average walking speed of a general person which is 1.25 m/s (B-line in Table 2). On the other hand, the actual travel time was attained through the nursing travel distance survey. In the simulation, movement within SS and NC was

		E-Ward							W-Ward						
Placement of Patients in Charge (Near: Near to SS or NC Far: Far to SS or NC)		E_D_1	E_D_2	E_DL_1	E_DL_2	E_L	E_N_1	E_N_2	W_D_1	W_D_2	W_DL_1	W_DL_2	W_L	W_N_1	W_N_2
		Near	Near	Near	Near	Near	Near	Near	Far	Near	Far	Near	Near	Near	Far
Simulation Results	A. Nursing Travel Distance(m)	2084	2203	4066	5138	3362	2796	4628	4152	3399	9412	4806	4711	4795	7878
	B. Nursing Travel Time(min.)	27.4	29.24	54.12	68.3	44.48	37.18	61.48	55.24	45.18	125.3	64.06	62.48	66.54	105
Surveyed Results	C. Number of visits to Patient Rooms (times)	79	54	130	80	93	107	135	72	80	206	123	136	166	197
	D. Actual Nursing Travel Time (min.)	29.07	41.23	48.04	67.42	46.13	44.28	65.48	45.33	43.38	108.13	58.59	52.21	62.51	91.05
Compare Analysis	E. Average Walking Speed (m/s)	1.19	0.89	1.41	1.26	1.21	1.05	1.17	1.52	1.3	1.45	1.36	1.5	1.27	1.44
	F. Nursing Travel Distance(m/hour)	238	252	313	395	384	233	386	415	340	672	343	483	399	657
	G. Number of Visits to Patient Rooms (times/hour)	9	6.2	10	6.2	10.6	8.9	11.3	7.2	8	14.7	8.8	13.9	13.8	16.4

Table 2. Simulation Results of Present Situation

neglected. In this analysis, the movement time in the nursing base was subtracted from the actual movement time revealed by the nursing travel distance survey. Thus, the travel time in corridor was defined and used as the actual travel time (D-line in Table 2). Furthermore, the time spent talking and working in the corridor was not added, so it could be said to be the pure travel time. In addition, the average walking speed of nurse (E-line in Table 2) was calculated from the actual travel time and the travel distance calculated in the simulation (A-line in Table 2).

Looking at the average walking speed of nurses (E-line in Table 2), comparing E-ward and W-ward, it could be seen that the walking speed of nurse was larger in W-ward. This may be due to the busyness of W-ward.

Comparing the distance traveled and average walking speed between day-shift nurses and day-long-shift nurses in Table 2, nurses (W_D_1, W_DL_1) whose bed room was far from nursing base were larger. From these, it could be inferred that nurses, who were in charge of far bed rooms from nursing bases, moved faster to cover the longer walking distance and to meet the increased workload.

(2) Number of Bed Rooms Visits and Travel Distance

Focusing on the number of visits to bed rooms (C-line in Table 2) and corridor travel distance (D-line in Table 2), and comparing the figures of day-shift nurses and day-long-shift nurses, nurses with fewer bed rooms visits walked longer travel distances, excepting W_DL_1 nurse who had most frequently visited bed rooms. In other words, nurses who were in charge of far bed rooms from nursing base tended to take a PC-cart to the front of bed room and to carry out the patient information input there, so that the reduction in number of bed rooms visits might have been possible.

6.2. Analysis of Nursing Travel Distance by Changing Nursing Base

As mentioned previously, in W-ward NC was only used for throwing trash and washing hands, and the actual nursing base was SS. Thus, it was assumed that such usage situation resulted in the increase of nursing travel distance in W-ward. The simulation was carried out in order to verify this and to predict the increase in nursing efficiency by the use of NC in W-ward. Moreover, in recent years, the computerization of medical equipment and medical record system realized the freer settings of number, locations, and spatial sizes of nursing bases during their planning and design stage as well as in usage stage in the existing hospital wards. Thus, it was considered that BIM simulations were highly useful by changing the physical features of nursing base.

		Day_Shift		DayLong_Shift		Night_Shift	
		W_D_1	W_D_2	W_DL_1	W_DL_2	W_N_1	W_N_2
Placement of Patients in Charge		Near NC	Near SS	Near SS	Near NC	Near NC	Near SS
Present Situation	Travel Distance(m)	4152.1	**3399.2**	9411.6	4806.1	**4795.0**	7878.3
	Travel Time(min.)	55.4	**45.3**	125.5	64.1	**63.9**	105.0
After Transfer	Travel Distance(m)	**3239.0**	3241.9	6939.1	4362.3	6168.4	**5746.4**
	Travel Time(min.)	**43.2**	42.2	92.5	58.2	82.2	**76.6**
Comparative Analysis	Reduced Distance(m)	-913.1	-157.3	-2472.5	-443.8	1373.4	-2131.9
	Reduced Time(min.)	12.2	3.2	27.0	5.9	19.3	28.4
	Reduced Rate(%)	-22	-5	-26	-9	29	-27

Table 3. Simulation Result of Transfer of Nursing Base

(1) Method of Changing Nursing Base Simulation

In this analysis, all functions other than the nurses' rest rooms in SS in W-ward were transferred to the NC. In this simulation, the original movement lines were displayed as much as possible based on the surveyed waypoint information such as the position of bed room's door where the target nurse came out. However, in the simulation all the waypoint information was deleted to automatically configure the most probable route. Thus, a nursing movement line may show the shortest distance, which is likely to be smaller than the actual travel distances.

(2) Simulation Result of Nursing Base Change

By changing the nursing base from SS to NC, the time and distance traveled by nurses in charge of closer bed rooms from NC were greatly reduced (Table 3). In particular, the nurses (W_DL_1, W_N_2) who were in charge of closer bed rooms from NC of day-long-shift

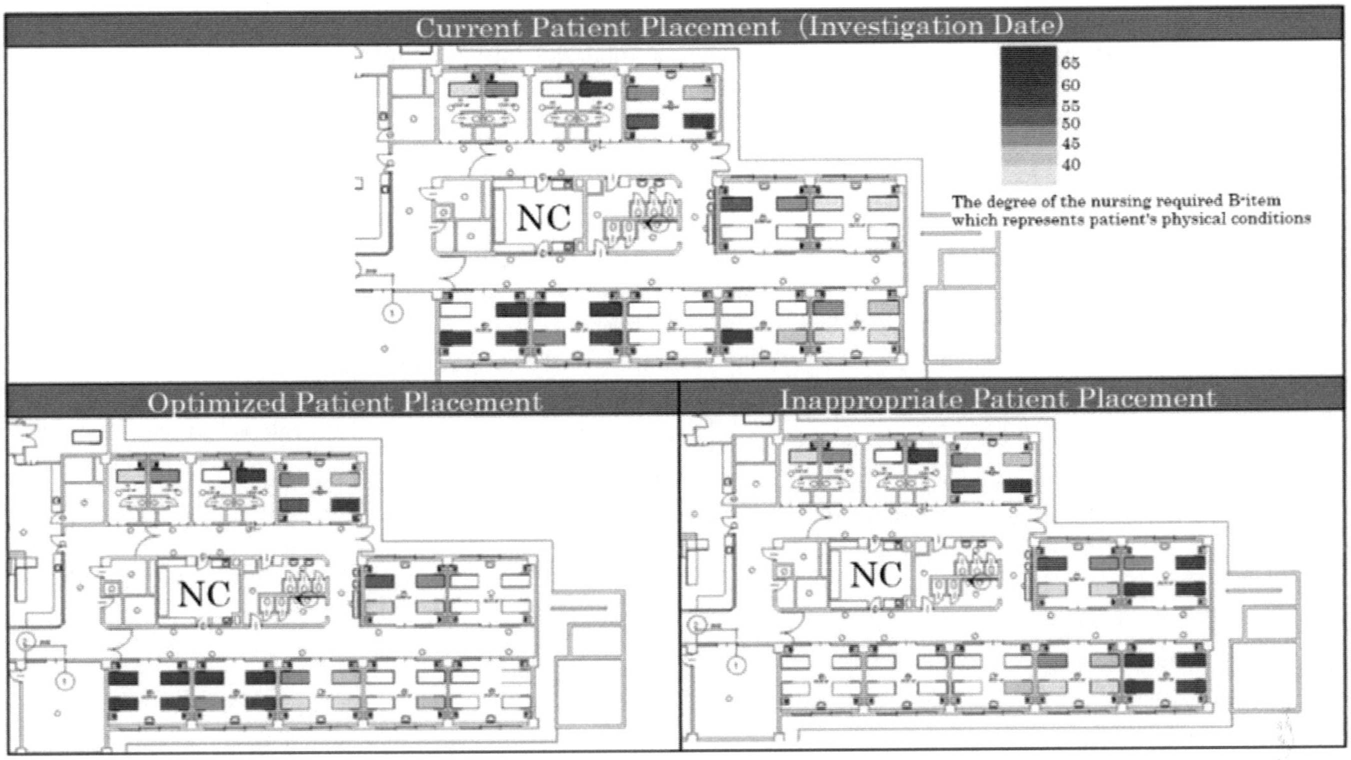

Figure 5. Patient Placement

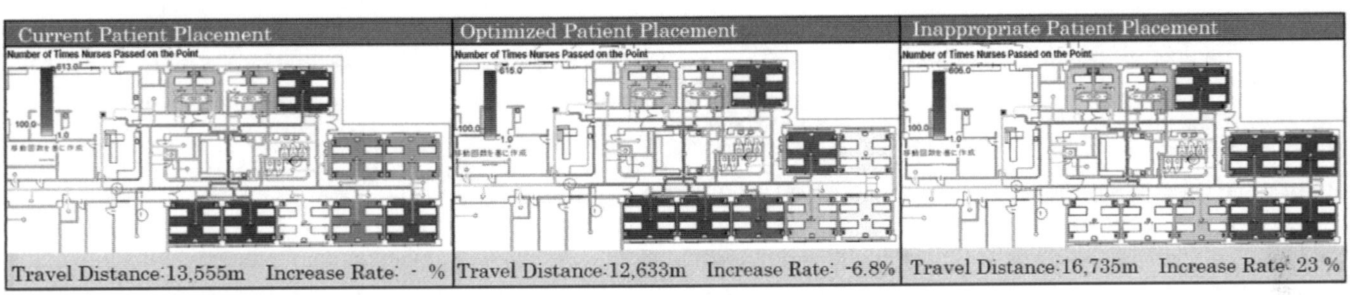

Figure 6. Simulation Results of Changing Patient Placement

nurses and night-shift nurses had reduced the travel distance of about 2 to 2.5 km, and the travel time of about 20 to 30 minutes (27% travel distance reduction). In addition, the reduction of the nursing travel time may result in the reduction of the workload, so that the nursing work time was reduced. On the other hand, nurses in charge of closer bed rooms to SS show little change. Furthermore, the night-shift nurse (W_N_1), who was in charge of close bed rooms to SS, increased the travel time by about 20 minutes (29% increase). In other words, it was not necessary to move the nursing bases for all nurses in the W-ward from SS to NC, but different nursing bases should be used depending on the positions of bed rooms in charge.

In addition, the current travel distance of nurses in charge of closer bed rooms from SS in W-ward after relocation of the nursing base was compared with the travel distance of nurses in charge of bed rooms closer to NC in E-ward. It was considered that the travel distance was almost the same. However, if a nurse W_DL_1 who frequently visited a number of bed rooms, that sample was excluded. Therefore, nurses in charge of closer bed rooms to NC were based in NC. Nurses in charge of closer bed rooms to SS could reduce the nursing travel distances by using SS as a nursing base. In addition, the amount of nursing work could be made comparable as E-ward. The use of hospital ward was considered closer to the ideal use of nursing base, which was the planning principle of original ward design shown in Figure 1 operating in three nursing units.

6.3. Estimation of total nursing travel distance by changing patient placements

Here, the hospital ward was evaluated from the viewpoint of total nursing travel distances using the nursing travel simulation. However, as shown in Section 5-2 (3), it was difficult to say that patients were always placed in the optimal room from the viewpoint of nursing travel distances. In fact, patients with infectious diseases were placed in a single bed room or in a 4-bed room far away from the nursing base, and those patients needing the ADL, activity of daily living, assistance were placed in closer bed room to the nursing base. Therefore, the hospital ward planning was evaluated by analyzing whether nursing travel distance changed with the appropriate or unsuitable bed room locations.

The simulation purpose to evaluate the hospital ward plan was performed using the total nursing travel distance, which was the sum of four day-shift nurses and day-long-shift nurses. (Figure 5).

(1) Method of Changing Bed Rooms Location

As the method of changing bed rooms location in this analysis, the optimal location was where the patient is moved to the nearest bed room to the nursing base, and the inappropriate location was where the patient was moved to the farthest bed room to the nursing base. The following describes the method to change bed rooms locations.

The optimal location was that the patient was transferred to near bed rooms to nursing base leaving the original bed room locations. For example, when one patient, referred as a-patient, in one 4-bed room (A-room) was transferred to a different 4-bed room (B-room) the number of bed room visits should be added to the original number of visits to B-room. However, the nurse movements in bed rooms were not recorded in the survey because of the patient privacy, and the number of visits for each patient in 4-bed rooms was not clarified. Thus, the proportional number of visits was calculated depending on the number of patients in 4-bed rooms, and this figure was added as related figure in patient transfer. Also, the transfer was not performed for single room patients to 4-bed rooms because the reasons for single room placement overweighed the simulation purpose and the repositions to other single rooms was found to be not significant from their placement. (Figure 4).

(2) Simulation Results of Bed Rooms Location Arrangements

As a result of the simulation, the total nursing travel distance by 4 nurses during the day-shift and day-long-shift in E-ward could be decreased by about 1 km (a 6.8% decrease). Previous studies show the hospital ward simulations with a reduction rate of about 5% (Taniguchi, et.al. 1984), confirming the planning and design principle in the surveyed hospital ward plan with the overall nursing efficiency issue.

Focusing on the range of rate of increase in nursing travel distances when the bed room was far from or near to nursing base, the optimal case was - 6.8% and the inappropriate case was + 23%, so the total of 29.8% difference was achieved. This showed the significant impact factor of patient placements on the nursing travel distances in the ward plan-type of the distributed nursing bases like the surveyed hospital. Moreover, following the downsizing trend of hospital beds for management efficiency and increasing demand for more patient privacy in Japan today, this plan-type feature should merit the renovation process to convert a 4-bed rooms to two single rooms in near future.

7. Conclusion

This study demonstrated the development and use of nursing travel distances simulation using BIM system. The analysis of nursing travel survey following BPE concept and the analysis of BIM simulation were conducted.

1) There was a strong correlation between Nursing Necessity Index B-item and the number of nurses' visits to patient beds room.

2) By improving the physical features and use of nursing bases, the nursing travel distances could be reduced by up to 27%.

3) The difference in nursing travel distance was clarified in the simulations for optimal bed rooms arrangement and unsuitable arrangement. There was a difference of nearly 30% in the surveyed hospital. As a result, it was analyzed that the nursing travel distances in a hospital ward plan-type of distributed nursing bases such as the surveyed hospital was significantly impacted by bed rooms configurations and patient locations.

In conclusion, by using and analyzing BIM simulation, not only the simulation results could be expressed visually, but also the ward planning and management could be analyzed from various viewpoints.

8. Future prospects

Some future focal issues include the significance of the view from nursing bases, the appropriateness of the bed rooms arrangement, and the impact of efficient material management as an evaluation axis for simulation in the hospital ward plan. The conventional Space Syntax analysis methodology might benefit the process. Moreover, these extra evaluation axes may be incorporated in the practice of evaluation of hospital ward floor plans.

In addition, the above BIM simulation system should merit the practice of planning design stage, construction management, and facility management following the BPE life-cycle process. Especially in the current stage of planning and design where there is only limited amount of time to carry out various simulations. The visual presentation from BIM simulations should improve the cooperation and coordination among building stake-holders including architects, construction managers, facility managers, project owners, as well as building users.

REFERENCES

Hendrich, A., Chow, M. P., Bafna, S., Heo, Y., and Skierczynski, B. A. (2009). Unit-Related Factors That Affect Nursing Time with Patients: Spatial Analysis of the Time and Motion Study, Health Environments Research & Design Journal Vol. 2, No. 2

Nuffield Provincial Hospitals Trust (1955). Studies in the Functions and Design of Hospitals, Oxford University Press

Preiser, W. F. E., Hardy, A. E., and Shramm, U. (2018). Building Performance Evaluation, From Delivery Process to Life Cycle Phases, Second Edition, Springer

Takeuchi, T. and Kato, A. (2018). Consideration on Simple Simulation Produced Visual Programming system of Building Information Modeling (BIM), Proceedings of Tokai Chapter Architectural Research Meeting, AIJ, Architectural Institute of Japan, (in Japanese)

Takeuchi, T., Yamaguchi, K., Takagi, M., and Kato, A. (2020). Study on Evaluation for Hospital Ward Planning and Management by Nursing Travel Distance Simulation Using BIM, Journal of Architecture and Planning, AIJ, to be published (in Japanese)

Taniguchi, G., Yanagisawa, M., Imai, S., Kato, A., Yamamoto, K. and Shida, K. (1984). A Study on Nursing Activities in Hospital Ward, Journal of Architecture and Planning, AIJ, Vol. 344, pp.116-125 (in Japanese)

Thompson, John D. and Golding, Grace (1975). The Hospital: A Social and Architectural History, Yale University Press

Toriyama, A., Watanabe, R., Nakayama, S., Kakehi, A. and Yamashita, T. (2008). Research on Function and Location of "Personal Nursing Bases" And Other Types of Nursing Bases: Reexamination studies on planning of acute care wards in the era of computerization of medical and patients' information Part II, Journal of Architecture and Planning, AIJ, Vol. 73, No. 625, pp.527-533 (in Japanese)

University of Fukui Hospital, Department of Nursing (2009). Introduction of PNS (in Japanese), https://www.hosp.u-fukui.ac.jp/kango/pns/ (accessed 2019-03-05)

Yanagisawa, M. (1978). Basic Planning for Hekinan Municipal Hospital, Nagoya University (in Japanese)

Yoshitake, Y. (1964). Nursing Work and Ward Planning in the Hospital. In: Study of Architectural planning. Kajima Institute Publishing Co., Ltd. (in Japanese)

Watanabe, R., Toriyama, A., Nakayama, S., Kakehi, A. and Yamashita, T. (2010). Relationship between Perioperative Period, Visit by Nurses and Nursing Care Needs in Acute Care Ward: Studies on a Wards Planning & Nursing Activity in the Hospitals part 8, Summaries of Technical Papers of Annual Meeting, AIJ, pp. 245-246 (in Japanese)

Zhou, Y. and Nagasawa, Y. (2013). Analysis on Nursing Activities in A Step-Down Care Unit, Journal of Architecture and Planning, AIJ (in Japanese)

Environmental Influence on Pro-Environmental Behavioral Intentions, Pro-Social Behavioral Intentions, and Donative Behavior

Micaela Lynne Nee (Connecticut College); Ann Sloan Devlin (Connecticut College)

1. ABSTRACT

The present research was designed to explore different types of environmental influences on pro-environmental behavioral intentions, pro-social behavioral intentions, and donative behavior. Participants were assigned to 1 of 5 environmental conditions: threatened nature, nature, urban, childhood home, or no environment. Participants included 252 individuals, 138 females and 114 males, recruited through Amazon Mechanical Turk. It was hypothesized that participants in the nature and threatened nature conditions would have the highest pro-environmental intentions and allocate the most money to charity, due to the effects nature has on pro-sociality, empathy, and giving. There were no significant differences between pro-social behavioral intentions or money donated in each condition. However, participants exposed to threatened environments had the highest pro-environmental intentions, and their scores were significantly different from those exposed to urban settings (p = .030). Exposure to threatened environments may prompt engagement in landscape preservation activities. Additional findings, limitations, and implications for future research are discussed.

1.1. Keywords

Pro-environmental behavior, nature, threatened environments, environmental influences

2. INTRODUCTION

The present research was designed to examine how exposure to different types of environments influences pro-social behavioral intentions, pro-environmental behavioral intentions, charity selection, and donative behavior. Spending time in certain types of environments, or simply being exposed to a type of setting, affects subsequent behavior (e.g., Arendt & Matthes, 2016; Klein & Hilbig, 2018; Weinstein, Legate, & Przybylski, 2013). In particular, environments influence types of donative behaviors (Arendt & Matthes, 2016; Klein & Hilbig, 2018). The potential of environments to trigger donative behaviors can be useful in understanding how to motivate charitable behavior that may encourage environmental protection. Different types of environments vary in their effects on charitable and donative behaviors (Weinstein, Przybylski, & Ryan, 2009a). Therefore, identifying and harnessing the potential of certain environments may encourage selfless behaviors.

Deep-rooted in the environmental literature is what Karmanov and Hamel (2008) refer to as the "urban vs. nature dichotomy" (p. 115). This dichotomy references the tendency in research to juxtapose beautiful natural scenes and heavily urbanized scenes to show the restorative effects of nature. Karmanov and Hamel point out the one-dimensionality of this approach to research, and show that well-designed urban spaces have the potential to be as restorative as natural spaces (2008). In an attempt to further break out of this dichotomy and expand our understanding of the impact of a greater diversity of environments, the present research includes a variety of settings: urban environments, nature environments, threatened nature environments, and childhood home environments.

2.1. Types of environments

Urban settings tend to be viewed more negatively, particularly in comparison to natural settings (e.g., Korpela, Klemettilä, & Hietanen, 2002; Weinstein, Przybylski, & Ryan, 2009a). When exposed to images of natural scenes, participants responded faster to a joyful tone than did those exposed to urban scenes; participants exposed to urban images responded faster to angry tones (Korpela et al., 2002). Urban settings also may promote selfish perspectives (Weinstein et al., 2009a). Viewing images of nature led to higher valuing of intrinsic aspirations, whereas viewing images of urban settings led to higher valuing of extrinsic aspirations, such as wealth and fame (Weinstein et al., 2009a). Relatedly, participants who viewed a nature documentary donated more money to an environmental NGO and recycled more than did participants who viewed urban New York architecture (Ibanez, Latourte, & Roussel, 2018).

Nature makes people more caring, generous

(Weinstein, Legate, & Przybylski, 2013), happy (Ibanez et al., 2018; Nisbet & Zelenski, 2011), and helpful (Zhang, Piff, Iyer, Kola, & Keltner, 2014). Simply placing plants in the lab setting for the research increased the amount of money participants decided to give to another participant (Weinstein et al., 2013). Relatedly, engaging in pro-social behavior may also increase pro-environmental behavior (Keenan, 2015). Those who wrote slogans for a poverty campaign were more likely to choose an eco-friendly product than were those who wrote slogans for an exercise campaign (control group), although those who wrote slogans for the environmental campaign were most likely to select the eco-friendly option (Keenan, 2015).

The present research expands on the literature that nature increases pro-social (e.g., Zhang et al., 2014) and pro-environmental behaviors (e.g., Arendt & Matthes, 2016), by including a threatened natural environment condition. Valtchanov and Ellard (2010) suggest that more research should be done on the physical properties of environments that contribute to the positive effects of nature. Research has historically looked at lush environments with abundant vegetation. Brooks, Ottley, Arbuthnott, and Sevigny (2017) extended the knowledge of the physical properties of environments by looking at the seasonal impact of the environmental setting.

The literature is not definitive regarding whether nature in a threatened state (e.g., deforestation) would result in more than fewer pro-social and pro-environmental behaviors. Therefore, one contribution of the present research is to further address that question. More beautiful natural scenes have led to more pro-social behaviors than have less aesthetically pleasing natural scenes (Zhang et al., 2014). Literature has also shown that the more concerned people are for the environment, the more connected to nature they feel, and the more likely they are to perform pro-environmental behaviors (Krajhanzl, 2010; Pereira & Forester, 2015). People who have greater environmental concern are more likely to engage in pro-environmental behavior, as a result of feelings of sadness and sympathy over environmental damage (Krajhanzl, 2010). The concept of "solastalgia", or "the distress caused by environmental change," has also emerged in the literature as a reaction to the emotions that stem from living in threatened environments (Albrecht et al., 2007). In the current experiment, the threatened environment condition was included to further explore the potential impact of such loss reflected in the physical condition of the environment, such as deforestation; including scenes of both categories of nature (threatened and unthreatened) permits an examination of their relative impact.

Klein and Hilbig (2018) recently assessed how threatened nature scenes may contribute to pro-environmental behavior. Participants who were shown scenes of littered beaches and poached animals had increased pro-environmental behavior; those who viewed scenes of aesthetically pleasing beaches or roaming animals did not. Klein and Hilbig (2018) further examined how pro-environmental behavior and cooperation were influenced when juxtaposed. Among non-selfish individuals, those in the destroyed nature group showed increased pro-environmental behavior (i.e., "invest money in environment"), but reduced cooperation (i.e., "invest money in public goods") while those in the social interactions video group showed increased cooperation. These results are pertinent to the current research in understanding how content contributes to charity selection, pro-social behavior, and pro-environmental behavior.

Childhood homes are typically viewed as nostalgic places (Low & Altman, 1992; Wilson, 2013), and these feelings increase empathy (Cordaro, 2011; Cheung, Sedikides, & Wildschut, 2017), prosocial tendencies (Stephan et al., 2014), and generosity (Ford & Merchant, 2010; Merchant, Ford, & Rose, 2011; Zhou, Wildschut, Sedikides, Shi, & Feng, 2012). Among the past spaces in people's lives, childhood homes are important because they serve as reminders of childhood memories (Low & Altman, 1992). The relationships, experiences, and memories in these spaces (Weinstein et al., 2013) may help transform houses into homes. The home or one's "spatial anchor" (Lewicka, 2011, p. 211) is often viewed as a special and nostalgic place (Wilson, 2013) that may provide individuals with a sense of comfort (Manzo, 2003; Wilson, 2013). Evoking nostalgia in participants increases social connectedness (Abeyta, Routledge, & Juhl, 2015; Stephan et al., 2014; Wildschut, Sedikides, Arndt, & Wildschut, 2010) and donative behavior (Ford & Merchant, 2010; Merchant, Ford, & Rose, 2011; Zhou et al., 2012). Weinstein et al. (2013) looked at how important spaces in childhood homes relate to nostalgia and found that "nostalgia links with aesthetic perceptions of the physical world, and may be responsible for the relation between basic physiological needs and perceptions of beauty" (Weinstein et al., 2013, p. 257). Inclusion of a childhood home condition in the current research helps further our understanding of the impact of thoughts of home on pro-social, pro-environmental, and donative behavior.

2.2. Environmental influences on behavioral intentions

According to Eisenberg and Mussen (1989), pro-social behavior is defined as, "voluntary actions that are intended to help or benefit another individual or group of individuals" (p. 3). In regard to environmental influences on pro-social behavior, natural settings seem to increase pro-social tendencies (Weinstein et al., 2009a; Zhang et al., 2014). More attractive natural environments have an even larger impact on pro-social behavior (Zhang et al., 2014). Less attractive threatened nature may be less influential on pro-social behavior than more attractive, unthreatened natural scenes.

In contrast to natural settings, urban settings are associated with greater emphasis on selfish aspirations (Weinstein et al., 2009a), and therefore may have a demotivating effect on pro-social actions. Childhood homes often trigger nostalgic feelings (Weinstein et al., 2013), which contribute to greater connectedness to (Stephan et al., 2014) and concern for others (Cordaro, 2011; Cheung et al., 2017). In order to assess different environmental influences on prosocial behavioral intentions, a childhood home condition was also included in the present research.

Pro-environmental behavior, or "behavior that consciously seeks to minimize the negative impact of one's actions on the natural and built world" (Kollmuss & Agyeman, 2002, p. 240) is needed given the current state of the environment. Despite climate change being "here" and "now" (Schwartz & Popovich, 2019, para. 3), there is a tendency for people to overlook the damages and impending threats of the warming climate frequently discussed in the literature (e.g., Gifford, 2011; Kollmuss & Agyeman, 2002; van Vugt, Griskevicius, & Schultz, 2014;Yale Climate Opinion Maps, 2018), even though 57% of adults think global warming is mostly caused by human activities (Marlon, Howe, Mildenberger, Leiserowitz, & Wang, 2018). Despite these threats, the minimal response to action noted by Gifford (2011) and others (e.g., Kollmuss & Agyeman, 2002) may be due to the psychological barriers, such as not thinking individual actions will make an impact.

Lack of engagement with the issues (Kollmuss & Agyeman, 2002) is one such barrier, and fear of environmental degradation may also prompt disengagement and a sense of helplessness (Gifford, 2011; Landry, Gifford, Milfont, Weeks, & Arnocky, 2018; Manolas, 2011), which may explain the inconclusive results of the use of fear appeals in climate communication (Landry et al., 2018). Donations to an environmental organization and interest in joining an environmental activist group were related to environmental concern only among individuals with low feelings of helplessness (Landry et al., 2018). In summary, despite the varying names for human biases/barriers to climate change, there is a consensus that encouraging environmental action requires people to see the issues as personally meaningful and salient, and to recognize themselves as capable agents of change. While there is some research about what aspects of exposure to environments may prompt salience and subsequent action, arguably more is needed.

Research has looked at ways to make environmental issues more immediate, apparent, and emotionally charged. Using the language from Wang et al. (2018), a major question for the current research was the extent to which different kinds of environments reflect personally relevant "objects of care" that encourage the intention to protect the environment. Pro-environmental messages that describe the specific benefits of engaging in a behavior have been successful in promoting pro-environmental behavior (Tate, 2014), such as the selection of an eco-friendly toilet paper (Keenan, 2015).

Encouraging pro-environmental behavior does not require direct experience of the effects of climate change. Spending time in and cultivating attachment to natural settings also lead to pro-environmental behavior (e.g., Walker & Chapman, 2003). Forming attachments to natural places may transform these environments into objects of care (Wang et al., 2018). Strong attachments to natural environments lead to greater protection efforts (Halpenny, 2010a; Najafi & Kamal, 2012; Walker & Chapman, 2003) and engagement in higher commitment behaviors to protect the natural place, such as donating or volunteering (Tonge, Ryan, Moore, & Beckley, 2015). Although these studies show the promise of engagement with natural settings, many people live in cities, or areas with limited access to nature. Considering the limitations of residence, as well as an increasingly tech-oriented society (Klein & Hilbig, 2018), the simplest way to expose masses of people to nature may involve media. Visuals are more impactful than text (Dixon, McKeever, Holton, Clarke, & Eosco, 2015; Lazard & Atkinson, 2015), and are more effective in pro-environmental communication (Lazard & Atkinson, 2015). Several studies have effectively exposed people to nature through visual stimulation, including photographs or videos (e.g., Arendt & Matthes, 2016; Berto, 2005; Brooks et al., 2017; Klein & Hilbig, 2018). Exposure to a nature documentary strengthened connectedness to nature and increased pro-environmental behavior (Arendt & Matthes, 2016; Ibanez et al., 2018; Klein & Hilbig, 2018). Participants who watched the nature documentary chose a nature-based charity more than did the participants who watched the documentary unrelated to nature (Arendt & Matthes, 2016). Photographs, used as heuristics, can overcome biases (e.g., low deference to science), which can be useful when representing information about global warming (Dixon et al., 2015). Photographs, considered a form of 'proof', have stronger effects on behavioral intentions when compared to iconic or text-based information, particularly among people unfamiliar with an event (Rickard, Schuldt, Esoco, Scherer, & Danziano, 2017), which has important implications for distant/invisible effects of climate change.

2.3. Relationship between pro-social and pro-environmental behavior

There is inconsistent evidence about the relationship between pro-social and pro-environmental behavior. Psychological and physical barriers to pro-environmental action may impede the formation of that relationship. For instance, a physical barrier may include a lack of refillable bottles at local grocery stores (Kurisu, 2015). Other research shows a correlation between pro-social behavior and pro-environment behavior; pro-social behavior was shown to foster pro-environmental behavior, and vice-versa (Neaman et al., 2018). On the other hand, Markowitz, Goldberg, Ashton, and Lee (2011) did not find pro-environmental individuals to have more pro-social traits, possibly because pro-environmental behavior may involve non-human recipients (nature and animals) (Nolan & Schultz, 2015). However, content may influence these kinds of behaviors. Klein and Hilbig (2018) show that exposure to destroyed nature increased pro-environmental behavior, at the cost of cooperative behavior. Within the context of exposure to different kinds of environments, the current research examined the relationship between pro-social and pro-environmental behavioral intentions.

2.4. Charitable intentions and donative behavior

Beyond pro-social and pro-environmental behavioral intentions, the present research also assessed how environments affect charitable intentions and donative behavior. There are many demands for donations from a variety of charitable campaigns; this competition (Krieg & Samek, 2016) leaves organizations wondering how to maximize donations. One of the goals of this paper was to see if charity selection (with possible implications for environmental protection) could be influenced by exposure to different kinds of environments.

People rate humanitarian causes as more specific, concrete, and making a more immediate impact than they do environmental causes (Hossain, 2014), consistent with the psychological barriers to climate change (Gifford, 2011; Kollmuss & Agyeman, 2002). When given different pairs of humanitarian and environmental organizations (e.g., UNICEF vs. WWF), participants chose the humanitarian organization slightly more frequently (de Groot & Steg, 2008). Similarly, this proclivity toward humanitarian causes emerged in attitudes toward cause and brand alliances. People indicated more positive attitudes to causes that directly impact people over animal or environmental causes (Lafferty & Edmondson, 2014). This preference for people-related causes is also reflected in Giving USA 2018: The annual report of philanthropy from 2017 (https://givingusa.org/see-the-numbers-giving-usa-2018-infographic/). While \$50.06 billion in donations went to human services, only \$11.83 billion in donations went to environment/animals.

3. RESEARCH OBJECTIVES

Given the reviewed literature, this research examined the effect of different environments on behavioral and donative intentions, employing exposure to a broader range of types of environments than has previously been assessed. Participants were exposed to one of four kinds of environments (childhood home, natural, urban, and threatened natural environments), or a control. Pro-social behavioral intentions and pro-environmental behavioral intentions were assessed after exposure to one of these conditions.

To assess charity selection and donative behavior, participants were given a choice to donate to Habitat for Humanity (HH; humanitarian-based as well as home-based) or The Environmental Defense Fund (EDF; nature-based). Increased feelings of social-connectedness among participants in the childhood home condition were expected to result in a larger number of person-related charity selections (HH), than nature-based charity (EDF) selections, because HH directly helps others. Conversely, exposure to the natural environments was expected to prompt greater selection of the EDF, in which the environment is the recipient.

Building on the literature, the current research examined whether exposure to nature not only influences participants' pro-environmental behavior intentions and charity selection, but also how it influences participants' generosity. Instead of specifically giving participants a dollar to donate, as in Arendt and Matthes' (2016) study, participants chose how much of the dollar they hypothetically wanted to allocate, which could be argued to be a situation that encourages pro-social behavior (and social desirability), and thus tests the limit of such generosity.

It was hypothesized that the five conditions would differentially impact behavior:

1) Viewing pictures of natural environments would increase not only pro-environmental behavioral intentions, but also pro-social behavioral intentions and

amount allocated to charity, when compared to the other environmental conditions.

2) Viewing images of urban environments would increase the amount allocated to self, based on the literature that suggests urban environments increase extrinsic aspirations (Weinstein et al., 2009a), when compared to the other environmental conditions.

3) Imagining childhood homes would increase pro-social behavioral intentions, when compared to the other conditions. No directional hypothesis was formulated regarding the effect on pro-environmental behavioral intention, given the lack of guidance in the literature.

4) Imagining childhood homes would cause participants to be more likely to select HH than the EDF, because helping other people supports nostalgia's social function.

5) Viewing the threatened environment condition would lead participants to select EDF more often and have higher intentions to engage in pro-environmental behavior than would participants in the urban, control, and childhood home conditions, but not in the nature condition.

4. METHODS

4.1. Participants and design

There were 252 participants in total in the following conditions: natural environment (n = 52), urban environment (n = 51), home environment (n = 41), threatened natural environment (n = 53), and control (no pictures) condition (n = 55). Two participants were excluded from the pro-social behavioral measure for leaving one unanswered question, n = 250; Nature (n = 51), Urban (n = 50), (other conditions remain the same as the original sample). One participant was excluded from the charity selection measure for not selecting an option, n = 251; Threatened nature (n = 52), (other conditions remain the same as the original sample). The participants were recruited through Amazon MTurk and compensated $.50 for their participation. There were 138 females and 114 males from the United States; participants had a record of at least a 97% job acceptance rate on Amazon MTurk. Mean age was 38.24 (range = 20-76 years). Self-reported race/ethnicity was 83.3% White, 5.6% Asian/ Pacific Islander, 5.2% Black and African American, 2.8% Latino, 1.6% biracial, and 0.04% American Idian and Native American.

4.2. Measures

Pro-environmental behavioral intentions were measured using Halpenny's Pro-environment Behavior Intentions Scale (Halpenny, 2010b) that consists of 11-items assessing intentions to engage in pro-environmental behavior; it employs a 5-point scale ranging from "Not probable" to "Highly probable." A sample item is: "Learn more about the state of the environment and how to help solve environmental problems." In this study, the scale had high reliability, $\alpha = .91$.

Prosocial behavioral intentions were measured using the Pro-social Intentions Measure (Pavey, Greitemeyer, & Sparks, 2012) that assesses intentions to engage in general pro-social behaviors through six items ($\alpha = .70$); it employs a 5-point scale ranging from "Definitely will not" to "Definitely will." A sample item is: "Go out of my way to help a friend in need." In this study, two items that dealt with helping a stranger and the community were presented as one question to yield five items; the scale had good reliability, $\alpha = .83$.

Level of environmental immersion was measured using The Player Experience of Need Satisfaction Physical Presence Scale (Adapted Version) (Weinstein, Przybylski, & Ryan, 2009b) that assesses immersion in the environments through three items ($\alpha = .84$); it employs a 5-point scale ranging from "A great deal" to "Not at all." A sample item is: "How much did the visual aspects of the environments involve you?" The scale was included to assess the level of involvement in the four environmental conditions.

To assess charity selection, participants were given a choice between an environment-focused charity (EDF; Environmental Defense Fund) and a humanitarian-focused charity (HH; Habitat for Humanity). Each charity was accompanied with a mission statement. The EDF was presented with, "Environmental Defense Fund's mission is to preserve the natural systems on which all life depends" (https://www.edf.org/our-mission-and-values), and HH was presented with, "Habitat for Humanity brings people together to build homes, communities and hope" (https://www.habitat.org/about/mission-and-vision).

Hypothetical donative behavior was assessed by informing participants they had a hypothetical dollar to split between the charity they selected and themselves. They were offered the choice to give the $1 to the charity, keep the $1 for themselves, or divide the $1 between the charity and themselves. There was a space to type in the amount they chose to give to charity and another space for the amount they chose to give themselves.

4.3. Stimuli

In the natural environment condition, participants viewed 50 images of nature scenes; in the urban environment condition, participants viewed

50 images of urban scenes. The images were taken from Berman et al. (2008) In the natural environment condition, six images from Berman et al. (2008) were removed for depicting any threat in a natural setting, such as fallen trees, and were replaced with non-threat nature photos similar to Berman et al.'s found online. In the threatened environment condition, there were 50 images of threatened natural environments that were carefully selected to avoid explicit signs of human intervention. The websites from which these images were drawn are available from the first author. In order to select a sample of 50 photos that represented the most threatened environments, 10 adults, ages 19-53, were asked to rate 70 images on a 1-5 scale based on how much they perceived the environment to be threatened. The 50 images that received the highest averaged ratings were selected for inclusion in the study. Image viewing was self-paced; a rating scale was provided with each image to encourage engagement. In all three image conditions, each image was shown with a 1-5 scale of how much participants liked each image, adapted from the procedure of Berman et al. (2008), who used a 1-3 rating scale. A 5-point scale was selected to provide the opportunity for more variability in response.

4.4. Procedure

Participants were recruited through Amazon MTurk and received $.50 for participation. The study was created on Qualtrics; participants were randomly assigned to one of the five conditions by the Qualtrics randomizer function. Participants first provided informed consent. Then, depending on the group to which the participants were randomly assigned, they were exposed to pictures of an a natural environment, urban environment, threatened natural environment, no pictures, or asked to immerse themselves in their childhood home through the following prompt: "Think back to your childhood home - where you spent the longest period growing up. When you close your eyes, imagine being fully immersed in the environment and describe each of the following rooms (living room, bedroom, kitchen, dining room) in detail." Then, excluding those in the control group, participants completed the Player Experience of Need Satisfaction Physical Presence Scale Adapted Version to assess their level of immersion in the environments. Participants in the nature, urban, and threatened nature conditions also completed a manipulation check to assess how much the images represented urban to nature-based scenes. All participants completed the Pro-social Intentions Measure and the Pro-environmental Behavioral Intentions Scale.

Following the completion of these two measures, they selected a charity (representing an environmental or a humanitarian cause) they would most like to donate to; the order of choices was randomized. After they made their selection, participants were informed they were receiving a hypothetical $1 to allocate between the charity they previously selected and themselves. They filled in their designated amounts in the text boxes provided. Participants then filled out their demographic information, including: gender, age, race, annual income, how often they donate per year, and where they choose to donate. At the end they were debriefed. The study took less than 10 minutes to complete.

Participants were excluded from the corresponding scale analyses if they left unanswered more than 15% of questions on a scale. On the Pro-environment behavior measure four participants each left unanswered one question; the questions for these participants were imputed with the overall mean of each question.

5. RESULTS

Participants reported different levels of preference among natural, urban, and threatened environments, $F(2, 141) = 111.78$, $p < .001$, $\eta p2 = 0.61$. Participants in each group rated all 50 images in their condition on a 1-5 scale (1 = like very much and 5 = dislike very much). Participants in the natural environment group reported a mean of 1.85, those in the urban environment group reported a mean of 3.12, and participants in the threatened natural environment group reported a mean of 3.96. Images of nature were more preferred than images of urban scenes, $p < .001$, 95% CI [0.94, 1.60], or threatened natural environments, $p < .001$, 95% CI [1.77, 2.44]. Images of urban scenes were also significantly more preferred than images of threatened natural environments, $p < .001$, 95% CI [0.50, 1.18].

The Player Experience of Need Satisfaction Physical Presence Scale Adapted Version (Weinstein et al., 2009b) had three questions that were rated 1-5. Lower scores indicated more immersion in the environment. The average of the three questions in each condition is as follows: threatened environment ($M = 2.04$), natural environment ($M = 1.74$), urban environment ($M = 2.35$), and home environment ($M = 1.82$). To evaluate levels of immersion in each environmental condition, a one-way analysis of variance was conducted. There were significant differences in reported levels of environmental immersion, $F(3, 193) = 4.96$, $p = .002$, $\eta p2 = 0.07$; participants in the nature, $p = .003$, 95% CI [1.51, 1.98], and childhood home, $p = .020$, 95% CI [1.56, 2.08], conditions reported higher levels of immersion than did participants in the urban condition, 95% CI [2.11, 2.58]. A manipulation check that asked participants to rate

Table 1. Means, SDs, and CIs of Pro-Environmental Behavioral Intentions Scores.

		Scale Score		
Environment	*n*	*M*	*SD*	*CI*
Threatened Nature	53	38.09[a]	8.67	[35.41, 40.78]
Nature	52	35.10	9.27	[32.39, 37.81]
Urban	51	32.39[a]	10.79	[29.66, 35.13]
Control	55	33.91	10.18	[31.28, 36.54]
Home	41	33.24	10.75	[30.19, 36.30]

Note. The maximum score is 55. Superscripts show significant differences.

1-7 how much their images were urban (1) to nature-based (7), confirmed that participants attended to and accurately categorized the images. The average rating was 2.20 in the urban condition, 6.85 in the nature condition, and 6.53 in the threatened nature condition. An ANOVA was conducted to evaluate the manipulation check and confirmed significant differences, $F(2, 153) = 435.47$, $p < .001$, $\eta p2 = 0.85$. The participants in the urban condition rated the images as significantly more urban-based, 95% CI [1.95, 2.44], than did participants in the nature group, $p < .001$, 95% CI [6.60, 7.10], and participants in the threatened nature group, $p < .001$, 95% CI [6.29, 6.77].

5.1. Pro-environmental behavioral intentions

An ANOVA was conducted to evaluate the different environmental conditions on pro-environmental intentions, and there was a significant effect, $F(4, 247) = 2.57$, $p = .039$, $\eta p2 = .040$. It had been hypothesized that the nature and threatened environment conditions would increase pro-environmental behavioral intentions. A Tukey post-hoc analysis revealed that exposure to threatened natural environments (95% CI [35.41, 40.78]) led to increased intentions to engage in pro-environmental behavior compared to exposure to urban environments (95% CI [29.66, 35.13]), $p = .030$ (see Table 1 for Means, Standard Deviations, and Confidence Intervals).

5.2. Pro-social behavioral intentions

To evaluate environmental influence on general pro-social behavioral intentions, an ANOVA was conducted. Environmental condition did not have a significant effect on the scores of the pro-social behavioral intentions measure, $F(4, 245) = 0.62$, $p = .649$, $\eta p2 = 0.01$ (see Table 2 for Means, Standard Deviations, and Confidence Intervals).

5.3. Pro-environmental and pro-social behavioral intentions

Higher pro-social behavioral intentions were associated with higher pro-environmental intentions, $r(250) = .44$, $p < .001$.

5.4. Donative behavior

An ANOVA was conducted to evaluate the environmental influences on giving behavior. There was no significant effect of environment on donation amounts, $F(4, 247) = 0.09$, $p = .985$, $\eta p2 = .001$ (see Table 3 for Means, Standard Deviations, Confidence Intervals).

Table 2. Means, SDs, and CIs of Pro-Social Behavioral Intentions Scores

		Scale Score		
Environment	*n*	*M*	*SD*	*CI*
Threatened Nature	53	18.53	4.01	[17.34, 19.72]
Nature	51	8.10	4.30	[16.89, 19.31]
Urban	50	17.34	4.50	[16.12, 18.56]
Control	55	17.64	4.55	[16.47, 18.80]
Home	41	17.51	4.50	[16.16, 18.86]

Note. The maximum score is 25.

Table 3. Means, SDs, and CIs of Amounts to Charity and Self in Each Environment

		Charity		*Self*		
Environment	*n*	*M*	*CI*	*M*	*CI*	*SD*
Threatened Nature	53	0.55	[0.44, 0.66]	0.45	[0.34, 0.56]	0.39
Nature	52	0.58	[0.47, 0.68]	0.42	[0.32, 0.53]	0.40
Urban	51	0.55	[0.44, 0.66]	0.45	[0.35, 0.56]	0.42
Control	55	0.56	[0.45, 0.66]	0.44	[0.32, 0.54]	0.36
Home	41	0.53	[0.41, 0.65]	0.47	[0.35, 0.59]	0.39

Note. Participants were given a hypothetical $1 to allocate between the charity they selected and themselves

5.5. Charity selection

To evaluate the role of the environment on charity choice, a chi-square analysis was conducted. Environment did not influence charity choice, $\chi2$ (4, N = 251) = 5.00, p = .288. Most participants selected HH (60.6 %) over the EDF (39.4%). However, the natural environment condition was the only group that had a higher number of participants choose the EDF than HH, and the only group where selection of HH dropped below 50% (see Table 4 for Percentages). To evaluate the influence of nature and home-based environments on choosing a charity that was either nature or home-based, a chi-square analysis was conducted between the nature and home environmental conditions. The influence of these two environments on charity choice approached significance, $\chi2$ (1, N = 93) = 3.82, p = .051. The influence of images of natural environments, compared to the control condition (no environmental manipulation), was trending, $\chi2$ (1, N = 107) = 3.29, p = .070. Participants in the natural environment condition chose the environmental charity more often than did participants who did not receive an environmental manipulation. To evaluate the influence of threatened and natural environments on charity choice a chi square analysis was conducted between the threatened nature and nature conditions. Although not significant, $\chi2$ (1, N = 104) = 1.90, p = .168, more chose EDF in the nature group (51.9%) than in the threatened nature group (38.5%).

6. CONCLUSIONS

This research suggests that viewing threatened natural environments may make people more likely to engage in pro-environmental behaviors, particularly more than when viewing urban environments. These results contribute to expanding beyond the "urban and nature dichotomy" (Karmanov & Hamel, 2008), and suggest that viewing threatened natural environments may be more effective in increasing pro-environmental behavior than viewing natural environments. Images of threatened environments may be used to motivate environmental engagement and action and, as stated earlier, may help give a face to climate change. Further, the scenes do not need to include 'graphic and catastrophic imagery' (Nicholson-Cole, 2005, p. 264) to prompt action. 'Negative landscape change' (p. 150) increases support for landscape conservation and preservation strategies (Walker & Ryan, 2008).

7. DISCUSSION

This study was designed to assess the impact of exposure to environments on pro-social and pro-environmental intentions and donative behavior and to test a wider variety of environments than has typically been the case. Participants' ratings of the different environments indicated that participants had a much higher preference for images of natural environments than they did of urban environments or threatened

Table 4. Percent of Participants Who Chose Each Charity by Condition

Charity	Threatened (n = 52)	Nature (n = 52)	Urban (n = 51)	Control (n = 55)	Home (n = 41)
EDF	38.5	51.9	39.2	34.5	31.7
HH	61.5	48.1	60.8	65.5	68.3

Note. EDF = Environmental Defense Fund; HH = Habitat for Humanity

natural environments, and that exposure to threatened environments has the capacity to significantly affect pro-environmental behavioral intentions, in comparison to viewing urban environments.

Exposing participants to threatened natural environments resulted in higher intentions to perform pro-environmental behaviors, such as talking to others about environmental issues, than did exposure to urban environments. This finding is noteworthy because it expands on literature that compares the effects of natural versus urban environments. Viewing threatened natural environments led to the highest intentions to participate in pro-environmental behaviors, in agreement with the findings by Klein and Hilbig (2018). The ability of still photographs to increase behavioral intentions has been documented in previous research (e.g., Dixon et al., 2015; Rickard et al., 2017), and has promising implications for pro-environmental communication. However, unlike the destroyed nature used in Klein and Hilbig's study (2018), the threatened nature images in the present study did not include overt signs of human intervention (e.g., poaching and littering), and were still able to impact behavioral intentions. When people think of climate change they often draw on emotional yet often distant imagery, such as polar bears on melting ice (Nicholson-Cole, 2005). Although climate change imagery should be emotionally arousing, often featuring people or animals, it is also recommended that the imagery be "easy to relate to both spatially and temporally" (Nicholson-Cole, 2005, p. 267). The potentially relatable content of the photographs may provide an explanation as to why the scenes influenced behavioral intentions.

The loss of valued landscape features, threatened through environmental disruption/change, evokes a sense of loss in land users. Residents who viewed their rural hometown as threatened by development were more willing to support and engage in landscape preservation activities (Walker & Ryan, 2008). Another explanation may be that photographs of threatened environments are used as a heuristic to overcoming contradictory beliefs (Dixon et al., 2015) about climate change, and may "represent abstract ideas- possible outcomes that may be (at present) unobservable in a physical sense" (Rickard et al., 2017, p. 5), which may strengthen behavioral intentions. Ultimately, it seems images of threatened environments help give a face to climate change, and therefore motivate pro-environmental behavior.

Although not statistically significant, the natural environment group reported the second highest intentions to engage in pro-environmental behaviors among the five conditions. The control and home groups reported similar pro-environmental intentions, closer to the mean of the urban group. This pattern may indicate that viewing threatened natural and natural environments leads to increased intentions to participate in environmental behaviors. The higher mean in the nature condition, compared to the urban, home, and control conditions, may lend support to the connection between nature exposure and environmental attitudes and behaviors (Halpenny, 2010a; Tonge et al., 2015; Walker & Chapman, 2003; Zelenski, Dopko, & Capaldi, 2015); furthermore, the demonstrated influence of exposure to threatened natural environments, expands on this connection between varieties of natural environments and pro-environmental behavior. The finding also reminds us that nature is not a singular construct; researchers need to acknowledge the variability in the content of nature.

Although the nature and threatened nature conditions had slightly higher means than the urban, control, and home groups, these differences were not statistically significant. This outcome does not support the literature that suggests exposure to nature increases pro-social tendencies and donative behavior (Weinstein et al., 2009a; Zhang et al., 2014) and viewing urban settings increases selfishly motivated aspirations. Nor did imagining childhood home environments heighten pro-social behavior, despite the social-connectedness and nostalgia elicited by home spaces (Weinstein et al., 2013).

People with higher pro-social behavioral intentions tended to have higher pro-environmental behavioral intentions. This positive relationship has been shown in previous research (e.g., Neaman et al., 2018). Future work may explore how this may contribute to increasing both types of behaviors, as suggested by Neaman et al. (2018). Viewing images of nature or threatened nature did not lead to selection of the environmental charity (EDF) at higher rates than for Habitat for Humanity. However, when individual conditions were tested against each other, there were trending results between the nature and home groups and the nature and control groups, indicating that exposure to nature may increase preference for an environmental charity over the home and control groups. The home group also had the highest percentage of participants who chose Habitat for Humanity, which was expected because this charity matched the content of that condition's environment. While not significant, the trending results of the influence of natural environments lend some support to Arendt and Matthes' (2016) finding that exposure to a natural environment leads to choosing to donate to an environmental charity more than another type of charity.

Contrary to the hypothesis, viewing threatened

environments did not lead to increased selection of the environmental charity. These results may be attributed to a preference for people-related causes (de Groot & Steg, 2008); people like to donate to people (Nolan & Schultz, 2015). Based on donor history, the majority of charities participants reported were under the category of "human services" (charitynavigator.org). These results are similar to those seen in Giving USA's annual report (givingusa.org). Furthermore, the type of natural environment (threatened vs. unthreatened) may influence selection differently. While not significant (p = .168), that level of the difference between the threatened nature and nature charity selections may suggest potential for different kinds of nature scenes to be used in different ways to benefit the environment. Exposure to threatened nature may remind people of the need for corrective pro-environmental behavior, such as joining in community clean up efforts, whereas exposure to unthreatened nature may serve to draw attention to preserving the natural systems, which is part of the EDF mission statement participants read.

This research built on earlier research (e.g., Arendt & Matthes, 2016) by not only looking at how exposure to natural environments influences charity selection, but also whether it contributes to donation amount. It suggests that viewing images of natural environments does not lead to significantly different hypothetical donations. Looking not only at the influence of selection choice, but also how much donation generosity is influenced is important, because in order to increase donations, people must first be motivated to donate.

In regard to charity selection, the degree of familiarity between the two charities could have confounded the environmental manipulations. However, both were selected from Charity Navigator's "Top 100 Nonprofits on the Web" (https://topnonprofits.com/lists/best-nonprofits-on-the-web/), compiled based on which charities get the most traction.

Another limitation of this study may have been the strength of the environmental manipulation. Although images have successfully been used to manipulate environments in other studies (Berman et al., 2008; Karmanov & Hamel, 2008), the length of viewing may not have been sufficient, although others (e.g., Berto, 2005) have found effects also using self-paced viewing. Additionally, others have found effects (e.g., Berman et al., 2008) using the same number of scenes. An alternative could also be exposure to environments through video as in Arendt and Matthes' (2016) design, which may be more immersive than the use of static images. Virtual reality is a promising, new way to simulate nature "without being there." Using virtual reality to simulate a restorative natural island scene was able to increase participants' positive affect and decrease negative affect compared to participants who were exposed to a simulation of an urban scene (Valtchanov & Ellard, 2010).

Future research might focus on the potential for threatened environments to increase pro-environmental behavior. In particular, different kinds of threatened nature should be examined, such as nature threatened by human intervention (e.g., cutting down trees) or natural intervention (e.g., lightning strike causing a forest fire). Future work may explore how increased pro-environmental behavioral intentions translate into actual pro-environmental behavior, through the use of an environmental task, such as recycling (Ibanez et al., 2018) or selecting eco-friendly products (Keenan, 2015). Research has explored how connectedness to nature contributes to pro-environmental behavior, however the potential for exposure to threatened natural environments to promote pro-environmental behavior could be further examined, particularly through immersive visual techniques that expose climate change.

REFERENCES

Abeyta, A. A., Routledge, C., & Juhl, J. (2015). Looking back to move forward: Nostalgia as a psychological resource for promoting relationship goals and overcoming relationship challenges. Journal of Personality and Social Psychology, 109, 1029–1044. doi:10.1037/pspi0000036

Albrecht, G., Sartore, G., Connor, L., Higginbottom, N., Freeman, S., Kelly, B., ...Pollard, G. (2007). Solastalgia: The distress caused by environmental change. Australian Psychiatry, 15(1), S95–S98. doi:10.1080/10398560701701288

Arendt, F., & Matthes, J. (2016) Nature documentaries, connectedness to nature, and pro-environmental behavior. Environmental Communication, 10, 453–472. doi:10.1080/17524032.2014.993415

Berman, M. G., Jonides, J., & Kaplan, S. (2008) The cognitive benefits of interacting with nature. Psychological Science, 19, 1207–1212. doi:10.1111/j.1467-9280.2008.02225.x

Berto, R. (2005). Exposure to restorative environments helps restore attentional capacity. Journal of Environmental Psychology, 25, 249–259. doi:10.1016/j.jenvp.2005.07.001

Brooks, A. M., Ottley, K. M., Arbuthnott, K. D., & Sevigny, P. (2017) Nature-related mood effects: Season and type of nature contact. Journal of Environmental Psychology, 54, 91–102. doi:10.1016/j.jenvp.2017.10.004

Cheung, W., Sedikides, C., & Wildschut, T. (2017). Nostalgia proneness and reduced prejudice. Personality and Individual Differences, 109, 89–97. doi:10.1016/j.paid.2016.12.045

Cordaro, F. (2011). The relationship between nostalgia, social exclusion, and empathy (Doctoral dissertation, University of Southampton). Retrieved from http://eprints.soton.ac.uk/187383

DeGroot, J. I. M., & Steg, L. (2008). Value orientations to explain beliefs related to environmental significant Behavior: How to measure egoistic, altruistic, and biospheric value orientations. Environment and Behavior, 40, 330–354.doi.org/10.1177/0013916506297831

Dixon, G. N., McKeever, B. W., Holton, A. E., Clarke, C., & Eosco, G. (2015). The power of a picture: Overcoming scientific misinformation by communicating weight-of-evidence information with visual exemplars. Journal of Communication, 65(4), 639-659.doi: 10.1111/jcom.12159

Dunfield, K. A., Best, L. J., Kelley, E. A., & Kuhlmeier, V. A. (2019). Motivating moral behavior: Helping, sharing, and comforting in young Children with autism spectrum disorder. Frontiers in Psychology, 10, 25. doi:10.3389/fpsyg.2019.00025

Eisenberg, N., & Mussen, P. H. (1989). The roots of prosocial behavior in children. New York, NY: Cambridge University Press. doi:10.1017/CBO9780511571121

Environmental Defense Fund. (2017). Retrieved March, 2018, from https://www.edf.org/our-work

Ford, J. B., & Merchant, A. (2010). Nostalgia drives donations: The power of charitable appeals based on emotions and intentions. Journal of Advertising Research, 50(4), 450–459. doi:10.2501/S0021849910091592

Gifford, R. (2008). Psychology's essential role in alleviating the impacts of climate change. Canadian Psychology, 49, 273–280. doi:10.1037/ a0013234

Gifford. R. (2011). The dragons of inaction: Psychological barriers that limit climate change mitigation and adaptation. American Psychologist, 66, 290–302. doi:10.1037/a0023566

Gifford, R., & Nilsson, A. (2014). Personal and social factors that influence pro-environmental concern and behavior: A review. International Journal of Psychology, 49(3), 141–157. doi:10.1002/ijop.12034

Giving USA Foundation Infographic. (2018). Retrieved from https://givingusa.org/wp-content/uploads/2018/06/GUSA-2018-Infographic-FINAL.png

Habitat for Humanity. (n.d.). Retrieved March, 2018, from https://www.habitat.org/about/mission-and-vision

Halpenny, E. A. (2010a). Pro-environmental behaviours and park visitors: The effect of place attachment. Journal of Environmental Psychology, 30, 409–421. doi:10.1016/j.jenvp.2010.04.006

Halpenny, E. A. (2010b). Pro-environment Behavioural Intentions Scale. PsycTESTS, doi:10.1037/t25866-000

Hossain, M. T. (2014). The effect of marketing communication on individuals' gift-giving: Role of cause type, appeal characteristics, and donor mindset. (Unpublished doctoral dissertation). The University of Texas at Arlington.

Ibanez, L., Latourte, J., & Roussel, S. (2018). Do people adapt their pro-environmental behavior according to their environment? An experimental investigation. Retrieved from https://faere2018.sciencesconf.org/210231/document

Karmanov, D., & Hamel, R. (2008). Assessing the restorative potential of contemporary urban environment(s): Beyond the nature versus urban dichotomy. Landscape and Urban Planning, 86(2), 115–125. doi:10.1016/j.landurbplan.2008.01.004

Keenan, E. (2015). Drivers of prosocial behavior and pro-environmental choice. UC San

Diego. ProQuest ID: Keenan_ucsd_0033D_14736. Merritt ID: ark:/20775/bb0274823w. Retrieved from https://escholarship.org/uc/item/1wt4k040

Klein, S., & Hilbig, B. (2018). How virtual nature experiences can promote pro-environmental behavior. Journal of Environmental Psychology, 60, 41–47. doi:10.1016/j.jenvp.2018.10.001

Kollmuss, A., & Agyeman, J. (2002). Mind the gap: Why do people act environmentally and what are the barriers to pro-environmental behavior? Environmental Education Research, 8, 239–260. doi:10.1080/13504620220145401.

Korpela, K. M., Klemettilä, T., & Hietanen, J. K. (2002) Evidence for rapid affective evaluation of environmental scenes. Environment and Behavior, 34, 634–650. doi:10.1177/0013916502034005004

Krajhanzl, J. (2010). Environmental and pro-environmental behavior. In Rehulka, E. (Eds.), School and Health (pp. 251-274). Retrieved from http://www.ped.muni.cz/z21/knihy/2011/36/36/texty/eng/krajhanzl.pdf

Krieg, J. & Samek, A. (2017). When charities compete: A laboratory experiment with simultaneous public goods. Journal of Behavioral and Experimental Economics, 66, 40–57. doi:10. 1016/j.socec.2016.04.009

Kurisu, K. (2015). Influential factors on PEBS. In Pro-environmental behaviors (pp. 27–46) Japan: Springer.

Lafferty, B., & Edmondson, D. (2014). A note on the role of cause type in cause-related marketing. Journal of Business Research, 67, 1455–1460. doi:10.1016/j.jbusres.2013.07.021.

Landry, N., Gifford R., Milfont, T. L., Weeks, A., & Arnocky, S. (2018). Learned helplessness moderates the relationship between environmental concern and behavior. Journal of Environmental Psychology, 55, 18–22. doi:10.1016/j.jenvp.2017.12.003

Lazard, A. & Atkinson, L. (2015). Putting environmental infographics center stage: The role of visuals at the elaboration likelihood model's critical point of persuasion. Science Communication, 37(1), 1-28. doi:10.1177/1075547014555997

Lewicka, M. (2011). Place attachment: How far have we come in the last 40 years? Journal of Environmental Psychology, 31(3), 207–230. doi:10.1016/j.jenvp.2010.10.001

Low, S. M., & Altman, I. (1992). Place attachment: A conceptual inquiry. In S. M. Low, & I. Altman (Eds.), Place attachment (pp. 1). New York: Plenum Press.

Manolas, E. (2015). Promoting pro-environmental behavior: Overcoming barriers. Aegean Journal of Environmental Sciences, 1, 13–21. Retrieved from http://www.env.aegean.gr/wp-content/uploads/2015/06/Manolas_AEJES_2015.pdf

Manzo, L. C. (2003). Beyond house and haven: Toward a revisioning of emotional relationships with places. Journal of Environmental Psychology, 23(1), 47–61. doi:10.1016/S0272-4944(02)00074-9

Markowitz, E. M., Goldberg, L. R., Ashton, M. C., & Lee, K. (2012), Personality and pro-environmental action. Journal of Personality, 80, 81–111. doi:10.1111/j.1467-6494.2011.00721.x

Markowitz, E., Slovic, P., Västfjäll, D., & Hodges, S. (2013). Compassion fade and the challenge of environmental conservation. Judgement and Decision Making, 8, 397–406.

Marlon, J., Howe, P., Mildenberger, M., Leiserowitz, A., & Wang, X. (2018). Yale climate opinion maps 2018. Retrieved from http://climatecommunication.yale.edu

Merchant, A., Ford, J. B., & Rose, G. (2011). How personal nostalgia influences giving to charity. Journal of Business Research, 64, 610–616. doi:10.1016/j.jbusres.2010.06

Najafi, M., & Kamal, M. (2012). The concept of place attachment in environmental psychology. Sustainable Architecture, 45, 7637–7641. Retrieved from http://www.elixirpublishers.com/articles/1350368123_45%20(2012)%207637-7641.pdf

Neaman, A., Otto, S., & Vinokur, E. (2018). Toward an integrated approach to environmental and prosocial education. Sustainability, 10(3), 583. doi:10.3390/su10030583

Nicholson-Cole, S. A. (2005). Representing climate change futures: A critique on the use of images for visual communication. Computers, Environment and Urban Systems, 29(3), 255-273. doi:10.1016/j.compenvurbsys.2004.05.002

Nolan, J., & Schultz, W. (2015). Prosocial behavior and environmental action. In (Ed.), The Oxford Handbook of Prosocial Behavior. Oxford University Press, doi:10.1093/oxfordhb/9780195399813.013.011

Pavey, L., Greitemeyer, T., & Sparks, P. (2012). Prosocial intentions measure. PsycTESTS, doi:10.1037/t36770-000

Pereira, M., Forster, P., & Darwin (2015), The relationship between connectedness to nature, environmental values, and pro-environmental behaviours. Reinvention: An International Journal of Undergraduate Research, 8(2), Retrieved from http://www.warwick.ac.uk/reinventionjournal/issues/volume8issue2/pereira

Pierre-Louis, K. (2018, November 9). Why does California have so many wildfires? New York Times. Retrieved from https://www.nytimes.com/2018/11/09/climate/why-california-fires.html

Rickard, L., Schuldt, J. P., Eosco, G., Scherer, C., & Daziano R. A. (2017). The proof is in the picture: The influence of imagery and experience in perceptions of hurricane messaging.Weather, Climate, and Society, 9(3). doi:10.1175/WCAS-D-16-0048.1

Schwartz, J., & Popovich, N. (2019, February 6). It's official: 2018 was the fourth-warmest yearon record. New York Times. Retrieved from https://www.nytimes.com/interactive/2019/02/06/climate/fourth-hottest-year.html

Sengupta, S. (2018, March 12). Hotter, drier, hungrier: How global warming punishes the world's poorest. New York Times. Retrieved from https://www.nytimes.com/2018/03/12/climate/kenya-drought.html

Stephan, E., Wildschut, T., Sedikides, C., Zhou, X., He, W., Routledge, C., ...Vingerhoets, A. J. J. M. (2014). The mnemonic mover: Nostalgia regulates avoidance and approach motivation. Emotion, 14(3), 545–561. doi:10.1037/a0035673

Tate, K. (2014). Influencing green consumer choice through environmental goal activation. (Unpublished doctoral dissertation). The University of Manchester, United Kingdom.

Tonge, J., Ryan, M. M., Moore, S. A., & Beckley, L. E. (2015). The effect of place attachment on pro-environment behavioral intentions of visitors to coastal natural area tourist destinations. Journal of Travel Research, 54, 730–743. doi:10.1177/0047287514533010

Valtchanov, D., & Ellard, C. (2010). Physiological and affective responses to immersion in virtual reality: Effects of nature and urban settings. Journal of Cybertherapy andRehabilitation, 3(4), 359–374. Retrieved from https://pdfs.semanticscholar.org/b577/3abeddf48fa5e5ffd1788686fa7a23c95975.pdf

Walker, G. J., & Chapman, R. (2003). Thinking like a park: The effects of sense of place, perspective-taking, and empathy on pro-environmental intentions. Journal of Park and Recreation Administration, 21(4), 71–86. Retrieved from https://js.sagamorepub.com/jpra/article/view/1492/1457

Walker, A. J., & Ryan, R. L. (2008) Place attachment and landscape preservation in rural New England: A Maine case study. Landscape and Urban Planning, 86(2), 141–152. doi:10.1016/j.landurbplan.2008.02.001

Wang, S., Leviston, Z., Hurlstone, M., Lawrence, C., & Walker, I. (2018). Emotions predict policy support: Why it matters how people feel about climate change. Global Environmental Change, 50, 25–40. doi:10.1016/j.gloenvcha.2018.03.002.

Weinstein, N., Legate, N., & Przybylski, A. K. (2013). Beauty is in the eye of the psychologically fulfilled: How need satisfying experiences shape aesthetic perceptions
of spaces. Motivation and Emotion, 37(2), 245–260. doi:10.1007/s11031-012-9312-7

Weinstein, N., Przybylski, A. K., & Ryan, R. M. (2009a). Can nature make us more caring? Effects of immersion in nature on intrinsic aspirations and generosity. Personality and Social Psychology Bulletin, 35, 1315–1329. doi:10.1177/0146167209341649

Weinstein, N., Przybylski, A. K., & Ryan, R. M. (2009b). Player experience of need satisfaction physical presence scale-adapted version. PsycTESTS, doi:10.1037/t19148-000

Wildschut, T., Sedikides, C., Arndt, J., & Routledge, C. (2006). Nostalgia: Content, triggers, functions. Journal of Personality and Social Psychology, 91(5), 975–993.doi:10.1037/0022-3514.91.5.975

Wildschut, T., Sedikides, C., Routledge, C., Arndt, J., & Cordaro, F. (2010). Nostalgia as a repository of social connectedness: The role of attachment-related avoidance. Journal of Personality and Social Psychology, 98, 573–586. doi:10.1037/a0017597

Wilson, G. D. (2013). Superfund evaluation: The families of Tar Creek. Dissertation Abstracts International Section A: Humanities and Social Sciences. Retrieved from http://scholarworks.uark.edu/etd/455/

Windsong, E. A. (2010). There is no place like home: Complexities in exploring home and place attachment. The Social Science Journal, 47(1), 205–214. doi:10.1016/j.soscij.2009.06.009

Zelenski, J. M., Dopko, R., L., & Capaldi, C., A. (2015). Cooperation is in our nature: Nature exposure may promote cooperative and environmentally sustainable behavior. Journal of Environmental Psychology, 42, 24–31. doi:10.1016/j.jenvp.2015.01.005

Zhang, J. W., Piff, P. K., Iyer, R., Koleva, S., & Keltner, D. (2014). An occasion for unselfing: Beautiful nature leads to prosociality. Journal of Environmental Psychology, 37, 61–72. doi:10.1016/j.jenvp.2013.11.008

Zhou, X., Wildschut, T., Sedikides, C., Shi, K., & Feng, C. (2012). Nostalgia: The gift that keeps on giving. Journal of Consumer Research, 39(1), 39–50. doi:10.1086/662199

D4- Transform Places with People and the Environment: Sustainable Planning, Design & Behavior I

Emergency Management and Signage System of Chubu International Airport: A Study on Facility Management of Airport Terminal Buildings

Natsuki Imai (Mie university)

1. ABSTRACT

The airport terminal is a place where passengers who use airplanes go through various procedures, meet-up and transfer to other transportation facilities. Therefore the safety, convenience, and economic efficiency are required. New airports and terminals have been built in various locations in response to the rapid increase in global aviation demand. In Japan, the number of tourists is increasing year by year and because international events are coming up in recent years, maintenance of airport terminals is essential. As countermeasures, the government is working on the maintenance of 27 airports and the private sectors are also working on various attempts like the construction of a new terminals and the introduction of an airport-indoor navigation applications that uses a mobile phone. Therefore, this study focuses on Chubu International Airport, which is one of the company-managed base airports, and clarifies the actual state of airport facility management. In recent years, Chubu International Airport has constructed an international exhibition hall, a LCC (Low Cost Carrier) terminal, and commercial facilities. In this study, we conducted a field survey at the current airport from the perspective of disaster prevention planning and signage planning as a basis for continuing verification of the linkage and facility continuity between the LCC terminal constructed in 2019 and the current terminal. The survey revealed 1) the current situation of the disaster prevention planning of the airport, 2) the current situation of the airport space composition, 3) the current situation and issues of the airport signage planning compared with other airports.

1.1. Keywords

Facility Management, Business Continuity Plan, Disaster Management Plan, Sign System, Wayfinding

2. INTRODUCTION

2.1. Recent airport and disasters

Japan is a disaster-prone country with various disasters occurring all over the country. When a large-scale disaster occurs, the land transportation network is severed, and it takes time to recover. On the other hand, because the air route in the sky is no way to be damaged, the air network will not be damaged unless the airport on land is not damaged. In order for the airport to fully function as a base for rescue activities in the event of a disaster, special operations different from daily airport operations need to be performed, and BCP (Business Continuity Planning) in preparation for it is also required to be updated as frequent as possible. In addition, lessons and knowledge gained from past disasters need to be utilized in the case of the Nankai Trough Earthquake, which is feared to occur in the near future.

2.2. Recent airport design

In recent years, due to the rise of LCC (Low Cost Carrier), LCC terminals have been built at multiple base airports. A terminal building plan that takes into consideration the characteristics of the LCC business model which is different from the conventional FSC (Full-Service Carrier) is required for the LCC terminal building. In particular, the ease of understanding and efficiency in signage planning and passenger flow planning is pursued, while realizing the overall low-cost planning.

The Ministry of Land, Infrastructure, Transport and Tourism announced that the number of foreign visitors to Japan has increased from 8.4 million in 2012 to 30 million in 2018. To further increase in the number of foreign visitors to Japan, the ministry is working to strengthen the regional airports. One of them is the approval of 27 airports for “airport supporting the increase of foreign tourists to Japan”, and diversification of tourists is also expected at conventional airports (Ministry of Land, Infrastructure, Transport and Tourism, 2017).

2.3. Previous research - Study on quantification of signage system

The ones that quantitatively grasp the signage information in the space and see the relationship with the wayfinding behavior could be found in studies by Watanabe, Kim and Akizuki. Watanabe divided the floor with a grid of 1.0 [m] x 1.0 [m] and set the subject stand at the center of the grid, extracted the visible grid range to understand the relationship between the wayfinding in the architectural space and the amount of signage information (guiding board, direction board, room name board). Furthermore, after weighning the degree of sign recognition, the sign information density of the entire space is calculated and quantified. However, with this method, it is not possible to grasp how much signage was set in the space, and the amount of obtained information includes human judgment whether or not it is visible. Because the amount of information is reduced subjectively, it is hard to say that it was judged objectively. The research by Akizuki has a novelty different from other researches in that it uses the solid angle projection rate in the visual field based on the ease of identification of evacuees as a method to grasp the signage installation status of evacuation routes.

3. RESEARCH OBJECTIVES

In this study, Chubu International Airport was analyzed from the two aspects; disaster prevention planning and signage planning in reference to the current situation of facility management at the airport. In the survey of the signage planning, the same survey is partly conducted at the Haneda Airport International Terminal as well as the Chubu International Airport and the results are compared. The paper aims to obtain better guidelines for airport planning, programming, design and management from the results.

4. METHODS

The first research method involved 3 investigations to understand the current situation of disaster management; an interview survey with the management company of the target facility, a facility tour, and participating in the drills for management operation.

The second research method also involved 3 investigations to understand the current situation of signage planning. As a preliminary survey of the signage planning, a drawing survey was conducted to understand the space composition in the terminal building and how the building was used. Further, a plot survey was conducted. The signage system in the non-restricted area was photographed, and the signage was quantitatively analyzed and considered from the plotted drawing or classification table. Finally, a passenger flow line survey was performed by walking the route that assumes the daily users to analyze and consider the actual situation and appearance of the signage.

5. COMPARISON OF DISASTER PREVENTION PLAN AND BCP FORMULATION AT EACH AIRPORT

In this study, as a basic study before conducting a survey of disaster prevention planning, the formulation of disaster prevention plans and BCP's at company-managed base airports were compared. Among the airports in Japan, the company-managed airports are Narita International Airport, Chubu International Airport, Kansai International Airport, and Osaka International Airport. (Ministry of Land, Infrastructure, Transport and Tourism) Items compared were disaster prevention work plans, disaster prevention drills, organizations of disaster prevention management departments, and BCP's within the scope of public information.

Regarding the variety of training, it was found that Narita International Airport had the most complete content in the public information, and Kansai International and Osaka International Airport had the most complete content in the public disclosures of BCP's. At Chubu International Airport, a revised disaster prevention manual has been issued annually for three years from the airport opening. However, the third edition was the final. Because actual disasters may occur unexpectedly, it is necessary to consider and update further variations by using the proactive management strategy, not just following the routines and training drills.

6. OVERVIEW OF CHUBU INTERNATIONAL AIRPORT

6.1. Overview of surveyed facility

The Chubu Centrair International Airport is a 24-hour international airport located on an artificial island in Ise Bay built off-coast of Tokoname City, Aichi Prefecture, in Japan and has a role as a gateway to the Chubu region. The distance from the center of Nagoya, the metropolis of the region, is about 35km, and it takes about 30 minutes by rail or highway. It is also the first airport in Japan that introduced the Universal Design Principals.

7. THE CURRENT SITUATION OF DISASTER PREVENTION PLANS

7.1. Survey method

Interview surveys with the management and operation company of the surveyed facility, participations in in-house disaster prevention management workshops and seminars, and facilities tours were conducted. The surveys were carried out on November 22 and 28, 2018.

Table1: comparison of BCPs in company-managed airports.

Company-managed airport	BCP
Narita International Airport	NAA Group large-scale earthquake countermeasure business continuity plan (BCP) NAA Group New influenza action plan (NAA : Narita Airport)
Chubu International Airport	Disaster prevention manual
Kansai International Airport	Evacuation policy in case of an earthquake with a Japanese seismic intensity scale of 5 or higher and a large tsunami warning is announced
Osaka International Airport	Evacuation policy in case of an earthquake with a Japanese seismic intensity scale of 5 or higher

Table2: Comparison of disaster prevention drill in company-managed airports.

Company-managed airport	Disaster prevention drill
Narita International Airport	Airport facilities (fire drill: twice a year, disaster prevention drill: once a year) Aircraft accident-fire fighting and rescue training: once a year Aircraft accident fire fighting and rescue comprehensive training: once a year/ practical desktop training Aircraft refueling facility (fire drill: twice a month) Maritime disaster drill (once a year) Joint training (3 times a year) / evacuation drill (tsunami), Joint comprehensive training (tank disaster) Pipeline disaster drill (once a year)
Chubu International Airport	Desktop training (twice a year) Partial training General training (once a year)
Kansai International Airport	Earthquake and tsunami disaster prevention drill Aircraft fire and rescue training
Osaka International Airport	Earthquake evacuation drill, earthquake (tsunami) disaster prevention drill Aircraft accident countermeasures comprehensive training

7.2. Tsunami and earthquake countermeasures (terminal facility management)

The basic movements (information transmission and evacuation routes) at the time of disaster were disseminated to employees as follows by distributing the Centrair Disaster Prevention Manual.

Information transmission: In the event of a large-scale earthquake, disaster information related to tsunami is transmitted. In the unlikely event that the Japan Meteorological Agency issues a "Great Tsunami Warning" to Ise/ Mikawa Bay, the evacuation sites to be guided within Airport Island are as shown in Table 3. Site evacuation: To facilitate evacuation in the event of a tsunami, high-rise buildings managed by the airport company and other high-rise buildings are defined as tsunami shelter sites. The tsunami shelter map shows the evacuation shelter sites and exit route along with the height.

7.3. Current status of disaster prevention measures (actual efforts and facilities)

During the two months from October to December of the survey, a safety seminar for all airport employees was held once, and workshops for managers were held once a week. In these workshops, there were attendees from various departments for each topic, and it was a workshop that deepened knowledge of individual jurisdictions and related duties. From this, it could be said that the goal is to raise awareness and deepen knowledge about the safety of the entire facility.

At the central operation and management facility, called Centrair Operation Center (COC), the main work is to check PC screens on a daily basis however, in case of an emergency, a meeting room has been set up at next room to hold a meeting with the Japanese Self-Defense Forces for disaster dispatch and air traffic controllers.

Table3: Disaster information transmission method by area

Zone	Area	Transmission method
PTB etc	Passenger Terminal Building (PTB), Access Plaza, International bus lounge, administration building, *multistory parking	Guards in disaster prevention center make emergency broadcasts * Employees call out to multi-story parking lots with the sound of an outdoor alarm siren.
Restricted area	Apron etc	Outdoor alarm siren
Cargo area	Cargo office building, domestic AL shed, domestic FD shed, first, second and third international sheds, *Machine Supplies Building	Along with outdoor alarm siren ringing, the third emergency guard station in the cargo office broadcasts in the emergency system. * Extension phone to the machine supplies building
Land area	Temporary parking lot, outdoor shelter (Centrair garden, hill with airplane view, green zone at north side of airport station)	Along with outdoor alarm sirens, fire trucks at airport fire stations patrol the land area.

8. THE CURRENT SITUATION IN SPACE COMPOSITION (PRELIMINARY SURVEY)

8.1. Survey method

A drawing survey was conducted based on the drawing plan data of the surveyed facility. The existing classification and the special characteristics in the non-restricted area, color-coded, calculated the area were classified, and the building use was investigated.

8.2. Space composition by existing division

Chubu International Airport: By floor, the proportion of the non-restricted area increases as you go up, and the proportion of the management area increases as you go down. This is because most of the management area is occupied by BHS (Passenger Baggage Handling System). In the top view (Figure 1), which is color-coded by area, you could see not only the plane but also the processing of vertical flow lines. As the airport is characterized by horizontal movement, smooth movement from each transportation means to non-restricted areas (departure lobby), international / domestic security checkpoints, or movement from arrival gates to non-restricted areas (arrival lobby) could be confirmed. On the other hand, if we focus on the movement only within the unrestricted area, it could be seen that the movement from the first floor and the second floor to the fourth floor is not performed vertically and linearly, and it takes time to transfer between international and domestic flights.

Haneda Airport International Terminal and comparison of two airports: Similar to Chubu International Airport, the proportion of the non-restricted area increases, and the proportion of the management area increases as you go down. The Haneda International Terminal is adjacent to two stations, a monorail and a train and the train station has a ticket gate on the arrival lobby on the second floor, and the monorail has a ticket gate on both the second and third floors. Compared to Chubu International Airport, the prospects for the lower floors are not good at the third floor and below, and it is assumed that sign system will be important in wayfinding. In addition, the security checkpoint is located in the center of the terminal, however both airports cannot be easily recognized the place when standing at the terminal entrance on the 3rd floor. As a one factor, a safety consideration could explain the arrangement of the security checkpoint.

8.3. Space composition by each space characteristic (unrestricted area)

Chubu International Airport: Focusing on the non-restricted area that occupies the most area in the terminal and considering the space composition, the first floor has a small number of customers while calm and quiet dedicated corner where you can stay for a certain time is well equipped. On the second floor, there are waiting areas for welcoming people and smoking areas. There is plenty of space for staying, at the same time there are also corners and facilities to help guests move, such as information counters. The third floor is a fluid space for departure guests to check in, and there are fewer dedicated corners and staying spaces than the second floor. The ceiling height is high and the prospects to the security inspection area on the fourth floor are well opened.

Being a central space of the building, it allows you to manage many people's administrative procedures smoothly without pile up. The fourth floor is a commercial facility floor, therefore there is a staying space in the center. However, there are few sign displays and dedicated corners. The space composition is different

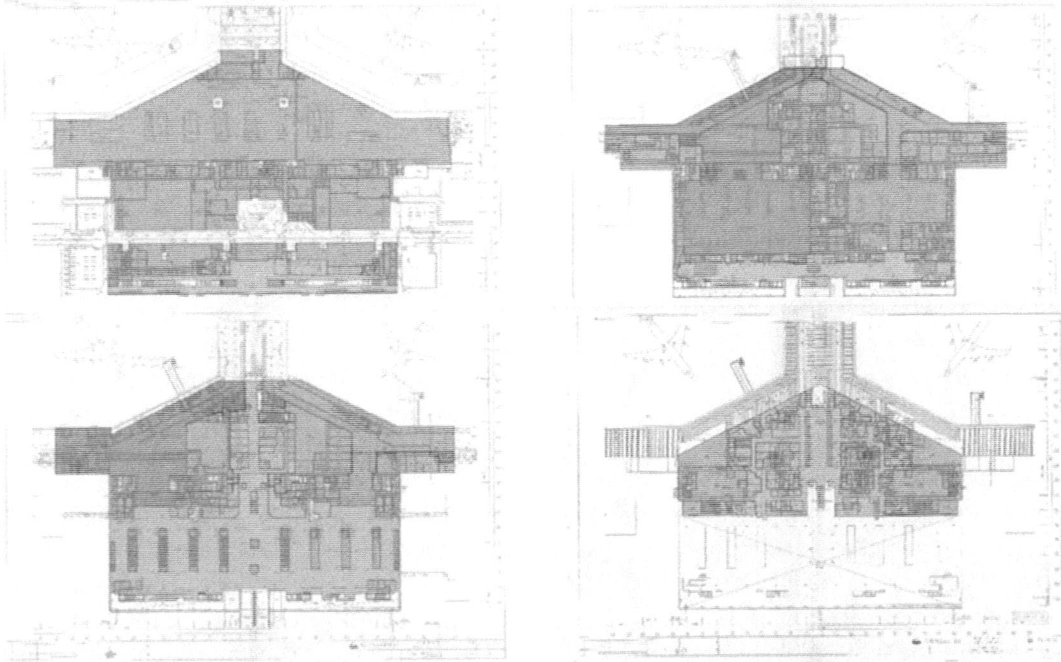

Figure 1. Floor plan by area (first to fourth floors in order of Z-shape from upper left)

from the first to third floors in order to encourage the use of commercial facilities.

Haneda Airport International Terminal and comparison of two airports: Similarly, focusing on the non-restricted area in Haneda Airport International terminal, the first floor had much more users than Chubu International Airport. As with Chubu International Airport, many people stayed for a certain period of time for charging mobile devices and connecting time, and it was almost full. The large number of people can be attributed to the demand for connections and the fact that there is less waiting space on the third floor than in the central area. On the 2nd floor, there are domestic transit check-in counters, shower rooms, ticket counters for buses, ticket gates for each station, making the movement flow after arrival relatively compact. Moreover, the floor can be recognized unconsciously using brown as the floor color compared to gray on the other floors. The third floor is a fluid space for departure guests to check in. In the Chubu International Airport, Access Plaza connected to each transportation system are connected by slopes from both the second and third floors, whereas Haneda Airport International Terminal is connected to each transportation system and transit check-in are provided on the second floor. Therefore, different from Chubu International Airport, there was no deviation in the number of people on each floor. The 4th and 5th floors are the floors of commercial facilities. Unlike the Chubu International Airport, the signs are the same as those on the other floors, and the stay spaces are distributed.

9. THE CURRENT SITUATION OF SIGN PLANNING

9.1. Survey method

The three types of surveys: plot survey, interview survey, and customer flow line survey were conducted. The survey date for the plot survey was on December 13, 2018, the interview survey was on November 22, 2018, and the customer flow line survey was on January 18, 2019. In this paper, the plot survey and customer flow line survey are focused. The same plot survey was conducted at Haneda Airport. The survey dates were September 21 and 22, 2019.

9.2. Sign system classification method (plot survey)

Signs were classified into four categories according to the transmission function from the viewpoint of wayfinding and plotted on the drawings shown in section 8 (Figure 1).

- ***Guidance sign:*** a sign that indicates the direction of the facility, etc.
- ***Location sign:*** a sign that announces the location of the facility
- ***Information map:*** a sign map that guides boarding / exiting conditions and location
- ***Regulation sign:*** a sign that regulates user behavior

9.3. Assumed route (customer flow line survey)

The following 12 routes were prepared. 6 are the departure flow lines, 2 are the arrival flow lines, and 4 are the flow lines for customers who come to pick-up or send-off someone:

1: International departure passenger flow line (DI) (direct to the boarding gate)

1-2: International departure passenger flow line (DI2) (direct to the boarding gate)

2: Domestic departure passenger flow line (DD) (direct to boarding gate)

2-2: Domestic departure passenger flow line (DD2) (direct to Boarding Gate)

3: International arrival passenger flow line (AI)

4: Domestic arrival passenger flow line (AD)

5: Send-off visitor (for International flights passenger) flow line (SI) (via Skytown on the fourth floor)

6: Send-off visitor (for domestic flights Passenger) flow line (SD) (via Skytown on the fourth floor

7: Pickup visitor (for international arrival Passenger) flow

8: Pickup visitor (for domestic arrival passenger) flow line (PD)

9.4. Regularity seen in signs

It was found that there are certain rules regarding the sign arrangement at Chubu International Airport. The sign changes from information maps to guidance signs as you go from the terminal entrance to forward (east-west direction). As you approach the end of the terminal (north-south direction), only the number of location sign is increased and the amount of information that could be observed is decreased. A symmetric composition could be seen in the north-south axis in the terminal facility layout and sign system layout. According to the classification of international and domestic flights, there are differences in the location signs. However, they are arranged in almost the same way. And the equidistant arrangement of the signs could be seen. In particular, on the third floor, by placing signs at regular intervals along columns and counters, it is possible to grasp the flow line distance and position information by the arrangement of signs. Among them, only the fourth floor has a different signing system arrangement from the other floors, and it seemed that priority was given to production as a commercial facility. Comparing the number of signs by floor, the number was proportional to each non-restricted area calculated by the space composition survey (session 8) except the fourth floor.

The same regularity was seen at Haneda Airport International Terminal like the sign changed from information maps to guidance signs as you go from the terminal entrance to forward. The difference from Chubu International Airport is that the same sign system is used on the fourth and fifth floors, and the number of signs on the second floor is larger compare to the area. This is because the domestic transit check-in counter, shower room, bus ticket counters and each train ticket gates are located on the second floor while these are provided in Access Plaza in Chubu International Airport. In addition, there are more clocks compare to Chubu International Airport. However there is no floor display sign. While the sign display uses fluorescence white letters on a black background in Chubu International Airport, Haneda Airport International Terminal uses black letters on a fluorescent white background. It seems to have differences in visibility.

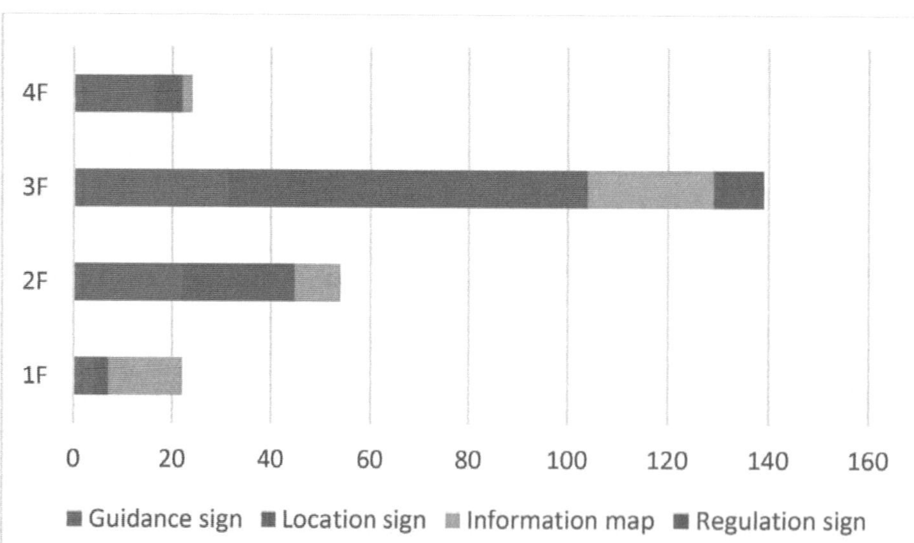

Figure 2. The number of signs by floor in Chubu International Airport

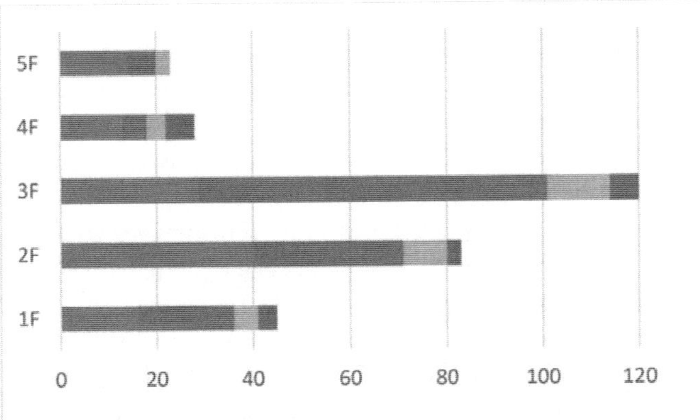

Figure 3. The number of signs by floor in Haneda Airport International Terminal

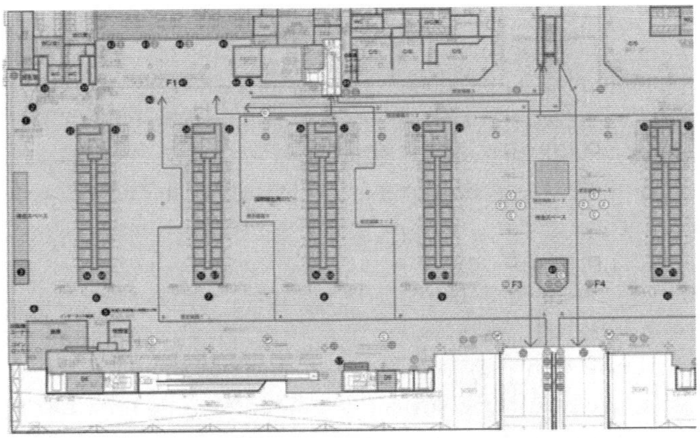

Figure 3. The number of signs by floor in Haneda Airport International Terminal

Figure 3. The number of signs by floor in Haneda Airport International Terminal

9.5. Problems in the current signage planning

In the signs used in practice, there are signs useful for wayfinding at many corners of the route. On the other hand, it also turned out that it is easy to lose the way in some open spaces, in the center of the building (Fig. 4), around the third and fourth floor escalators, and at the end of the third floor.

10. CONCLUSIONS

10.1. The current situation of disaster prevention plans

In this study, the actual situation of Chubu International Airport from the viewpoint of disaster prevention planning and signage planning was investigated. As a result, it was found that for the disaster prevention planning, BCP formulation was voluntarily carried out according to the characteristics of each airport.

A comparison of the disaster prevention plan and BCP formulation focusing on the four base airports managed by the private sector shows that Narita International Airport has the most extensive training content in the public information, and Kansai International Airport and Osaka International Airport have the most substantial content in the BCP disclosure. From the frequency and purpose of the seminars and workshops in which I participated, it could be said that the goal is to raise overall awareness of safety and deepen their knowledge in Chubu International Airport. At the Centrair Operation Center (COC), a meeting room is set up next to the Japanese Self-Defense Forces and air traffic controllers in order to hold meetings in case of an emergency.

However, since an actual disaster may cause an unexpected situation, it is important not only to formulate and train, but to use imagination, consider and update further variations, and provide guidance. In the future, it will be necessary to have a framework for management and cooperation that could be applied to new company-managed airports.

10.2. The current situation of space composition

In Chubu International Airport, which features horizontal travel, it could be confirmed that smooth travel from each transportation to unrestricted areas (departure lobby), international / domestic security checkpoints, or arrival gates to unrestricted areas (arrival lobby) and each transportation. On the other hand, focusing on the movement within the unrestricted area, it could be seen that the transfer from the first floor and the second floor to the fourth floor is not performed vertically and linearly, and it takes time to transfer between international and domestic flights. In addition, compared to the Chubu International Airport, it is assumed that at Haneda Airport, on the third floor and below, the prospect to the lower floor is not good, and it is assumed that a wayfinding is performed relying on a signage. In the Chubu International Airport, Access Plaza that connects to each transportation system is connected by slopes from both the second and third floors, while in Haneda airport transit and all transportation systems are connected on the second floor causing no bias in the number of people in each floor.

10.3. The current situation of signage planning

As for the signage planning, as a result, it was found that there was some regularity in the sign arrangement from the first floor to the third floor. Further, there were some places that were actually overlooked by the arrangement plan alone, and there were places where the arrangement plan worked well. Specifically, it was found that it was easy to lose track in some open spaces, in the center of the building (Fig. 5), around the escalator on the third and fourth floors, and at the end of the third floor. From these surveys, it will be possible to reconsider the necessity of the present signs and propose signs with new regularity where the arrangement plan is not working well by comparing the actual survey with the sign-plotted drawings. Through comparison of the spatial composition and signs with Haneda Airport International Terminal, the differences and similar points were identified. The size of the area was almost same, however features in the sign systems, space composition, and usage have been seen in each airport terminal. Based on the results, we will proceed with further analysis of the points that are working well and the points that are not working well and show the points that should be shared in the international airport space composition, and the points to be differentiated.

In addition, at the space composition survey, it was confirmed that it would take time to connect between international and domestic flights. Although it was not investigated in this survey, it is expected that transit demand will increase as the number of foreigners visiting Japan increases in the future, and it will be necessary to proceed with development. Providing them and verifying the cooperation and the continuity between the new LCC terminal constructed in October 2019 and the current terminal from the perspective of disaster prevention planning and signage planning, it will be expected to enable better airport operation.

10.4. The current situation and outlook of airports due to COVID-19

In November 2019, the first occurrence of COVID-19 was confirmed near Wuhan City, Hubei Province, China, and it was reported to WHO in December. As of April 2020, the outbreak has spread to the world, and the EDRA51 was also affected, so that the conference was carried out virtually. The following is a brief summary of how border measures have been implemented at airports in Japan from January to April 2020. The information is based on related news articles after the discovery of "COVID-19" from the keyword search of the Asahi Shimbun (major newspaper in Japan).

In Japan, the infection of a Chinese man in his 30s living in Kanagawa was confirmed for the first time in mid-January 2020. After that, in late January the government implemented thorough quarantine, distributing questionnaires to passengers on direct flights from Wuhan, and on transit flights, submitting to quarantine officers at airports, and requested in-flight announcements to each airline company to prevent secondary infections and waterfront measures. Quarantine officers originally check thermographically at airports where there are international flights. As measures against COVID-19, quarantine was strengthened by newly distributing and submitting questionnaires, in-flight announcements, and calling on panels and posters. At airports where Wuhan flights arrive, measures were taken to carefully observe the booths that passengers pass through by reducing the two booths to one at the arrival time of Wuhan flights (Narita International Airport). At the same time the United States responded by limiting the number of airports that can enter from Wuhan. Quarantine was strengthened at three airports: JFK Airport in New York, Los Angeles and San Francisco, and two airports in Chicago and Atlanta were strengthened next. Passenger acceptance from Wuhan was limited to these five airports. At the port where regular ferry arrives, from the last bird flu, a quarantine officers originally go in and check the body temperature using thermography when the previous port was in China. In addition, measures such as posters have been strengthened to alert people.

As for the main measures, quarantine strengthening / restriction, announcements, posters, panels, etc. were issued at all airports. Due to the reduction of air flights and suspension of operations, airports are currently closing airport terminals and runways, and operating scale and operating time are being reduced.

At Singapore's Changi Airport, IT is being boldly promoted, such as check-in with unattended counters and automatic departure control at terminals 4 and 5. At the moment in Japan, infections of airport staff have been confirmed. The difference from the introduction example in Japan to see how it would affect the situation by IT system other than saving time should be verified in future. With the hope that the future COVID-19 will be calmed down soon and that each facility will be restored, this paper concludes with a view to keeping track of the current status of the airport terminal support.

REFERENCES

Akase, T. (2013) Signology for Public Spaces. (in Japanese) Kajima Institute Publishing

Akihiko Watabe, Kazuhiko Mori, The relation between signage and way-finding behavior a cognitive study in way-finding in architectural space, No.1, Journal of Archit, Plan, Environment, Engng, AIJ, No.437, July.,1992

Centrair Council (2013). Centrair Disaster Prevention Manual (Earthquake and Tsunami)-Chubu International Airport Tsunami Evacuation Plan (in Japanese)

Central International Airport.co (n.d.). Centrair Operation Center (in Japanese)

Central International Airport.co (n.d.). Universal Design (in Japanese) Retrived September 25 from https://www.centrair.jp/corporate/csr/universal-design/

Hyunsuk Kim, Satoshi Asano, Kunie Kajisjima, Yoshiaki Horikoshi, Haruhiko Goto, Koichi TonumaAn investigation into the stage of important plans fir public signs -proposals for and themes of public signs in Setagaya Ward, Tokyo-, Journal of Archi, Plan, Environ, Engng, AIJ, No.415, Sept., 1990

Ipek Kastas, Didem Kan Kilic, Wayfinding Behavior of Airport Users in Relation to the Five City Elements of Lynch, edra43Seattle Conference Proceeding, P.241

Kansai Airports (n.d.). Safety and disaster prevention plans (in Japanese) Retrived September 25, 2019, from http://www.kansai-airports.co.jp/efforts/safety/disaster-prevention-plan.html

Kato, A., Tsunekawa, K., Taniguchi, G. Universal Design Approach for Planning, Design and Management of Terminal Building of Chubu International Airport, Japan, edra37Atlanta Conference Proceeding, P.287

Kato, A., Taniguchi, G., Yamada, A., Mizutani M., Hara, T., Morisaki, K. (2007). Effects of universal design research for facility management of Chubu International Airport Terminal Building, The 2nd International Conference for Universal Design in KYOTO 2006 P.351-358.

Lori.McGilberry, Scott Gorenc, Jonathan Massey, Predictive Analysis and Passenger Experience in Terminal Design, edra47Raleigh Conference Proceeding, P.323

Ministry of Land, Infrastructure, Transport and Tourism (2017). The Tourism Vision in Support for Tomorrow Japan; Designating 27 Regional Airports as Welcoming Facilities of Foreign Visitors (in Japanese), Retrived September 25, 2019, from http://www.mlit.go.jp/report/press/kouku19_hh_000070.html

Ministry of Land, Infrastructure, Transport and Tourism (n.d.). International base airport (in Japanese), Retrived September 25, 2019, from http://www.mlit.go.jp/koku/15_bf_000304.html

Narita international airport corporation (n.d.). Initiatives for safety (in Japanese), Retrived September 25 from https://www.naa.jp/jp/airport/safety_torikumi.html

Simon J. Buechner, Christoph Hoelscher, Peter Eckart, Bernd Hilpert, Jan Wiener, Evaluation of Signs Using Eye-Tracking, edra41Washington, DC Conference Proceeding, P.209

D7- Transform Places with People and the Environment: Work Environments

Examining Police Officer's Workplace in Police Box and Residential Police Box in Mie Prefecture, Japan

Ryota Kobayashi (Mie University)

1. ABSTRACT

The current study explores two types of police facilities, police boxes and residential police boxes with new functions, and the characteristics of the physical space of the facilities were clarified. The police box is a facility where multiple police officers work on a 24-hour shift basis, while a residential police box is a facility where a police officer work while living with their families. Prefectural police is required to create an "individual facility plan" that describes the current state of the facilities they own and future renovation plans. According to the "Individual Facility Plan", police boxes and residential police boxes were built concentrated during specific era. If this situation continues, the timing to renovate these facilities will be concentrated, and it is expected that the financial burden will be increased. Therefore, it is necessary to accurately grasp the current situation and to make an appropriate maintenance plan. As a result of investigating the characteristics of physical space, it was found that police box has a layout that ensures the safety of police officers,however, the residential police box did not have the same safety features. The results of a survey on satisfaction with the facilities indicated that police officers working in the residential police box were dissatisfied with safety. In the layout of the police box, two types were confirmed. In addition, it has become clear that in future facilities it will be necessary to provide facilities that take visitors into consideration, such as barrier-free toilets.

1.1. Keywords

Police facility; Police box; Facility Management;

2. Introduction

In Japan, the Tokyo Olympics were held in 1964, and the urban infrastructure was newly developed intensively in the 1960s. Currently, facilities built around this time period are now at least 60 years old and in general, the useful life of buildings is 60 years. Thus, many facilities are needed to be rebuilt. Also, a number of buildings were built during the bubble economy period in the late 1980s; those buildings need to be repaired after 30 years. Many public facilities such as government buildings and schools are also constructed during these time periods, and these buildings need to be renewed and repaired.

The National Government Buildings Division Managers Meeting was held in June 2014, and the survey was conducted on the buildings used by the nation, prefectures and major cities designated by government-ordinance, and managed by the department in charge of the conference. According to this survey, the total area of concerned public buildings was about 320 million square meters, or 3.2 billion square feet. The details are shown in Table 1. In addition, the ratio of total floor area by age for these buildings is shown in the top graph in Figure 1. More than 45% of the facilities (by floor area ratio) accounts for more than 30 years after construction and they are needed to be renovated. In 10 years, the percentage will be 65%, more than half. The bottom graph in Figure 1 shows the total floor area by construction year. The area of the facilities constructed around 1970-2000 is large, and it can be said that there are a number of buildings that are in the period for rebuilding or large-scale renovation.

In recent years, the declining birthrate and aging population, and the decline of population decline has become a serious problem in Japan. Local governments have difficulty in investing in the maintenance of public facilities due to the decrease in tax revenues and increased welfare costs arising from these problems. Under these circumstances, there is a demand to develop a method to manage facilities efficiently and effectively. The situation surrounding public facilities development is severe, and it is important to make a plan for renovation and renewal of facilities from a strategic Facility Management perspective.

Police facilities are also public facilities, and it is necessary to systematically improve future facilities. In particular, there are more than 12,000 police boxes and residential police boxes nationwide. Since there are many facilities that are small in size, it is necessary to grasp the situation of the facilities as soon as possible and to make appropriate plans such as renovations.

Table 1 The detail of the total area of public buildings subject to survey (Unit: 1,000m^2) (National Government Buildings Division Managers Meeting, 2014)

Facility classification	Office*	Board of Education facilities	Public Safety Commission facilities	Public housing	Total
National	48,466	-	-	-	48,466
Prefectures (47)	43,290	58,840	11,631	63,801	177,562
Government-ordinance-designated cities (20)	33,933	33,200	-	29,929	97,062
Total	125,689	92,040	11,631	93,730	323,090

*"Office" includes all facilities (libraries, hospitals, etc.) not included in other categories, in addition to the general office building

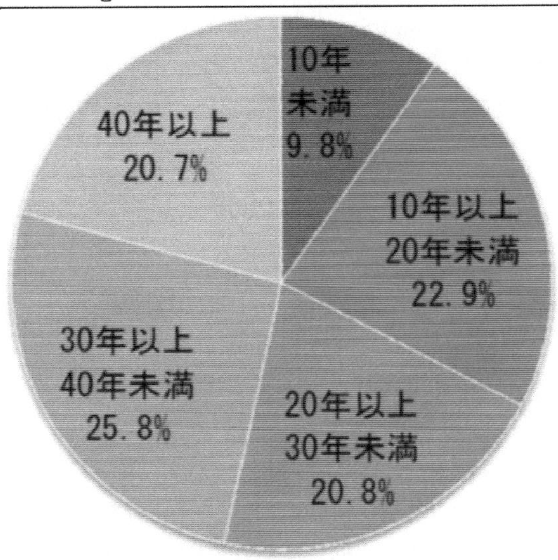

On the other hand, in recent years there have been incidents of attacking police officers working in police boxes for the purpose of robbing handguns or killing police officers. Although police officers work to keep citizens safe, they work in a hostile environment due to the nature of their job. Therefore, it becomes necessary to also consider the safety of police officers.

3. Objectives

Police facilities have rarely been studied in the field of architectural planning and environmental behavior from the viewpoint of regarding information leaks and escape incidents of suspects. When designing these facilities, the architects simply follow the design of currently existing facilities, and the

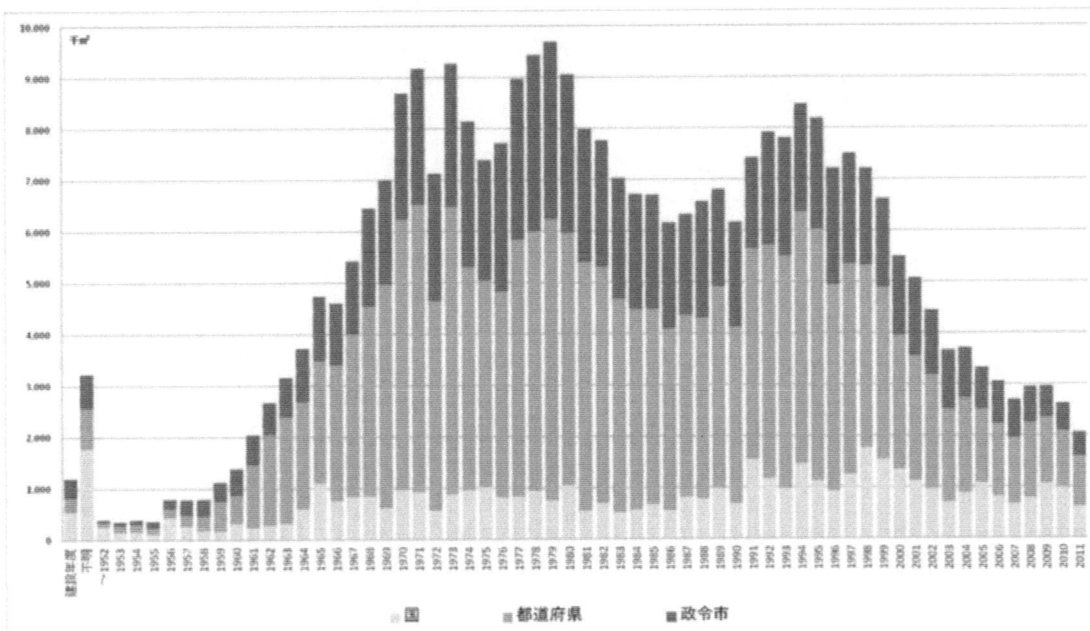

Figure 1 The ratio of total floor area by age for public buildings subject to survey and the total floor area by construction year (National Government Buildings Division Managers Meeting, 2014)

design methodology is not well-developed, making it is difficult to propose a new and better design. Thus, the purpose of this research is to clarify:
1) The facility situation regarding police box.
2)The physical and spatial features of police boxes.

4. Method

In this study, the following surveys were conducted in order to understand the facility situation regarding police box as well as the physical and spatial features:

1) Facility maintenance survey
Investigate the degree of aging of facilities based on police facility maintenance policies published by prefectural police and local governments.
2) Police box layout survey
Investigate the characteristics of physical space of police boxes by measuring the corresponding space and developing detailed drawings.
3) Satisfaction survey
Investigate how police officers who actually work in police boxes feel about their facilities.

4. The situation of police facilities

4.1. Aging police facilities

In order to clarify the direction of medium- to long-term efforts including grasping the current situation and issues of public facilities and planning for maintenance, the ministry and local governments are required to formulate an infrastructure life extension plan (referred to as "action plan"). In addition, they are required to formulate a "long-life plan for each individual facility (referred to as" individual facility plan "), which defines a more specific policy for each individual facility.

In response, the National Police Agency formulated the "National Police Agency Infrastructure Life Extension Plan" in March 2015 as an "Action Plan". The target facilities are roughly divided into the following three facilities.

1) Police facilities: A) All police facilities that are financed by the national expenditure (referred to as "national expense facilities") However, rented facilities and police communication facilities such as wireless relay stations are excluded and B) Facilities that are subsidized by the national expenditure (referred to as "subsidized facilities")
2) Traffic safety facilities: Traffic lights
3) Police communication facilities: Wireless relay station (excluding rented facilities)

Then, the current situation and issues of police facilities are confirmed from the "National Police Agency Infrastructure Life Extension Plan". National expense facilities are mainly police school facilities and riot police facilities. Subsidized facilities are police office buildings, police boxes and residential police boxes that are managed by prefectural police. Tab 2 shows the aging status of major national expense facilities as of April 1, 2014, and Table 3 shows the aging status of major subsidized facilities. Facilities over 30 years old are approximately 53% of the main school building, 62% of the student dormitory, and 56% of the police station. So, more than half of each facility type requires renovation. As for the riot police facilities, the office buildings and waiting dormitories are close to half, 48% and 40%, respectively, are over 30 years old, and the facilities are aging as a whole. The police school facilities were intensively developed in response to a significant increase in the number of local police officers in the 1960s. The riot police facility was intensively developed in the event of a security event due to mutual cooperation and the conclusion of a security treaty between Japan and the United States in 1960. These facilities are currently undergoing renovations. It is necessary to accurately grasp the state of the facility through inspection and diagnosis, and to appropriately plan the timing of large-scale renovations by performing preventive maintenance.

4.2. The situation such as police box

Prefectural police is expected to be responsible for maintaining police boxes, which are subsidized facilities, so each prefectural police prepare "individual facility plans" to grasp the current situation. Gunma Prefectural Police and Hiroshima Prefectural Police have organized the construction status of police boxes and residential police boxes according to the construction year in their respective "individual facility plans". Figure 2 shows the situation in Gunma Prefecture. In Gunma Prefecture, there are 91 facilities constructed between 1986 and 1995, accounting for about 48% of all police boxes and residential police boxes, so renewal and renovation are expected to occur within a certain period of time. Figure 3 shows the situation in Hiroshima Prefecture.

The Hiroshima Prefectural Police uses the useful life as one of the indicators in grasping the current situation of facilities. The useful life span is determined in the prefectural property evaluation standard (hereinafter referred to as "useful life"), and is determined by the structure as follows:

Table 2 The aging status of major national expense facilities (National Police Agency, 2015) and year (National Government Buildings Division Managers Meeting, 2014)

		Over 40 years	30-39 years	Less than 30 years
Police school facilities	main school building	39.68%	14.29%	46.03%
	(cumulative percentage)	39.68%	53.97%	100.00%
	student dormitory	38.71%	22.58%	38.71%
	(cumulative percentage)	38.71%	61.29%	100.00%
Riot police facilities	office buildings	25.40%	22.22%	52.38%
	(cumulative percentage)	25.40%	47.62%	100.00%
	waiting dormitories	22.41%	17.24%	60.35%
	(cumulative percentage)	22.41%	39.65%	100.00%

Table 3 The aging status of major subsidized facilities (National Police Agency, 2015)

	Total	Over 40 years	30-39 years	Less than 30 years
Police station	1,169	325	329	515
Percentage	-	27.8%	28.1%	44.1%
(Cumulative percentage)	-	27.8%	55.9%	100.0%

1. Reinforced concrete construction (RC) 60 years
2. Concrete block construction (CB) 45 years
3. Steel frame construction (S) 35 years
4. Wooden construction (W) 25 years

Police boxes of concrete block construction were built between 1951 and 1965, and those of reinforced concrete construction were built between 1988 and 1997. It can be seen that police boxes were intensively built during these two periods. There are 62 wooden residential police boxes built between 1991 and 2001, accounting for 44% of the residential police boxes. This wooden residential police boxes will be renewed. The status of the facilities is shown in Table 4.

It was confirmed that there were times when the renewal of facilities was concentrated in both Gunma and Hiroshima prefectures. Since each prefecture has different circumstances, it is necessary for the prefectural police to quickly grasp the current state of the police facilities they have and make an appropriate maintenance plan.

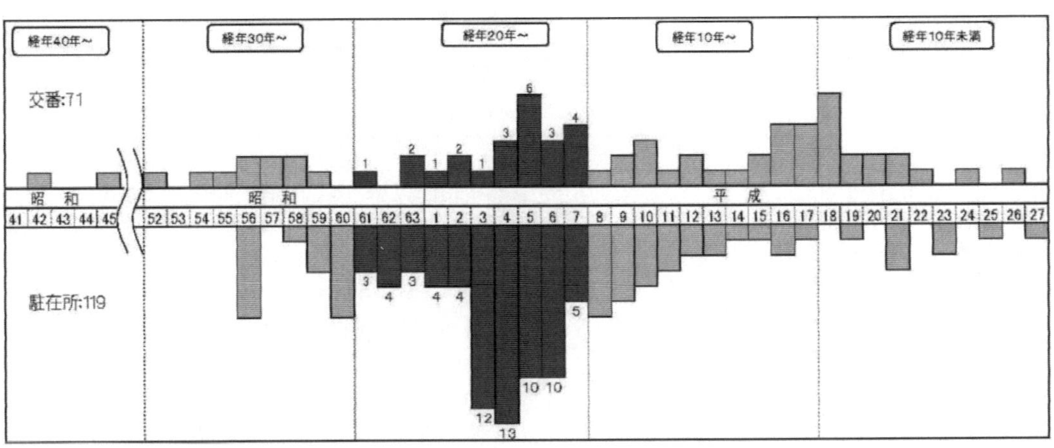

Figure 2 The situation according to construction year of police box and residential police box in Gunma Prefecture (Gunma Prefectural Police Headquarters, 2016)

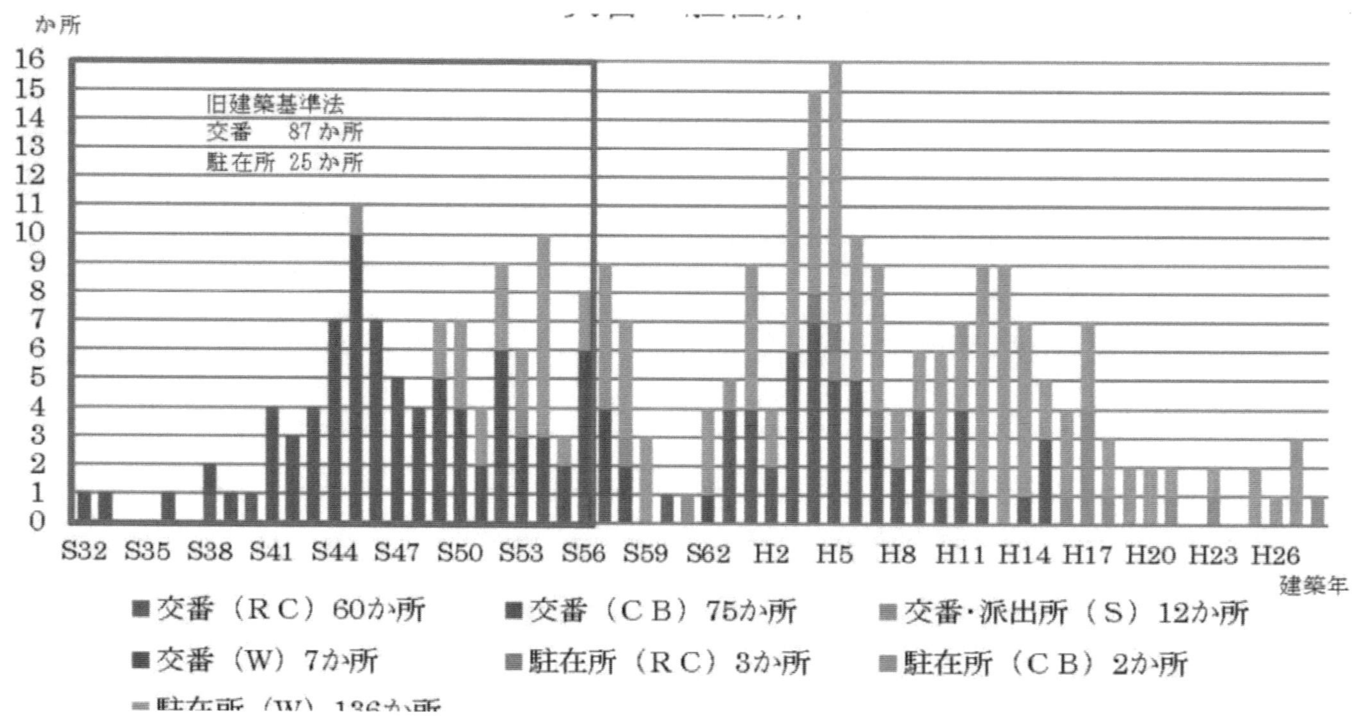

Figure 3 The situation according to construction year of police box and residential police box in Hiroshima Prefecture (Hiroshima Prefectural Police Headquarters, 2017)

Table 4 Situation of police box and residential police boxes (Hiroshima) (Hiroshima Prefectural Police Headquarters, 2017)

	Structure (Number of facilities)	Passed the useful life	Old Building Standard Act	Passed the useful life and old Building Standard Act
Police box (154)	RC (60)	0	8	0
	CB (75)	38	74	38
	S (12)	1	1	0
	W (7)	6	4	4
Residential police box (141)	RC (3)	0	0	0
	CB (2)	1	1	1
	W (136)	52	24	24
Total (295)		98	112	67
Percentage (%)		33.2	37.9	22.7

5. Physical space features

5.1. Facility overview

In April 1992, the National Police Agency repositioned the external work section to the regional section, and newly established “Regional Police Management Rules” clarifying the responsibility to maintain security (regional responsibility) in each area under the jurisdiction of the Hasyutsujo (police boxes) and the residential police boxes. The Hasyutsujo (police boxes) and the residential police boxes are positioned as a life safety center in the local community. It is thought to have changed the facility function of the police boxes by this.

Therefore, a facility maintenance survey was conducted on three police boxes and three residential police boxes in Mie Prefecture that were built after 1992. At the target facilities, plan drawings were created by measuring with a laser rangefinder. The “office room” is a place where a user performs a procedure for obtaining a notification. In addition, the room where the computer stations are placed at the back of the office is called the "working room".

The results of the surveyed facilities are shown in Table 5 and Figure 4. The ratio of the office room to the total floor area of the surveyed facilities is 26% for police boxes A and C, and 18% for police box B. Only the B police box has a consultation room and a barrier-free toilet, so it is considered as the reason for a lower ratio when compared the other types. In the residential police box category, residential police boxes D and E account for 14% and residential police box F accounts for 10%. The residential police box F also has a consultation room and a barrier-free toilet. The size of office room of the residential police boxes is 10 to 14m^2, which is smaller than the office room of the police boxes (20 to 30m^2). As a general rule, two or more police officers are assigned to the police box, but only one person is assigned to the residential police box. So the office room of the residential police box is smaller than the police box.

Table 5 The results of the surveyed facilities

	Police box			Residential police box		
	A	B	C	D	E	F
Building year	27-Mar-84	25-Mar-10	1-Aug-00	16-Mar-85	23-Mar-92	23-Mar-06
Site area (m²)	155.00	352.37	-	330.00	266.36	436.00
Total floor area (m²)	98.00	169.91	75.92	71.21	102.00	96.00
Office room area (m²)	25.73	30.32	20.08	10.14	14.22	9.70
Office room area / Total floor area (%)	26.26	17.84	26.45	14.24	13.94	10.10
Consultation room	×	○	×	×	○ (7.52m²)	○ (7.42m²)
Barrier-free toilet	×	○	×	×	×	○

Figure 4 Police box and Residential police box layout

5.2. Police box layout

In the police box layout survey, it was confirmed that police officers and visitors are divided into areas using counters and swing doors. The security of police officers is ensured by creating boundaries and preventing visitors from easily approaching police officers.

Two types were found in the office layout of the police box (Figure 5). The first was a "screen type" (Figure 5). This is a layout where desks are placed parallel to the counter so that a screen is made at a desk in the police officer area. This layout is designed to prevent visitors from entering the police officer area. Police officers sit at the entrance, making it easier to notice the presence of visitors. The second is the "island type" (Figure 5). This is characterized by creating a large free space in the police officer area by arranging the desks in the police officer area in an island shape. By securing a large free space, the police officers can move freely, and it is thought that the layout was easy to handle the quick movements of visitors. The police officers sit face-to-face in daily work, making communication easier.

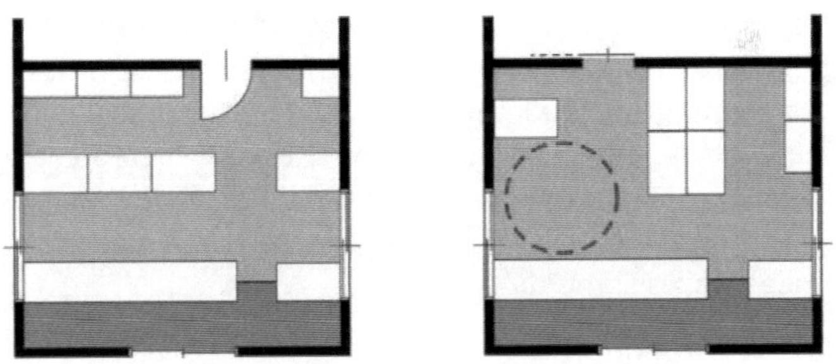

Figure 5 Two types of office room layout

5.3. Residential police box layout

There were only minimum desks in the office because of space limitations. Also, police officers and visitors cannot be divided into areas due to the business characteristics of affinity with local residents. Therefore, considering the fact that the residential police box is for one person, it becomes easier to attack than the police box. Thus, it is necessary to consider improving safety measures.

The residential police box is different from the police box, and the office room is attached to the residence. Police officers live and work with the family in the residential police box. Therefore, the office room and the residence are directly connected. When police officers are not in the residential police box due to patrols, some member of the family responds. Visitors can rest assured that there is always someone. However, in this open relationship between the office room and the residence, suspicious people may enter the residence. It may be safer to connect to the consultation room instead of the office room where anyone can see the entrance to the residence as in the F residential police box.

5.4. Use of counter

Police officers basically respond to visitors across the counter. The height of the counter is about 900mm, which may be difficult to get over. Desks and counters with a height of around 700mm are always provided for wheelchair users. These counters prevent visitors from easily approaching police officers. However, when a visitor stands in front of the counter, the area around their waist is a blind spot for police officers. Care should be taken because it is difficult to notice whether the visitor is in possession of a weapon .

Due to the height of the counter, both police officers and visitors are basically facing each other in a standing condition. When police officer fills out a document at the counter, he can't see the visitor's face. Therefore, when a police officer fills in a document, it is necessary to fill in at the desk in the police officer area, not at the counter. It should be considered changing the way one responds. For example, if the visitor was to sit at a desk height of 700mm and the police officer was to stand on 900mm counter side. The position of police officer standing and visitor sitting, will create an advantage positioning for the police officer. It may be easier for visitors to sit down because the registration process takes some time.

5.5. New function

In the new facility, visitors-friendly facilities such as consultation rooms and barrier-free toilets are installed. This is thought to be due to the policy that police box has been positioned as a life safety center in the local community since 1992. The consultation room is used when people who do not want to see where they are consulting or who are victims. By protecting the visitor's privacy and creating an environment where it is easy to consult, it can be handled before it develops into a major incident. The consultation rooms should not use a counter as in the office room, but should use relaxing furniture such as a sofa and a low table.

Barrier-free toilets can be accessed from the office room or consultation rooms, and none can be used directly from outside. It is difficult to use because one has to enter the police box first to use it. In addition to the use of ordinary people, it is also used for a drunk person. In that case, it would be more convenient if the entrance was connected to the office room.

5.6. Satisfaction survey

In order to investigate the evaluation of the actual facilities of police officers who worked there, a satisfaction survey for police officers who work at the target facilities was conducted.

In the police box, the level of satisfaction with the workplace functionality was low. In particular, it was found that there was a problem with the accessibility of information. There is no computer station in the office room, so police officers can't immediately find out what the inquirer is asking for. In the case of route guidance, one of the main tasks of the police box, is handled by paper maps, but this task may not be necessary now that smartphones are popular.

In the residential police box, the result showed low satisfaction with the functionality, safety and common space of the workplace. As confirmed in 5.3, there is no working room in the residential police box, and there is only one desk in the office room, so this may be the reason for the low satisfaction of functionality and safety. In the residential police box F, there were places where about five people met about 15 times a year as a place to exchange information with local residents. To remedy this, it is considered that the dissatisfied shared space is not a shared space between police officers but a shared space between police officers and local residents. For this, instead of securing a large space at the representative office for the meeting, the meeting should be chosen to be held in a local community center.

6. Conclusion

This study examined the physical and spatial features of police facilities, especially police boxes and residential police boxes categories. Although the situation of maintaining police boxes varied from one prefecture to another, there were cases where they were intensively maintained at specific times. For this reason, it is expected that the time to renovate and renew facilities will also be concentrated. Since the financial burden also increases, it is necessary to determine the priority of repairs based on the degree of aging of the facilities and the situation of the area under the control of the facilities, and to prepare an appropriate repair plan.

In the case of police box, it was revealed that police officers and visitors were divided into areas using counters and desks to ensure the safety of police officers. At that time, paying attention to the desk layout, two types, "screen type" and "island type" were confirmed. In the police box where the number of police officers working is small, it is considered that the "screen type", which makes it difficult for visitors to approach police officers, is suitable. On the other hand, in the police box with many police officers, the "island type", which is easy to communicate in daily business, is considered suitable. However, there is a problem that a blind spot can be made for police officers by a counter functioning as a safety measure.

In the residential police box, the space was not divided to ensure the safety of police officers like police boxes. Both the police box and the residential police box should be equipped with facilities with consideration for visitors such as consultation rooms and barrier-free toilets in the new facilities. This is considered to be a change due to the fact that the dispatched station and the representative office were positioned as a life safety center in the local community in the "Regional Police Management Rules" of 1992. In future renovations, these new facilities will be necessary, but it should be discussed how to use these facilities with the police officers who actually work, and be careful not to create unused spaces.

REFERENCES

FM Promotion Liaison Council (2018). The Official Guide of Facility Management, in Japanese, Nikkei Publishing Inc.

Gunma Prefectural Police Headquarters (2016). Gunma Prefecture Police Facility Management Plan, in Japanese, https://www.police.pref.gunma.jp/keimubu/08kouhou/15koukai/keimubu/kaikei/kaikei_28.pdf. Accessed September 30, 2019.

Hiroshima Prefectural Police Headquarters (2017). Police facility maintenance policy, in Japanese, https://www.pref.hiroshima.lg.jp/uploaded/attachment/234946.pdf. Accessed September 30, 2019.

Liaison conference of related ministries and agencies regarding promotion of infrastructure aging measures (2013). Action plans for extending the lifespan of infrastructure, in Japanese, www.cas.go.jp/jp/seisaku/infra_roukyuuka/pdf/houbun.pdf. Accessed September 30, 2019.

National Government Buildings Division Managers Meeting (2014). Case examples pertaining to aging measures for public buildings, in Japanese, www.mlit.go.jp/common/001060850.pdf. Accessed September 30, 2019.

National Police Agency (2015). Action plans for extending the lifespan of infrastructure, in Japanese, https://www.npa.go.jp/seisaku/koutsu/kisei/H270319/02_honbun.pdf. Accessed September 30, 2019

National Police Agency (2004). 2004 police white paper. In Japanese, Gyosei.

D11 - Transform Places with People and the Environment: Children, Youth & Environment II

Forms of Nature in Play-Space Making

Sarah Khalid Malak (Arizona State University)

1. ABSTRACT

Today, more than half of the world's children are urban dwellers who have limited access to outdoor and natural play areas, whose indoor environments are injected with technological attractions that keep them occupied in a sedentary life. In History of Children's Play and Play Environments, Joe L. Frost categorizes 'play' in the United States into six eras beginning with the wilderness era where play happened in the wilderness of nature and the open spaces, and streets of old towns. With the industrial revolution of the late 19th century, play-spaces evolved into structured, built, and planned play-spaces. Today, with the growing research on the power of play and especially in nature, we find that play-space design is finding its way back to nature in the form of natural playscapes, nature-inspired playgrounds, and dynamic free-play designs.

Since play appears to have started in nature and is heading towards it, this paper questions the influence of nature on the built play-space and whether this influence can help lead to better built-play environments. This paper explores the research question: What are the forms of nature in built-play environments that are conducive to children's play, development, and wellbeing? By reviewing the literature on the development of play theories and play environments, this paper concludes that nature encourages play because it aligns the emergent, nonlinear, and dynamic aspects of play. Nature-inspired play-spaces differ in the way they bring nature into the built environment. Only those that address the emergent, nonlinear and dynamic aspects of play can provide play opportunities similar to nature in the way that it encourages child play, development, and wellbeing.

1.1. Keywords

Play-space, play theories, nature play, urban play, nonlinear dynamic theory

2. THE STATE OF PLAY

Today, where more than half of the world's population lives in urban environments (UN, 2018), fewer children are playing outdoors. Children in the United States spend less time playing outdoors than their mothers did as children. A survey done in 2014 records a significant decrease in creative play, pretend play, and child-initiated street games, which are critical for children's social and cognitive development (Clements, 2014, p. 72-73). Additionally, Stephen Kellert, a social ecologist, states that most American children spend 90% of their time indoors; most of them spend 30 minutes daily in unstructured outdoor play; in contrast, the generation before them spent more than 4 hours outdoors (Kellert, 2015, p. 288-89). The deprivation of the basic right to play is not limited to the United States. Similar studies around the world show that the decrease in outdoor and physical play is a global problem that has negative consequences on children (Else et al.,2014).

Research on the decrease in outdoor play explains that this issue results from many interrelated factors (Carr & Luken, 2014, p. 79; Frost, 1997,54-57; Frost, 2009, p.214; Kellert, 2015, p. 288-89; Louv, 2015, p. 11-13; Tremblay et al., 2015, p. 6485). With industrialization came urbanization, digital technology, air conditioning, and the modern lifestyle. The focus on transportation in an urbanized world is limiting direct access to children's play-spaces. Digital technology is providing children with indoor entertainment in the comfort of air-conditioned spaces. The modern lifestyle of an over-scheduled professional life is pushing parents to invest in their children's academic success by focusing on passive learning and organized sports. Parents have become cautious about outdoor un-supervised play and from risky play that they would sue manufacturers and designers of apparently dangerous play features (Tremblay et al., 2015, p. 6483). People are now living as nuclear families with a fewer number of children; this means fewer peers to play and socialize with (Heseltine & Holborn, 1987, p. 72; Brett, Moore, & Provenzo, 1993, p. 61). This growing list shows that we have a problem in the making of play-spaces for children.

Why is the lack of outdoor play a concern? Recent studies such as the Canadian study, "Position Statement on Active Outdoor Play" (Tremblay et al., 2015), and "A Movement for Movement" (M4M) (Sigman, 2019) clarify that playing outdoors, and especially in an unstructured environment, have many benefits on children's health, wellness, and development. Active and outdoor play are conducive to children's cognitive, social, communicative, psychomotor, and emotional development. The increase in screen-time, decrease in outdoor play, inactivity, and the increase in sedentary behavior can cause obesity and other non-communicable diseases like diabetes, sleep disorders, and mental health issues. Due to these negative impacts, the American Academy of Pediatrics (AAP) is now recommending pediatricians to prescribe play for children up to age two (Yogman, Garner, Hutchinson, Hirsh-Pasek, & Golinkoff, 2018). The Position Statement suggests that "access to active play in nature and outdoors—with its risks—is essential for healthy child development." It also recommends "increasing children's opportunities for self-directed play outdoors in all settings—at home, at school, in childcare, the community, and nature" (Tremblay et al., 2015, p. 6476). Concerns about the decrease in child play are being addressed locally and globally by childhood organizations that work towards what the play advocate, Joe Frost, calls the "Contemporary Child-Saving Movement." According to him, this movement requires collaborative efforts on different levels to save children from a "dual crises": decrease in outdoor play and alienation from nature. Both the early child-saving movement which took place after industrialization in the late 19th century and the contemporary child-saving movement include a sub-movement called "back to nature" (Frost, 2009, 242-248), or as the nature and childhood researcher, Richard Louv, calls it, "the new nature movement: not going back to nature, but forward to a nature-rich civilization" (Louv, 2017). Louv aims to save children from "Nature-Deficit-Disorder" and revive their connection to nature (Louv, 2015, p. 34, 260-61). Both, Frost and Louv's advocacies for play suggest that play in the urban and built environment needs to be reconsidered, and that play in nature as a rich and stimulating environment needs to be revived.

3 . RESEARCH OBJECTIVES

Guidelines for the design of play environments vary between those focused on the built environment (Brett, Moore, & Provenzo,1993; Heseltine & Holborn, 1987; Moore, Goltsman, & Iacofano, 1987) and those focused on natural play-scapes as designed landscapes with natural elements intended for play (Moore & Wong, 1997; Moore et al., 2009; Moore, 2014; Carr & Luken, 2014). The line between built-play and nature-play environments is blurry, and in this blurry line, there is a gap for research. Maybe, nature-play guidelines do not have to be separate from built-play guidelines? Compared to the power of the natural environment in stimulating play, urban and built play settings are limiting children's opportunities to play. Therefore, nature could be our guide and solution to better the state of play in built environments. We need to observe children's interaction with natural play elements to distinguish new play-design strategies from nature. We also need to explore how to apply these new strategies in the built environment. These observational and analytical explorations are part of a dissertation currently in progress. This paper, however, reviews the literature that explores: What are the forms of nature in built-play environments that are conducive to children's play, development, and wellbeing? By asking this question, this paper aims to find the playful aspects of nature that can be replicated in the built environment to make it as playful and beneficial to children as nature.

4. METHODS

This study aims to find the forms of nature in play-space design by conducting an interpretative-historical review of the development of western play theories, the development of play-environments in the United States, and their relationship to nature. First, western play theories that explain the meaning of play are reviewed mainly from Doris Bergen's perspective (2014), then the influence of nature on them is highlighted. Then, the development of play environments in the United States is analyzed mainly based on Joe L. Frost's classification of the play and playground eras (2009), then the influence of nature on them is distinguished. According to Tremblay et al. (p. 6486), natural play environments "include natural elements such as plants, soil, and water. These may be human made (e.g., gardens, nature playgrounds and urban parks) or wild and naturally occurring (e.g., wooded areas, meadows and beaches)." In this study, the focus is on the naturally occurring environments as nature-play settings to learn and be inspired from.

5. PLAY & PLAY ENVIRONMENTS

5.1. What is Play?

Plato is one of the first philosophers to express his perspective on play and its importance in learning, gaining experience, and training for adulthood. Early western play theories describe play from developmental, sociocultural, and biological perspectives. According to the educational psychologist Doris Bergen, most of the early play theories describe play by its characteristics and purposes instead of defining it (Bergen, 2014, p. 9). In fact, many writers express that play is difficult to define (Brown, 2010, p. 74; Eberle, 2015, p. 214-33) and developing a "unifying play theory" is challenging (Hakkarainen, 2017, 1-2). This review of play theories examines the theories that shape the meaning of play for children.

Early developmental theorists like Fredrich

Frobel and Maria Montessori have acknowledged the importance of play and applied it in their educational settings. However, Bergen notes that these theorists do not provide a clear definition of play. She explains that the period between the mid to late 20th century is when developmental theories start to shape the meaning of play (Bergen, 2014, p. 10-11). In his developmental stages theory, Jean Piaget (1896-1980) explains how children's understanding of meaning develops from concrete to abstract ideas as they play. Play is an "assimilation process" that allows children to relate new experiences to their current cognitive understanding of the world. Lev Vygotsky (1896-1934) emphasizes the role of play in imagination and socialization. While playing, children pretend, solve problems, create conversations, regulate their emotions, and feel challenged to perform to higher levels of skills (Bergen, 2014, p.14-16; Mooney, 2013, p. 78, 101).

Early Biological play theories, Bergen explains, describe play by its purposes, such as teaching survival skills, relieving from stress, stimulating curiosity, and fostering social skills (Bergen, 2014, p. 11-14). From a sociocultural aspect, Johan Huizinga (1872-1945) described play as a cultural process that forms the base of civilization, language, philosophy, music, poetry, and art. To Huizinga, play has an aesthetic quality that transcends psychological and physiological explanations, the "fun-element." He defines play by its "formal characteristics" as (Huizinga, 2014, p.1-13): "a free activity standing quite consciously outside ordinary life as being not serious, but at the same time absorbing the player intensely and utterly. It is an activity connected with no material interest; a no profit can be gained by it. It proceeds within its own proper boundaries of time and space according to fixed rules and in an orderly manner. It promotes the formation of social groupings which tend to surround themselves with secrecy and to stress their difference from the common world by disguise or other means."

The accumulation of developmental, biological, and sociocultural theories of play shows that play is essential for children's social, intellectual, emotional, and physical wellbeing. Play is vital for the development of imagination and creativity, for socialization, finding a purpose, understanding the world, problem-solving, and creating art. Collectively, these theories form a foundation for the contemporary and interdisciplinary understanding of play. Contemporary play theorists including David Elkind, Stuart Brown, Karen Vander Ven, and Scott Eberle, explain play as a complex behavior that is paradoxical in its nature, as an essential element for learning, development, happiness, and wellbeing of children and adults.

David Elkind describes play based on psychological and educational theories. To him, "play is our need to adapt the world to ourselves and create new learning experiences," and "it is only when we integrate play, love, and work that we, as children and adults, can live happy, healthy, and productive lives." (Elkind, 2007, p. 2-5, 218) Similar to Elkind, Stuart Brown links play to happiness in that play is the opposite of depression, not work. Brown also states that play and work should happen together, because both involve creativity. In his book "Play: How it Shapes the Brain, Opens the Imagination, and Invigorates the Soul", Brown states that play is inherently too complex and that it must be experienced to be understood. Instead of providing a defining statement, he chooses three ways to describe play: His list of "properties of play," "process" of play, and Huizinga's "formal characteristics" of play (Brown, 2010, p. 15-20, 126-37).

In his search for a clear definition of play, Eberle states that the difficulty of defining play lies in its dynamic nature. He builds his definition on two main

Table 1. Defining Play by Brown's Properties, Eberle's Process & Huizinga's Characteristics

Properties of Play (Brown)	Process of Play (Eberle)	Formal Characteristics of Play (Huizinga)
- apparently purposeless - voluntary - inherent Attraction - freedom from time - diminished consciousness of self - improvisational potential - continuation desire	- anticipation - surprise - pleasure - understanding - strength - poise and balance	- voluntary and free - superfluous - not ordinary nor real - limited to a specific time or location - tense - comes with fixed rules - includes secrecy and differentiation of the group

concepts: (1) Play is a "function" of physical, social, emotional, and intellectual development; (2) Play is an "emergent system" with "adaptive variability" and "self-organization." Adaptive variability suggests that play is a dynamic process that includes change and movement with time. Self-organization means that play can happen anywhere at any time, even at "in-between" spaces like imagination and cyber-spaces. Based on these concepts, and based on understanding play as a "process," Eberle states that (Eberle, 2015, p. 214-33):

"Play is an ancient, voluntary, 'emergent' process driven by pleasure that yet strengthens our muscles, instructs our social skills, tempers and deepens our positive emotions, and enables a state of balance that leaves us poised to play some more."

Parallel to Eberle's theory, Karen Vander Ven addresses play as a nonlinear dynamic system. She explains that this nonlinear dynamic aspect of play prepares children for the dynamic, complex, and chaotic future of the 21st century (Vander Ven, 2015, p. 409-16). She criticizes the linear approach in developing educational programs. She acknowledges the importance of the environment in facilitating play, and by changing the environment or its characteristics, she suggests a nonlinear dynamic environmental solution for nonlinear dynamic play experiences that encourage creativity. The same critical opinion could be directed towards the design of playgrounds and play environments today.

5.2. Play Theories and Nature

Although not directly mentioned by Eberle or Vander Ven, the non-linear, dynamic, and emergent processes that characterize play also characterize nature and natural play environments. Earlier theories, on the other hand, do recognize and mention nature as a stimulator of curiosity, imagination, and creativity in play.

According to Frost, Comenius is one of the first philosophers to relate play during childhood to nature by comparing children's learning and development to a growing tree that needs nurture and care when young ,and will eventually grow with branches and ripe fruits. Frobel adopts this concept in his kindergarten, the garden where children grow like flowers (Frost, 2009, p.20). Frobel emphasizes the importance of play in nature by designing an outdoor nature play space, which, according to Frost, is the first place to be named "playground" (Frost, 2009, p. 28-30). In the kindergarten, Frobel applies a nature principle: children learn from active, hands-on, self-directed experiences. He translates this principle to the indoor classroom, where he designed "gifts" and "occupations" as objects and activities for self-directed learning. In that context, it seems that Frobel was able to extract and observe a principle from nature and that nature is the origin of learning-by-doing and self-learning tools that are still being used and emphasized in early education environments today.

In contemporary play theories, Elkind mentions five things that spark children's curiosities in this world. Three of those are related to nature: (1) natural elements: earth, water, air, and fire, (2) plants and animals, and (3) cosmos. The other two are (4) playacting, and (5) life issues. Elkind explains that self-directed play experiences guided by children's curiosity are one of the best ways to learn and mature (Elkind, 2007, p. 199-35). Moreover, the contemporary play theories of Eberle and Vander Ven explain play as a non-linear dynamic process that is complex, emergent, and adaptive in its nature. Play needs an environment that fits these characteristics like the natural environment with its transformative landscapes and cyclic processes of day and night, in addition to four seasons. The state of play in traditional model-based playgrounds is fixed and structured, thus not fitting the non-linear dynamic aspects of play.

5.3. Where Do Children Play?

"Where children are is where children play" (Opie and Opie, 1969). Children play anywhere and everywhere. In natural and built environments, they play outdoors and indoors. Historians who document the history of play and playgrounds explain that it has a similar history in most parts of the world (Eriksen, 1985; Frost, 2009). Frost categorizes the history of play and playgrounds in the United States into six eras (Frost, 2009). Nature play dominates some of these eras and built play features dominate most. The history of play-spaces begins in the "natural and wilderness era." For centuries, rural children played outdoors in the wilderness and farms, and city children played in streets and plazas. The appearance of the "child saving movement" and the attention to designing and organizing play-spaces begins with industrialization in the late 19th century. In both Europe and the United States, cities got crowded, streets became dangerous and filthy, and children lost their access to nature-play and street-play (Frost, 2009, p. 174).

As Frost explains, in the late 19th century nature was brought to the urban setting in the form of sandboxes. Parallel to the sandbox projects was the outdoor gymnasiums that led to the "manufactured appliance era" in the early 20th century. This era shows a detachment from nature in its emphasis on manufactured appliances that structured the play experience. After World War II, the era of adventure playgrounds brought nature into the play-space in the form of sand-play, gardening, and petting areas. It also provided built-play areas with loose construction tools and materials (Frost, 2009, p. 30, 92-95, 174, 183-190). The adventure playground is a playground in the making; it is dynamic and ongoing like play, like nature.

In the "novelty era" of the 1950s and 60s, artistic and architectural approaches dominated. Aldo van Eyck's playgrounds in Amsterdam and Richard Dattner's playscapes in the United States used abstract and concrete forms, and others like the Japanese octopus-slides, replicated forms from nature to attract children to play. In spite of being criticized for their fixed character and high expense (Eriksen, 1985, p. 35; Frost, 2009, p. 180-181), some playground designers of this era applied novel ideas that influence today's play-space design. According to Liane Lefaivre, Aldo van Eyck created a poly-centric system of playgrounds as place-makers and activators of city voids like block corners, wide sidewalks, and temporary spaces between buildings (Lefaivre, 2002, p. 27). On the other hand, Richard Dattner carries a different ground-up approach in the design of play-scapes. He prefers to call them play-scapes because he involves children in the design process (Dattner, 1969, p.56-70,137).

In the 1970s and 80s, standardization and modularity characterized the "modern playground era," a continuation of the "manufactured appliance era." Frost explains that as designers reconsidered the manufactured appliance era with novel innovative designs, playground manufacturers reconsidered their products according to safety and child development. Due to their affordability, low maintenance, and commitment to safety standards, standardized "modern playgrounds" are abundant today. However, Frost criticizes them for being "light years away from the developmentally focused playgrounds of America's best child development centers" (Frost, 2009, p. 190-93). In, "How NOT to Cheat Children," Simon Nicolson, criticizes modern playgrounds and other everyday children environments like schools, hospitals, and museums for their fixed and static characters. Nicholson discovered the "loose parts" theory that "in any environment, both the degree of inventiveness and creativity and the possibility of discovery, are directly proportional to the number and kind of variables in it." With this theory, he aims to break the barrier between "education and recreation" in children's everyday environments (Nicholson, 1971, p. 30-34).

Today, with the decreasing number of children playing outdoors, the modern playground is being questioned. Frost explains that the present era, "the postmodern era" of play is characterized by moving towards nature and natural playscapes (Frost, 2009, 190-93). The expanding research on the developmental benefits of nature-play is motivating the "new nature movement" (Louv, 2017) to strengthen children's connection to nature. The modern playgrounds that still exist today are also following the nature-play trend by decorating slides, swings, and post-deck structures with natural themes. However, these model-based nature-inspired playgrounds do not function like nature because they do not reflect its openness, versatility, and loose-parts.

Since research has proven that "children are more curious about, and interested in, natural spaces than prefabricated play structures" (Tremblay et al., 2015, p. 6486), some designers and child development specialists advocate for using natural elements as play features. The architect Robin Moore has produced multiple guidelines for playground design early in his career (Moore, Goltsman, & Iacofano, 1987; Brett, Moore, and Provenzo, 1993). Realizing the importance of nature play on child development, Moore's guidelines shifted to focus on naturalizing the playground either by either by retrofitting it or designing it with natural features from scratch (Moore & Wong, 1997; Moore et al., 2009; Moore, 2014). Similarly, the child development specialists, Victoria Carr and Eleanor Luken, propose a nature based "play-scape paradigm." They suggest replacing prefabricated playgrounds with all-natural playscapes to encourage outdoor play, prepare children to advance academically, and strengthen their relationship with the environment (Carr & Luken, 2014, 73-74).

Model-based playgrounds are being criticized for structuring and limiting the play experience, and nature is proving to be open-ended and conducive to learning and development. Do we have to replace built play environments with natural elements? Perhaps natural play features can be integrated with urban and built play features. Probably built-play features can apply the principles of play inherent in natural environments and features without having to use the natural features themselves. Parallel to the nature-movement characterizing the "postmodern era" of play, another contemporary approach to play-space design is growing It aims for deconstructing the playground and decentralizing it so that play can happen anywhere and everywhere.

Influenced by the ground-up approaches of the novelty era, Lefaivre proposes the PIP ground-up principle of play-space design. PIP stands for "Participation, Interstitial (the in-between) and Polycentric (network)" (Lefaivre, 2007, p. 80-109). She collaborated with Doll Atelier to observe children's favorite play spots in two selected neighborhoods in Rotterdam. Then they designed a decentralized system of interconnected playgrounds to transform these neighborhoods into playgrounds. Like Dattner, Lefaivre adopts a ground-up approach that involves observing and taking the opinions of children. Gunter Beltzig is another contemporary designer who recommends this approach. He suggests that designers behave like anthropologists and respect the "otherness" of children by observing them and discovering their needs (Beltzig, 2015, p. 288).

Decentralizing the playground and allowing play to happen anywhere is a yearly challenge that KaBOOM! has launched in 2018 to promote play everywhere in the United States. They also collaborate with Imagination Playground and Rigamajig to provide pop-up transportable playgrounds to kids anywhere, indoors and out. The Imagination Playground, designed by architect David Rockwell, is composed of large pieces of blue foam with different shapes and sizes. On the design team of the Imagination Playground is Cas Holman ("Imagination Playground," n.d.), who designed Rigamajig based on similar concepts to the Imagination Playground. She uses wooden boards, wheels, pulleys, nuts, bolts, and ropes as constructive play materials ("Brain Powered" n.d.). Both of these pop-up playgrounds use 'loose parts,' a concept found in nature. Cas Holman acknowledges the relationship between her product and nature play in the documentary film Design in Practice: "I think the ideal toy is a tree or a pile of sticks and sand and water. Rigamajig is a kind of a similar play value" (2016).

6. PLAY-SPACES AND NATURE

Nature has its "gifts." This is how Louv describes the physical, cognitive, social, and emotional benefits of nature for children. We need to connect to nature, preserve it, and live in harmony with it. Yet, this does not mean that digital technology and urban development are monsters. They are human-made, evolutionary, and young on this old and thriving planet. They do have the potential to nurture child development through play. We only need to understand how to make this possible. The Imagination Playground adopts a nature principle, the "loose parts." A new version of it is currently exhibited at the National Building Museum, Washington DC. It uses the "loose parts" principle in a built indoor space, enhanced with light features and interactive screens with moving images and sounds ("Play Work Build," n.d.). Here indoor technology stimulates the play experience. Perhaps we do not have to "turn off the tech-toys" as Frost states (Frost, 2009, p.158). Instead, we could bring the tech-toys in an interplay with nature.

Nature can refer to the woods, beaches, lakes, rivers, tunnels, deserts, mountains, snow, sand, and trees. Unfortunately, not all children have the same opportunity to access these natural features. Children around the world live in different climate zones and geographic regions. Some children live in poor conditions where nature is too expensive to maintain, and others live in extreme climates where outdoor play is limited to a few months a year or a few hours of a day. Therefore, urban design and technology can bring the unreachable, stimulating, and developmentally beneficial experiences to those who cannot access them. How can the built environment translate, replicate and import the benefits of nature-play to young urban dwellers? Based on the previous review of play-spaces, we can synthesize that until today, nature has influenced the making of children's play-spaces in three ways: 1) Importing natural elements into the play-space: Froebel's playground – sandboxes - natural playscapes; 2) Mimicking natural forms: Some novelty playgrounds - nature-themed modern playgrounds and 3) Mimicking nature-play principles: Adventure Playgrounds, Imagination Playground, Rigamajig.

These three approaches to nature-inspired play show that play-spaces can range on a spectrum of all-natural to all-artificial. The all-natural, mixed, or all-artificial play-spaces can be developmental if they address the non-linear dynamic character of play. The historically successful play-spaces such as natural playscapes, adventure playgrounds, and the Imagination Playground all share the aspect of a dynamic manipulatable open-ended environment. It also happens that nature, with its cyclical processes, is the most dynamic environment. The urban environment is static when compared to the dynamism of nature. Could this be the element that allowed play to start in the wilderness and the alleys of old towns? Could this be the element that is activating the movement towards nature-play today?

The dynamic element of play and nature can lead to a new understanding of the play-space. The common term for the traditional play-space is the playground with its structured play equipment. In 1985, Aese Eriksen proposed the term playscapes referring to the "carefully designed and landscaped outdoor environment that supports and suggests activities that are an essential part of child's learning and development (physical, emotional, social, and intellectual)" (Eriksen, 1985, p. 56). In 2014 Carr and Luken defined natural play-scapes as "all-natural playgrounds...dedicated spaces for children to play within an engaging outdoor environment that reflects the local landscape" (Carr & Luken, 2014, p. 73). The process of understanding play and the play-space brings the playground, the play-scape, and the natural play-scape under one umbrella: the play-space, a space that affords opportunities for play activities that promote children's social, emotional, cognitive and physical skills. It can be outdoor or indoor, natural or built, found, or intentionally designed. It is a space that transforms into a place by the dynamic, transformative, and emergent play opportunities it provides.

This review of play theories and environments shows that nature's influence on play-space design can be more developmentally beneficial if we as designers mimic nature's principles of play-space making instead of mimicking natural forms. These principles are abstract design strategies that are conducive to play, such as loose parts and open-endedness. By understanding those principles, we can create environments conducive to children's play and wellbeing, whether we use natural features, built play features, or integrate between the two. Having the option to choose between natural and built features makes play-spaces maintainable, affordable, and accessible to more children. The next step is to go into the field to observe and find the play-space principles that allow natural environments to provide children with play opportunities that are conducive to their development and wellbeing. But where can we find children playing in nature?

The dynamic aspect of play is the same reason that makes modern early education models a step ahead of structured modern playgrounds (Frost, 2009, 110). Froebel's emphasis on nature transformed into Montessori's learning-by-doing paradigm (Mooney, 2013, p. 27-35). Therefore, the natural-artificial spectrum works in education systems, too. Since nature and child-centered early education settings have proven to influence developmental play, they could be of significant influence on play-space design. Coincidentally, forest schools represent both these settings at once. They are early education settings that take place in natural environments that lie on the natural edge of the nature-artificial spectrum of play. They are schools with no buildings, because the natural environment is their school. Nature schools make a suitable setting to observe children and explore the new research question.

Since 1971, Nicholson, who discovered the nature principle of "loose parts" also anticipated that schools and education systems are heading towards "de-schooling" and that the line between education and recreation is diminishing (Nicholson, 1971, p. 30-34). Therefore, it makes sense to look at educational settings to develop play-spaces, and it also makes sense to look at playful environments to develop educational spaces. Parallel to Aldo van Eyck's analogy that the leaf's relationship to the tree is similar to the house's relationship to the city (1962, p. 102):

"Tree is leaf and leaf is tree – house is city and city is house – a tree is a tree but it is also a huge leaf – a leaf is a leaf but it is also a tiny tree – a city is not a city unless it is also a huge house – a house is a house only if it is also a tiny city."

We can also make a new analogy: Tree is leaf and leaf is tree – School is playground and playground is school – a tree is a tree but it is also a huge leaf – a leaf is a leaf but it is also a tiny tree – A school is not a school unless it is a huge playground– a playground is a playground only if it is a tiny school.

At any forest school, the natural environment is both the school and the playground. In built and urban settings, however, there is a separation between the school and the playground. The natural environment teaches us that both can be the same, a play-space that nurtures children's development and wellbeing. The theoretical concepts explored in this paper tell us that nature can guide and inspire us to design play-spaces on every scale. We can learn from the affordances of trees as climbable features, and sand and water as mixable materials. We can learn from nature settings that urban environments could become offshoots of schools and learning environments in their own right. Urban environments need new infrastructures that encourage the play-life of children and increase opportunities to learn, creating experiences that engage the body and senses, and spaces that are conducive to their wellbeing.

REFERENCES

Bergen, D. (2014). Foundations of Play Theory. In Brooker, L., M. Blaise, and S. Edwards (Eds.), SAGE Handbook of Play and Learning in Early Childhood (pp. 9-20). SAGE Publications. ProQuest Ebook Central.

Beltzig, G. (2015). Early Child Development: Body of Knowledge. In Kiser, B. (Ed.), Early Child Development: Body of Knowledge. Nature 523(7560): 286–89. https://doi.org/10.1038/523286a.

Brett, A., R. C. Moore, and E. F. Jr. Provenzo (1993). The Complete Playground Book. Syracuse, NY: Syracuse University Press.

Brown, S. L., and Vaughan, C. C. (2010). Play : How It Shapes the Brain, Opens the Imagination, and Invigorates the Soul. New York: Avery.

Carr, V., and Luken, E. (2014). Playscapes: A Pedagogical Paradigm for Play and Learning. International Journal of Play 3(1), 69–83. https://doi.org/10.1080/21594937.2013.871965.

Clements R. (2014). An Investigation of the Status of Outdoor Play. Contemporary in Issues Early Child. 5(1), 68-81. Retrieved from: https://doi.org/10.2304/ciec.2004.5.1.10.

Dattner, R. (1969). Design for Play. First edit. Cambridge,

Mass: MIT Press.

Eberle, S. G. (2014). The Elements of Play Toward a Philosophy and a Definition of Play." American Journal of Play 6(2), 214–33. http://www.journalofplay.org/sites/www.journalofplay.org/files/pdf-articles/6-2-article-elements-of-play.pdf.

Elkind, D. (2007). The Power of Play : How Spontaneous, Imaginative Activities Lead to Happier, Healthier Children. Cambridge, MA: Da Capo Lifelong.

Eriksen, A. (1985). Playground Design: Outdoor Environments for Learning and Development. New York: Van Nostrand Reinhold.

Eyck, Aldo van. The Child, the City and the Artist: An Essay on Architecture: The in-between Realm: Written in 1962. Edited by Vincent Ligtelijn and Francis Strauven. Amsterdam: SUN, 2008.

Frost, J. (1997). Child Development and Playgrounds. Parks & Recreation 32(4), 54–57. http://search.ebscohost.com.ezproxy1.lib.asu.edu/login.aspx?direct=true&db=eih&AN=9705163596&site=ehost-live.

Frost, J. (2009). A History of Children's Play and Play Environments: Toward a Contemporary Child Saving Movement. New York, NY: Routledge. ProQuest Ebook Central.

Hakkarainen , P. (2017). Introduction from editors. In Bruce, T., P. Hakkarainen and M. Bredikyte (Eds.), The Routledge International Handbook of Early Childhood Play (pp. 1-2). New York: Routledge. Kindle.

Heroes Will Rise. (n.d.). Imagination Playground. Retrieved from: https://casholman.com/projects#/imagination.

Heseltine, P., and Holborn, J. (1987). Playgrounds : The Planning, Design and Construction of Play Environments. London: Mitchell.

Huizinga, J. (2014). Homo Ludens : A Study of the Play-Element in Culture. Mansfield Center, CT, USA: Martino Fine Books. (original work published 1955)Imagination Playground. (n.d.). Our Mission. Retrieved from http://www.imaginationplayground.com/mission/index.html.

KaBOOM! (2018). Play Everywhere Playbook (An expanding online playbook). Retrieved from https://kaboom.org/playbook.

Kellert, S. (2015). Build Nature into Education, In Kiser, B. (Ed.), Early Child Development: Body of Knowledge. Nature 523(7560), 286–89. https://doi.org/10.1038/523286a.

Lefaivre, L. (2002). Space, Place and Play. In de Roode, I. and Lefaivre, L. (Eds.), Aldo van Eyck: The Playgrounds and the City (pp. 25-45). Netherlands: Stedelijk Museum.

Lefaivre, L. and Döll-Atelier. (2007). Ground-up City : Play as a Design Tool. Rotterdam: 010 Publishers.

Louv, R. (2005). Last Child in the Woods: Saving Our Children from Nature-Deficit Disorder. 1st ed. Chapel Hill, NC: Algonquin Books of Chapel Hill.

Louv, R. (2017). The New Nature Movement, 2017: Not Going Back to Nature, but Forward to a Nature-Rich Civilization. Children and Nature Network. https://www.childrenandnature.org/2017/01/12/the-new-nature-movement-its-not-about-going-back-to-nature-but-forward-to-a-nature-rich-civilization/

Mooney, C. G. (2013). Theories of Childhood: An Introduction to Dewey, Montessori, Erikson, Piaget, and Vygotsky. Second ed. St. Paul, MN: Redleaf Press. ProQuest Ebook Central.

Moore, R. C., Goltsman, S. M., & Iacofano, D. S. (1987). Play for all Guidelines: Planning, Design and Management of Outdoor Play Settings for All Children. Berkeley, CA: MIG Communications.

Moore, R. C., & Wong, H. H. (1997). Natural learning: the life history of an environmental schoolyard. Berkeley, Calif: MIG Communications.

Moore, R. C., Cosco, N., Sherk, J., Bieber, B., Varela, S., Gurina, N., Vickery, H., Murphy, J. (2009). Creating and retrofitting play environments: Best Practice Guidelines. Published in collaboration between PlayCore and the Natural Learning Initiative (NLI).

Moore, R. C. (2014). Nature play and learning places: Creating and managing places where children engage with nature. (With A. Cooper). Raleigh, NC: Natural Learning Initiative and Reston, VA: National Wildlife Federation.

Nicholson, S. (1971). How NOT to Cheat Children: The Theory of Loose Parts. Landscape Architecture 62, 30–34. https://media.kaboom.org/docs/documents/pdf/ip/Imagination-Playground-Theory-of-Loose-Parts-Simon-Nicholson.pdf.

Opie, I., and Opie, P. (1969). Children's Games in Street and Playground. Chasing, Catching, Seeking, Hunting, Racing, Dueling, Exerting, Daring, Guessing, Acting, Pretending. Oxford: Clarendon P.

Plato, C. H. A. Bulkley, and B. Jowett. (1876). Plato's Best Thoughts: Compiled from Prof. Jowett's Translation of the Dialogues of Plato. New York: Scribner, Armstrong. https://babel.hathitrust.org/cgi/pt?id=uc2.ark:/13960/t5v69ck6g&view=1up&seq=152.Rigamajig. (n.d.). Brain Powered. No Batteries Needed. https://rigamajig.com/about-us.

Rockwell Group. (n.d.). Play Work Build," https://www.

rockwellgroup.com/projects/play-work-build. An ongoing exhibition at the National Building Museum, Washington DC.

Sigman, A. (2019). A Movement for Movement Screen Time, Physical Activity and Sleep: A New Integrated Approach for Children (Report). https://www.api-play.org/wp-content/uploads/sites/4/2019/01/API-Report-A-Movement-for-Movement-A4FINALWEB.pdf.

Tremblay, M. et al. (2015). Position Statement on Active Outdoor Play. International Journal of Environmental Research and Public Health 12(12), 6475–6505. https://doi.org/10.3390/ijerph120606475.

United Nations. (2018). World Urbanization Prospects: The 2018 Revision (Report). United Nations, Department of Economic and Social Affairs, Population Division. Retrieved from: https://population.un.org/wup/Publications/Files/WUP2018-KeyFacts.pdf.

Vander Ven, K. (2015). Protean Selves, Trading Zones, and Nonlinear Dynamical Systems: The Role of Play in Future Progress. In Fromberg , D. P., and Bergen, D. (Eds.), Play from birth to twelve: contexts, perspectives, and meanings, 3rd ed (pp. 409-418). London: Routledge. ProQuest Ebook Central.

Weber, R. and Zeig-Owens, N. (Directors). (2016). Design in Practice: Cas Holman [Motion Picture] USA: Architecture and Design Film Festival, A/D/O. https://casholman.com/about.

Yogman, M. et al. (2018). The Power of Play: A Pediatric Role in Enhancing Development in Young Children. Pediatrics 142(3), e20182058. https://doi.org/10.1542/PEDS.2018-2058.

D17 - Transform Places with People and the Environment: Health I

Farmers Markets As Restorative Environments

Kelsea Dombrovski (University of California Davis); Angela Sanguinetti (University of California Davis); Stephen Wheeler (University of California Davis)

1. ABSTRACT

Restorative environments provide stress relief and renewed attention capacity. Pristine natural environments are most commonly considered restorative, however a growing body of work has been extending the concept to built settings. This study explores farmers markets' potential as restorative environments. We analyzed site features at two farmers markets and surveyed patrons to assess perceived restorativeness and market features' influence on perceived restorativeness. Results indicated farmers markets can function as restorative environments. Site features related to restorative qualities were Market Activities, Sensory Elements, Market Structure, Patron Characteristics, and Groups and Density. Perceived restorativeness did not differ between a market in a park or in a hardscaped setting, implying that markets do not need to be in park-like settings to provide respite from urban living. Results highlighted social interaction and sense of community as important features contributing to restorative farmers market experiences. Social aspects of restorative environments are ripe for further study.

1.1. Keywords

Attention Restoration Theory, farmers markets, restorativeness, restorative environments

2. INTRODUCTION

2.1. Farmers Markets as Restorative Environments

Restorative environments are sites that provide relief from stress and accumulated strains on attention (Kaplan, 1995; Ulrich et al., 1991). Traditional conceptualizations of restorative environments and most empirical research on the topic focus on pristine nature (wild or manicured) as the iconic setting for restoration. For example, parks, nature preserves, and gardens have been widely studied as restorative sites (Berto, 2005; Han, 2003; Hartig, Kaiser, & Bowler, 1997; Hartig, Korpela, Evans, & Gärling, 1997; Korpela, Ylén, Tyrväinen, & Silvennoinen, 2008; Laumann, Gärling, & Stormark, 2001; Ulrich et al., 1991). However, a growing body of work is extending the concept of restorative environments, for example, to built spaces like monasteries, museums, and plazas (Abdulkarim & Nasar, 2014; S. Kaplan, Bardwell, & Slakter, 1993; Ouellette, R. Kaplan, & S. Kaplan, 2005).

This study explores the potential for farmers markets to function as restorative environments. Identifying and promoting spaces that can restore mental fatigue is increasingly important as citizens become more harried in an unpredictable society. Investigation of new types of restorative environments, the features that make them restorative, and how to increase their restorativeness are research topics that could help inform placemaking efforts in cities worldwide. make restoration more accessible to more people. This research adds to that effort.

2.2. Theories of Restorative Environments

Two prominent theories describe the mechanisms underlying restorative environments: psychophysiological stress recovery theory (Ulrich et al., 1991) and attention restoration theory (ART; R. Kaplan, S. Kaplan, & Ryan, 1998). The former describes the role of restorative environments in recovery from stress and illness via impacts on the parasympathetic nervous system. ART describes the role of restorative environments in renewing capacity to direct attention via cognitive processes.

A theme in empirical research approaching restorative environments from either of the above theoretical approaches is that vegetated natural environments (especially wild nature, but also cultivated nature) are more restorative than hardscaped urban environments (Han, 2003; Hartig, Kaiser, et al., 1997; Hartig, Korpela, et al., 1997; Korpela et al., 2008; Laumann et al., 2001). Korpela and Ylén (2007) noted that it may be specific elements within a setting, not the natural setting per se, that contribute to restorative effects. Hunter and Askarinejad (2015) identified three categories of physical environment attributes (structure, content, and landscape) that influence restorativeness, based on environmental psychology theory and empirical research; for example, horizontal line position, framing tree count, and circulation boundary, among other features, contribute to Coherence.

These objective environmental features correspond to subjective qualities underlying the restorative experience. ART specifies four such qualities: Being Away, Fascination, Extent, and Compatibility (R. Kaplan, S. Kaplan, & Ryan, 1998; S. Kaplan, 1995). Being Away refers to physical or mental distance from daily stressors. Extent refers to the range of the site and the degree to which components of the environment make sense together (sometimes called out as a related aspect called Coherence, and containing the concept of Legibility, or the extent to which the environment is understandable and interpretable; Hartig, 1997b). Compatibility refers to the degree to which the environment matches the user's needs. Fascination, or Soft Fascination, gives people the opportunity to focus their attention with little effort, allowing their mind to wander and relax.

2.3. Restorative Built Settings

Objective environmental attributes and subjective restorativeness qualities can apply to built settings as well as natural ones, but relatively few studies have considered built settings as restorative environments. Built settings that have been studied as potential restorative environments include museums (S. Kaplan et al., 1993), monasteries (Ouellette et al., 2005), and public plazas (Abdulkarim & Nasar, 2014). For example, Abdulkarim and Nasar (2014) explored perceived restorativeness of a public plaza in relation to three of William Whyte's (1980) elements of successful urban spaces: seating, food, and a triangulating feature promoting interaction between passersby (e.g., a sculpture). Presence of a triangulating feature or a triangulating feature and seating were associated with greater perceived restorativeness.
Kaplan et al. (1993) studied museum visitors. They found that individuals who had little experience with museums felt restored after visiting the space, but not as restored as those more familiar with museums. This suggests that the restoration benefit of some settings, perhaps built settings in particular, may require experience with the setting.

Ouellette et al. (2005) surveyed over 500 retreat attendees at a Benedictine monastery. Attendees' motivations mapped onto ART's four perceived qualities of restorative environments. However, an additional factor, spirituality, was also apparent.

Similarly, Runnerstrom (2008) suggested self-actualization via experiences in symbolic environments as another path to restoration in built settings. Altaher and Runnerstrom (2018) also connected restoration to the positive psychology concept of flow, which involves participating in an engaging activity (e.g., sports, socializing, art; Csikszentmihalyi, 1990). Most respondents in their study, which examined the practices and environments sought out by college students during stressful times, indicated that they pursued restoration by participating in a leisure activity in a built environment - most often being with friends - even though there were natural areas nearby. During such social and leisure activities, attention is diverted from daily demands and stressors.

Other studies have highlighted the role of activities in restoration, particularly in urban settings. For example, Laumann et al. (2001) found that the ART quality of Fascination in urban settings pertained more to the activities occurring rather than physical characteristics of the environment, as was the case in more natural settings they studied. Scopelliti and Giuliani (2004) found that socializing, especially for middle-aged adults (mean age 42) more than young adults (mean age 28) or older retired adults (mean age 68), was an important component of a restorative environment.

The role of social components in perceived restoration has been found by other researchers as well, despite the lack of a specific social component in the original conception of ART. Peschardt and Stigsdotter (2013) defined and then assessed human-perceived dimensions of environments. They found that the 'social' perceived sensory dimension (PSD) was significantly associated with the perceived restorativeness among average urban greenspace users, with more stressed users finding the 'nature' PSD to be very important. The authors advocate for the use of their findings in design and creation of new urban greenspaces, which could easily include farmers markets.

Thus, the subjective qualities of restorative environments may differ or manifest differently in built settings compared to natural settings, and more research is needed to expand theories of restoration to be inclusive of these experiences. The role of spirituality (Ouellette et al., 2005), self-actualization (Runnerstrom, 2008), and flow (Altaher & Runnerstrom, 2018) in restoration via built settings seem related to the concepts of awe and transcendence as discussed in the context of experiences in wild natural environments (Davis & Gatersleben, 2013), which have also not been clearly tied to the dominant theories of restorative environments. Psychophysiological stress recovery theory focuses on the body, ART on the "mind", and perhaps a third level is needed describing higher-level restoration of the "spirit".

Another possibly complementary theory is that of connectedness to nature (Mayer & Frantz, 2004) and to others/community (Aron, Aron, & Smollan, 1992). Connectedness to nature may be a path to restoration in natural settings, whereas built settings that enable socializing and more complex experiences like self-actualization and flow could lead to restoration through social connectedness, with oneself or community. This theory is particularly relevant to the current research, which explores the potential for farmers markets, urban settings with lots of social opportunities, to function as restorative environments.

2.4. Restorative Potential of Farmers Markets

While the restorative potential of farmers markets has not been previously studied, other research on farmers markets has relevant implications. For example, markets may often provide a sense of Being Away, per ART. Although markets are typically located in the neighborhoods they serve (Brown, 2002), patrons' familiarity with the market sites' use when the market is not occurring (such as streets or parking lots) may heighten the sense of Being Away. Temporary spaces like this are called urban ephemera (Colomb, 2013; Schuster, 2001).

Farmers markets may also provide Soft Fascination via rich sensory experiences, such as colors, odors, sights, smells, and, sounds. Ng (2003) and Sommer (1980) discuss international differences in farmers markets' sensory stimuli. Smithers and Joseph (2009) argue that patrons use such sensory factors plus localness of producers to determine authenticity of markets.

ART qualities of Extent may be supported by the open-air setting of most farmers markets, creating a connection to the local natural environment as the source of the produce being sold (Ng, 2003). Further emphasizing this connection is the fact that market booths and overall layout are influenced by "the scale and requirements of the pickup truck" (Sommer, 1980, 15; Stephenson, 2008). This also relates to the idea of connectedness to nature. Extent in time can also be relevant for restoration (S. Kaplan, et al., 1993). McGrath, Sherry, and Heisley et al. (1993) described the felt authenticity of markets as a deliberate re-creation of a past, more simple time, with stronger human connections and less omnipresent capitalistic controls. This manifestation of extent relates to connectedness to community.

In terms of Compatibility, farmers markets provide opportunities for socializing and enabling patrons to express their values. For example, McGrath, et al. (1993) described how people are drawn to farmers markets to enjoy social connections. Sommer (1980, 1989) described aspects of these social connections, which include interactions with the 'real' farmers and a wide range of ages and backgrounds of people at the market, all contributing also to a fun experience. Sommer et al. (1981) compared behavior at farmers markets and supermarkets and found 2.5 times as many social interactions at farmers markets, with 75% of customers arriving in groups. Furthermore, markets enable patrons to demonstrate their political and social values (McGrath, et al., 1993), most notably supporting local growers. For example, Alkon (2008) compared patrons' motivations at two California Bay Area farmers markets with different representations of the environment: patrons of a North Berkeley market with an ecological sustainability focus concentrated on supporting local agriculture as a means to preserve pristine wilderness, whereas patrons of a West Oakland market with an environmental justice and equality focus emphasized the goal of enriching urban spaces.

3. METHOD AND RESULTS

To explore the potential for farmers markets to function as restorative environments, we analyzed site features and surveyed patrons at two farmers markets in Central California. This study was approved by the institutional review board of the authors' university. We wanted to know if patrons found these markets restorative, and what specific features were associated with restorativeness. The study proceeded in two stages: site selection and analysis, and market patron survey. Method and results for each of these stages is presented in turn; results of the site analysis are used in analysis of patron survey data.

3.1. Site Selection and Analysis

We chose the Davis Farmers Market and the Sacramento Central Farmers Market to represent a range of market features as these markets are quite different (Figure 1). These markets were also selected to compare restorativeness of a market setting high in natural elements versus a less vegetated, more hardscaped market setting. The Davis Farmers Market is in that city's Central Park, and the Sacramento Central Farmers Market is located beneath a concrete freeway overpass and bordered by busy city roads.

Site Analysis Method

We assessed market features at each site to aid hypothesis formation regarding relationships between site features and restorative potential, thus informing our market patron survey design and analysis. This site analysis used objective, observational post-occupancy evaluation methods as outlined by Cooper Marcus and Francis (1997), such as initial exploration and assessment of sub-areas, behavior mapping, and use-analysis. The lead researcher spent at least one hour at each site on two separate occasions. The assessment focused on objective site features, including physical environment attributes as well as occupant characteristics and activity observations. Physical features analyzed included site

Figure 1. Davis (left) and Sacramento (right) farmers markets (2016). Photos by the author.

layout (vendor orientation, booth setup, congregation sites, location, amenities, and layout), music, weather, crowd density, and noise. Activity observations consisted of noting how patrons moved around the setting and what they did in different spaces. Occupant observation focused on the users of the space themselves, e.g. their physical characteristics and demeanor.

Site Analysis Results

We identified nine site feature types present in both farmers markets. These categories, which represent components common to most farmers markets, were distilled by grouping similar site features within and across the two markets. The sections below define each feature type, characterize each of our farmers markets in terms of the features, and consider hypotheses about how the feature types related to the four ART qualities: Being Away, Fascination, Coherence/Extent, and Compatibility.

1) Market Structure. Market structure features pertain to the path dictated by the market layout and the physical structural features defining and delimiting the market. The Davis market is structured along a central corridor with flexible open space at both ends. The Sacramento market has a wider main aisle that wraps around the market location. We hypothesized that Market Structure features that help patrons navigate the space (e.g., walkways, landmarks, demarcations) would have implications for Coherence and Compatibility--particularly legibility. Shopping in and experiencing the open-air structure could also support a feeling of Being Away.

2) Programming. Programming features demonstrate control and planning by a person or team, for example, the presence of ATM machines and attractions other than produce stalls. Both markets had such features. Programming at Davis was more extensive, including park features and market-specific activities, such as a bicycle-powered merry-go-round. Attractions could potentially contribute to a sense of Being Away or Compatibility, but we did not have strong hypotheses about these relationships.

3) Stall Characteristics. These features characterize the appearance, setup, size, and produce/product inventory of market stalls. Sacramento had more stalls overall and fewer stalls with elaborate décor compared to Davis, but otherwise Stall Characteristics were similar at both markets. Stall Characteristics could potentially impact Fascination or Compatibility by appealing to the style and interests of different patrons.

4) External Infrastructure. These features affect patrons' access to the market and amenities available but are not part of the market grounds or structure. They include parking, transit lines, and restrooms unmanaged by the market. The Davis market is near multiple sets of public restrooms but lacks parking; the Sacramento market has plentiful parking but there are few restrooms. External infrastructure (or lack thereof) might have implications for Being Away and Coherence, but we did not anticipate a strong role in perceived restorativeness.

5) Market Activities. Market activities always include produce shopping, but may also include eating, shopping for other goods, and entertainment. The Davis market had more activities related to entertainment, including jugglers, musicians, and a carousel. Market Activities has clear implications for Compatibility via opportunities for person-environment fit.

6) Sensory Elements. Those features sensed using one or more of the five senses, e.g. cool temperature under the market structure vs. warmth on the lawn, sounds of music or sounds of traffic. Davis market's park location and Sacramento market's underpass location engage different senses in positive and negative ways. We hypothesized that Sensory Elements would contribute to Fascination via positive sensory distractions, as well as Being Away via out of the ordinary or new sensory experiences.

7) Uncoordinated Features. Those features that organically arise and/or are not planned by market management, e.g., concrete berms at the Sacramento market used by street musicians, and the lawn edges at the Davis market where patrons set up personal areas (usually blankets) for picnicking, lounging, and socializing. Such features might have implications for Compatibility by providing flexibility for more activity preferences, but we did not have strong hypotheses about this.

8) Patron Characteristics. Patron characteristics include demographics like age and ethnicity, and attributes like preparedness for market (measured as having carts, bags, baskets). Market demographics may or may not represent the diversity of the host city. Sacramento patrons came fully prepared for large shopping trips - many patrons had wheeled carts, large woven baskets, and reusable grocery bags - whereas Davis patrons seemed more prepared for browsing or small purchases. We hypothesized that Patron Characteristics would contribute to Fascination via people-watching and to Compatibility via social opportunities.

9) Groups and Density. Groups and Density features include crowd size (number of patrons in relation to spaces they occupy), as well as the tendency for patrons to attend in groups. These features vary based on time of day, with fewer groups and smaller crowds early in the morning, and larger groups and crowds later in the day. We hypothesized that Groups and Density would contribute to Fascination via people-watching and to Compatibility via social opportunities.

In sum, we hypothesized that Sensory Elements and Market Structure would relate to Being Away, as might Programming and External Infrastructure to a lesser extent. Stall Characteristics, Sensory Elements, Patron Characteristics, and Groups and Density would relate to Fascination; Coherence would be affected by Market Structure, and to a lesser extent, External Infrastructure. Market Structure, Stall Characteristics, Market Activities, Patron Characteristics, and Groups and Density, would relate to Compatibility, as might Programming and Uncoordinated features, to a lesser extent.

3.2. Market Patron Survey

We conducted a survey of market patrons at each site to assess perceived restorativeness and explore how the objective site features outlined above influence perceptions of the four environmental qualities described in ART. The lead researcher visited the markets on three successive spring weekends, surveying patrons from opening until closing. She intercepted patrons on their way into the market near entrances, approaching every patron or group during low traffic times and every third patron or group during high traffic times. Approaching patrons as they arrived rather than as they left encouraged them to reflect on all previous visits collectively rather than only the current visit, to capture more experiences and in case their current visit was anomalous in some way relative to other visits. Rigorous data on response rate were not collected, but approximately one quarter of patrons approached agreed to participate.

After introducing herself, the researcher asked patrons if they had visited the market at least five times previously. Only those who answered in the affirmative were recruited to take the survey. Kaplan et al. (1993) noted that it is rare to experience meaningful restoration during initial visits to a museum setting, likely because becoming oriented to a new place is intensive in terms of cognitive resources. We anticipated a similar first-time experience for farmers market patrons, and thus only included relatively experienced patrons.

We used an augmented version of Hartig et al.'s Perceived Restorativeness Scale (PRS; 1997b), which consists of a series of statements corresponding to the four qualities of restorative environments per ART. The Extent subscale is labeled Extent/Coherence in the instrument, but hereafter is referred to as Extent. Participants indicate their agreement with each statement on a seven-point scale (Table 1 shows the items that comprise the PRS). We augmented the PRS by appending an open-ended question at the end of each subscale (When responding to the above, what specific features of the farmers market came to mind?). These open-ended questions were intended to tease out narratives of how site features contribute to or detract from perceived restorativeness.

To identify aspects of the social and experiential dimensions of farmers markets not covered by the PRS, we added an open-ended question assessing the effect of people and crowds (In what ways, if any, does the amount of people at the market affect your experience?), as well as three additional questions asking patrons their likes and dislikes about the market as well as why they visit. Additional closed-ended questions assessed regularity of visit, group size, gender, age, race, socioeconomic status, and distance traveled. The researcher recorded weather and time.

Across the two markets, a total of 71 patrons completed the survey. Participants were relatively similar at both markets in terms of gender balance, ethnic mix, visit frequency, age, and approximate income (Table 2). However, several Sacramento patrons declined to participate because of a language restriction, so a more complete sample at that location might have shown greater diversity, and results of this study should be

Table 1. Items comprising the PRS as developed by Hartig et al. 1997b.

Being Away	
	Being here is an escape experience.
	Spending time here gives me a break from my day-to-day routine.
	It is a place to get away from it all.
	Being here helps me to relax my focus on getting things done.
	Coming here helps me to get relief from unwanted demands on my attention.
Fascination	
	This place has fascinating qualities.
	My attention is drawn to many interesting things.
	I want to get to know this place better.
	There is much to explore and discover here.
	I want to spend more time looking at the surroundings.
	This place is boring.
	The setting is fascinating.
	There is nothing worth looking at here.
Coherence/Extent	
	There is too much going on.
	It is a confusing place.
	There is a great deal of distraction.
	It is chaotic here.
Compatibility	
	Being here suits my personality.
	I can do things I like here.
	I have a sense that I belong here.
	I can find ways to enjoy myself here.
	I have a sense of oneness with this setting.
	There are landmarks to help me get around.
	I could easily form a mental map of this place.
	It is easy to find my way around here.
	It is easy to see how things are organized.

viewed as limited to English-speaking market patrons.

We conducted descriptive analysis of the PRS to explore the degree to which patrons perceived farmers markets to be restorative, and a one-way MANOVA to identify any significant differences between the Davis and Sacramento markets in terms of overall PRS score and each subscale: Being Away, Fascination, Extent, and Compatibility. Content analysis of open-ended survey data involved coding responses by relevant site feature category and perceived restorativeness subscale, as well as inductive coding for other emergent themes.

In line with our overall hypothesis, patrons perceived farmers markets as somewhat restorative. Overall and subscale mean PRS scores were in the restorative range, indicating that, on average, patrons at both locations experience, to some degree, the four main elements associated with restorativeness according to ART (See Figure 2). The sense of Being Away was not as frequently perceived as the other ART qualities.

We hypothesized that due to its park setting the Davis Farmers Market would be perceived as more restorative. However, there were no significant differences between market locations in terms of the PRS overall [$F(4, 52) = 0.46$, $p = .765$; Wilk's $\Lambda = 0.966$, partial $\eta 2 = .03$.] or any subscale, including Being Away: $t(67) = 1.34$, $p = 0.18$; Compatibility: $t(64) = 0.11$, $p = .92$; Fascination: $t(61) = -0.87$, $p = 0.39$; Extent: $t(64) = -0.31$, $p = 0.76$. Open-ended responses following each PRS subscale revealed more insight into patrons' experiences and indicated relationships between restorativeness and our identified site feature categories.

1) Fascination. Open-ended responses following the Fascination subscale indicated that site features in the categories of Market Activities, Patron Characteristics, Stall Characteristics, and Sensory Elements strongly contributed to a sense of Fascination, with a meta-theme of diversity, as these selected quotes illustrate:

"The beauty of the flowers, the lavender, wonderful taste of cheese; all of the different languages spoken."

"I love the mixture of cultures, socio-economic communities. Hustle and bustle."

"Variety of produce, flowers, people."

"The variety of products - so many different kinds. Farmers/sellers to interact with, shoppers to see and interact with, [and the] new mural!!!"

"Live music, people watching, playing kids, Lots of different people!"

2) Being Away. As previously mentioned, patrons overall did not express a strong sense of Being Away. In particular, many Sacramento market patrons explicitly noted here that the market is routine for them and not an escape. e.g., The experience of the market is great but more a part of our weekly life than an escape; and Part of my routine. Not relaxing, but I don't come here to relax. Davis market patrons' responses were more expressive of a sense of Being Away (e.g., Not having it be part of my weekly routine makes it more special). A role for Sensory Elements was again implied, e.g.: Open air, music, smell of good food; Fresh air away from traffic.

Market Activities, Patron Characteristics, and Groups and Density were also evident in responses, contributing to an overarching theme of social connection. In this sense, Being Away may represent a contrast to more socially isolated daily routines, as the following quotes illustrate:

"I get a sense of community at the market, and it is nice to feel like I am a part of the community."

"Different people that you don't see everyday and don't know everything about you."

"People are friendly, vendors. Beautiful garden. People are not just buying but come with family to just hang out or have breakfast."

"Sitting with friends. Grass, beer, blankets, conversations, fresh fruit".

"Kids running around, old people having picnics, the park, the vibe."

3) Extent. Open-ended responses following the Extent subscale indicated relevant site features in the categories Market Structure and Groups and Density. Specifically, the paths and layout were generally perceived as organized, contributing to Extent, e.g. Easy layout. One participant alluded to the importance of experience with the setting for perceived Extent: I would've agreed with these statements [Extent subscale] when I first came to this market, but I know my way around now.

Crowding was sometimes perceived as inhibiting Extent, e.g., Many, many people! In the park it's fine, but between the stalls is difficult to navigate; and Mostly just that it can be very crowded and hard to get through, particularly in the summer. Crowding was also the most common theme in responses to the open-ended question about general dislikes, particularly in Davis. However, in

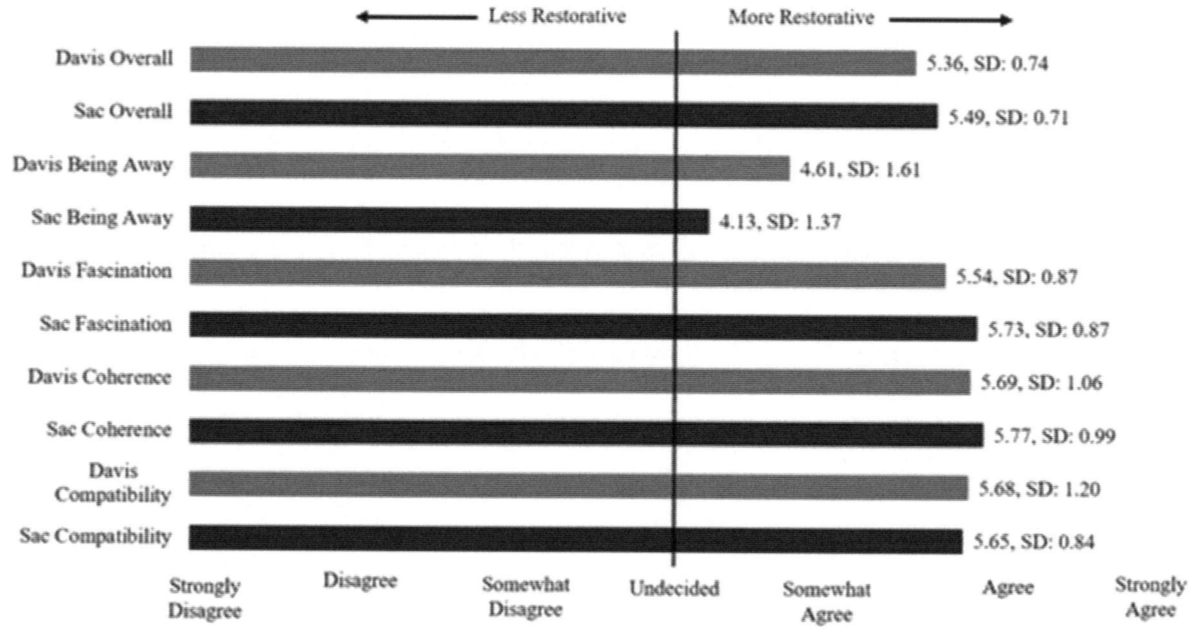

Figure 2. Response means and standard deviations of overall Perceived Restorativeness Scale (PRS) scores and subscale scores per market (2016). Diagram by the authors, inspired by Tenngart Ivarsson & Hagerhall (2008).

the context of the Extent scale crowding was much more often perceived as positive, and comments implied a synergy between crowds, structure, and activities that perhaps actually contributed to Extent, e.g.:

"It's structured chaos."

"There is a bit of distraction, but it's a good thing - the fact that there is so much good stuff!"

"It is how a good city functions".

"Very organized, method to the madness."

"Being chaotic is what makes it. You get to see your city come alive."

4) Compatibility. Open-ended responses following the Compatibility subscale also highlighted Market Structure features, including more mentions of experience with the setting as an important factor:

"I'm familiar with this market, so I have a good sense of the layout. It does feel overwhelming at times, nonetheless."

"It was intimidating the first few times but I am much more comfortable after 4 years."

"There's tape on the ground to distinguish different stalls."

"I thought about how the entire market is structured. The food up front and everything else in the back part of the market."

"How the tables are all set up, starting with fruits and vegetables, to food, then accessory items."

"Large tree, jungle gym, playground, loud, sea of people."

"There is the park as an easy way to help with a mental map."

Since the last four questions of this subscale measure environmental legibility rather than broader conceptions of person-environment fit, participants may have been primed to consider objective Market Structure features in their subsequent open-ended response more so than subjective experiences of social connection or conduciveness of the market setting to actualize their values and engage in preferred activities. Nevertheless, there were some mentions of these topics, e.g.:

"I feel a part of the Davis community when at the market."
"For me, liking and belonging are pretty separate; feel

Table 2. Respondent demographics.

	Davis	Sacramento
sample size	32	38
gender		
Female	69%	63%
male	31%	34%
non-binary	0%	3%
median frequency of visits		
every week	25%	32%
2-3 times per month	44%	37%
about once per month	28%	13%
less than once per month	3%	18%
distance traveled		
1-2 miles	62%	34%
3-5 miles	31%	37%
6 or more miles	6%	29%
mean age (standard deviation)	44 (18)	45 (15)
household income		
< $20,000	25%	6%
$20,000-49,999	14%	39%
$50,000-99,999	22%	42%
$100,000-199,999	21%	14%
$200,000 or more	18%	0%
race/ethnicity		
White	61%	63%
Latino/a	16%	18%
Asian	10%	0%
African American	0%	8%

like to belong would have to make more conversation or know people."

"I love farmers markets! They always make me feel happy :)"

Responses from the open-ended questions What do you like about the farmers market? and Why do you visit the farmers market? provided more insight into these more subjective aspects of Compatibility. Notions of community and social atmosphere were the strongest theme across these responses with the exception of themes related to the food (fresh, local, and variety). On the other hand, a theme in general dislikes was not being able to bring dogs to the market, which points to an aspect of incompatibility for some people.

5) Participant Characteristics. Finally, we explored correlations between PRS or subscale scores and each participant variable in Table 2. Only one relationship was found, which was a significant negative correlation between age and Being Away: $r(67) = -0.29$, $p = 0.02$, and age and overall perceived restorativeness: $r(56) = -0.29$, $p = 0.03$. Reasons for this are unclear, but perhaps older patrons have grown habituated to the markets, and perhaps they use the markets in different ways.

4. DISCUSSION

Results support the hypothesis that farmers markets can function as restorative environments. However, there were no statistically significant differences in overall PRS or subscale scores between patrons at the two markets we compared - one located in a park, the other under an overpass. This suggests that the quality and quantity of natural scenery at farmers markets is not a significant factor for restorative outcomes. Rather, it appears that it is the social and functional elements of a farmers market that contribute most to restoration. Measurable site feature types mentioned by survey respondents in relation to restorative qualities were Market Activities, Sensory Elements, Market Structure, Patron Characteristics, and Groups and Density. Regarding the latter, it is important to note that although many patrons felt crowd density had a positive impact on the experience, others noted that crowding can also interfere, particularly with Compatibility (i.e., legibility and being able to shop easily).

The themes of civic engagement and sense of community in farmers market patrons' discussions of restorativeness align with Runnerstrom and colleague's suggestions that restoration can occur via socializing and leisure, as well as self-actualization in symbolic urban environments (2008; Altaher & Runnerstrom, 2018). There was evidence that patrons perceive farmers markets as symbolic of many ideals, e.g., support of small farmers, local foods, community, alternative economies. Patrons also highlighted the importance of various types of socializing at markets in the context of perceived restorativeness.

Theories and measures of restorative environments require further development to provide a better account of social and activity-based factors. There have been systematic assessments of physical and perceived features that relate to restorativeness (Hunter & Askarinejad, 2015; Peschardt & Stigsdotter, 2013), but no similar attempts for social and activity-based features. As a start, this study suggests diverse crowds can contribute to a sense of Being Away and Fascination. Opportunities for socializing and a sense of community are clearly related to Compatibility. A better understanding of non-physical environmental features contributing to perceived restoration holds implications for new approaches to designing for health-promoting environments, perhaps along the line of Francis and Griffith's (2011) view of farmers markets as centers of social life and contributors to healthy cities. These authors note that markets help satisfy a need for lively civic spaces, and potentially promote mental health and well-being. This conceptualization ties in with the notion of third places (Oldenburg, 1989; Rosenbaum, 2006). Farmers markets meet the definition of third places – alternative community gathering spaces where people can engage in less formalized city life. Future research and theory could better tie together theories of restorative environments with third spaces.

Rather than seeking a green or pastoral site, managers can locate markets in neighborhoods that lack third places and enhance restorative potential via market features that increase diversity of produce, patrons, and activities. These social and functional aspects contribute to all four qualities of restorative environments.

Although our participants emphasized specific market characteristics, results could be different in other market settings. Managers of existing markets should engage their patrons to determine motivations for visiting and experiences related to restoration, and then consider incorporating or enhancing features identified in this research as promoting restoration. Future research should replicate this study at markets of varying sizes, locations, and specialties. The opinions and experiences of market managers, volunteers, and vendors could also be explored to more fully understand how and for whom farmers markets provide restorative benefits. Social and lively environments not focused on commerce should be investigated as well, to determine the extent to which the market environment itself plays a role. Finally, within all proposed topics it is important to obtain truly representative samples, and care should be taken to have translation services available for those participants who do not speak or read English.

5. CONCLUSION

This study was an initial look into the potential for farmers markets to function as restorative environments. It provides evidence that farmers markets can indeed provide restoration. Farmers market patrons were surveyed, using the PRS, with supplemental open-ended questions, and the PRS measured patrons' experience of the four environmental qualities articulated in ART. Compatibility and Fascination were expressed most saliently, and open-ended responses revealed relationships between these qualities and a variety of objective physical market site features, especially Market Activities, Sensory Elements, Market Structure, Patron Characteristics, and Groups and Density.

Farmers markets have spread rapidly across North America in the past several decades and form an important economic outlet for local farmers as well as an important restorative and social space for citizens. Given the lack of many other human-scale, informal civic spaces within our communities that are not dominated by corporate culture, these markets appear to be fulfilling an important niche in postmodern society and an important restorative role for individuals. Better understanding of how they do these things can help maximize their potential as well as inform creation of other such spaces.

REFERENCES

Abdulkarim, D., & Nasar, J. L. (2014). Are livable elements also restorative? Journal of EnvironmentalPsychology, 38, 29–38. http://doi.org/10.1016/j.jenvp.2013.12.003

Alkon, A. (2008). Paradise or pavement: the social constructions of the environment in two urban farmers'markets and their implications for environmental justice and sustainability. Local Environment, 13(3), 271-289.

Altaher, Y., & Runnerstrom, M. G. (2018). Psychological Restoration Practices Among College Students. Journal of College Student Development, 59(2),

227-232.

Aron, A., Aron, E. N., & Smollan, D. (1992). Inclusion of other in the self scale and the structure of interpersonal closeness. Journal of personality and social psychology, 63(4), 596.

Berto, R. (2005). Exposure to restorative environments helps restore attentional capacity. Journal of Environmental Psychology, 25(3), 249–259. http://doi.org/10.1016/j.jenvp.2005.07.001

Brown, A. (2002). Farmers' market research 1940–2000: An inventory and review. American Journal ofAlternative Agriculture, 17(4), 167-176.

Colding, J., & Barthel, S. (2013). The potential of "Urban Green Commons" in the resilience building of cities. Ecological Economics, 86, 156–166. http://doi.org/10.1016/j.ecolecon.2012.10.016

Colomb, C. (2013). Pushing the urban frontier: Temporary use of space, city marketing, and the creativecity discourse in 2000s Berlin. Journal of Urban Affairs, 34(2), 131–152.

Cooper Marcus, C., & Francis, C. (Eds.). (1997). People places: Design guidelines for urban open space.

John Wiley & Sons.Csikszentmihalyi, M. (1990). The domain of creativity.

Davis, N., & Gatersleben, B. (2013). Transcendent experiences in wild and manicured settings: The influence of the trait "connectedness to nature". Ecopsychology, 5(2), 92-102.

Feagan, R. B., & Morris, D. (2009). Consumer quest for embeddedness: a case study of the Brantford

Farmers' Market. International Journal of Consumer Studies, 33(3), 235–243. http://doi.org/10.1111/j.1470-6431.2009.00745.x

Francis, M., & Griffith, L. (2011). The Meaning and Design of Farmers' Markets as Public Space An Issue-Based Case Study. Landscape Journal, 30(2), 261–279. http://doi.org/10.3368/lj.30.2.261

Han, K. T. (2003). A reliable and valid self-rating measure of the restorative quality of natural environments. Landscape and Urban Planning, 64(4), 209–232. http://doi.org/10.1016/S0169-2046(02)00241-4

Hartig, T., Korpela, K., Evans, G. W., & Gärling, T. (1997a). A measure of restorative quality in environments. Scandinavian Housing and Planning Research, 14(4), 175–194. http://doi.org/10.1080/02815739708730435

Hartig, T., Kaiser, F. G., & Bowler, P. A. (1997b). Further development of a measure of perceived environmental restorativeness. Institute of Housing Research, Working Paper, 5.

Hunter, M. R., & Askarinejad, A. (2015). Designer's approach for scene selection in tests of preference and restoration along a continuum of natural to manmade environments. Frontiers in Psychology, 6.

Kaplan, R., Kaplan, S., & Ryan, R. (1998). With People in Mind: Design and Management of Everyday

Nature. Island Press.

Kaplan, S. (1995). The restorative benefits of nature: Toward an integrative framework. Journal of Environmental Psychology, 15(3), 169–182. http://doi.org/10.1016/0272-4944(95)90001-2

Kaplan, S., Bardwell, L. V., & Slakter, D. B. (1993). The Museum as a Restorative Environment. Environment and Behavior, 25(6), 725–742. http://doi.org/10.1177/0013916593256004

Korpela, K. M., Ylén, M., Tyrväinen, L., & Silvennoinen, H. (2008). Determinants of restorative experiences in everyday favorite places. Health & Place, 14(4), 636–652. http://doi.org/10.1016/j.healthplace.2007.10.008

Laumann, K., Gärling, T., & Stormark, K. M. (2001). Rating scale measures of restorative components of environments. Journal of Environmental Psychology, 21(1), 31–44. http://doi.org/10.1006/jevp.2000.0179

Mayer, F. S., & Frantz, C. M. (2004). The connectedness to nature scale: A measure of individuals' feeling in community with nature. Journal of environmental psychology, 24(4), 503-515.

McGrath, M. A., Sherry, J. F., & Heisley, D. D. (1993). An ethnographic study of an urban periodic marketplace: Lessons from the Midville farmers' market. Journal of Retailing, 69(3), 280-319.

Ng, C. F. (2003). Satisfying shoppers' psychological needs: From public market to cyber-mall. Journal of Environmental Psychology, 23(4), 439–455. http://doi.org/10.1016/S0272-4944(02)00102-0

Oldenburg, R. (1989). The great good place: Cafes, coffee shops, community centers, beauty parlors, general stores, bars, hangouts, and how they get you through the day. New York: Paragon House.

Ouellette, P., Kaplan, R., & Kaplan, S. (2005). The monastery as a restorative environment. Journal of Environmental Psychology, 25(2), 175–188. http://doi.org/10.1016/j.jenvp.2005.06.001

Peschardt, K. K., & Stigsdotter, U. K. (2013). Associations between park characteristics and perceived restorativeness of small public urban green spaces. Landscape and Urban Planning, 112, 26-39.

Rosenbaum, M. S. (2006). Exploring the Social Supportive Role of Third Places in Consumers' Lives. Journal of Service Research, 9(1), 59–72. http://doi.org/10.1177/1094670506289530

Runnerstrom, M. (2008). Testing a Theory of Psychological Restoration Through Self-Actualization Processes (PhD Dissertation). University of California, Irvine, Irvine, CA.

Schuster, J. M. (2001). Ephemera, Temporary Urbanism, and Imaging. In L. J. Vale & S. B. Warner Jr. (Eds.), Imaging the City - Continuing Struggles and New Directions. New Brunswick, New Jersey: CUPR Books.

Scopelliti, M., & Vittoria Giuliani, M. (2004). Choosing restorative environments across the lifespan: A matter of place experience. Journal of Environmental Psychology, 24(4), 423–437. http://doi.org/10.1016/j.jenvp.2004.11.002

Smithers, J., & Joseph, A. E. (2009). The trouble with authenticity: separating ideology from practice at the farmers' market. Agriculture and Human Values, 27(2), 239–247. http://doi.org/10.1007/s10460-009-9250-5

Sommer, R. (1980). Farmers markets of America: a renaissance. Capra Press.

Sommer, R. (1989). Farmers' Markets as Community Events. In I. Altman & E. H. Zube (Eds.), Public Places and Spaces (pp. 57–82). Springer US. Retrieved from http://link.springer.com/chapter/10.1007/978-1-4684-5601-1_4

Sommer, R., Herrick, J., & Sommer, T. R. (1981). The behavioral ecology of supermarkets and farmers' markets. Journal of Environmental Psychology, 1(1), 13–19. http://doi.org/10.1016/S0272-4944(81)80014-X

Stephenson, G. (2008). Farmers' Markets. Amherst, NY: Cambria Press.

Tenngart Ivarsson, C., & Hagerhall, C. M. (2008). The perceived restorativeness of gardens – Assessing the restorativeness of a mixed built and natural scene type. Urban Forestry & Urban Greening, 7(2), 107–118. http://doi.org/10.1016/j.ufug.2008.01.001

Ulrich, R. S., Simons, R. F., Losito, B. D., Fiorito, E., Miles, M. A., & Zelson, M. (1991). Stress recovery during exposure to natural and urban environments. Journal of Environmental Psychology, 11(3), 201–230. http://doi.org/10.1016/S0272-4944(05)80184-7

Whyte, W. H. (1980). The Social Life of Small Urban Spaces. Retrieved from https://trid.trb.org/view.aspx?id=521122

D24 - Transform Places with People and the Environment: Cities & Globalization / Residential Environments

Living in Transition: Displacement and Temporary Living Arrangements after the Woolsey Fire in Southern California

Mariana Junqueira (University of California Irvine)

1. ABSTRACT

The Woolsey Fire ignited on November 8, 2018, in Los Angeles and Ventura Counties of the U.S. state of California. Among the unprecedented destruction, at least 400 structures were homes that burned down completely. Naturalistic field research conducted over the Summer of 2019, included several households who were still relying on temporary living arrangements at the time (September of 2019). This paper focuses on the role of temporary living arrangements in the lives of three of these Southern California families displaced by the fire who are still living in transition. There are hardly any studies on what people's lives look like in these types of living arrangements: how they view and cope with living in temporary housing while managing the loss of their homes. Observations (including walkthroughs) and unstructured, open-ended interviews with three families inform this paper. The lesson from this study is that despite significant variability among types of temporary living arrangements made by families displaced by the Woolsey Fire, the experience of living in temporary housing seems to hold strong parallels. Three emerging themes are: (1) a temporary living arrangement might be a house, but not always a home; (2) living in a temporary arrangement may influence one's well-being; (3) undesired spatial characteristics and the absence of objects in a temporary living arrangement contribute to intrusive and repetitive thoughts about past experiences. All these seem to play a role both in the quality of people's daily lives and in the management of their recovery process.

1.1. Keywords

Displacement, Temporary living arrangements, Temporary housing, Transition, Woolsey Fire.

2. INTRODUCTION

"I can't live in limbo for two years. Because that's what we are. We are living in limbo here." (Patricia[1], 7/12/2019). This is how Patricia, a Southern California resident who had her house destroyed by the Woolsey Fire, in November of 2018, described how she felt about her temporary living arrangement in July of 2019. Patricia was not experiencing displacement alone. Rita, an accomplished writer, who lost not only her house but also her manuscripts and book inventory, said:

"I am mainly waiting... I have gone back to work a few months ago. But I am going to be getting some more equipment when I move to a new place. Right now, I've lived in 12 places, and I'm going to my 13th to do some house and pet sitting tomorrow. And I'll maybe be in 14 places before I can actually settle into a dwelling. So, at that point, I'll be back up to speed. But it's really been an experience to work out of so many different places... living out of a shoebox... or a suitcase." (Rita, 8/13/2019)

The Woolsey Fire ignited on November 8, 2018, and burned 96,949 acres (39,234 hectares) of land in Los Angeles and Ventura Counties of the U.S. state of California. The fire destroyed 1,643 structures, killed three people, and prompted the evacuation of more than 295,000 people (Los Angeles County Fire Department, 2018). Among the 1,643 structures destroyed, at least 400 were homes that burned down, displacing its dwellers for months, and in some instances, will continue to do so for years to come.

Patricia and Rita shared more than the experience of losing their homes to the Woolsey Fire; Patricia, Rita, their families, and most of their neighbors are all still experiencing displacement some ten months after the fire. They are among the participants of a broader study in the aftermath of the Woolsey Fire, who are still living in temporary living arrangements. There are hardly any studies on what their lives look like while living in transition.

Disasters usually leave a tremendous amount of damage in their wake. However, we are likely to classify that damage as an "aftermath" and assume that both the physical wreckage and the human wreckage will be repaired over time (Erikson, 1976). What this means, for practical purposes, is that disasters rupture peoples' sense of normalcy and continuity in their lives. When leading to displacement, the experience of disaster can be even worse, as the loss of home may hold multiple meanings and constraints to one's recovery. In the process of managing and adapting to such loss, living in a precarious environment (such as a temporary housing) may pose daily limitations. This paper focuses on understanding how the lives of displaced people develop in temporary living arrangements, and how these environments may play a role in people's recovery processes.

2.1. Displacement: a rising issue in people-environment relationships

Displacement is defined as the situation in which people are forced to leave the place where they usually live. (The Cambridge Academic Content Dictionary) By the end of 2017, according to a report by United Nations High Commissioner for Refugees (UNHCR), there were 68.5 million forcibly displaced people in the world. The number includes 25.4 million refugees, and about 40 million internally displaced people, who were forced to leave their homes but are still in their home countries (UNHCR, 2017).

Displacement is also described as a four-stage process — push factors to relocation mark the antecedent phase. The uprooting phase involves the loss of place, of possessions, and connections with people. Relocation and rapid changes to one's physical and social environments mark a transition phase. Lastly, the resettlement phase concludes the process when people forge new ties and connections in a new place (Greene, Tehranifar, Hernandez-Cordero, & Fullilove, 2011). Thus, an inherent part of displacement is relocation to a new environment, and often involves a transitional period of living in a temporary or improvised environment. These environments are often characterized by their "less-than-ideal conditions" and are referred to as "precarious." In this study, a precarious living environment is considered unstable because it is temporary or improvised, or fraught with uncertainty, as danger or inability to change conditions (either due to lack of resources or to lack of ability or options).

With over 68 million people known to be displaced worldwide, instances of groups of people living in precarious environments are countless. An extremely brief list includes informal housing (e.g., Favelas and Palafitas in Brazil, Comunas in Colombia, and Colonias in the U.S.-Mexico border), refugees and migrant camps, post disasters' shelters and temporary housing, among others. Two types of scenarios seem to cause the phenomenon (i.e., displaced groups of people living in precarious environments) – the inability to deal with: (a) a persisting condition, such as poverty, and (b) a disturbing event, such as a natural or human-made disaster. This is a study of displacement caused by an event: the Woolsey Fire. Though displacement can have many forms, one thing that makes these events analogous is the human experience of loss for which definitions or categories are not always precise. It is not uncommon for the media to speak of people's displacement as a "diaspora," and for people displaced by a disaster to describe themselves as "refugees." (Erikson, 1976)

2.2. Experiencing displacement

Displacement has been primarily recognized as a traumatic experience in which the loss one experiences have been found to impact the health and well-being of individuals, with significant short and long-term consequences. Many researchers have explored such links between well-being and displacement. Commonly explored have been the impacts of reduced social, cultural, and economic capital in one's health during the process of managing displacement (Greene et al., 2011). Individuals who are displaced are especially psychologically injured as they experience various kinds of personal and collective losses (Fullilove, 2016). Because some of these losses can be prolonged or even permanent, ramifications of displacement for the well-being of those affected are profound. For instance, per Diaz & Dayal (2008, p.1174), the most catastrophic impact of a disaster is one's feeling of "loss of place." From this perspective, place is both the material and social environment for anchoring a sense of self and a sense of self-in-relation, through memory and meanings invested in that environment through repeated interactions.

Further, displacement interrupts the tacit experiencing of a seamless narrative of 'who' and 'what' we are concerning 'where' we are (Milligan 1998; Casey 1993). For instance, Fullilove (2016) investigates, from a social psychiatry standpoint, both individuals and communities suffering in the aftermath of displacement. She states that displacement causes emotional pain and a reaction similar to that of a "root shock," which she defines as a "stress reaction to the loss of all or part of one's emotional ecosystem." (Fullilove, 2013, p.293)

Displacement is thus inherently trauma-inducing as it leads to a sudden disruption of both ties to a geographical place and the social relationships fostered in that place. From a social psychological standpoint, Fried (1966, p.377) describes the psychological costs of displacement as reactions of grief for a lost home, in which such affective reaction to the loss of a place shows "...most of the characteristics of grief and mourning for a lost person." Especially regarding forced displacement, Fried (1966) describes it as a highly disruptive and disturbing experience.

Therefore, the human experience aspect of displacement, mainly caused by disasters, is not yet fully understood from a perspective other than a psychological one – with a few exceptions (e.g., Erikson, 1976). The study of the loss of a home, as it pertains to people's relationships to their places, could gain relevant insight by including considerations of an ecological component to such loss, which touches on broader extra-psychological constructs.

2.3. Individual and community responses to displacement

If the displacement process ends successfully or ends at all, it can be affected by the appropriateness of the response efforts carried out by the various agents involved (e.g., governments, various kinds of organizations, concerned citizens, and even the displaced themselves). Most response efforts are carried out by formal organizations (e.g., the State) and by organizations of the third sector (e.g., non-profits). Nonetheless, this study aims to gain insight into responses to displacement at the level of the displaced people themselves. This is because efforts made to provide people with housing solutions after displacement become useless if they are inappropriate. Besides the money and efforts spent, from a cultural-ecological perspective, when a response to displacement fails, "...it becomes yet another disaffection that needs to be overcome." (Mazumdar, 2012, p.1904) It becomes yet another trauma. Thus, there is a need for further investigation on responses to displacement at the level of the displaced people themselves.

Previous studies have shown that despite the ruptures caused by displacement, commitment to remake place connections remains strong, and people press on to remake their lives and start again (e.g., Fullilove, 2013, 2016; Greene et al., 2011; Mazumdar & Mazumdar, 2016). In order to better understand the meaningful adaptation efforts of people displaced by disasters, we draw a parallel through examples of immigrant groups who look to create meaningful spaces for themselves in a new country.

An ethnographic study at the residential unit level, by Mazumdar & Mazumdar (2012), looks at immigrants from India, Vietnam, Indonesia, Philippines, Iran, China, and Taiwan, living in Southern California. They found that home gardens play an important role in the lives of those immigrants. For these persons, their home gardens can be a religious space, a culture space through plants, fruits, and flowers, an ecological space, and a family memorial space. The addition and maintenance of these home gardens "enable immigrants to engage with, personalize, and experience their new environment in deeply meaningful ways." (Mazumdar & Mazumdar, 2012, p.258)

Several other studies have documented homemaking efforts by displaced people in a new context. Despite the challenges faced by forcibly displaced groups being different from those of voluntary immigrants (Fullilove, 2016; Greene et al., 2011; Mazumdar & Mazumdar, 2016), empirical evidence has shown that meaningful adaptation efforts to the new environment are still present. For instance, Mazumdar & Mazumdar (2016) examined how Vietnamese immigrants displaced by war in their home-country engaged in creating meaningful spaces for themselves in the U.S. – a setting with vastly different customs, values, buildings, and institutions.

Further, at a community level, Mazumdar, Mazumdar, Docuyanan, & McLaughlin (2000) examined the physical, social, symbolic, and emotional significance of the Little Saigon in the lives of Vietnamese immigrants. Their findings highlight how the built environment interacts with the immigrants' social, commercial, and ritual activities "to create and sustain a sense of place, foster community identity, and structure social relations." (Mazumdar et al., 2000, p.319) Similarly, in a study of displaced families forcibly displaced from various countries for distinct reasons, Greene et al. (2011) found that families were actively engaged in adapting to their new communities by confronting barriers and mourning their losses. Survival was the primary goal.

In a study of community recovery after displacement from the McLure fire in Canada, Cox & Perry (2011) concluded that the collective disruption caused by the fire exposed the roles of place and home that were taken for granted in the identity formation of residents, triggering a collective identity crisis. In this way, the McLure fire acted as an insertion point: a time and place in which routine social practices were disrupted and the smooth functioning of those affected interrupted. The authors suggest that the direct encounter with our materiality and with suffering calls forth more than coping. They posit that through this "pause," a discursive opening is created, and in this opening, a process of navigating, negotiating, and reconstructing identity, or reorientation, is initiated.

2.4. Living in transition: a post-disaster temporary housing

Definitions of what constitutes a disaster have varied in the literature. They have been categorized based on damages associated with an event (Baum, 1987), duration (Bolin, 1985), or the subjective psychological responses (Davidson & Baum, 1986). In social science usage, as well as in everyday speech, the term disaster refers to a sharp and furious eruption of some kind that splinters the silence for one terrible moment (Erikson, 1976). A disaster is an event with a distinct beginning and a sharp end, and it is, by definition, extraordinary – a disturbance of the natural processes of life. So, the two distinguishing properties of a disaster are: first, that it does a good deal of harm, and second, that it is sudden, unexpected, acute (Erikson, 1976).

Disasters tend to inflict injuries on the minds of at least a few of the people who survive them, and these injuries are often diagnosed as "psychic trauma," which means an assault on the person (so sudden and so explosive) that it smashes through one's defenses and does damage to the sensitive tissues underneath (Erikson, 1976, p.253). Accordingly, some have defined disasters as collective traumas (Silver & Garfin, 2016), encompassing a wide variety of natural and human-made adversities that affect large segments of the population.

As previously introduced, findings of the effects of displacement demonstrate that many people navigate sudden changes in their home environments with great difficulty (Brown & Perkins, 1992; Fried, 1963). In developing the concept of habitus, Bourdieu (1986) challenged clear-cut distinctions between psychological, social, and physical processes, arguing that they are engrained with one another in the course of everyday life. However, in the specific context of disasters, some scholars have criticized an apparent overemphasis on social capital in the theorizing of managing the loss. They argue that the repeated focus on social capital minimizes other vital factors, such as the built environment and its meanings to the well-being of people (Cox & Perry, 2011). In doing so, some have drawn tacit connections to place. In the specific context of disasters, planning and design efforts for housing the displaced have been known to be unsuccessful, and even considered to be failures (Mazumdar, 2012). Evaluations of the wide variety of responses to disasters have allowed them to evolve, leading to improved interventions. However, response failures are still registered worldwide, especially in the provision of temporary housing to the displaced.

By analyzing several cases of success and failure in design through time, Petroski (2006, p.51) emphasizes that designers can often "overlook the details that make the difference between success and failure," and that "a building need not collapse catastrophically to be considered a failure." Consequently, most failures and our responses to them are not matters of life and death. "Ignorable failures," are treated as unremarkable (Petroski, 2006, p.62) and thus often understudied and unaddressed. The practical implication of this gap in studying and addressing failures is the fact that these, when in fact ignored or inappropriately addressed, become a disaffection (or a constraint) that needs to be overcome repeatedly by the user of the designed product (e.g., temporary housing), hindering the user's experience.

Despite identifying distinct stages of housing that are often employed in post-disaster situations, Quarantelli (1995) defines temporary housing as a physical structure with the function to shelter people when a disaster occurs until they are permanently resettled. However, several solutions provided to house those displaced temporarily have registered a low level of acceptance, despite being structurally sound and functional. Oliver (2003, p.130) stresses that a house is "built to serve a variety of functions, but one of the most important is to create living conditions that are acceptable to their occupiers;" these conditions refer to many aspects of life. This raises questions about the nature of failure in post-disaster response to displacement. By shifting our focus from the house as a product of design to homemaking as a design process, we may gain valuable insight into the potential role of people-environment relationships, especially after traumatic experiences. Selecting the "person-in-environment" as the unit of analysis (rather than the environment as solely a design product) may inform new directions in post-disaster research, especially concerning design assessments and the provision of more appropriate temporary housing to those displaced.

3. ENVIRONMENTAL DESIGN RESEARCH: Approaches and Trajectories

Prior research examining some of the issues concerning people-environment relationships is revealing for what has been covered and how, for what has been overlooked, and for providing ideas, concepts, and direction.

3.1. Early scholarship

Early scholarly works dealing more closely with people-environment relationships are described in the Environment Psychology (EP) literature. An important conceptualization provided by Stokols (1978) that is essential to this proposed study is the identification of "modes of human-environment transaction," which are described as four distinct sets of processes. The "interpretive mode" corresponds to individuals' cognitive representation or construction of the environment, and it involves research on the cognitive representation of the spatial environment and personality and the environment. The "evaluative mode" corresponds to an individual's evaluation of the situation against predefined standards of quality and involves research on environmental attitudes and environmental assessment. The "operative mode" corresponds to an individual's movement through or direct impact on the environment and involves research on the experimental analysis of ecologically relevant behavior and human spatial behavior. Last, the "responsive mode" corresponds to the environment's effects on the individuals' behavior and well-being and involves research on the impact of the physical environment and ecological psychology.

These modes allowed for the development of an essential way of seeing man-environment relationships referred to as "people-environment transactions." "Transactionalism" was further explored by Saegert & Winkel (1990), who organized the qualities of this type of study in five dimensions - the unit of analysis is the "person-in-environment" (p.443); person and environment are dynamically defining and changing each other; stability and change coexist; change is emergent; and it is vital to look for the sources of change.

From EP's formative years, most researchers were aware of the extent to which environmental issues were a function of socio-cultural and economic factors. However, efforts to develop theoretical and research perspectives linking macro-scale issues to social and individual level environmental concerns only started to systematically appear by the time of Saegert & Winkel's (1990) review. The authors emphasized the fact that many experiences in and uses of environments should be understood in the context of broader physical, economic, historical, and political forces. For instance, what the authors called a paradigm of "sociocultural forces" encompasses the individual as a social agent who seeks and creates meaning in the environment. It also treats the environment as a benign medium and resource for group development and maintenance.

Psychologists tend to characterize meaning primarily in terms of the categories that people use to organize their physical worlds mentally, but they ignore the sources of these categories. Moreover, "[a]lthough there are theoretical differences regarding the ontology of meanings, there is a consensus that neither meanings nor actions are individual acts." (Saegert & Winkel, 1990, p.458) Furthermore, meanings are not just constructed but are also given by the culture and social structure within which the person operates. Meaning thus stems from various levels of nonverbal communication from the environment to people (Rapoport, 1982). Moreover, the extent to which the environment works as a source of cues for appropriate behavior is culture-specific and not necessarily cross-culturally transferable. The main contribution of this turn is the acknowledgment that environmental meanings and actions are not solely individual constructions and that the individuals both define and are defined by the group in which they participate.

Mazumdar's (2008, pp.12-14) Environmental Psychology in the USA: a reflective view, posits that some characteristics encompassing all these approaches are an emphasis on people, a neglected social dimension, a focus on "behaviorist environmental psychology," and a view of individual and environment as separate (Mazumdar, 2008, p.14). This characterization drives the author to question the multidisciplinary nature of EP, and to conclude that the term EP appears to designate "both a specific discipline (psychology) as well as a larger multidisciplinary field" (Mazumdar, 2008, p.15). The author considers that EP, in the U.S., has the potential to be an even stronger discipline using the existing modes, methods, and approaches, but engaging in the use of different perspectives or lenses.

Specifically, Mazumdar (2008, p.18) calls for the use of different lenses to further explore understudied areas, such as the physicality of the environment, as it "can be deconstituted and be examined in greater detail;" the sensorial environment, as it is "multimodal;" the emotion-laden environments, as "emotional connections and reactions to environments" can be further investigated; the designers' and planners' minds; and specific populations and their environments. In sum, emphasis should be given to a definition that puts both humans and the environment at an equal standing.

3.2. A cultural-ecological approach to EDR

EP has been moving towards the psycho-social perspective since its inception. However, Bonnes & Secchiaroli (1995) are the ones who describe this as the most appropriate approach for enabling EP to answer "external" questions and to become a theoretically relevant reference. Accordingly, they defend the use of the key concept of "transaction," and even further, propose the term "transactional-contextual." (Stokols & Altman, 1987) This is an essential turn in EP. The intent by various authors to not only bring to light an extra-psychological construct (i.e., that of "place") into EP, but also, to "assign it a central position as the main construct characterizing the entire disciplinary area." (Bonnes & Secchiaroli, 1995, p.162) Interestingly, the architectural field had already been the first to be oriented in this direction, shifting from a "behavior setting" (Barker, 1968) to place.

Emphasis is given to human intentionality

through which people assume a "sense of place," or the realization that places are a material and symbolic product of human action (Stokols & Shumaker, 1981). More specifically, places are not direct results of physical characteristics of the environment, but products of intentional human acts turned towards the 'creation of places.'" (Tuan, 1974) The growing attention to this shift can be traced back to two main factors: the possibility of collaboration between the phenomenological approach and the psychological research, and the fact that the limits of the constructs being used were increasingly noted. Many authors contributed by calling for a critical review of the popular "behavior setting" construct directed to what they called "place perspective" or "place theory" (Bonnes & Secchiaroli, 1995, p.168)

Accordingly, the cultural-ecological view sees people in environments as an integrated whole, rather than as a loose collection of relatively independent parts. To take a cultural-ecological approach to design means to assume that cultures exist in space and time and cannot be free of either and that cultures have interrelationships with space. The latter implies that cultures are affected by the quality of the environment and that cultures affect their environments (Mazumdar, 2012).

There is a substantial and complex body of literature devoted to defining culture. Rapoport (1969) suggests that these various definitions are seen as complementary, rather than contradictory. Three general views give an idea of what culture is: (1) culture as a way of life typical of a group; (2) culture as a system of symbols, meanings, and a cognitive schemata transmitted through symbolic codes; and (3) culture as a set of adaptive strategies for survival related to ecology and resources (Rapoport, 1969). Building on this broad characterization of culture, this study adopts Mazumdar's (2000, p.161-162) working definition, which recognizes culture as "the negotiated, shared understandings, world view, values, ideals, and norms of a group of interacting people."

3.3. Cultural-ecological considerations of Place

Culture is thus herein broadly defined as a group of mutually interacting persons with negotiated shared understandings, ideals, values, a way of looking at the world and their place in it. This broad definition of culture includes national, geographic, regional, local as well as sub-local cultures. The "groundedness" of people in a particular place, and the physicality of cultures are seen as necessary (Mazumdar, 2012). Cultural products include the ideational constructs, the socially negotiated and developed ideals, as well as the constructed environment, including buildings, cities, plazas, parks, among others. The place component is not devoid of people; it is the sum-total interactive effects of both (Mazumdar 2011).

Some definition of place is also needed. In this study, a place has three features (Gieryn, 2000, p.464-465): a geographic location, (i.e., a place is "a unique spot in the universe"); a material form (i.e., a place "has physicality... place is stuff"); and an investment with meaning and value (i.e., without naming, identification, or representation by ordinary people, a place is not a place). Places are interpreted, narrated, perceived, felt, understood, and imagined (Soja, 1996).

The simplest definition of place-making is that of a process of creating quality places that people want to live in. Place-making is thus a means to an end – the creation of an ideal place. Place-making is a concept that transcends the boundaries of the potential effects of urban cultural policies on space. Its original formulation could be attributed to an encounter between the practical orientation of planning and community development studies, mainly concerning the material handling of space and built environment (Rapoport, 1997), and the humanistic geographical thought, which has eventually appropriated and reformulated the meaning of place-making.

Bringing together the concepts of 'self' and 'place,' which are considered mutually constitutive, humanistic geography stresses the role of the subject as an autonomous actor who, interacting with other actors and nature, society, and culture, transforms environments (Entrikin & Tepple, 2006). Therefore, place-making is more than a spatial process in which the actor locates in a given place in order to exploit some local externalities without any stable relationship to an existing path-dependency. In this process, elements such as individual and collective experiences, representations, and intentions acquire increasing importance as powerful engines of environmental transformation. Indeed, spontaneous cultural practices can be powerful catalysts of these elements, both supporting their production and leading them to coherence.

Homemaking involves the personalization of space with meaningful objects (Mazumdar & Mazumdar, 2016). These may remind people of their past – places, friends, family, and religion, and significant "personal mementos." Particularly in a context of displacement, Parkin (1999, p.317) emphasizes that such meaningful objects, "under the conditions of rapid and sometimes violent flight... may take the place of interpersonal relations as a depository of sentiment and cultural knowledge." For some, displacement can lead to a greater sense of attachment to these cherished possessions and places (Feld & Basso, 1996).

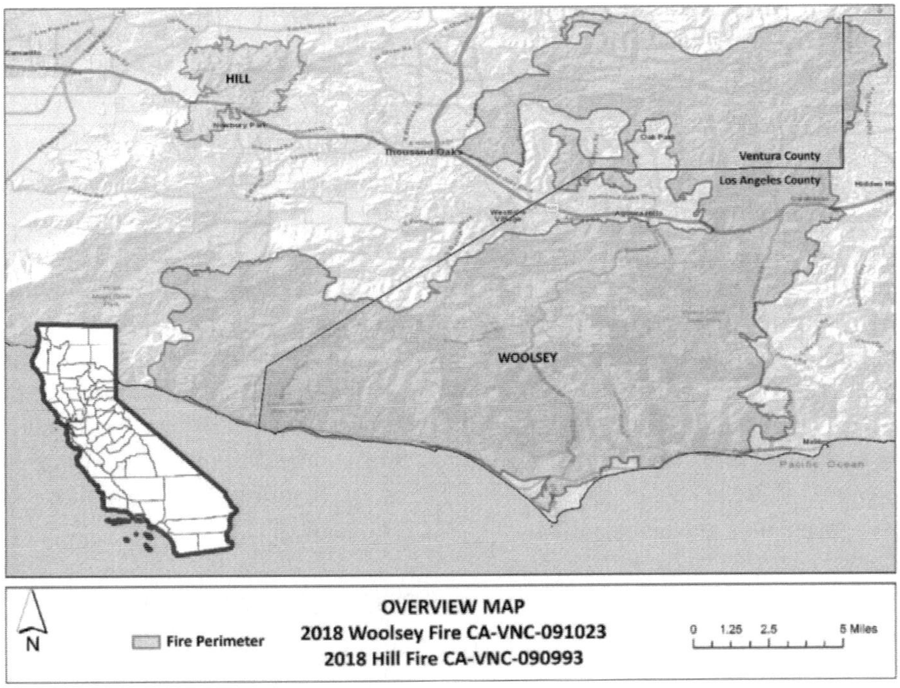

Figure 1. The Woolsey Fire perimeter in Ventura and Los Angeles County, California (2018). Public domain image from the State of California Watershed Emergency Response Team.

Cultures develop an attachment to places and physical as well as cultural and social communities (Low & Altman, 1992). For instance, place attachment can lead people to reconstruct prior environments they were familiar with and attached to, instead of building in ways that would reduce the problems of the design of their old environment (Mazumdar, 2012).

Critical concluding thought is that a people-in-place perspective includes "encounter, spirit, impressions, is open to and includes poetics, aesthetics, romance, imaging and imagination, and people's feelings and emotions, including unique experiences, and intersubjectivity." (Mazumdar, 2009, p.9) This perspective on people-environments relationships accepts that there may be variations in cultures; it accommodates culture and its effects on experience.

4. RESEARCH AIMS AND QUESTIONS

By using a cultural-ecological approach to Environmental Design Research that seeks to understand people in environments as an integral whole, this study investigates how people conceptualize and deal with living in transition (e.g., in precarious environments) in a context of displacement. In other words: What do peoples' lives look like while living in temporary living arrangements?

The goal of this research is to gain an in-depth understanding of how people view and interact with precarious environments occupied as a result of displacement caused by the Woolsey fire. This study is interested in understanding, from an emic perspective, how processes, such as homemaking unfold in precarious environments, and if such processes of meaningful adaptation play a role in the pursuit of well-being (i.e., as a coping mechanism) when dealing with trauma.

5. THE INTERPRETIVIST FRAMEWORK

A methodology for learning about cultural-ecological relationships must focus on the people and their environment (Mazumdar, 2012). A qualitative interpretivist framework was best suited for this study. Interpretivists view experience and perspective as valuable sources of knowledge and seek to uncover the interpretations of reality that people have based on their understanding of the world (Hesse-Biber & Leavy, 2011; Roth & Mehta, 2002). Further, interpretivists focus on understanding how meaning is created through the interaction between people, and between people and their environments (Hesse-Biber & Leavy, 2011).

Studying meaning-making in context through a displaced person's understanding of the process of making and remaking place in a new environment is needed to address this study's questions. One of the underlying assumptions in this study is that multiple realities exist (and coexist). Cultures are unique, but so are disasters (Mazumdar, 2012). Obtaining "as complete a picture as possible" of the phenomenon is to understand how these people see the world, the way they perceive constraints, and how they choose to act (Mazumdar & Geis, 2001, p.262).

5.1. Setting & Subject

The Woolsey Fire impacted neighborhoods in various cities of both Ventura and Los Angeles counties, California. Though social media (i.e., Facebook), a group created and managed by victims of the event was identified. The group administration, who played an important role in the disaster relief efforts put in place in the aftermath of the Woolsey Fire, identified a few victims that were willing to participate in this study. All of the initial participants actively engaged in identifying other potential participants, allowing for a snowballing strategy. This paper focuses on three of the initial families (four participants), who are still living in temporary housing.

5.2. Data collection techniques

Data collection strategies included observations and interviews. Visiting a home (in any form of living arrangement, e.g., temporary housing) can tell us a lot about its residents as our surroundings often reflect a great deal about who we are (Bailey, 2007). Thus, observations were a significant form of data collection. While obtaining people's accounts about activities within a setting through interviews in interactions is fundamental to field research, so too "is seeing with one's own eyes." (Bailey, 2007, p.79) A strategy for initial observation was to conduct a walkthrough. A walkthrough is somewhat a combination of observation and interview that allows for the descriptive identification of negative and positive aspects of the investigated environments, in which the physical aspects of the environment elicit participants' reactions to it (Preiser, Rabinowitz, & White, 1988).

Usually, a walkthrough relies on checklists and is conducted by the professional or researcher. In this study, the goal was to facilitate an understanding of the environment through the resident's eyes. So this was somewhat an itinerant conversation that encompassed all spaces of the house and was supplemented with photographs, audio, and video recording. The amount – and especially the quality – of the information offered was in the hands of the participant, and it was very telling of what was foremost in their minds. Observations were also conducted at different times, focusing on people's expressions, expressed feelings, interactions with each other, and with their environments.

Interviews were also used to collect data. Unstructured, open-ended interviews were conducted. Questions attempted to learn about the lives of people displaced by the Woolsey Fire that are living in temporary arrangements from an emic perspective.

5.3. Data analysis

For data analysis, a "discovery of grounded theory" (Glaser & Strauss, 1967) approach was used to capture emergent themes arising from the data. This process involved the conversion of the data into text form through transcription, reading, and familiarization with the data. Portions that revealed participant's experiences of living in temporary housing arrangements were identified, coded for emergent aspects, ideas, and themes. These were then organized into broader thematic categories. Three themes that emerged are described next.

6. PRELIMINARY FINDINGS

6.1. A temporary living arrangement can be a house, but it is not always a home

Patricia (earlier introduced) and her husband lost their home to the Woolsey Fire. About eight months later, seated in the living room of the guest house they are renting from a friend, the couple flips through the pages of a booklet portraying the still-under-construction condo they are buying in the following weeks (Field observation, 7/12/2019). "It [the condo] will be home. That's what we need... a home. Because this is not." (Patricia, 7/12/2019)

Home is an ambiguous concept, and it cannot be taken for granted. There is a lack of consensus about the uses of terminology, including "housing," "dwelling," and "home" Lawrence (1987; 1997). The distinction between these terms has been discussed by vast scholarship (for an overview, see Lawrence, 1987). From the perspectives of economics and politics, the tenure status of housing with its implications for personal control is the critical variable that defines what makes a house a home (Johnson, 1971). The tenured status of housing is somewhat restrictive and cannot account for a large portion of the population in many countries that choose to rent housing units. However, often used to explain the meaning of home (e.g., Saunders, 1990), ownership seems to be especially relevant in the U.S. cultural context.

To Cox & Perry (2011), within the dominant Western discourse, home is constructed as the site of the most significant power and control. Further, homeownership is seen as a necessary, desirable, and healthy milestone, which constructs the subjectivity of the homeowner as a better citizen. The notion of home thus has multiple layers of meaning and engages both a physical discourse and a psychological, symbolic one.

Hence, the loss of home has powerful and multiple meanings for people. Rita, who also lost her home to the fire, compares her lack of ownership (of a house and all of the objects contained in it) to the status of a dead person:

"It's been eight months, and I don't own a thing. Now, that is a real interesting concept. I mean, you know, normally, when somebody is in a situation like mine, it's after they're dead, you know, it's that... you die, and you don't take it with you. And then it's up to your family to distribute your belongings or do what they want with them or hang on to them. But in my case, who knows... if I live another 30 years or more, what is it going to be like, for me, to keep living without anything? I mean, that is a new one. Not many people get to experience that." (Rita, 8/13/2019)

To the conversation of ownership, Rita added: "everything that I actually own, no joke, is what I can fit in my car." Cecilia also shares the estrangement of not owning a thing. She is yet another person displaced by the Woolsey Fire, and who is still unsure about when she can rebuild. Cecilia expressed her conflicting feelings about the apartment she rented to live with her 18-year-old son, after weeks of both of them sleeping on friends' couches, and living by other arrangements made by her church leaders:

"When I first got here, I felt great because I had one place to roost. Now that I've been here a while... I'm miserable here. It's depressing. This isn't home. There are no memories [cries] There is no... We don't have any of our stuff. And, I mean, not that stuff is that much, but you have a certain comfort to having your bed, and your sheets, and your towels, and your 'this,' your 'that'... And there's just like, nothing." (Cecilia, 8/6/2019).

When asked about the pieces of furniture filling every corner of the apartment, Cecilia clarifies:

"This was a friends' table [touches the dining table], like most of stuff. There are very few things that we bought. Like, a lot of people helped as far as giving us stuff. The only [thing] I think we bought was that couch there [motions to the living room]. That was the only thing we bought was that couch there [motions to the living room]. That was the only thing we bought and our beds. I can't remember which [Goodwill and Salvation Army] did which... One gave us gift cards and stuff. They gave us vouchers. I got a $200 voucher for furniture, so we went in, and we bought small things that we could... like, we got this nesting set of tables [points to a set of small tables by the wall], so we just use those as side tables and, you know... my son, he just has a little storage thing and a big table... so we just haven't gone crazy... I don't want to buy stuff until I totally go back home. This isn't home. That's why nothing is on the walls and stuff... really, very barren, which is very depressing [cries] And I don't want to nest here, because I'm getting out of here, as soon as I can." (Cecilia, 8/6/2019).

It has been posited that household living, along with its daily patterns and use, is what turns a house into a home (e.g., Turgut & Ünügür, 1995). That might be true for conventional housing arrangements and circumstances in which relocation is optional. However, in this case, household patterns and use seem to be only one part of a more complex process of adaption to forced displacement and relocation to precarious environments. Ownership status of the house and of all things that are contained in it (both the objects themselves and the memories that may be engrained in them) seem to play a role in the process of homemaking for Southern California residents displaced by the Woolsey Fire.

6.2. Living in a temporary arrangement may influence one's recovery process

Preliminary findings suggest that, beyond playing a role in the process of homemaking through a sense of ownership, spatial characteristics of a temporary living arrangement may influence one's well-being. Following the literature of trauma presented earlier, participants in this study were emphatic when describing their physical and mental health is impacted not only by the disaster itself but also by the daily frustrations of living in temporary housing.

"The living room is very eerie, it's very dark in there. You know, this is okay [motions to the dining room, which opens to a small patio]. The bedrooms are brighter, but it's just different, and I have a lot of anxiety... I've been really suffering from anxiety, and I have more when I'm here than anywhere else. I wake up in the morning, like borderline-panic attack, almost every morning. And I go away like, to my brother's [house], or anywhere. I don't wake up like that. So, it's just being here... this is not a comfortable situation. And then, because it's so eerie down here, I do most of my work upstairs. And I guess, because I sleep in the room, and I'm waking up, and, and doing my work, and... it just snowballs." (Cecilia, 8/6/2019).

Similarly, Patricia related spatial features of their temporary house to health issues faced by her husband, Ralph. When asked about how was the bedroom working for them, she was quick to evaluate: "The bedrooms are really tiny. He [Ralph] has had an injury because he banged his head on a shelf in the bedroom. Because it's an unfamiliar place. So, he has been dealing with a concussion... since March." (Patricia, 7/12/2019). "A brain bruise," Ralph added to his wife's account. Ralph continued by pointing out that he injured himself three times after the fire. The first two injuries were related to the being on their lot after the fire. In the first case, Ralph stepped on a nail when wearing tennis shoes to walk through the debris and ended up in a hospital. In the second incident, Ralph "smashed his hand" when attempting to clear the debris out of the lot. However, to Ralph, his physical injuries are connected to his state of mind at the time, as he added:

And then, the third one, of course, was my head concussion, which was... you know, if I probably wasn't suffering from the PTSD, and I wasn't all emotional that night, I probably wouldn't have done that (Ralph, 7/12/2019).

Ralph was open about his PTSD diagnosis. Cecilia also openly discussed her depression and anxiety. These are not rare diagnosis after a disaster. Garfin & Silver (2016, p.42) emphasize that given the "robust findings linking natural disasters to psychosocial problems that may persist long after the structural debris has been cleared," policies are essential to meet the short and long-term needs of the impacted population appropriately.

In the case of the Woolsey Fire, it seems this type of relief is continuously offered by the various agents involved (e.g., local government, third sector organizations, and even local mental health professionals). Cecilia, for instance, was offered free individual therapy sessions through a local foundation, which she was happy to use for a couple of months. Nonetheless, she mentioned that when a former therapist she had seen, maybe a year before the fire, noticed that the Woolsey Fire had impacted her area, he contacted her offering a few free therapy sessions, and then a discounted plan, so she currently sees him once a month. Identifying populations most in need and appropriate timing for the offering of services is essential to maximize limited relief funds (Garfin & Silver, 2016). In this case, group therapy sessions and support group meetings have been widely advertised on social media, targeting specific groups, such as the Facebook group created and managed by Woolsey Fire victims (Social media observations, 7/10/2019).

Further, independent accounts, voluntarily offered by participants, indicate that some of the displaced people passed away shortly after the fire. The voluntary mention of deaths (no questions were asked in this topic) seems to indicate that part of the participants in this study perceives the impacts of the disaster on one's well-being to be extreme. With no evidence that could confirm such a connection, it is reasonable to say that participants, to some extent, perceive the trauma imposed by the disaster as a potential contributor to one's death.

6.3. Living in transition: the nexus of space and time perceptions

The discussion involving the nature of time and its role in the development of human consciousness, self, and identity is not a new one. Over the last century, psychologists have suggested that perceptions of time are part of the foundation from which conscious thought and behavior emerge. As such, temporal perceptions have been considered the primary context through which humans understand and make sense of their life experiences (for an overview, see Holman & Silver, 1998). In a cross-sectional and longitudinal study of traumatized individuals, which included one sample consisting of two Southern California communities devastated by fire, Holman & Silver (1998) looked at the relations between an individual's temporal orientation and long-term psychological distress. Results indicated that individuals with a past temporal orientation – when one focuses attention on prior life experiences – was associated with elevated levels of distress long after the trauma (or the traumatic event) had passed, even when controlling for the degree of rumination reported.

Rumination is an interesting concept in this study. Also described as intrusive or repetitive thoughts about past experiences, rumination may occur in response to many different types of trauma, such as natural disasters (Holman & Silver, 1998). Intrusive or repetitive thoughts about past experiences were frequently mentioned by participants in this study and on various occasions. Interestingly, spatial features of the lost home and lost objects appear to be an evident trigger to such thoughts. Not having the objects that one would expect a house to have, or not knowing what one has at that new house seem to cause similar reactions.

"You're moving into somebody else's home. It's all fairly furnished, which is great, but it's not your stuff. It's not your house... I got groceries, and I was going to make something, and I knew that there was stuff in the kitchen. There is like, you know, there's pots and pans, and plates, and everything. But little things, like a baking pan. You would assume that you would have a baking pan to heat something up. They didn't have a baking pan... oh, okay, well... it's like we don't even know what was here." (Patricia, 7/12/2019).

For Rita, who has lived in about sixteen different hotels even since the night if the fire, the absence of her things is noticeable, as she pointed out: "When somebody's lost everything like this, and I've talked to some of my other neighbors, we end up having these flashbacks and, and sometimes you know, where people are just kind of doing their day and then all of a sudden you think in your mind 'Oh, yeah, that dish...' and you go to reach for something, and you say, wait a minute, I don't have that anymore. Or if you want to look at a photograph of summit anything 'Oh, that album that had the picture.' And then 'Oh, I don't have that.' Or you want to listen to some music that you might have had somewhere on something. It's gone. And it's... you don't think about it all at once. It sorts of goes in little increments" (Rita, 8/13/2019).

The nexus of space and time perceptions that may trigger rumination seem to be aggravated by living in a precarious environment, in which daily frustrations caused by inappropriate spatial features or the absence of known objects lead to comparisons to what one used to have in their home. For Patricia, for example, the "little things that we take for granted" would make her think of her lost home: the lack of garbage disposal, the absence of an ice maker, the newly-renovated deck that overlooked a reservoir that is now gone.

It seems reasonable to say that, based on preliminary analyses of the data collected, living in temporary living arrangements may contribute to ruminative or intrusive thinking in the context of coping with trauma. Patricia concluded the walkthrough of her temporary house in a big patio overlooking the mountains by saying, "Beautiful view, but I personally like water views. You know, I like water," referring to the view her lost home had of a reservoir. Ralph noticed the instant sadness in Patricia's face, apparently brought by the comparison, and added, "Someday, dear... It [the reservour] was beautiful, but I think... we are going to settle for a little water fountain on the new condo." (Field observations, 7/12/2019)

7. CONCLUDING DISCUSSION

This study assumed that the sudden and unforeseen nature of a disaster, such as the Woolsey Fire, could uniquely contribute to the precariousness of temporary housing arrangements obtainable by the displaced people. Preliminary analyses seem to confirm the presence of such a perceived relationship (i.e., a link between the sudden nature of the fire and the precariousness of housing arrangements available) in the context of Southern California's extended and increasing housing crisis (Legislative Analyst's Office, 2015).

This lack of options is evident is most of the participant's accounts. Cecilia made her choice of where to live temporarily based on location because it was close to her neighborhood. Nevertheless, she was categorical in adding that "It was the only thing in our budget that wasn't terrible. I mean, you would be shocked what they want to rent you for that amount of money in this area" (Cecilia, 8/6/2019).

Patricia and Ralph told a similar story to explain how they ended up renting their friend's guest house. They were living in a hotel for a month while engaged in disaster relief efforts in their neighborhood. According to Patricia, they were so busy with the disaster relief, that by the time they started looking for a place to live, the cheapest option they could find was about $3,000 a month, about double of their budget (their insurance Alternative Living Expense money). That price, Patricia pondered, without utilities or furniture, so when their friend offered them to rent his empty guest house at a ranch, they felt they were one of the fortunate ones, among all the displaced families they knew.

Okay, well, we're going to go ahead and do this, because money was so tight. And at least, you know, this was all furnished, we didn't have any money to buy furniture or anything. We didn't want to buy temporary furniture. We didn't want to have to pay for a moving truck later. So, we thought okay, we will just live here for now.

There were many different temporary housing arrangements made as a response to displacement to people impacted by the Woolsey Fire. This study focused on the experiences of some of those whose temporary arrangements have lasted at least ten months after the fire. Analysis of the data collected throughout the Summer of 2019 sought an understanding of what their lives look like when living in transition and recovering from the loss of home. The preliminary analysis points to three emerging themes that seem to play a role in the participants' lives, despite their different types of temporary living arrangements.

First, the recurring distinction implied by participants that a temporary living arrangement can be a house, but it is not always a home. In many instances, participants tried to distinguish their temporary living arrangements from their lost homes by pointing out an absence of meaning and memories embedded in the temporary house itself and its objects. In contrast, when referring to their lost homes, spatial features and objects were frequently related to people, meanings, and memories.

Second, aspects of one's well-being were consistently related to the spatial characteristics of the temporary home. Participants, without being asked, frequently made a point on how their (or their family members') physical and mental health were impacted not only by the disaster itself but also by the daily frustrations of living in temporary housing.

Lastly, temporal perception showed to play a role in participants' speech (e.g., they often recounted past events using the present tense) and by the information they shared (e.g., many focused on past or present experiences, while few talked about plans for the future). These indicators of temporal orientation indicate what is foremost in their minds. In this case, most participants seem to focus their attention on the past. Previous studies have linked a past orientation to difficulty in recovering from trauma. For the participants in this study, undesired spatial characteristics and the absence of objects in the temporary housing arrangement seem to notably contribute to intrusive or repetitive thoughts about past experiences (i.e., rumination).

At some point, Patricia, Ralph, and Rita decided to move on and remake a home somewhere else. Fortunately, although being underinsured, they were able to make arrangements to buy a house somewhere else. In both cases, there is additional waiting time as their houses get built. Also, in both cases, it seems like the decision-making happened through a compromise: they are downsizing, relocating to another area, but hoping to settle sooner rather than later. For Cecilia, displacement is still an ongoing concern. Her neighborhood will not be ready for rebuilding for another year or so, as the required debris removal program carried out by the State lasted nearly six months.

Currently, the new infrastructure needed for the whole neighborhood is still in the planning process. It seems that the impacts of the Woolsey Fire will keep playing a role in Cecilia's living arrangements, as she claimed that, even after rebuilding her house, she might not be able to resettle anytime soon emotionally. "I may want to get back there just because I don't want to be here. But I don't plan on living there forever. When I rebuild, I want to like, stay there for a few years, and move. I don't want to live in a canyon anymore. I'm tired of fires" (Cecilia, 8/6/2019).

---

[i] All names are pseudonyms.

REFERENCES

Bailey, C. A. (2007). A Guide to Qualitative Field Research. (2nd ed.). Thousand Oaks, CA: Pine Forge Press.

Barker, R. (1968). Ecological Psychology: Concepts and methods for studying the environment of human behavior. Stanford, CA: Stanford University.

Baum, A. (1987). Toxins, technology, and natural disasters. In G. R. VandenBos, & B. K. Bryant, Cataclysms, crises, and catastrophes: Psychology in action (pp. 5-49). Washington, DC: American Psychological Association.

Bolin, R. (1985). Disaster characteristics and psychosocial impacts. In B. J. Sowder, Disasters and mental health: Selected contemporary perspectives (pp. 3-28). Rockville, MD: U.S. Department of Health and Human Services.

Bonnes, M., & Secchiaroli, G. (1995). Environmental Psychology: a Psycho-Social Introduction. (C. Montagna, Trans.). London, UK: Sage Publications.

Bourdieu, P. (1986). The forms of social capital. In J. Richardson, Handbook of theory and research for the sociology of education (pp. 241-258). New York, NY: Greenwood Press.

Brown, B. B., & Perkins, D. (1992). Disruption in place attachment. In I. Altman, & S. M. Low, Place attachment. (pp. 279-304). New York, NY: Plenum.

Casey, E. (1993). Getting back into place: Toward a renewed understanding of the place-world. Bloomington, IN: Indiana University Press.

Cox, R. S., & Perry, K. E. (2011). Like a Fish Out of Water: Reconsidering Disaster Recovery and the Role of Place and Social Capital in Community Disaster Resilience. American Journal of Community Psychology, 48, 395-411.

Davidson, L. M., & Baum, A. (1986). Chronic stress and posttraumatic stress disorders. Journal of Consulting and Clinical Psychology, 54, pp. 303-308.

Diaz, J., & Dayal, A. (2008). Sense of place: A model for community based psychosocial support programs. The Australasian Journal of Disaster and Trauma Studies, 1, 1174-4707.

Entrikin, J. N., & Tepple, J. H. (2006). Humanism and democratic place-making. In S. Aitken, & G. Valentine, Approaches to Human Geography (pp. 30-41). Thousand Oaks, CA: Sage Publications Ltd.

Erikson, K. T. (1976). Everything in Its Path: Destruction of Community in the Buffalo Creek Flood. New York: Simon & Schuster Paperbacks.

Feld, S., & Basso, K. H. (1996). Senses of Place. Santa Fe, NM: School of American Research Press.

Fried, M. (1966). Grieving for a Lost Home: Psychological Costs of Relocation. In J. Q. Wilson, Urban Renewal: the Record and the Controversy (pp. 359-379). Cambridge, MA: The MIT Press.

Fullilove, M. T. (2013). The frayed knot: What happens to place attachment in the context of serial forced displacement? In L. C. Manzo, & P. Devine-Wright, Place attachment: Advances in theory, methods, and applications (pp. 292–316). Boston, MA: Taylor & Francis.

Fullilove, M. T. (2016). Root Shock: How Tearing Up City Neighborhoods Hurts America, and What We Can Do About It (2nd ed.). New York, NY: New Village Press.

Garfin D.R., & Silver R.C., (2016) Responses to Natural

Disasters. In: Howard S. Friedman (Editor in Chief), Encyclopedia of Mental Health, 2nd edition, vol 4, Waltham, MA: Academic Press, pp. 35-46.

Glaser, B., & Strauss, A. (1967). The discovery of grounded theory. New York: Aldine de Gruyter.

Gieryn, T. F. (2000). A space for place in Sociology. Annual Review of Sociology, 26, pp. 463-496.

Greene, D., Tehranifar, P., Hernandez-Cordero, L. J., & Fullilove, M. T. (2011). I Used to Cry Every Day: A Model of the Family Process of managing Displacement. Journal of Urban Health: Bulletin of the New York Academy of Medicine, pp. 404-416.

Hesse-Biber, S. N., & Leavy, P. (2011). The practice of qualitative research. Los Angeles, CA: Sage Publications.

Holman, E. A., & Silver, R. C. (1998). Getting "Stuck" in the Past: Temporal Orientation and Coping With Trauma. Journal of Personality and Social Psychology, 1146-1163.

Johnson, M. (1971). Household Behaviour: Consumption, income and wealth. Harmondsworth, UK: Penguin.

Lawrence, R. (1987). Housing, Dwelling, and Homes: Design Theory, Research and Practice. Chichester, UK: John Wiley.

Lawrence, R. (1997). Culture and Space in the Home Environment. In S. M. Ünügür, O. Hacihasanoğlu, & H. Turgut (Ed.), Culture & Space in the Home Environment (Introduction). Istanbul, Turkey: Dünia.

Legislative Analyst's Office. (2015) California's High Housing Costs: causes and consequences. Retrieved July 20, 2019.

Los Angeles County Fire Department. (2018). Woolsey Fire Incident Update. Retrieved November 22, 2018.

Low, S. M., & Altman, I. (1992). Place attachment: A conceptual inquiry. In S. M. Low, & I. Altman, Place attachment: Human behavior and environment (pp. 1-12). New York, NY: Plenum Press.

Mazumdar, S. (2000). People and the Built Environment. In P. Knox, & P. Ozolins, Design professions and the built environment: An introduction (pp. 157-168). New York, NY: John Wiley Ltd.

Mazumdar, S. (2008). Environmental Psychology in the USA: a reflective view. MERA, 21, pp. 11-22.

Mazumdar, S. (2009). From the material to the immaterial in Architecture: An experiential approach. MERA Journal, 12 (1), 5-13.

Mazumdar, S. (2011). Yearned crowding, desired closeness: Hakata Gion Yamakasa. Journal of Asian Urbanism, 4, 12-15.

Mazumdar, S. (2012). A Cultural Ecological Approach to Disaster Planning. 9th International Conference on Urban Earthquake Engineering/ 4th Asia Conference on Earthquake Engineering (pp. 1901-1906). Tokyo, Japan: Tokyo Institute of Technology.

Mazumdar, S., & Geis, G. R. (2001). Case study method for research on disability. In B. N. Altman, & S. N. Barnartt, Research in Social Science and Disability: Exploring theories and expanding methodologies (vol.2) (pp. 255-275). New York: NY: Elsevier.

Mazumdar, S., & Mazumdar, S. (2012). Immigrant home gardens: Places of religion, culture, ecology, and family. Landscape and Urban Planning, 105, 258-265.

Mazumdar, S., & Mazumdar, S. (2016). Interiors and Homemaking in Displacement: A Study of Vietnamese Americans in Southern California. Journal of Interior Design, 41 (2), pp. 23–37.

Mazumdar, S., Mazumdar, S., Docuyanan, F., & McLaughlin, C. M. (2000). Creating a Sense of place: The Vietnamese and Little Saigon. Journal of Environmental Psychology, 20, 319-333.

Milligan, M. J. (1998). Interactional past and potential: The social construction of place attachment. Symbolic Interaction, 21 (1), 1-33.

Oliver, P. (2003). Dwellings: The vernacular house worldwide. New York, NY: Phaidon.

Parkin, D. J. (1999). Mementos as transitional objects in human. Journal of Material Culture, 4 (3), 303–320.

Petroski, H. (2006). Success through failure: the paradox of design. Princeton, NJ: Princeton University Press.

Preiser, W., Rabinowitz, H., & White, E. (1988). Post-Occupancy Evaluation. New York: Van Nostrand Reinhold.

Quarantelli, E. L. (1995). Patterns of sheltering and housing in US disasters. Disaster Prevention and Management, 4, 43-53.

Rapoport, A. (1969). House Form and Culture. New Jersey, NJ: Prentice-Hall.

Rapoport, A. (1982). The Meaning of the Built Environment: A Non Verbal Communication Approach. Beverly Hills, CA: Sage.

Rapoport, A. (1997). The nature and role of neighborhoods. Urban Design Studies, 3, 93-118.

Roth, W. D., & Mehta, J. D. (2002). The Rashomon effect combining positivist and interpretivist approaches in the analysis of contested events. Sociological Methods & Research, 31 (2), 131-173.

Saegert, S., & Winkel, G. H. (1990). Environmental Psychology. Annual Review of Psychology, 41 (1), pp. 441-478.

Saunders, I. (1990). A Nation of Horse – Owners. London, UK: Unwin Hyman.

Silver, R. C., & Garfin, D. R. (2016). Coping with disasters. In J. C. Norcross, G. R. VandenBos, D. K. Freedheim, & N. Pole, APA handbooks in psychology: Psychopathology and health (pp. 597-611). Washington, DC: American Psychological Association.

Soja, E. (1996). Postmodern Geographies: The Reassertion of Space in Critical Social Theory. New York, NY: Verso.

State of California Watershed Emergency Response Team. (2018). Woolsey-Hill Fires.

Stokols, D. (1978). Environmental Psychology. Annual Review of Psychology, 29, pp. 253-295.

Stokols, D., & Altman, I. (1987). Handbook of environmental psychology. New York, NY: Wiley.

Stokols, D., & Shumaker, S. A. (1981). People in place: A transactional view of settings. In J. H. Harvey, Cognition, social behavior, and the environment (pp. 441-488). Hillsdale, NJ: Lawrence Erlbaum.

Tuan, Y.-F. (1974). Topophilia: A Study of Environmental Perceptions, Attitudes, and Values. New York, NY: Columbia University Press.

Turgut, H., & Ünügür, S. M. (1997). Changing Households Patterns and the Use of Home: A Case Study of the Traditional Turkish Houses. In S. M. Ünügür, O. Hacihasanoğlu, & H. Turgut (Ed.), Culture & Space in the Home Environment (pp.1-42). Istanbul, Turkey: Dünia.

United Nations. (2017). United Nations High Commissioner for Refugees (UNHCR). United Nations.

D24 - Transform Places with People and the Environment: Cities & Globalization / Residential Environments

Designing Inclusive SMART Cities: The Opportunities and Challenges

Prakash Kamtam (UN Global Compact Cities Programme)

1. ABSTRACT

The twenty-first century is the age of cities. The growth and development of a country depends on the success and failure of its cities. Therefore, today there is an increased and renewed focus on urban governance and city development models. Cities are experimenting with new and innovative solutions for their urban problems.

The smart city concept which is the dominant model even though still at an evolving stage has likely potential to change the cities towards sustainable development and inclusive communities. But 'smart cities' still is an elusive concept and often dominated by tech solutions, where technology is only a means to achieve larger outcomes and greater impacts and not an end in itself. Appropriate technology innovations no doubt have unlimited possibilities. The renewed focus on smart urbanisation and use of innovative tools and smart applications may address the challenge of diversity, equity and inclusion in modern cities. Is smart city an inclusive and holistic approach and can it benefit all stakeholders including the vulnerable, disadvantaged and excluded sections – esp. the elderly, the differently-abled, the children, the women, the youth, the poor and the migrants? People-centric smart city design is an essential element for successful outcomes of any smart city project.

1.1. Keywords

Inclusive Smart Cities, Inclusion, Sustainable development, Inequality, Equity

2. Introduction

Increasing urbanisation is no doubt setting a new global order with more than half of the total population residing in cities and exerting pressure on already strained resources. This trend will further change in the coming decades, where nearly seventy per cent of the total global population is expected to live in cities by 2050 (UN,2014). The increasing urban population is looked at both as a crisis and an opportunity!

It is a crisis because it will further strain the deteriorating global environment by increasing ecological footprint of cities (Agenda 21 'the Earth Summit' and UN Sustainable cities programme); where the increasing size and population of cities will adversely affect the social, economic and environmental dimensions of the development, unless it is corrected by cities with increased emphasis on environment and sustainability (Rogers,2008). On the other hand, increasing urbanisation is looked at as an opportunity because it ushers in the new era of modernisation, mega conglomerations, new opportunities and livelihoods, improved quality of life, better management of resources, economies of scale and so on.

Goal 11 of United Nations Sustainable Development Goals (SDG) states to: "Make cities inclusive, safe, resilient and sustainable". According to this goal, cities are considered to be the main hubs of human settlements with ideas, culture, commerce, productivity, social development etc; whereas many challenges exist in making the cities a better place to live, to create job opportunities and welcome prosperity. Today cities across the globe are encountering several issues and problems. The SDG 11 aims to mitigate these problems and work towards livable and resourceful cities. Similarly, the new urban agenda sets out goals to ensure that future cities, towns and basic urban infrastructures and services will be – "more environmentally-friendly, accessible, user-friendly and inclusive of all people's needs".

Today cities are facing two conflicting issues (Francesco Bifulco et.al, 2016): 1. Complex nature of problems (due to increasing population, overexploitation of resources, demand for more services and so on) and 2. Need for sustainable development

However, managing cities has become one of the very crucial development challenges of the current century. The focus and importance of urban governance is increasing day by day with cities struggling to adapt to workable solutions for urban problems. Governments are struggling with new ways and means and investing millions and millions to improve the quality and standard of living in cities. These emerging urban development models across the globe are termed as- Smart Cities,

Resilient Cities, Sustainable cities and so on.

On the other front, the new digital technologies are bringing in greater connectivity across the world with interconnected people. The national boundaries are shrinking with raising worldwide web and the internet. The force fueling this rapid transformation of silo islands to inter-connected superhighways is Information and Communication Technology (ICT). This new ICT is revolutionizing how fast communications are made, how customers are served, how tasks are performed, and how organisations are run.

This fast-changing technology world is influencing the urban governance sector also. At present smart, intelligent and cognitive cities have become new city development models, all across the globe. There is a renewed interest among all to address challenges of urban management using technology solutions (Nasrin Khansari, et.al,2013). The new technology is being looked at providing solutions to emerging urban challenges. The city managers are expected to leverage the benefits of ICT and use the same for efficient delivery of services, for the better quality of life thus achieving sustainability goals (Franscesco Bifulco, et.al 2016).

ICT is also recognized as an integral part of UN-sustainable development goals. Wherein this modern technology is accepted as an enabler which helps measures the progress and implementation of Sustainable Development Goals (UN-HABITAT,2015). According to UN-HABITAT, the advantages of using ICT tools/solutions are reduction in administrative cost, improving access to healthcare, education etc, wider participation of citizens, smart buildings and infrastructure, e-Governance and real-time data monitoring.

Based on the fact they are using technology tools, there are multiple self-proclaimed smart cities all across the world (Hollands, 2008). No doubt technology is critical to developing smart cities but does this state of art technology solutions transform them into truly smart and sustainable cities? ((Tan Yigitcanlara et.al 2019). Does ICT have the potential to change our cities with a better quality of life, bridge the digital divide and eliminate social, economic and physical exclusion and existing inequalities? Are smart cities inclusive and beneficial, especially for vulnerable and disadvantaged population groups like the -elderly, differently-abled, children, women, youth, poor and migrants?? Whether cities could use this opportunity to become smart and enable a parallel shift towards greater inclusivity in all respects?

How to build inclusive smart cities? These questions need to be addressed in the development and implementation of smart cities projects. Smart city planning is not just a data-based linear practice, it is an integrated, collaborative, interactive analysis and visualizations of data sets and functions and the ongoing processes including the local history and stories (Susa Eräranta and Aija Staffans,2015).

People-centric smart city design is an essential element for successful outcomes of any smart city project. Technology can offer new and innovative solutions, but it should always remain a means to an end; with the outcome being the service and quality of life of citizens. However, increasing smart cities projects the world over has posed more questions than answers!

The main objectives of this paper are:

- To understand the concept of smart urban governance
- To explore the potential of smart applications and innovations in enhancing sustainability and urban inclusion and thus building inclusive smart cities

This study is based on desk research method; where the literature reviewed includes research articles and reports on- the concepts of smart governance, the concept of smart city, sustainability in smart urbanisation, urban inclusion and different smart city models from across the globe.

3. Smart governance leading to Smart city

Smart urban governance is a wide concept and includes terms such as- smart city, digital city, virtual city, international city, knowledge city, broadband city, broadband metropolis, wireless city, mobile city, ubiquitous city, eco-city and so on. Where the key focus is on the use of ICT and the term smartness is the combination of different components which lead to working together, quick action and making an efficient decision (Tunc Karadag, 2013). Currently, the global urban development process is influenced by globalization (World Bank,2008) and emerging technologies (Demirkan, et.al, 2011). As the cities are facing multiple challenges leading to the use of technology in every aspect of the urban ecosystem to provide a better quality of life and therefore the concept of the smart city emerged (Francesco Bifulco et.al, 2016). The concept of a smart city began in the 1970s when urban context started adopting digital configuration (Ishida and Isbister, 2000). Since then it has been tried and developed in different forms with an expectation to find a solution to urban problems. In this fast-changing world, the cities across the globe are transforming a smart concept like- smart TV, smartphone and so on (C Moreno Alonso et.al,2016).

Based on global experience it is found that there are four types of smart governance concepts (Albert Meijer,2013):

- Right policies and right implementation without any structural and procedural changes aimed at efficiency and effectiveness of city administration
- Innovation in decision making and implementation
- Smart administration by use of information technology, integration of information processes, interconnections of institutions & physical infrastructure for better service to citizens
- Rearranging the position of government within the urban systems like–collaboration across departments, community-based interconnected governance, open-minded and participatory governance and so on. This is aimed at maximizing the socio-economic and ecological performances of city governments

The key strategies for smart urbanization include (Neha Bansal et.al, 2015): use of renewable energy, smart Grid, smart urban planning, eco-cities, intelligent cities, smart policies, smart transportation, inclusive and livable cities and urban integration.

Today the concept of smart cities has become a global buzz word, and everyone agrees to make cities smarter by use of - smart technologies, smart collaborations (among the educated population) and with an effective institution (to address the challenges of modern cities) (Albert Meijer,2013). However, the felt need for smart cities is different for different regions-for example in case of South East Asia, smart cities are used as a vehicle to create national/international identity, boost local economy and technology innovations, test new solutions on large urban projects and so on; in contrary, the smart city projects in North America, Oceania and Europe are usually in small scale with focus on quality of life and sustainability (Tan Yigitcanlara et.al 2019).In emerging economies smart city is viewed more as a new urban development model with the possibility of solutions to all urban issues!

Among all different perspective's ICT is at the core of the smart city. The new technology is being looked at providing solutions to emerging urban challenges. The city managers are expected to leverage the benefits of Information and Communication Technology and use the same for efficient delivery of services, for a better quality of life thus achieving sustainability goals (Franscesco Bifulco, et.al 2016).

According to UN-HABITAT, the benefit of using ICT in urban development is to achieve transparency, two-way accountability in governance, digital literacy and better civic engagement of people and use of data and collaborative platforms in designing better infrastructure. The term smarter in the smart city means more efficient, livable and sustainable (Alawadhi et.al, 2012). There is renewed interest and enthusiasm at all levels especially among the public who expect their city life to become a more easy, comfortable and happy-with expectation of improved basic services, reduced pollution, minimal road congestion,24x7 supply of safe drinking water, affordable electricity and use of other resources appropriately. The smart city no doubt is a big hit across the globe in terms of raising expectations of the people who look forward towards- "economically vibrant, inclusive, efficient and sustainable urban habitats. If this is applicable to all sections of society and if everyone reap these benefits equally is a big question in search of the right answer!

4. Smart City Vs Sustainability

Sustainable development has no doubt become the new path of future development. Initially, the worry of environmental degradation due to an increase in population, rapid urbanisation, use of motor vehicles, unplanned industrialization posed serious concerns to human well-being. Accordingly, a lot of eco-friendly concepts of sustainable development are driving the current city development models across the globe (Yigitcanlar, 2015).

At the same time due to the increasing population in cities, urban governance has become the key challenge for attaining sustainability. There is an increasing demand for technology solutions for urban challenges. ICT is mainly used for better management of resources for improvement in the quality of life, to allow participation of all stakeholders and to achieve sustainable development (Anthopoulos and Tougountzoglou,2012)

Urban governance and sustainability are becoming critical issues of city management across the world. Urban issues are cross-cutting and therefore requires strategic decisions and accountability. Due to the all-pervasive nature of technology, during recent times the focus increased on achieving the goal of sustainability by use of ICT as a tool for smart governance (Meijer and Rodriguez Bolivar, 2015).

Further with the advent of technology, there can be several smart solutions for improving the resource efficiency and management of scarce resources. Today the smart city is considered as an efficient, green, technologically advanced and socially inclusive city (Vanolo, 2014). A smart city is expected to develop technology solutions for societal, economic, management and ecological challenges (Yigitcanlar, et .al, 2017). Smart cities give importance to digital technologies to become competitive and attain a sustainable future (Schuurman et.al,2012).

Where smart governance includes citizen's participation and co-creation of new services for improvement in the quality of life (Kolsaker and Lee Kelley, 2008), increased data exchange, service integration and collaboration (Maltby, 2013). Smart living includes all aspects of quality of living like – health, housing, tourism, culture and high levels of social cohesion (Francesco Bifulco et.al, 2016). The smart city focuses on mobility, resource efficiency, social sustainability, climate mitigation and adoption in addition to ICT (Arvin Ghasemi, 2015).

For example, in the case of Copenhagen, smart city approach is a climate strategic approach using Wi-Fi and Bluetooth tracking smart solutions for climate change (Arvin Ghasemi, 2015). A smart city aims not only at the sustainability of physical systems like- environmental, social and the economic world but also unphysical elements like climate strategy and planning including digital infrastructure. Where digital sustainability should balance and complement the physical sustainability (Arvin Ghasemi, 2015).

A smart city does not constitute mere development of a business park or industrial estates with largescale IT enterprises. In the smart city, all structures such as water, power, mobility etc are designed, constructed and maintained using technology like sensors, networks and electronics which are used to collect data, regular tracking and decision support system (Bowerman et.al, 2000). The smart city is expected to monitor its condition and self-repair if needed (Bowerman et.al, 2000).

A smart and sustainable city has the following goals to be achieved in adaptable, reliable, scalable, accessible, and resilience way (Dhingra and Chattopadhyay,2016):

- Improve the quality of life
- Ensure economic growth & employment opportunities
- Well- being of citizens
- An environmental approach to development
- Efficient delivery of services
- Address climate change and environmental issues
- Effective local governance ensuring equitable policies.

Where sustainability, quality of life, urban aspects and intelligence become the four attributes of smart and sustainable cities which become the core of all smart city projects (Dhingra and Chattopadhyay,2016). Therefore, the smart city is being considered a powerful tool towards achieving urban sustainability.

5. The need for Inclusive Smart cities for sustainable future

Today cities are the key agents of social integration and sustainability-social, economic and environmental (World Economic Forum,2016). The success or failure of SDG's will be decided in Urban areas, therefore national development strategies should be aligned with urban development to guarantee all citizens, especially the vulnerable, have equal access to quality basic services and means of production. (Development Bank of Latin America,2016). Hence smart city must recognise the diversity of citizens, struggle against segregation of minorities, eliminate both physical and digital barriers. Then it is called Inclusive Smart City (Neto and Kofugi, 2016).

"Ensuring that smart initiatives are inclusive is not always an easy task; it requires that these initiatives be conceived with inclusion as a number one priority. This is where shared values, strong leadership and a common vision become crucial," states Thomas Müller, Co-Founder of bee smart city. Unsustainable planning practices can have a significant impact on the lives of city inhabitants who are marginalised, either through identity or economic status. These practices can lead to the exclusion of citizens who cannot meet the requirements of participating in the city as it exists, for example, those with limited access to transport such as those on the periphery of urban areas (Xavier Lemaire and Daniel Kerr,2017).

Cities, in particular, have large numbers of older inhabitants and are home to 43.2% of this older population (OECD,2018). However, many cities lack old age-friendly planning and policy, have an insecure and inaccessible built environment. Similarly, cities continue to be the greatest obstacles for people with disabilities, let it be in terms of physical access or digital access (Neto and Kofugi, 2016).

Globally most cities are utterly unfriendly to people with disabilities. Barriers can range from blocked wheelchair ramps to buildings without lifts, to inaccessible toilets, to shops without step-free access. For example, US law doesn't specify how municipalities should design and implement digital services for disabled people. As a result, cities sometimes adopt new technologies that can end up causing, rather than resolving, problems of accessibility. Touch screen kiosks are one such example (MIT Study).

A quarter of the world's children live in poor urban settlements. They are subject to social exclusion and lack of access to basic services and opportunities for a better future (UNICEF, 2012). The other challenge of new-age communities is that children in cities are spending most of their time indoors playing on smartphones/gadgets and watching TV. There are very few cities or urban settlements who provide innovative ways and means to keep children outdoors- healthy, active and social.

In cities, millions of women, lead highly insecure life and experience violence in a multitude of forms every day (IDRC,2017). According to WHO figures from 2013,

35 per cent of women worldwide have experienced some form of sexual violence. Economic and financial crisis effects such as increased unemployment rates have a gender dimension (Maria Sangiuliano). Women and girls are both direct and indirect victims of the lack of basic services in slums (UN-HABITAT,2010).

Another most critical problem is mainstreaming youth development in city development planning. Youth make up 41% of the world's unemployed, (ILO) or 74 million young people living in urban areas. By 2030,60 per cent of the world, the population will live in cities, and that 60 per cent of these urban residents are likely to be under the age of 18(UN-HABITAT). How and where to utilise this population? What is there competency and where are jobs? Growing youth population and fast urbanisation are two phenomena with deep implications on any country's economy and growth. Otherwise, youths that are educated but excluded and disgruntled are susceptible to make societies insecure and unsafe.

One billion people, which is almost one-third of worlds urban populations currently live in slums (UN-HABITAT). Hundreds of millions of people living in poor conditions lack access to basic amenities, shelter, clean water and sanitation. Poverty has multiple contributing factors such as low educational attainment, lack of access to proper healthcare, high housing cost, unemployment or working in low and menial jobs, low income/pensions and inadequate social benefits, barriers to access to basic urban services including discrimination and low level of participation in community and public life.

Urban migration is a huge problem in any city. The cities which are the major economic hubs of any society always attract migrant populations from within and outside the country. This has put constant pressure on the city administration to develop innovative integration policies for migrants thus building a welfare society (Costa and Ewert, 2014).

As per World Urban Campaign -City, We Need 2.0 (2017) the technology revolution offers new opportunities for the efficiency and responsiveness of urban services. It proposes ten principles and ten drivers of change for the achievement of sustainable development where the principles (1), (2), (3), (7), (8) and (9) emphasise the need for inclusion and equity. Similarly, the principles (2), (3), (4) and (5) of International Guidelines on Urban and Territorial Planning (UN HABITAT,2015) aims to promote renewed urban governance striving for local democracy, holistic participation and inclusion, transparency and accountability ensuring sustainable urban development and spatial quality.

G3ict and World Enabled launched 'Smart Cities for All' (2017) with a focus on ICT accessibility and digital inclusion of older people and persons with disabilities. Each tool addresses a priority challenge /barrier to the digital inclusion of persons with disabilities and older persons in Smart Cities as identified by global experts. Cities that effectively foster inclusion will offer greater opportunities for their citizens and reap widespread economic benefits (Deloitte Insights 2019). The movement of immigrants to cities always transformed the urban areas around the globe – demographically, politically, culturally and economically and evidence suggest that when cities practice inclusive migration they function better and prosper faster (UN-HABITAT, 2012).

The Urban Institute conducted a study to analyse the relationship between inclusion and economic health of 274 large cities in the United States over the past four decades. The study revealed that, with few exceptions, the more inclusive cities have better economic health indicators than those that are not (Urban Institute,2018). As per OECD Report (2018) key policy dimensions that matter for cities are- inclusive education, inclusive housing and the urban environment, inclusive jobs and labour markets and inclusive infrastructure and public services. The UN-New Urban Agenda (2017) provides a first global platform to exchange best practices and to enable and enhance, review and reforms to increase the performance of urban spatial planning. As per UNESCO report (2016), smart cities call for new urban governance models integrating both top-down and bottom-up approaches including citizen-driven innovations and co-creation. Smart city applications must be inclusive in terms of providing opportunities for all citizens. The needs of vulnerable groups such as the elderly, women and persons with special needs etc., should be integrated into smart cities strategies and planning. Promoting participating governances is essential and pre-requisite for developing inclusive smart cities.

Some of the relevant key policy instruments recommended to achieve inclusion include:

- Formalising informal sectors through smart applications, as data on informal sectors/ settlements are often missing from many cities.
- Providing affordable smart infrastructure such as piped water or electricity for informal sectors
- Making smart cities gender-inclusive by providing gender-inclusive smart infrastructure
- Developing smart infrastructure for all vulnerable groups like the elderly, people with special needs, children etc.
- Encouraging local innovations to drive smart city applications that ensure inclusive development.

6. Findings

A smart city using technology and mechanization can inspire, share knowledge and culture with its citizens thus giving them a motivating environment (Rio's, 2008). However, there is a lack of consensus on what is the smart city (Angelidou 2015). In spite of their commitment to sustainable outcomes, smart cities are criticized for just being a buzz word (Kunzmann 2014; Shelton et.al, 2014; Yigitcanlar 2016). Recently, smart interventions in cities have been criticized for their fragmented inclusiveness (Walravens,2014), split urbanism (Vanolo, 2014) use of smart just as a label (Hollands, 2008). Smart cities are also throwing similar challenges as of e-government studies like overemphasis on technology or social structure with limited understanding of emerging socio-techno practices, emphasis on governance transformation without conserving institutional and organizational form, lack of knowledge for drivers of smart city with assumption that this is driven by social or economic need and training of smart city issues as managerial or technical but not looking into politics of technical choices. There is a need to understand the relationship between smart cities and sustainable urban development. The smart city is facing a serious problem in terms of whether it can be termed as a suitable strategy for integration of physical and digital sustainability in the society or if so how it approaches the change and its solution (Arvin Ghasemi, 2015).Most of the current literature is on smart issues (smart technology), human capital (smart people) or a combination of the socio-technical issue (smart cities). There is very limited analysis of smart cities as socio-technical systems which combines both the social and technical domains and studies the effect of one on the other including the interrelationships between them. There is a need for theories to highlight the complex interactions between the social system and technology (Albert Meijer,2013). The technical and social analysis will provide a theoretical and empirical understanding of the socio-technical dynamics of a smart city. There is a need for more such analysis for better understanding between government/social structures and new technologies (Albert Meijer, 2013). In urban setting 'inclusion' takes different meanings such as inclusion of-informal sector/ informal settlements, migrants, poor and vulnerable, different religious groups and disadvantaged communities (old, disabled, children, women, youth, migrants etc.,) (UN-HABITAT). The 'urban inclusion' concept is a comprehensive and integrated development approach and includes the physical inclusion (access to land, housing and infrastructure), the economic inclusion (opportunities for all) and the social inclusion (rights and participation) (World Bank,2015).

A coherent, smart, inclusive and green growth strategies must be adopted so that conflicts and contradictions between different objectives can be overcome and accomplishment of one objective may not be detrimental to meeting others (EU,2011). Freedom of speech and accessibility to public information and services increases with the use of ICT which becomes a key component of smart city framework (Partridge,2004). New planning and development approach helps urban administrations to see and appreciate the big picture view of sustainable urban development and helps them to take necessary actions for wider and successful implementation of the sustainability agenda, and consequently achieve desirable sustainability outcomes (T. Yigitcanlar and S. Teriman, 2015). Further to fully understand the sustainability process, it is essential to know why a project is happening, what are the benefits, how it fits into the long-term strategies of city development plans. Therefore, it is required to have continuous study on the impact of technological solutions on the city development which will help in understanding the sustainable city development approach (Arvin Ghasemi, 2015).

7. Conclusion

Urban inclusion is a key aspect of sustainable development. The use of technology innovations can create immense possibilities towards the achievement of physical, economic and social inclusion of most vulnerable and disadvantaged sections like -elderly, differently-abled, women, children, migrants, youth and poor etc. This paper investigates the smart city approach towards inclusive and sustainable urban development-the opportunities and challenges. People-centric smart city design is an essential element for successful outcomes of any smart city project. Where the city shall provide an opportunity for all (without excluding any sections) and make them feel part and parcel of the city, with access to basic amenities and opportunity to participate in the decision-making process. Future research is needed to identify key indicators to measure the impact of smart applications and innovations including ICT and their contribution towards enhancement of urban inclusion. There is need to develop an Inclusive Smart Cities framework that could aid in the transition towards sustainable urban development enhancing access, participation and livability of most vulnerable and disadvantaged groups, who are often neglected and left behind.

REFERENCES

Sustainabledevelopment.un.org.(2017)/https://sustainabledevelopment.un.org/content/documents/1684SF_-_SDG_Universality_Report_-_May_2015.pdf

The Role of ICT in the Proposed Urban Sustainable Development Goal and the New Urban Agenda: UN-HABITAT

Bifulco, F., Tregua, M., Amitrano, C. and D'Auria, A. (2016). ICT and sustainability in smart cities management. International Journal of Public Sector Management, 29(2), pp.132-147

Claire Quinn & Carolyn Snell (2008) Peter Rogers, Kazi Jalal, & John Boyd, An Introduction to Sustainable Development, Sustainability: Science, Practice and Policy

Nasrin Khansari, Ali Mostashari and Mo Mansouri, (2013). Impacting Sustainable Behaviour and Planning in Smart City

Tunc Karadag (2013)- An Evaluation of the Smart city approach

EU (2010), Cities of tomorrow Challenges, visions, ways forward

TWorld Bank (2013)-Building Sustainability in an Urbanising World

Arce-Ruiz, Rosa., Baucells, Neus., Moreno Alonso, Concepcion.,(2016)- Smart Mobility in Smart Cities

Journal article -Albert Meijer and Manuel Pedro Rodríguez Bolívar, (2013). Governing the Smart City: Scaling-Up the Search for Socio-Techno Synergy.

Neha Bansal, Vineet Shrivastava, Jagdish Singh (2015)-Smart Urbanization – Key to Sustainable Cities

Arvin Ghasemi, 2015, Sustainability and Data Collection in the Smart City- A Case Study of the Wi-Fi and Bluetooth Tracking Project in Copenhagen

Suha Alawadhi, Armando Aldama-Nalda, Hafedh Chourabi, J. Gil-Garcia, Sofia Leung, Sehl Mellouli, Taewoo Nam, Theresa Pardo, Hans Scholl, ShawnWalker(2012) -Building Understanding of Smart City Initiatives

de Oliveira Neto, J.S. and Kofuji, S.T., (2016), July. Inclusive Smart City: an exploratory study. In International Conference on Universal Access in Human-Computer Interaction (pp. 456-465). Springer, Cham.

Yigitcanlar, T. and Kamruzzaman, M. (2015). Planning, Development and Management of Sustainable Cities: A Commentary from the Guest Editors. Sustainability, 7(11), pp.14677-14688.

Yigitcanlar, T. and Teriman, S. (2015). Rethinking sustainable urban development: towards an integrated planning and development process. International Journal of Environmental Science and Technology, 12(1), pp.341-352

Meijer, Albert., Rodríguez Bolívar, Manuel Pedro(2015)- Governing the smart city: a review of the literature on smart urban governance

Vanolo, Alberto (2014)- Smartmentality: The Smart City as Disciplinary Strategy

Sangiuliano, M., Smart Cities and Gender: main arguments and dimensions for a promising research and policy development area.

Trindade, E., Hinnig, M., da Costa, E., Marques, J., Bastos, R. and Yigitcanlar, T. (2017). Sustainable development of smart cities: a systematic review of the literature. Journal of Open Innovation: Technology, Market, and Complexity, 3(1).

Kitchin, R., (2019) Toward a Genuinely Humanizing Smart Urbanism. In The Right to the Smart City (pp. 193-204). Emerald Publishing Limited.

Lemaire, X. and Kerr, D., (2017). Inclusivity in urban planning practices.

Korngold, D., Lemos, M. and Rohwer, M., (2017). Smart Cities for All: A Vision for an Inclusive, Accessible Urban Future. BSR and AT&T Inc.

Bronstein, Z., (2009). Industry and the smart city. Dissent, 56(3), pp.27-34.

Meijer, A. and Bolivar, M., (2013)- Governing the smart city: Scaling-up the search for socio-techno synergy. In 2013EGPA Conference Proceedings.

Batty, M., Axhausen, K.W., Giannotti, F., Pozdnoukhov, A., Bazzani, A., Wachowicz, M., Ouzounis, G. and Portugali, Y., (2012)- Smart cities of the future. The European Physical Journal Special Topics, 214(1), pp.481-518.

UN-HABITAT (2017)-UN Sustainable Development Knowledge Platform accessed at Sustainabledevelopment.un.org

UN-HABITAT (2009)-Planning Sustainable Cities

UNESC (2016)-Smart cities and infrastructure

UNESC (2015)-Towards a sustainable, inclusive and resilient urban future for Asia and the Pacific

EUSMARTCITIES @http://EU-SMARTCITIES.EU

Hall, R..,Bowerman, B.,Braverman, Joseph., Taylor, J.,Todosow, Helen., Wimmersperg, (2000)- The vision of a smart city

UN-HABITAT (2008)- State of the World's Cities Report 2008/2009: Harmonious Cities,

UN-HABITAT (2010)- GENDER EQUALITY FOR SMARTER CITIES-ChallengesandProgress

UN-WorldUrbanizationProspects-(2014)

UNEP-Sustainable, Resource-efficient cities – Making it Happen

OECD (2015)-Ageing in Cities-Policy Highlights

FICCI & NCPEDP (2016)-Structural Framework for Accessible Urban Infrastructure in Smart Cities'

AT&T Report-Smart Cities for All: A Vision for an Inclusive, Accessible Urban Future

Hollands G Robert, (2008). Will the real smart city please stand up? Intelligent, progressive or entrepreneurial?

Yigitcanlara, Tan et al,(2019)- The making of smart cities: Are Songdo, Masdar, Amsterdam, San Francisco and Brisbane the best we could build?

Eräranta, Susa and Staffans, Aija,(2015)-From Situation Awareness to Smart City Planning and Decision Making

World Urban Campaign (2017)-THE CITY WE NEED 2.0 Towards a New Urban Paradigm

UN New Urban Agenda (2017)

UN-HABITAT (2015)-International Guidelines on Urban and Territorial Planning

Deloitte Insights (2019)- Inclusive Smart Cities –Delivering digital solutions for All

An OECD White Paper for the Urban 20 (U20) –(2018) Inclusive Growth and Social Cohesion in Cities

Lee-Kelley, Liz., Kolsaker, Ailsa (2004)- E-government: the "fit" between supply assumptions and usage drivers

Schuurman, Dimitri., Baccarne, Bastiaan., Marez, Lieven., Mechant, Peter (2012)- Smart Ideas for Smart Cities: Investigating Crowdsourcing for Generating and Selecting Ideas for ICT Innovation in a City Context

Dhingra, Mani., Chattopadhyay, Subrata (2016)-Advancing Smartness of Traditional Settlements-Case Analysis of Indian and Arab old cities

Rios, Patrice(2008)-Creating 'the smart city'

Angelidou, Margarit (2015)- Smart cities: A conjuncture of four forces

Walravens, Nils (2014)- Mobile city applications for Brussels citizens: Smart City trends, challenges and a reality check

Partridge, Helen(2004)- Developing a Human Perspective to the Digital Divide in the Smart City

E5 - Transform Processes and Outcomes for Equity: Cities & Globalization / Participation

Fuhais Community Identity As Defensibility Mechanism Against Threats of Urban Sprawl

Majd Al-Homoud (German Jordanian University); Dina Hawash (German Jordanian University)

1. ABSTRACT

Continuation of urban sprawl towards suburbs has become a key issue threatening identity and solidarity of rural communities. Urban sprawl is perceived as a negative urban form that needs to be controlled to limit its impacts on host communities. The most challenging issue is solidarity and cohesion of the host community, which are indicators of communal stability. It is phenomenological that Amman's suburbs developed strong social cohesion based on kinship, ethnicity, shared values, and homogeneity. Recently, Fuhais town witnessed motions of urban sprawl from Amman city that caused major problems to the local community. This paper investigates community solidarity mechanisms that are used to defend Fuhais town from urban sprawl. The major objectives of the study are: (1) to pinpoint defense mechanism against urban sprawl; and (2) to provide reference to help decision makers and planners to deal with urban sprawl. The hypothesis of the study states that control of urban sprawl can be enhanced through sense of community identity, as well as its components: social affiliation and political affiliation. A field survey was conducted using a structured questionnaire disseminated among Fuhais town residents. The outcomes indicated a strong relationship between sense of community identity, social affiliation, and political affiliation and control of urban sprawl. Such outcomes suggest that community solidarity can resist the urban sprawl of Amman city towards the surrounding local communities. Therefore, such planning decisions should consider the involvement of the communities and their will to be part of the urban expansion before they are taken.

1.1. Keywords

Community, Identity, Jordan, Urban, Sprawl.

2. INTRODUCTION

The phenomena of the rapid and random spatial growth of the urban peripheries of Amman, the capital city of Jordan, into adjacent towns and villages located west and northwest threatens the socio-spatial structure of such communities. Such rapid growth at the peripheries represents a clear example of neoliberal practices and social restructuring of the city (Tewfik and Amr, 2015). High status sectors moved west-ward and inflated land prices. Consequently, upper-income districts showed an ideally prestigious lifestyle tendency growing in a sectorial wedge-shaped pattern. In addition, most representatives of the state choose to live near the symbolic and political landmarks such as Royal Palaces, King Hussein Medical Center, and Jordan's General Intelligence Center (Daher, 2013). However, it is phenomenological how in Amman's suburbs, residents had developed a strong social cohesion based on factors such as: kinship, ethnicity, shared values and homogeneity (Al-Asad, 2004).

Suttles (1972) discusses the concept of defended neighborhoods, where residents are the main players in defending their communities. Local residents tend to have similar interests that promote solidarity fostered by their homogeneity. Community solidarity is a social issue and yet the centrality of this issue is often masked identity (Durkheim, 1995). To hold to one's true identity is a crucial connection to other people beyond the family. Sharing identity with other people is to feel in solidarity with them, to owe them something special, and to count on them (Gotham, 1998; Durkheim, 1995).

The term urban sprawl is just another name for excessive metropolitan decentralization or suburbanization. It takes place when larger percentage of residential and business activities take place outside the city centre. Urban sprawl is usually driven by population and economic growth allowed by urban expansion into rural areas. It takes the form of low-density development, predominantly single-family residential subdivisions and commercial strips. Further, land prices in the centre of the city force developers to seek lower prices at the periphery. This international phenomenon affects both developed and developing countries causes negative impacts at the socio-spatial, economic, and environmental levels (Al-Asad, 2004; Small, 2000; Downs, 1999). Several studies tended to deal with urban sprawl as a universal problem, and most of the proposed spatial solutions and strategies were general without focus on the specificity of the place (Freilich, 1999).

Suburban communities can be viewed in

contrast to cities because they tend to be homogenous in socio-spatial character and their residents represent a self-selected group having the same preferences for social activities and lifestyle (Wassmer, 2002; Knox, 1987). Suburban communities differ from core cities in both physical and socio-spatial characteristics. Such communities are seen as potential providers for communal ties where residents practice their common activities, locality, and solidarity (Forsyth, 2012; Wilson and Baldassare, 1996; Gober and Behr, 1982).

Locally, the phenomenon of the rapid and random spatial growth at the urban peripheries of Amman calls for an immediate reconsideration of regulation to limit this uncontrolled growth (Tewfik and Amr, 2015). Due to the high population density and continued expansion of urban development in Amman city, the adjacent low-density peripheral suburban towns had to absorb some populations and potential developments. By 1990, Amman's growth from the west swallowed the towns of Sweileh and Wadi Al-Sir, which were transformed into characterless, shoddy satellite extensions to Amman. Such sprawl raised the concerns of neighboring towns' residents to preserve their communities from the foreseen negative impacts of such growth (Makhamreha and Almanasyeha, 2011).

Due to its strategic location close to the Royal Palaces and Al-Hommar and Dabouq upper class neighborhoods, which has abundance of resources and lack restriction legislations over urban development (Al-Asad, 2004), Fuhais is threatened by this urban sprawl. Fuhais is located in Shafa Al-Balqa 20 kilometers northwest of Amman, with an area of 16km2 and a population of 13,000. Fuhais town has the potential to be swallowed by this urban sprawl and may become incorporated into the Directorate of Amman city. It witnessed continuous mixed urban expansion from Amman city that threatens Fuhais town as a community and jeopardizes its solidarity. In addition, this act of rapid urban sprawl from Amman led to relocation of major services and facilities within Fuhais town.

Fuhais is considered an ethnic town where the majority of residents are Christians known as the young horse riders, which culturally refers to strong people who defend their territory. Further, a heritage awareness movement in Jordan made this town to be a cultural, historical, touristic, and religious destination and put Fuhais on the map for Christian pilgrimage. Fuhais residents have developed a strong social fabric that depends on kinship, religion, familial networks, and shared destiny, which solidify their community and intensify their inclusive identity. At the economic level, Fuhais was originally an agricultural village, but today its economy is fueled by non-agricultural activities such as remittances from expatriates in the United States, and income generated from micro businesses because of its closeness to the capital city, Amman. 2008 statistics show that Fuhais Municipality is one of the richest and self-sufficient municipalities in Jordan, with an annual budget of 1,983,936 JD. The majority of Fuhaisi residents are upper-middle class. The natural resources of vegetation, green fields, and rich soil make Fuhais an attractive destination for the surrounding neighbors. Lack of restricted legislation of investments in Fuhais has attracted various developers to construct mixed-use developments. In addition, the municipality lacks planning management and policies, making the purchase and sale of agricultural lands easy and contributing to premature, out of control growth (Makhamreha and Almanasyeha, 2011). Consequently, property values have risen and have contributed to the enrichment of landowners in the hills adjacent to Amman. These issues raised Fuhaisi residents' fear for their town to be swallowed by Amman. The major concern is the increased number of new commercial businesses along the major corridor that connects the town to Amman. The other concern is the new construction of residential apartments to accommodate the increased number of new-comers to the town. As a result, residents fear losing the communal character and solidarity that Fuhais was known for (Sweis, 2011; Abu-Dayyeh, 1997). The fear from urban sprawl was dominated by the residents' fear of the alteration to their community solidarity, which led them to organize campaigns against Amman's intervention under the slogan of "no residential apartments any more". The goal of the protest was to stop the phenomenon of spatial invasion of their town, which has caused a distortion of the natural landscape and the cultural heritage, and put pressure on their ethnic cohesion and on the existing infrastructure (Sweis, 2011). Additionally, a number of social, political, and voluntary organizations and associations in Fuhais town demanded that the Ministry of Municipal Affairs stop issuing building permits for investors coming from Amman and those Fuhaisi investors living outside the country. Most Fuhaisi residents insist on having a regulatory structure for the town to maintain its distinctive spatial and communal characters, claiming that they have the right to determine their town's future (Sweis, 2011).

2.1. Urban Sprawl

Urban sprawl is one of the key issues facing cities today (Harvey and Clark, 1965). The term urban sprawl was first used by planners in the 1950's to indicate an urban development that generates negative social

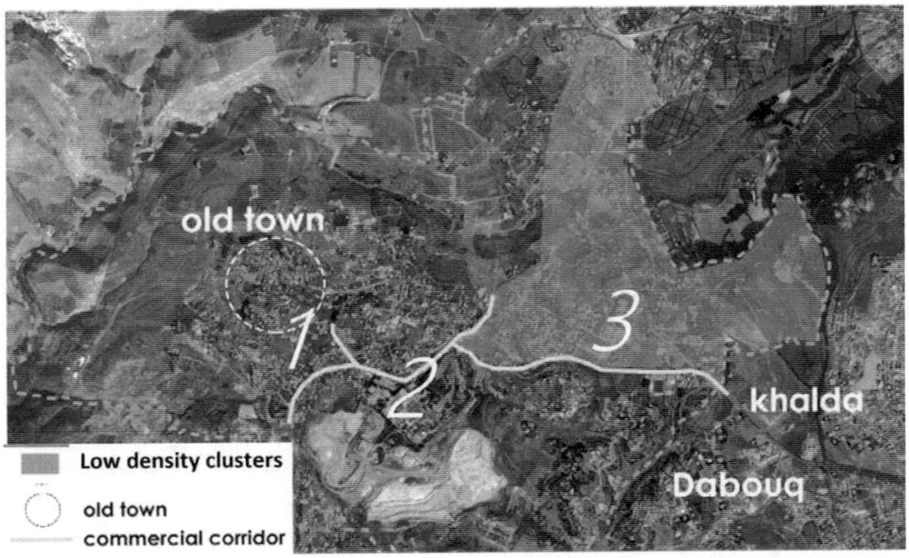

Figure 1. Map showing the Three Clusters of the Study Area. Source: authors, 2018.

impact. Later, urban economists adopted the term to deal with economic impact generated by sprawl. In the 1990's the phenomenon of urban sprawl was adopted by other disciplines (Wassmer, 2002). Studies suggest that where unplanned development dominates, sprawl will automatically occur in a mechanistic way. In contrast, where growth around the periphery of the city is implemented by tough urban policies, more compact forms can be secured (EEA, 2006; Wassmer, 2002). Researchers define urban sprawl as an uncontrolled haphazard growth resulting from real-estate development at the outskirt of the city. It is characterized by low-density, dispersed, auto-dependent development, and specialized land use that has environmental, social, and economical impact (Al-Asad, 2004; Wassmer, 2002; Small, 2000; Downs, 1999; Ewing, 1994). Urban sprawl is opposite to compactness. With the advent of the automobile, there is no distance long enough for urban growth and urban centers may extend almost indefinitely (Al-Asad, 2004). Such pattern of low density expands into large urban areas on the expense of agricultural land and show inefficient and uncontrolled development (EEA, 2006). Smith (1990) argued that urban sprawl has become a general term surrounding a wide range of urban forms. There is little agreement on the characteristics, drivers, and impacts of urban sprawl. Policymakers often see urban sprawl as a disease identified by undesirable symptoms that need to be cured (EEA, 2006; Al-Asad, 2004; Small, 2000; Downs, 1999).

In terms of form, there are three patterns of urban sprawl: (1) Secondary urban center, where development takes place alongside the boundaries of the existing town center, characterized with low density continuous development. (2) Ribbon-linear-development, which occurs along the major transportation network that connects urban area, leaving undeveloped lands without direct access to the network. (3) Leapfrog development, a discontinuous pattern of urbanization in which developed areas are widely scattered with lots of undeveloped lands (Ghani, Abidin, and Khalid, 2014; Glaster et al., 2001; Harvey and Clark, 1965).

In terms of land use, it takes the form of low density, single land use development with spatial segregation, and lack of integration and accessibility to commercial, residential, and retail developments. Such patterns create isolated pockets and fragmented masses. It usually consists of single-family homes on large lots or commercial strips (Small, 2000; Taylor and Harrell, 1996).

In terms of density, i.e., the relationship between the number of people living in an area and a given land area, it takes the form of low-density developments that weaken social capital, social interaction, and informal social control, and increase travel distance and cost (Franz, Maier, and Schröck, 2006; Brueckner, 2000).

The drivers of urban sprawl are macro and micro socio-economic forces that can be summarized as follows:

(1) Macro-economic: driven by population and economic growth allowed by urban expansion into rural areas that take the form of low-density development, including single-family residential developments and commercial strips. People want to live outside cities to avoid traffic, noise, and to have homes with more square meters, and private green spaces. Such developments expand in geographic size faster than they grew in population size; they impact health, economy, and social life (EEA, 2006; Small, 2000; Downs, 1999; Ewing, 1994).

(2) Micro-economic: driven by land prices as developers and investors seek lower prices in the peripheral agricultural areas. In addition, competition among municipalities for new income generating jobs and services is great fuels for urban sprawl. Many municipalities are tempted to relax controls on the development of agricultural land, and to offer tax benefits to commercial and industrial developments to invest in their municipalities (EEA, 2006; Ewing, 1994).

(3) Housing preferences: driven by households demand to make a residential location selection based on decentralized locations with cheaper land to build larger house, against longer commuting times and less urban amenities (EEA, 2006; Wassmer, 2002; Ewing, 1994).

(4) Demographics: driven by household lifecycle. Families with fewer children are most likely to move to suburban and rural areas outside the urban core, in comparison to the elderly and single people who are least likely to move out to the periphery (EEA, 2006).

(5) Inner city problems: driven by factors like poor environment, social problems, safety issues, and lack of green open spaces (EEA, 2006; Ewing, 1994; Harvey and Clark, 1965).

(6) Regulatory framework: driven by less controlled and more loosely regulated land use; as developers find these lands more suitable for new constructions (Harvey and Clark, 1965).

Impact and Threats of Urban Sprawl on Suburban Communities: Urban sprawl may have both positive and negative impacts; but because urban sprawl is often uncontrolled or uncoordinated, negative impact overweighs positive impact (Luther, 2005). Luther (2005) and Smith (1990) argued that threats can be recognized at three levels: socio-spatial, economic, and environmental as follows:

Socio-spatial threats include: (a) Socio-spatial fragmentation and segregation: refers to the lack of social integration between the different socio-cultural and economic classes and their large lots that in turn tend to separate neighbors, and lack of collective urban memory and sensitivity to the place. (b) Degradation of local identity: refers to the lack of the authentic sense of place and the rural character. (c) Lack of perceived safety: refers to the decrease in resident perception of safety when they perceive a large inflow of newcomers. (d) Population mobility: refers to the frequent change of residents who stay and leave, which creates community instability and less informal social control (Altınok and Cengiz, 2008; EEA, 2006; Franz, Maier, and Schröck, 2006; Brueckner, 2000; Ewing, 1994). This is the most threatening factor of urban sprawl, which degrades the social fabric and turns the community from saved into lost (Luther, 2005).

Economic threats include: (a) Infrastructure and public services inefficiency: refers to the inability to provide and maintain urban infrastructure and public services that requires linking the existing with the new developments by providing roads, water, electricity, sewage, hospitals, schools, fire stations, and recreation areas. (b) Local business inefficiency: refers to the growing competitive economic activities that negatively affect residents' local business. (c) Decay of old urban center: refers to the decay of old urban center because of the transmission of commercial activities to the new center. (d) Gaps in housing prices: refers to the reduction in housing prices in the urban center in comparison to the new developments, leading to concentrations of economically underprivileged groups in the downtown area. (e) Spatial segregation in land use: refers to the segregation in land use for housing, business, recreation, and shopping, which affect travel distance and cost, as well as land prices (Altınok and Cengiz, 2008; EEA, 2006; Franz, Maier, and Schröck, 2006; Bekele, 2005; Brueckner, 2000; Ewing, 1994; Harvey and Clark, 1965).

Environmental threats include: (a) Consumption of natural resources: the increase in use of agricultural lands and consumption of water and raw materials associated with growing population and dispersed developments. (b) Transport-related air pollution: refers to the diminishing air quality because of automobile dependence. (c) Transport-related energy consumption: refers to the increase in energy consumption associated with road networks, inefficient public transport, and car dependency. (d) Conversion of agricultural land use: refers to the transformation of farmlands into urban uses as a result of low price of agricultural lands (EEA, 2006; Brueckner, 2000; Ewing, 1994; Harvey and Clark, 1965).

2.2. Defended Neighborhood

Defensibility mechanism enables residents to take control over their community, restore their sense of community, and keep their neighborhood stable (Glasze, Webster, and Frantz, 2006). Defensibility takes two forms:

(a) Socially Defensibility: Suttles (1972) defined socially defended neighborhood as a spatial unit that serves day-to-day living needs for its members, who are able to exert formal and informal social controls on local residents and on access of outsiders to the neighborhood. Suttles (1972) and Flanagan (1990) reported that there are four necessary elements that contribute to a successful socially-defended neighborhood: fear, reputation, organization/disorganization, and social control. Such neighborhoods starts with a primary networks that

ties individuals to each other like ethnic villagers who developed a strong social cohesion based on kinship, ethnicity, shared values, homogeneity, and identity (Gans, 1991). The social life of residents in such communities is rich and solid. Residents of such communities develop defense mechanism to protect their own community. They become suspicious of anyone and anything outside their neighborhood and are insulated from life outside (Fischer, 1984).

(b) Physical Defensibility: Newman (1972) argued that defensible space restructure the physical layout of communities to allow residents to control the areas around their homes, including the streets and grounds outside their buildings, and the lobbies and corridors within them. The manipulation of spatial configurations introduces areas controlled by residents themselves, creating a social fabric that protects their community (Al-Homoud and Tassinary, 2004; Newman, 1972). The defensible space concept preserves areas in which local residents can realize their common values and lifestyle (Newman, 1972; Glasze, Webster, and Frantz, 2006). It has the ability to bring people of different backgrounds together in a mutually beneficial way. Defensible space relies on self-help rather than on government intervention (Newman, 1972).

2.3. Community Solidarity

Social solidarity provides a protected sphere for individuals, preventing others from penetrating without permission (Horațiu, 2014; Arnsperger and Varoufakis, 2003). It is a societal characteristic and an individual quality (Durkheim, 1995). Schuermans, Spijkers, and Loopmans (2013) reported three types of solidarity: local, social, and global. At the individual level, it is defined as bonding or bridging relationship among other people, or as feelings of commitment to other people, or as an empathic response to a condition distressing others independently of their personal or social character (Komter, 2005; Arnsperger and Varoufakis, 2003; Durkheim, 1995).

Durkheim (1995) argued that when tension develops, solidarity offers its social-psychological structure, where there becomes a need for a deep feeling of social belonging enabling intimacy, promoting effective exchange and alliance, enabling mutual defense, and facilitating a greater range of social and cultural experience. He suggested the existence of two types of tension: (1) concentration: the tendency to hold onto one's familiar ground, stick with one's own kind, and consolidate the richness of one's heritage, and (2) incorporation: to expand one's horizons, take on as much of the world as one can, and to try to locate the source as economic and communications systems allow to impose on one's home.

Human geographers describe the sympathies and loyalties between underprivileged people involved in struggles against oppression, injustice, or inequality. The Dictionary of Human Geography mentions that race, class, gender, and other differences can provide the basis for solidarities (Johans, 1998). From a spatial perspective, geographers emphasize that such forms of solidarity are generally rooted in a sense of community and a strong feeling of shared territory. While this territory is often linked to the national homeland, it can also be realized at the scale of the city and the neighborhood (Routledge, 2012). Komter (2005) reported two major concepts that make the spatial perspectives of solidarity: social cohesion and community identity.

Social Cohesion: According to Hollinger (2006) social cohesion is multi-dimensional in nature. It is described as one's inclusion of and the participation in the economic, cultural, social, and political life of their community. It is a sense of belonging to society based on an effective enjoyment of citizenship. It is the shared values and commitment in the recognition of a collective goal or good. It is the opposite of conflict over resources between individual and common interests. At this level, social sustainability reveals that the common goal of neighborhood is to raise quality of life.

Community Identity: The term includes political and social roles of individuals. It is to flag expectations about just who will make common cause with whom (Hollinger, 2006). Gotham (1998) pointed that to share an identity with other people is to feel in solidarity with them: to owe them something special and to believe one can count on them in ways that cannot count on others. Identification with one's community of descent is to transform that community into solidarity capable of advancing the interests of the community (Smith, 1990). The more we recognize the contingency of the process of identity/solidarity formation, the more respect we might develop for individual volition in deciding what one's identity is, where one belongs, and how we defend our identity (Hollinger, 2006; Gotham, 1998).

Hollinger (2006) and Korte (2006) define community identity as a cognitive construct that answers the question who the person is. Generally, the link between the individual and the group starts with the idea of the self as a bounded cognitive schema, a sort of implicit identity structuring of complex, rich, affectively charged, interrelated concepts about the self (Hogg and Terry, 2000; Gotham, 1998). Hollinger (2006) reports community identity functions in the assigning of the overall sense of community, social, and political roles of individuals to flag the expectations about just who will make common cause with whom. McMillan and Chavis (1986) state that overall sense of community is a feeling that members have of belonging, of mattering to one another and to the group, and a shared faith that members' needs will be met through their commitment to each other. Wilson and Baldassare (1996) found that the overall sense of community was greater when there was higher satisfaction with the amount of local participation in the community and with privacy in the residential setting. He also found that residents living in higher density and more ethnically diverse cities had lower overall sense of community.

Researchers discussed the relation between community identity and affiliation, claiming that a deep understanding of community identity is primarily about the process of affiliation, while its dynamics of affiliation are programmed by descent as registered by physical characteristics (Gotham, 1998). An individual comprises numerous roles and identities, each of which has its own set of goals. In any given situation, an individual may call into play multiple identities (household member, town resident, CEO, parent, member of religious organization). Even when the various identities may conflict with each other an individual decides which identity is most relevant in a situation. It is the strength of affiliation that someone feels toward other members of a group that can determine which identity that person chooses to apply in a particular situation (Hollinger, 2006; Gotham, 1998). Two types of affiliations are discussed:

Social Affiliation: The concept of affiliation can be translated as acceptance and open social behavior which is characterized by seeking presence of others, interest in cooperation, communication, and social contacts. Affiliation is the base of the human's motive to establish, keep, and develop instrumental social relationship with other people, achieving the necessary goals through their mediation in situations when people are mutually interchangeable. Affiliation motivation is stimulated by causes necessary for human survival; their cognitive picture of the world, tendency to compare self with others and to get feedback from them. In situations of uncertainty and danger, the tendency to cooperate increases in most people representing a precondition of more certain material, informational, and emotional benefit (Gotham, 1998).

Suburban communities differ from urban communities regarding affiliation, which makes the first homogenous and the second heterogeneous, as there are significant degrees of variation in relation to sense of place, familial/social network formation, civic engagement, and collective action. Highly affiliated suburbs exhibit high levels of place identification, are amenable to family life, and are characterized by a significant degree of social embedding and social activism(Hogg and Terry, 2000; Gans, 1991).

Group affiliation may influence whether an individual decides to cooperate in a group decision or not for several reasons: active social goals, reward for individuals, exertion of stronger influence on individuals of group norms (McMillan and Chavis, 1986). Factors that contribute in the process of social affiliation of members in community are: (1) Membership; which is the feeling of belonging or of sharing a sense of personal relatedness and it includes a sense of identification and social investment. (2) Integration and fulfillment of needs; which reinforces the idea that your community needs to solve problem for their members and vice versa. (3) Shared emotional connection; which is the kind of experiences one can manufacture for the members of your community. It is translated through members' expectation and feelings about their community (Hollinger, 2006; Gotham, 1998). According to Hollinger (2006) and McMillan and Chaves (1986), the concept of social affiliation refers to the following: (1) Seeking presence of others: refers to (a) informal affiliations such as one's affiliation with other residents in order to exchange instrumental resources like knowledge, information, and news, as well as socio-emotional resources such as experiencing positive mood and social connection in the presence of others, and (b) formal affiliations such as one's affiliation with local institutions of community whether political, social, or voluntary. (2) Group membership refers to social investment of residents in their time and money to become a member of the formal community organizations and have the right to belong to. (3) Group boundaries: refer to (a) the ability to identify community group members, and (b) the ability to identify and stop outsiders from interfering or threatening the community. Such boundaries provide members with emotional safety necessary for needs and feelings to be exposed and for intimacy to develop while stopping outside threats. (4) Power and influence: refer to the influence of the group on its members while providing them with a space to influence decision-making process. Such sharing of influence and power leads to greater participation, satisfaction, and greater cohesion in the community. (5) Integration and fulfillment of needs refer to the ability of group to bring members together so that they meet others' needs, while meeting their own.

Political Affiliation is defined as a social identity with political relevance, as it typically inherits an identity as part of a group with norms concerning shared political beliefs. There is agreement that political cohesion revolves centrally around social identity, when group identification involves a subjective sense of membership, which takes the form of social identity and perceived common fate. It forms when social identities are transformed into political ones to derive a political affiliation (Hollinger, 2006; McMillan and Chaves, 1986). McMillan and Chavis (1986) argued that political affiliation is defined by empowerment of residents' role, voting behavior social class, and influence. It is the members feeling like they have influence over the community, their voices can be heard, and their beliefs in community leadership. It is defined by their political role; political view point; and community leadership as in the following (Hollinger, 2006; McMillan and Chaves, 1986): (1) Empowerment refers to residents' feel of having (a) political voice, (b) community involvement, and (c) actions that they can themselves such as elections, protest, persuasion, noncooperation, and intervention in order to initiate the needed changes. (2) Political meaning of group membership: refers to residents having a political viewpoint through norms and beliefs that connect group membership to specific political attitudes and actions. (3) Existence of symbolic and realistic threats: refers to residents' feeling a sense of symbolic common fate and shared interests when political threats exist. (4) Effective Community Leadership: refers to residents' belief of having an influential leader who advocate certain beliefs, policy positions, and skills that help him to develop a vision for the community.

3. RESEARCH OBJECTIVES

The general hypothesis of this study states that control of urban sprawl can be enhanced through sense of community identity. Sub hypothesis 1 states that social affiliation of residents affects control of urban sprawl. Sub hypothesis 2 states that political affiliation of residents affects control of urban sprawl. The research method was a quantitative design using a survey for data collection. The study population was Fuhais town residents. The study audiences are planners, municipality decision makers, concerned sociologists, and policy makers.

The survey was carried over three major clusters: the first cluster is Al-Balad which is the old town, characterized by its mixed buildings, churches, residential housing of types "C" and "D", old schools, and historic sites. The second cluster was the new commercial spine which is part of the major corridor associated with Amman (Figure 1). The corridor is considered as the new center for the town with mixed developments. The third cluster was the east part of Fuhais town, characterized by low density residential developments of types "A" and "B" thriving towards Amman.

The main objective of this research is to test community identity as a defense mechanism to control urban sprawl. Additional objectives are: (1) to understand the defense mechanism that are used to control urban sprawl; (2) to highlight the importance of preserving the community identity to control urban sprawl; (3) to provide reference to help decision makers and planners to deal with urban sprawl issues that threatens communities and their identity.

4. METHODS

A quantitative research design was utilized to conduct a survey and collect data about the study variables through a questionnaire. The target population was residents of Fuhais town. The unit of analysis was the head of households and passers-by. The sampling technique was random sampling. The final sample size was 100 subjects. Subjects were selected randomly from three main zones of the town that was indicated in the map earlier. Selection system started from Al-Balad (town center) (cluster 1) by choosing houses alternatively; then moving to the commercial strip (cluster 2) and choosing passers-by randomly, if selected subjects were not residents from the town they were dropped; then moving to the new low density clusters close Amman (cluster 3) by choosing houses alternatively. Subjected were briefed and asked if they are willing to participate. After filling the self-administered questionnaire they were thanked.

4.1. Research Hypothesis

The general hypothesis of the study states that control of urban sprawl can be enhanced through overall community identity. Community identity was defined by social affiliation and political affiliation. The following sub-hypotheses were tested: (1) Social affiliation of residents affects control of urban sprawl. (2) Political affiliation of residents affects control of urban sprawl.

4.2. Constructs and Variables

- **The Independent variable:** Control of Urban Sprawl, which is defined by the ability to control urban sprawl from Amman city to Fuhais City. It is measured by dichotomous scale: (1) No, (2) Yes.
- **The Dependent variable:** Community identity, which is defined by overall sense of community identity. Its components are social and political affiliation of residents with their community during conflicts and uncertainty. They are defined and measured as follows:
- **Variable 1-** Overall Sense of Community Identity: It is defined by the average of the sum of political and social affiliation. It is measured using a 5-point Likert scale: (1) Strongly Disagree, (2) Disagree, (3) Not Sure, (4) Agree, and (5) Strongly Agree.

- **Variable 2-** Social Affiliation: It is defined by the following sub-variables: seeking presence of others (affiliation with others); Interest in cooperation (social role); cognitively consensual picture of the group (identification with members); cognitively consensual picture of the world (Identification with outsiders); communication and social contacts (social investment); and achieving the necessary goals (benefits). Each is measured using a 5-point Likert scale: (1) Strongly Disagree, (2) Disagree, (3) Not Sure, (4) Agree, and (5) Strongly Agree.
- **Variable 3-** Political affiliation: It is defined by the following sub-variables: Political Role; Political View point; and Community Leadership. It is measured using a 5-point Likert scale: (1) Strongly Disagree, (2) Disagree, (3) Not Sure, (4) Agree, and (5) Strongly Agree. The following question is assigned for this variable:
- **Confounding Variables:** The confounding construct of the study is the personal information of the subjects that might influence the results of the study. They included: gender, place of origin, current residence, length of your residence, place of work, ownership, ethnicity, age, education level, and income level. The confounding constructs were measured using a nominal scale.

4.3. The Instrument of the Study

A written structured questionnaire was developed to collect the data. The instrument was administered by the researcher. The instrument included three sections: Demographics: This section was designed to collect data about personal information of the subjects that might influence the results of the study. It will be measured using a nominal scale. Data was collected through questions (Q1-10).

- The Dependent Variable: This section was designed to collect data about ability to control urban sprawl from Amman city to Fuhais City. It is measured by dichotomous scale: (1) No, (2) Yes. Data was collected through question (Q11).
- The Independent Variables: overall sense of community identity, and social and political affiliation of residents with their local community. It is measured using a 5-point Likert scale: (1) Strongly Disagree, (2) Disagree, (3) Not Sure, (4) Agree, and (5) Strongly Agree. Data was collected through questions (Q12-21).

5. RESULTS

5.1. Demographic Analysis

Table 1 shows that about 71% of the sample was male subjects. More than half were originally from Fuhais. Most of the subjects are currently living in Fuhais. About half of the sample has lived more than 20 years in Fuhais. Two thirds of the interviewed subjects work in town. More than half of the subjects own homes. Two-thirds indicated that they are Jordanian. About two thirds of the sample is younger than 50 years. Two-thirds of the subjects got higher education. About half of the sample has less than 500JD, which is the limit for low-income level. About half of the sample indicated that they have control over the sprawl of neighboring capital city over their town, Fuhais.

5.2. Hypotheses Testing

Community Identity attributes that enhance control over urban sprawl by the local community at Fuhais town is tested through the general hypothesis and the sub-hypotheses:

General Hypothesis: control of urban sprawl can be enhanced through overall community identity: A One-way ANOVA test was carried out to test the difference in the mean of scores of control of urban sprawl with the mean of scores of sense of community identity and its components: social affiliation and political affiliation. Results are presented in Table (2). The test indicated a significant effect of control of urban sprawl by sense of community identity and its components in the order of their strength as follows: political affiliation [$F(1, 99) = 22.86$], $P < 0.00$, overall sense of community identity [$F(1, 99) = 18.08$], $P < 0.00$, and social affiliation [$F(1, 99) = 6.58$], $P < 0.01$. This suggests that solidarity is influential, especially when it is based on political identity.

A regression test was carried out for overall sense of community identity and its components as an interactive model to see their influence on control of urban sprawl in Fuhais. The interactive relationship of variables with control of urban sprawl is reported significant in Table (3). This suggests that overall sense of community identity and its components social affiliation and political affiliation all together influence control of urban sprawl in Fuhais town.

However, individual contribution of attributes of sense of community identity to control of urban sprawl in Fuhais town within the interactive model was significant only by overall sense of community identity [$t(1, 99) = 2.45$], $P < 0.016$ and overall political affiliation [$t(1, 99) = 3.52$], $P < 0.00$, Table (4). Also here political affiliation was more powerful.

Table 1. Distribution of the Demographic Variables

		Frequency	Percent	Valid Percent	Cumulative Percent
Gender	Male	71	71.0	71.0	71.0
	Female	29	29.0	29.0	100.0
	Total	100	100.0	100.0	
Origin	Fuhais	55	55.0	55.0	55.0
	Other	45	45.0	45.0	100.0
	Total	100	100.0	100.0	
Current Residence	Fuhais	78	78.0	78.0	78.0
	Other	22	22.0	22.0	99.0
	Total	100	100.0	100.0	
Length of Residence	<5 Years	21	21.0	21.0	21.0
	5-9 Years	19	19.0	19.0	40.0
	10-19 Years	15	15.0	15.0	55.0
	>20 Years	45	45.0	45.0	100.0
	Total	100	100.0	100.0	
Place of Work	Fuhais	69	69.0	69.0	69.0
	Amman	24	24.0	24.0	93.0
	Other	7	7.0	7.0	100.0
	Total	100	100.0	100.0	
Ownership	Owner	58	58.0	58.0	58.0
	Renter	42	42.0	42.0	100.0
	Total	100	100.0	100.0	
Ethnicity	Jordanian	75	75.0	75.0	75.0
	Other	25	25.0	25.0	100.0
	Total	100	100.0	100.0	
Age	<20 Years	6	6.0	6.0	6.0
	20-29 Years	28	28.0	28.0	34.0
	30-39 Years	26	26.0	26.0	60.0
	40-49 Years	17	17.0	17.0	77.0
	50-59 Years	13	13.0	13.0	90.0
	>60 Years	10	10.0	10.0	100.0
	Total	100	100.0	100.0	
Education	High School	33	33.0	33.0	33.0
	Bachelor	40	40.0	40.0	73.0
	Higher Education	14	14.0	14.0	87.0
	Other	13	13.0	13.0	100.0
	Total	100	100.0	100.0	
Income	<200 JD	15	15.0	15.0	15.0
	200-499 JD	41	41.0	41.0	56.0
	500-749 JD	15	15.0	15.0	71.0
	750-1499 JD	16	16.0	16.0	87.0

	1500-2999 JD	10	10.0	10.0	97.0
	>3000 JD	3	3.0	3.0	100.0
	Total	100	100.0	100.0	
Control of Urban Sprawl	No	55	55.0	55.0	55.0
	Yes	45	45.0	45.0	100.0
	Total	100	100.0	100.0	

Table 2. ANOVA of Overall Sense of Community Identity and its Components with Control of Urban Sprawl

		Sum of Squares	Df	Mean Square	F	Sig.
Overall Sense of Community Identity	Between Groups	37.83	1	37.83	18.08	0.00
	Within Groups	205.13	98	2.09		
	Total	242.96	99			
Social Affiliation	Between Groups	4.95	1	4.95	6.58	0.01
	Within Groups	73.71	98	0.75		
	Total	78.66	99			
Political Affiliation	Between Groups	22.50	1	22.50	22.86	0.00
	Within Groups	96.46	98	0.98		
	Total	118.96	99			
Social Affiliation Components						
Affiliation With Others	Between Groups	1.63	1	1.63	0.54	0.47
	Within Groups	297.68	98	3.04		
	Total	299.31	99			
Social Role	Between Groups	8.85	1	8.85	5.84	0.02
	Within Groups	148.59	98	1.51		
	Total	157.44	99			
Identification with Members	Between Groups	0.15	1	0.15	0.13	0.72
	Within Groups	113.81	98	1.16		
	Total	113.96	99			
Identification with Outsiders	Between Groups	3.17	1	3.17	2.79	0.09
	Within Groups	111.35	98	1.14		
	Total	114.51	99			
Social Investment	Between Groups	9.15	1	9.15	6.27	0.01
	Within Groups	143.04	98	1.46		
	Total	152.19	99			
Benefits	Between Groups	15.29	1	15.29	13.61	0.00
	Within Groups	110.11	98	1.12		
	Total	125.39	99			

Political Affiliation Components						
Political Role	Between Groups	32.36	1	32.36	19.71	0.00
	Within Groups	160.88	98	1.64		
	Total	193.24	99			
Political Viewpoint	Between Groups	28.91	1	28.91	13.25	0.00
	Within Groups	213.84	98	2.18		
	Total	242.75	99			
Community Leadership	Between Groups	10.02	1	10.02	5.69	0.02
	Within Groups	172.73	98	1.76		
	Total	182.75	99			

Table 3. Regression Model Summary - Overall Sense of Community Identity and its Components with Control of Urban Sprawl

Model	Sum of Squares	Df	Mean Square	F	Sig.
Overall Sense of Community Identity and its Components					
Regression	6.25	3	2.06	10.82	0.00
Residual	18.50	96	0.19		
Total	24.75	99			
Social Affiliation Components					
Regression	3.42	6	0.57	2.49	0.03
Residual	21.33	93	0.23		
Total	24.75	99			
Political Affiliation Components					
Regression	5.14	3	1.71	8.38	0.00
Residual	19.61	96	0.20		
Total	24.75	99			

Table 4. Coefficients for Regression Model Summary - Overall Sense of Community Identity and its Components with Control of Urban Sprawl

Model	Unstandardized Coefficients		Standardized Coefficients		
	B	Std. Error	Beta	T	Sig.
Overall Sense of Community Identity and its Components					
(Constant)	0.74	0.22		3.31	0.00
Overall Sense of Community Identity	0.10	0.04	0.31	2.45	0.02
Overall Social Affiliation	-0.03	0.07	-0.06	-0.47	0.64
Overall Political Affiliation	0.15	0.04	0.34	3.52	0.00
Social Affiliation Components					
(Constant)	0.73	0.25		2.90	0.01
Affiliation With Others	-0.03	0.03	-0.10	-0.84	0.40
Social Role	0.05	0.05	0.12	0.96	0.34
Identification with Members	-0.02	0.06	-0.03	-0.25	0.81
Identification with Outsiders	0.02	0.07	0.04	0.28	0.78
Social Investment	0.01	0.06	0.02	0.13	0.90
Benefits	0.14	0.06	0.31	2.39	0.02
Political Affiliation Components					
(Constant)	0.87	0.15		5.68	0.00
Political Role	0.12	0.04	0.32	3.09	0.00
Political Viewpoint	0.05	0.04	0.14	1.24	0.22
Community Leadership	0.04	0.04	0.12	1.19	0.24

Sub-Hypothesis 1: *Social affiliation of residents affects* control of urban sprawl: A One way ANOVA test was carried out to test the difference in the mean of scores of control of urban sprawl with the mean of scores of social affiliation components: affiliation with others; social role; identification with members; identification with outsiders; social investment; and benefits. Results are presented in Table (2). The test indicated a significant effect of control of urban sprawl by social affiliation components in the order of their strength as follows: benefits [F (1, 99) = 13.61], P < 0.00, social investment [F (1, 99) = 6.27], P < 0.01, and social role [F (1, 99) = 5.85], P < 0.01. Meanwhile, affiliation with others, identification with members, and identification with outsiders showed no effect. This suggests that their social influence on control of sprawl occurs when they have proactive gains. A regression test was carried out for social affiliation components as an interactive model to see their influence on control of urban sprawl in Fuhais. The interactive relationship of variables with control of urban sprawl is reported significant in Table (3). This suggests that social affiliation components: affiliation with others, social role, identification with members, identification with outsiders, social investment, and benefits together influence control of urban sprawl in Fuhais town. However, individual contribution of attributes of social affiliation components to control of urban sprawl in Fuhais within the interactive model was significant only by benefits [t (1, 99) = 2.39], P < 0.01, Table (4). This suggests how powerful it can be to gain benefits from the local community.

Sub-Hypothesis 2: Political affiliation of residents affects control of urban sprawl: A One way ANOVA test was carried out to test the difference in the mean of scores of control of urban sprawl with the mean of scores of political affiliation components: political role; political viewpoint; and community leadership. Results are presented in Table (2). The test indicated a significant effect of control of urban sprawl by all political affiliation attributes in the order of their strength as follows: political role [F (1, 99) = 19.71], $P < 0.00$, political viewpoint [F (1, 99) = 13.25], $P < 0.01$, and community leadership [F (1, 99) = 5.69], $P < 0.01$. The strongest of which was political role. This suggests that political involvement at the local community empowers residents to defend their communities from urban sprawl and its impact. It seems when they are politically involved they contribute more.

A regression test was carried out for political affiliation components as an interactive model to see their influence on control of urban sprawl in Fuhais. The interactive relationship of variables with control of urban sprawl is reported significant in Table 3. This suggests that political affiliation and its components political role, political viewpoint, and community leadership together influence control of urban sprawl in Fuhais town. However, individual contribution of attributes of political affiliation and its components to control of urban sprawl in Fuhais within the interactive model was significant only by political role [t (1, 99) = 3.09], $p < 0.00$, Table 4. This suggests also the empowerment of local communities can be enhanced by proactive political stands.

Demographic Characteristics Test: A Chi-square test was carried out to investigate the individual association of control of urban sprawl with the demographic characteristics: gender, origin, current residence, length of residence, place of work, ownership, ethnicity, age, education, and income. Table 5 shows a significant association of control of urban sprawl with length of residence [χ^2 (3, 99) = 14.54, P = 0.00] and origin [χ^2 (1, 99) = 4.5, P = 0.03]. It makes sense that the more people reside in a place, the more they become attached, defensible, and protective to their community.

A regression test was carried out to test control of urban sprawl with the demographic characteristics as an interactive model. The interactive relationship of the set of variables with control of urban sprawl is reported none significant. Also none of the sub-components contributed individually to control of urban sprawl at Fuhais Town within the interactive model.

6. CONCLUSIONS

For the model of control of urban sprawl to be enhanced through overall community identity, the following relationships must be confirmed:

- Attributes that affect control of urban sprawl in the order of their strength: political affiliation, overall sense of community identity, and social affiliation.
- In addition, the interactive relationship of the set of variables with control of urban sprawl was reported significant. Therefore, overall sense of community identity and its components social affiliation and political affiliation together influence control of urban sprawl in Fuhais town.
- The contribution of each attribute to control of urban sprawl in Fuhais town was significant only by overall sense of community identity and overall political affiliation.

For the model of social affiliation of residents affects control of urban sprawl:

- Attributes that affect control of urban sprawl by social affiliation components in the order of their strength as follows: benefits, social investment, and social role. Meanwhile, affiliation with others, identification with members, and identification with outsiders showed no effect.
- In addition, the interactive relationship of the set of variables with control of urban sprawl was reported significant. Therefore, social affiliation components:

Table 5. Pearson Chi-Square Tests - Control of Urban Sprawl with Demographic Characteristics

	Pearson Chi-Square	Df	Asymp. Sig. (2-sided)
Gender	0.00	1	0.98
Origin	4.50	1	0.03
Current Residence	5.82	2	0.05
Length of Residence	14.54	3	0.00
Place of Work	1.02	2	0.60
Ownership	2.52	1	0.11
Ethnicity	1.09	1	0.30
Age	10.07	5	0.07
Education	6.87	3	0.08
Income	7.64	5	0.18
N of Valid Cases	100		

affiliation with others, social role, identification with members, identification with outsiders, social investment, and benefits all together influence control of urban sprawl in Fuhais town.

- The contribution of each attribute to control of urban sprawl in Fuhais within the interactive model was significant only by benefits.

For the model of political affiliation of residents affects control of urban sprawl:

- Attributes that affect control of urban sprawl by political affiliation components in the order of their strength as follows: political role, political viewpoint, and community leadership.
- In addition, the interactive relationship of the set of variables with control of urban sprawl was reported significant. Therefore, political affiliation components political role, political viewpoint, and community leadership together influence control of urban sprawl in Fuhais town.
- The contribution of each attribute to control of urban sprawl in Fuhais within the interactive model was significant only by political role.

For the demographic characteristics relationship control of urban sprawl:

- Results indicate a significant association of control of urban sprawl only with length of residence. It showed no relationship with gender, origin, current residence, place of work, ownership, ethnicity, age, education, and income. Also there was no interactive relationship of demographic attributes with control of urban sprawl. Further, none of the attributes contributed individually to control of urban sprawl at Fuhais Town within the interactive model.

Fuhais town, like the other towns in Jordan has witnessed a continuation of urban sprawl that threatens its solidarity and affects its community identity and cohesion. In order to reach suitable solutions to the gowth of such local community in harmony with its identity, this research built on the existing situation the suggestion to take into consideration community engagement, listening to residents' opinions, needs, and their proposed solutions in any urban or planning developments. Such communities provide haven for its own residents, and it is worth to protect their seclusive urban growth away from the growth of the adjacent urban centers.

The study has some limitations; the researcher faced many defensible residents who were cautious and not willing to participate.

Future studies might want to incorporate questions on the socio-spatial integration of Fuhais town with the surrounding communities at the regional level, such as Mahis and Al-Salt. During the field survey, the researcher heard many side stories by the elderly who witnessed the process of Fuhais growth, such stories have the potential to be researched in an expanded ethnographic study. Comparative studies with other similar communal towns in Jordan that witness urban sprawl should be conducted. Finally, studies on the relationship between the socio-cultural context and the spatial configuration of settings in Fuhais town could be carried out using GIS mapping.

Recommendations: This study provides decision makers and planners with useful evidence to help them deal with the issue of urban sprawl when it threatens solidarities and community's resources. Based on the study outcomes, recommendations for planners and officials are presented as follows:

- **Officials of Fuhais Municipality:** Fuhais Municipality officials are advised to: (1) involve multiple stakeholders in the urban planning processes, including the public and private sectors, community's key residents, key business owners in order to address the needs and priorities of different stakeholders and allow better quality decision. (2) Establish investment strategy, which prioritizes infrastructure, and involves investing in Fuhais commercial corridor, residential clusters, and social services to ensure local residents' prosperity and job opportunities. (3) Establish restricting legislations to limit the scattered settlements at the edge of community. (4) Revise the existing zoning and regulations to ensure that the development vision is on the right track.
- **Planners:** (1) Develop special policies related to the management of land use for local communities and their surroundings, taking into account the current and future expansion, and establishing control legislations. (2) Redefine town's density by focusing on making better use of Fuhais lands with policies that enable more houses to be built in a given area to accommodate the town's rising population. This strategy could deliver many benefits to local residents, better and more diverse local services, and improve social and transport infrastructure. (3) Integrate urban agriculture as a vital tool in planning for agricultural communities. (4) Adopt a green belt around the town to control urban growth. This tool will enhance the aesthetic value of the environment and slow down the rate of spatial expansion as well.

REFERENCES

Abu-Dayyeh, N. (1997). Village Air for Urban Elites. Expedition Magazine, 44(2), 10-17.

Al-Asad, M. (2004). Center for the Study of Built Environment, Retrieved May 15, 2015, from CSBE: http://www.csbe.org/publications-and-resources/urban-crossroads/urban-sprawl/

Al-Homoud, M., &Tassinary, L. (2004). Social Interactions at the Neighborhood Level as a Function of External Space Enclosure. Journal of Architecture and Planning Research, 21(1), 10-23.

Altinok, E., & Cengiz, H. (2008). The Effects of Urban Sprawl on Spatial Fragmentation and Social Segregation in Istanbul, Conference on 'Urban Growth without Sprawl: A Way Towards Sustainable Urbanization', Istanbul, Turkey, September, 8, 2008, 19-23.

Arnsperger, Ch., & Varoufakis, Y. (2003). Toward a Theory of Solidarity. Erkenntnis, 157-188.

Bekele, H. (2005). Urbanization and Urban Sprawl, A Master Thesis in Masters of Science in Geography, Virginia: Virginia Polytechnic Institute and State University.

Brueckner, J. (2000). Urban Sprawl, Diagnosis and Remedies. International Regional Science Review, 23(2), 160-171.

Daher, R. (2013). Neoliberal Urban Transformations in the Arab City: Meta-narratives, Urban Disparities and the Emergence of Consumerist Utopias and Geographies of Inequalities in Amman. Environmental Urban, Urban Environment, 7(3), 99-115.

Downs, A. (1999). Some Realities about Sprawl and Urban Decline. Housing Policy Debate, 10(4), 955-974.

Durkheim, E. (1995). The Elementary Forms of the Religious Life. New York: Free Press, 45-72.

European Environment Agency (EEA). (2006). Urban Sprawl in Europe - The Ignored Challenge, A Report on 'Urban Sprawl', Luxembourg: European Commission.

Ewing, R. (1994). Characteristics, Causes and Effects of Sprawl: A Literature Revie, Environmnetal and Urban Issues, FAU/FIU Joint Center, Washington D.C.

Fischer, C. S. (1984). To Dwell among Friends: Personal Networks in Town and City. Chicago: University of Chicago Press.

Flanagan, W. G. (1990). Urban Sociology: Images and Structure. USA: Rowman & Littlefield.

Forsyth, A. (2012). Defining Suburbs. Journal of Planning, 27(3), 270-281.

Franz, G., Maier, G., & Schröck, P. (2006). Urban Sprawl: How useful is this Concept? Journal of Economic Perspectives, 18(4), 177-200.

Freilich, R. H. (1999). From Sprawl to Smart Growth-Successful Legal, Planning, and Environmental Systems. Planning Practice and Research, 16(3), 271-279.

Gans, H. (1991). The Levittowners: Ways of Life and Politics in a New Suburban Community. New York: Pantheon Books.

Ghani, A., Abidin, Z., & Khalid, A. (2014). Urban Sprawl Shape Description. Malaysian Journal of Computing, 2(1), 27-35.

Glaster, G., Hanson, R., Ratcliffc, M., Wolman, H., Coleman, S., & Freihage, J. (2001). Wrestling Sprawl to the Ground: Defining and Measuring an Elusive Concept. Housing Policy Debate, 12(4), 681-718.

Glasze, G., Webster, Ch., & Frantz, K. (2006). Private Cities Global and Local Perspectives. USA : Routledge.

Gober, P., & Behr, M. (1982). Central Cities and Suburbs as Distinct Place Types: Myth or Fact? Economic Geography, 58(4), 371-385.

Gotham, K. F. (1998). Political Opportunity, Community Identity, and the Emergence of a Local Anti-Expressway Movement. Oxford: Oxford University.

Harvey, R. O., & Clark, O. W. (1965). The Nature and Economics of Urban Sprawl. Land Economics, 9-41.

Hogg, M., & Terry, D. (2000). Social Identity and Self-categorization Processes in Organizational Contexts. Academy of Management Review, 25(1), 121-140.

Hollinger, D. A. (2006). From Identity to Solidarity. The American Academy of Arts, 23-31.

Horațiu, R. (2014). Social Change Review: Measuring Social Solidarity. Some Research Notes. The Journal of Lucian Blaga University of Sibiu, 71-90.

Johans, R. A. (1998). Bridging the Gap between Class and Space: US Worker Solidarity with Guatemala. Economic Geography, 252-271.

Knox. P. (1987). Urban Social Geography: An Introduction. Liverpool University Press, 54(1), 85-129.

Komter, A. E. (2005). Social Solidarity and the Gift. Cambridge: Cambridge University, 189-225.

Korte, R. F. (1977). A Review of Social Identity Theory with Implications for Training and Development. Journal of European Industrial Training, 31(3), 166-180.

Luther, T. (2005). Private Forest, Retrieved May 15, 2015, from privateforest.org: Private Forest, http://www.privateforest.org/forest101/sprawl.htm

Makhamreha, Z., & Almanasyeha, N. (2011). Analyzing the State and Pattern of Urban Growth and City Planning in Amman Using Satellite Images and GIS. European Journal of Social Sciences, 24(2), 252-264.

McMillan, D., & Chavis, D. (1986). Sense of Community: A Definition and Theory. Journal of Community Psychology, 6-20.

Newman, O. (1972). Creating Defensible Space. USA: U.S. Department of Housing and Urban Development.

Routledge, P. (2012). Sensuous Solidarity: Emotion, Politics and Performance. In: The Clandestine Insurgent Rebel Clown Army, 28-55.

Schuermans, N., Spijkers, F., & Loopmans, M. (2013). Solidarity in Human Geography: Responsibility, Care, Place and Encounter. USA: DieGem.

Small, K. (2000). Urban Sprawl: A Non-Diagnosos of Real Problems. In: Metropolitan Development Patterns, Lincoln Institute of Land Policy, Annual Roundtable: Cambridge, MA.

Smith, T. (1990). Metropolis on the Geographers Look at Urban Sprawl. New York: John Wiley and Sons.

Suttles, G. D. (1972). The Social Construction of Communities. The University of Chicago Press, 142-180.

Sweis, M. (2011). Fuhais Town is in Danger, Retrieved on September 11, 2011 from Jalaad.com : http://www.jalaad.com/index.php?module=participations&id=1929&category

Taylor, R. B., &Harrell, A. (1996). Physical Environment and Crime, NIJ Research Report, Washington: National Institute of Justice.

Tewfik, M., & Amr, A. (2015). Growth of Urban Peripheries with Reference to Inconsistent Spatial Planning Policies: South-east Amman as Case Study. European International Journal of Science and Technology, 4(3), 50-65.

Wassmer, R. (2000). An Economic Perspective on Urban Sprawl, Working Paper for California Senate Office of Research, 1-21.

Wilson, G., & Baldassare, M. (1996). Overall "Sense Of Community" In A Suburban Region: The Effects of Localism, Privacy, and Urbanization. Environment and Behavior, 27-43.

E6 - Transform Processes and Outcomes for Equity: POE/ Programming

Using Post-Occupancy Evaluation to Analyze the Effective Use of Photovoice and Focus Groups in Developing a Successful Program for an International Service Learning Design Project

David Boeck (Univeristy of Oklahoma)

1. ABSTRACT

Architectural programming has always been an integral part of building design and design education. Before the Pyramids or the Parthenon or the Pantheon were built, someone gathered information about what would go into designing and building that project. Architectural programming has transformed over the centuries but has always maintained its importance in helping the project owners and architects to understand the scope of the project and to know what needs to be included in its development from plan and use to design detailing and materiality.

In developing an architectural program for an other-culture educational studio project, the author introduced photovoice and focus group methodologies to help students better understand how important cultural influences are in the success of an educational facility design.

1.1. Key Words

Programming, Photovoice, Focus Groups Education

2. PROJECT BACKGROUND

As a practicing architect who focused on educational projects in the first 15 years of my practice, the importance of programming was paramount in developing successful educational projects. Interviews, in tandem with reviewing state and local building codes, we thought, ensured the success of our projects. And for the most part it seemed that we had a successful formula for developing these programs.

For the last fifteen yeas as an Associate Professor in the College of Architecture at the University of Oklahoma I have been actively involved in teaching design studio at all levels. In each of these courses, design programming was a given for each design project. None of these programming activities used any participatory action research methodologies to help with programming.

In 2014-2016 I took students from the College of Architecture to Zambia to participate in a service learning course, working with an American ministry -Family Legacy Ministries International (FLMI)-to develop educational and living facilities for homeless and poor children in the capitol, Lusaka. The service learning project developed in 2015, a prototypical primary school design for up to 1,000 students, was built in 2017 and occupied in 2018.

A post occupancy evaluation was implemented in the fall of 2018 to determine how successful the programming was in developing this new primary school concept, and to evaluate how the program and compared to the desired outcome in supporting the educational goals.

3. DEVELOPING THE PROGRAMMING PROCESS

All architects and architecture students understand how important programming is in ensuring the development of a quality building design that will work, incorporating the specific functional requirements of each project. Successful design happens when the architect's program takes into consideration all aspects of the building to be designed-budget, code, use, structure, size, environmental considerations, systems, building materials, finishes, construction schedule, long-term maintenance (Cherry and Petronis 2016)

In the university setting, architecture students are given a pre-written program with the information needed to develop a successful building design. Every studio project includes a program to organize the project and allow the students to engage properly in the design process. They are given the size of the building, and the functions of the spaces and are required to figure out the siting, the code issues, environmental issues and connections, orientation, views, systems layouts, accessibility issues, materiality and form. There are no interviews with clients or owners at the early levels. That information is given. As said, programming is the most important part of the predesign process. The design process will not be completed in a holistic manner without adequate information as to what the building is about. As students move through the studio sequence, the projects become larger and more complicated. In each case, the first activity is reviewing the program, and as they mature, actually engage in writing at least part of the program.

After visiting Lusaka to evaluate the potential projects, it was evident that teaching our students the process of programming used in the U.S. would not adequately prepare them to develop a building program that would help Family Legacy Ministries International (FLMI) in designing and building the kind of schools they needed. With experience in using photovoice and conducting focus groups, we decided to incorporate these methodologies into the course to help the student team develop an understanding of the culture, the communities, the neighborhoods, the school system, the pupils, and the teachers and the administrators, and prepare them to apply what we learned to developing an appropriate program to meet their needs.

4. HISTORIC BACKGROUND OF ZAMBIA

Zambia was been a British Colony until 1964, when it gained its independence. The original city of Lusaka , the capitol of Zambia, , was built for a population of 150,000 mostly British residents. As was typical in the African colonies, cities were built for the British-not the Africans-and Lusaka was no different. The service jobs that were performed by Zambians did not include provisions for housing. Africans who worked in the city were forced to live on the outskirts of town. By the time the country gained independence, Africans were allowed to live in the city but mostly in the slums or compounds as they are commonly referred to in Zambia. These compounds were developed on fringe properties owned by wealthy individuals to house the people coming into the city to find work (Mwakikagile 2010).

4.1. THE HISTORY OF FAMILY LEGACY MINISTRIES INTERNATIONAL

Family Legacy Ministries International (FLMI) grew out of a desire to provide quality education for the children of the compounds of Lusaka, whose families cannot afford to invest in a quality education. FLMI has set up schools in existing buildings, renovated existing schools acquired from other ministries, and has constructed new schools on land they purchased. With the help of donors, Family Legacy has been able to cloth, feed, and provide books, learning supplies, and equipment for the compound kids and the schools to use.

Approximately 75% of the Lusakan population living in the compounds is considered unemployed by the Zambian government. For the formally unemployed and undereducated, the inability to pay for uniforms and books means that those children will not attend governments schools. Because of this, private church-based schools have proliferated in the compounds, where the churches find sponsors to pay for the uniforms, books and meals.

Over the last 14 years Family Legacy Ministries International as developed 20 primary schools with total enrollments of around 20,000 students. Although they have been very successful in developing a strong educational system in Lusaka, they have been plagued with poor programming and design, substandard construction and numerous maintenance issues.

5. USING PHOTOVOICE IN DEVELOPING A PRIMARY SCHOOL PROGRAM FOR FAMILY LEGACY MINISTRIES INTERNATIONAL

Photovoice is a participatory action research method which enables people to record their community's strengths and concerns through photographs and narrative; to promote critical dialogue of personal and community issues through large and small group discussions based on the photographs; and to be used as a tool to reach policy makers (Wang and Burris 1997). Photovoice is a qualitative method used for community-based participatory research to document and reflect reality. (Nykiforuk et al. 2011) It is an empowering and flexible process that combines photography with grassroots social action and is commonly used in the fields of community development, international development, public health, and education.(Strack et al. 2004) Participants include community members of all ages and status including those who are discriminated against due to language, gender, race, class, disability, etc. (Nykiforuk et al. 2011) These narratives are then used to promote dialogue to mobilize and help change-makers (i.e. policymakers) better understand and change the community, and develop effective solutions and programs that address the issues and needs (Wang 2003).

The idea of using photovoice as a research method to get a better understanding of the needs of schools in Zambia emerged from research done 10 years ago in Norman when faculty investigated the community's ability to support aging in place. By using photovoice and focus groups, researchers were able to give these seniors a voice in examining what was age friendly and what was perceived as a barrier to their independence.

This understanding of photovoice informed the

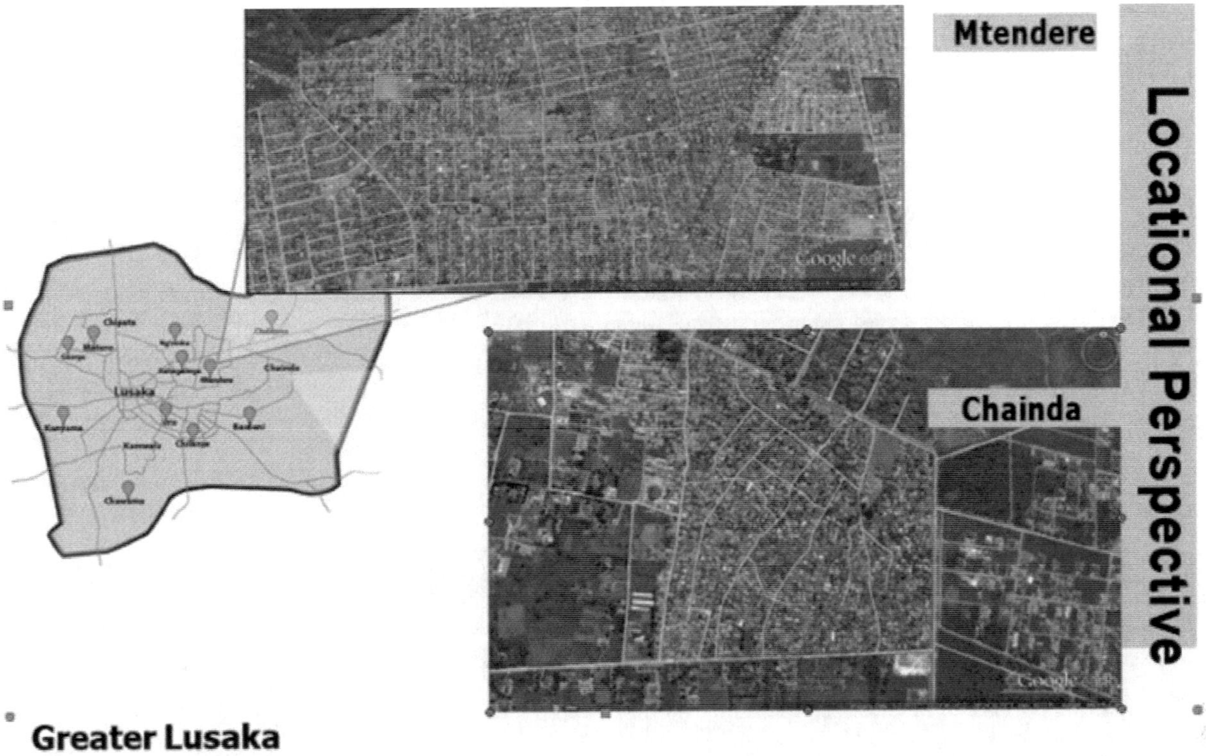

Figure 1. Locations of the compounds where Family Legacy has developed their primary schools

development of the service learning courses we used in Zambia. It facilitated the gathering of information and cultivation of an understanding of the local students we would be designing for, the culture these schools were to be developed in, and the education system that is influenced by this immediate cultural experience. We visited the neighborhoods or "compounds" where the Family Legacy students lived in an effort to bring back photos of the respondents' favorite places, social spots, and least favorite place or places they were afraid to go. We would use the photos and narration from the respondents to begin to identify key assets and challenges within the student's environment and see how these would influence the school design and programming. This qualitative research and student empowerment would provide a powerful groundwork for the expansion of Family Legacy support structure and personal development for the students.

In the process of developing the course, we looked at photovoice and what it was we wanted to accomplish using it in helping us develop the program for the 3 project. We saw that the greatest gap in understanding the educational system FLMI had developed and in meeting the needs of the Zambian students they were educating, was understanding their culture and social system. We came up with 3 questions we needed them to answer as we walked around their communities with camera and notebook in:

1) Show us and tell us about the places or things that help you in your day to day life. Show us places or things that are challenging in your day to day life.

2) Show us and tell us about the places where you and your family rely to support themselves
(This is a place where they can make money or buy things they need for their daily life).

3) Show us and tell us about the places where you spend time doing social activities

Out of this photovoice exercise we were able to determine what are called Community Assets and Community Problems. The Community Assets (CA) include the churches, homes, and schools (figures 2-7) and the Community Problems (CP) included the bars, homes and schools shown in figures 8-12.

Figure 2 (Left), 3 (Right). **Church**- Churches were considered CA because they are places to meet and cultivate important and supportive relationships.

Figure 4 (Left), 5 (Right). Homes are considered CA because they are the places where they can connect with friends and family.

Figure 6 (Left), 7 (Right). Schools are considered CA because this is a social place where they learn and meet friends- some even considered the library as the most important school place.

Figure 8 (Left), 9 (Right). Bars are considered CP because of their proximity to homes and schools and markets. The noise they produce makes it hard to study and the students are harassed by people at the bar while walking to school and home.

Figure 10 (Left), 11 (Right). Homes are considered CP because they are small and crowded. Many families have 6-8 family members in a 2 room small structure making it hard to study. Homes are built poorly and often deteriorate in the rainy season and fall down. There are no codes and require the manufacturing of consistent quality of building materials.

Figure 12 (Left), 13 (Right). School are considered CP because not all children can afford to go to schools so there is conflict between those in school and those not going. There is a lack of educational programs not only for children but also for their parents and other adults in the communities. Often the schools do not have the spaces for active learning like computers and home economics and science or technology classes.

6. PHOTOVOICE AND HOW ITS USED IN THIS PROJECT

The Zambia program was developed with the intent to produce several design projects over a 3-year period with a multidisciplinary team of students from architecture, landscape architecture and, regional and city planning. The faculty, including the author and a faculty member from Regional and City Planning, decided to incorporate these methodologies into the planning process. In 2015 we were asked to develop a prototypical primary school project with flexibility to be placed on different sized site. The course structure was organized around the use of photovoice, and focus-group design charrettes to gather the information needed to develop a program that would address all the needs of the children from the compounds where the school was going to be built, the teachers, the administrators, and the community coordinators (PV-OU- COA 2015).

The primary school project for 2015 was chosen as the focus for evaluation the photovoice and focus group's use in design programming, because it was the only project that had been built.

7. FOCUS GROUPS AND DESIGN CHARRETTES

The idea and structure of focus groups came about as social science researchers looked for ways to engage the locals in the conversation and the development of solutions to the issues being evaluated (Greenbaum 2000). By making the primary school student population, teachers, and administrators, a part of a collaborative conversation, a wider variety of solution ideas could be generated. Through the use of tours within both schools and compounds, focus groups, as well as design charrettes, the University of Oklahoma team began deliberation on the most efficient use of limited land and resources to give the students of Lusaka the best education possible (DP-OU-COA 2015).

The programming report written at the conclusion of the service-learning project was developed around the ideas that we needed to address the following:

- Address the layout concerns noted through meetings with students, teachers and Family Legacy Staff
- Provide a framework and template for future Family Legacy schools
- Offer solutions on the challenges which are not conducive to the learning environment.

Figure 14 (Left), 15 (Right). Photos of the Focus Group Session

Figure 16 (Left), 17 (Right). Photos of the Focus Group Session

The team identified the following structural and environment issues that need to be addressed in the building program:

1) Classrooms: We would need to address the conditions of the classrooms-this is the most important unit of the school layout directly effecting the learning process. Our programming investigation brought out the importance of club and extra-curricular activity space needs as well as technology and shop and home economics and science labs. This included school day programs as well as after school activities.

2) Exterior Activity Spaces: In discussing outdoor spaces with local pupils, teachers, and administrators we found out that there is a primary need for multi-use, covered space not connected to the classroom area, for instruction, assembly, eating, and recreation.

3) Administration Block: The administration block will include offices, work area, lounge and teacher dining, secured storage for testing materials as required by the government, and toilets.

4) Community Outreach and Development Block:
This area will house all community programs, counseling, outreach including staff and space for after-school and weekend outreach, as well as continuing education.

5) Toilets: There is a need for durable and secure toilet areas for the students, as current toilet areas are require significant and frequent repairs and replacements.

6) Environmental Requirements, including. natural light, moisture control and ventilation, acoustical issues and sound control, site security, water management and recycling.

7) Infrastructure: Plumbing and water management and use.

The design charrettes became a part of the program development as a natural outcome of the Photovoice and Focus Group processes. In discussing the input from these methods, it seemed that allowing the pupils as well as teachers to be involved in design process, graphically developing aspects of the school they saw as important, would help us to develop the program considering how important the cultural input was coming out of the photovoice activities (PV-OU-COA 2015).

8. THE PRIMARY SCHOOL PROGRAM

Armed with information gathered during the photovoice, focus group, and design charrette activities, the Oklahoma team started developing a program by setting up a hierarchy of design issues and elements which included the proximity matrix (Figure 19-21). The prototypical floor plan developed was organized around the data collect from the photovoice, focus group and design charrette activities carried out in face to face meetings at different schools (PV-OU-COA 2015).

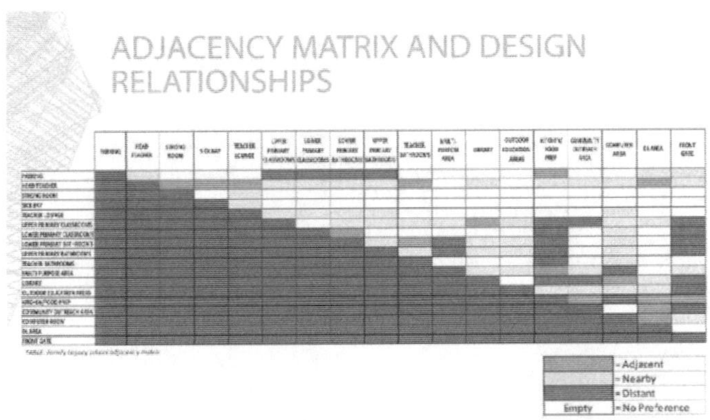

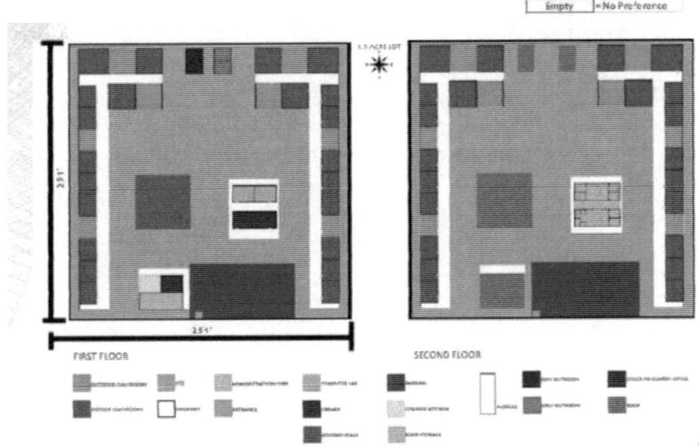

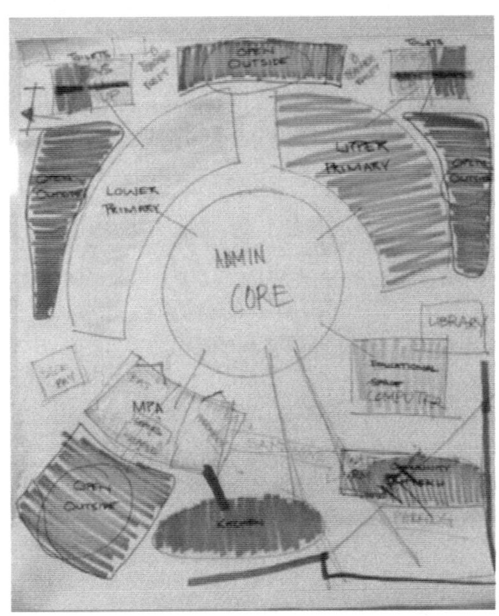

Figure 19, 20, 21. Photos of the program drawings of proximity matrix and plan.

9. POST OCCUPANCY EVALUATION PROCESS

The importance of focusing on the primary school project in this discussion of the value of using photovoice and focus groups in the development of educational facility programming is that this project is the only one to have been built. To determine how successful our process was in improving the programming and therefore the design process, I returned to Lusaka in October 2018 to evaluate the new school using a post occupancy evaluation methodology.

The goal for the anticipated focus group discussions was to use a more traditional facility evaluation process- Post Occupancy Evaluation(POE)- to compare the work done two years earlier and the resultant building configuration developed by FLMI and their architect for the Chainda school. Through the POE format I would be able to look at aspects of environmental concerns, building function concerns, safety concerns, and special interaction issues at both schools. In the most common definition of the process, "A POE is generally defined as the process of systematically evaluating the degree to which occupied buildings meet user needs and organizational goals. Post Occupancy Evaluations for educational facilities generally cover items, which are organized in six areas: the school site, structural and mechanical features·, plant maintainability, school building safety and security, educational adequacy and environment for learning" (Sanoff et. Al. 1977).

This study aimed to produce a comprehensive evaluation of the program overall, tying the outcomes of the 2015 primary school designed during the service learning course to understand how the Photovoice and Focus Group processes influenced and affected the outcome of the primary school design. The data collection and analysis took two weeks during October 2018, during which I spent a number of days with students, teachers, and administrators as well as Family Legacy executive staff.

To get a good understanding of how our process worked, we set up these interviews not only with the Chainda school but also with those groups at Bauleni, another FLMI primary school complete about two years earlier. It is the most current new school programmed and designed before the service learning course activities in 2015. It is also a two story school constructed with the same materials as Chainda. The project was developed by a local architect who had been involved in a number of their school projects using traditional programming activities. The focus group activities included using traditional programming methodologies to gather information determining the schools overall function (Colucci 2007).

The form used to structure the focus group discussion with the students, teachers and administrators is included below:

9.1. POST OCCUPANCY EVALUATION OF THE NEW CHAINDA AND EXISTING BAULEINI PRIMARY SCHOOLS

To evaluate the new primary school effectively we need to consider the following elements as they relate to the facilities where they taught and studied before moving into this new school

1. Durability

How has the new construction improved the durability of materials and their ability to withstand wear and tear.

- Roofing system
- Walls-Interior and exterior
- Doors and windows
- Paving
- Plumbing
- Electrical

2. Systems

How has the new construction improved the function of the following systems

- Ventilation
- Lighting- natural and artificial
- Sound control/acoustics
- Comfort
- Durability and continued function

3. Building Function

How has the design and construction of the new school improved the following?

- School room function
- Circulation
- Toilets
- Security

4. Grounds Function

How has the design and layout of the school grounds improved the following?

- Play areas
- Sound control
- Meals
- Outdoor classrooms spaces and use

5. Security

How does the design of the school support overall site and facility security?

- Perimeter security
- Entrance to the site

- Building security at entrances
- Building function organization to support building security

Included in the POE focus group were some design Charrette activities inspired by our initial programming focus groups, aimed at obtaining some insights as to current functional successes and additional needs.

We first visited with the Bauleni School people. The process was broken down into three separate meetings. We met the students first starting with the POE questions and then engaging in the design charrette activities.

Through this process we found that the feedback from students, teachers, and administrators at both schools was consistent. The premise for this research was that the traditional ways that architecture students learn to do programming, and thus the way architects also do programming for educational facilities design, is limited in its understanding of the programming needs of any or all facilities because it does not include input and feedback from all the user groups. The research statement that programming can be improved with an interdisciplinary approach was also being tested. Typical programming is centered around an architectural structure and that an interdisciplinary structure would ensure a more holistic and successful outcome.

Figure 21. Pupils design charettes in Bauleni

Figure 22. Pupils design charettes in Chainda

Figure 23. Pupils design charettes in Bauleni

Figure 24. Pupils design charettes in Chainda

Figure 25. Touring the school with the pupils of Bauleni to get feedback on issues they recognized in the design and construction of their facilities.

Figure 26. Touring the school with the pupils of Chainda

Figure 27. Meeting with the teachers of Bauleni

Figure 28. Meeting with the teachers of Chainda

Figure 29. Bauleni design charrette with teachers

Figure 30. Chainda design charrette with teachers

Figure 31. Meeting with the teachers of the school in Bauleni

Figure 32. Meeting with the teachers of the school in Chainda

The following is a breakdown of the information gathered during these four days of post occupancy evaluation interviews:

9.1.1 BAULENI

This school was completed two years prior to our visit and was designed by a local architect. We could not find out how the programming process was done or if there even was a programming process. Through our interviews we were able to develop aza list of deficiencies. In the POE focus groups, it became evident that all three groups were consistent in understanding what was missing or not done right as well as what worked and supported educational activities:

- Library was too small for the program of each class visiting the library at a given time. Classes designed to hold 25-30 pupils while the library would only hold 15-18 pupils. This area also needs to have a separate space for computers and other technology
- There was no administration office block developed for the three administrative staff plus assistants. This space should include a reception area, three separate offices, a work room, a secure storage room for all testing materials, toilets, and storage. This area also needs to include a faculty break/ work/ lunch room with kitchenette, and toilets.
- There was no development and outreach block included in the design. This space should include 3 offices for discipleship, community outreach and development staffer, and counselor. It needs group meeting rooms for community groups during the week and on weekends, as well as private counselling rooms. It needs a nurse's room to allow for sick children to be kept during the day. This area should have its own entry and be accessible from outside the classroom part of the building.
- This building design did not have enough toilets and the numbers of toilets and location made it difficult for any large groups to access them during breaks. Because of the weather and environment, these rooms should have a dressing and shower area that can be private. Also the type of toilets used are easily broken and so are often out of order and needing to be replaced or repaired.
- The modular classrooms are uncomfortable. They are not designed for adequate ventilation and sun control.
- One of the most used spaces is the School Hall. This space is currently uncovered and gravel, but if it were
- covered, paved and connected to the classroom areas, it could be used for assemblies, lunch, and recreation in all weather conditions.

These findings illustrate many shortcomings and significant area omitted from the design. The researchers found no evidence of interviews and could not find out if anyone reviewed and approved the program.

9.1.2 CHAINDA

Focus group and charrette activities revealed that all three groups for the most part were unified in talking about both the positive aspects as well and the limited deficiencies of the building design. Through the focus POE groups, we were pleased to find out how consistent and positive the groups were in discussing how much they loved the way the school functioned and how they enjoyed the way it worked and supported the education process (Figures 45-46).

Classrooms

Consistently, everyone agreed the classroom designs support the education process.

- The room sizes, lighting, and ventilation are good.

- The desks have places for student books and utensils.
- There were no disruptive noise issues.
- There is enough space for learning and other activities.
- The only need was for a private space for teachers to meet with students and do class preparation while still have connection to the room
- Too many classrooms were built. The school was designed for 1,000 but they are determining that that is too large.
- Adequate and consistent classroom display continues to be an issue.
- The location of the teacher work office outside the classrooms is an issue. This location was not supported in our program.

Corridor/Circulation

Everyone agrees that the circulation works and bathroom and stair locations minimize noise disturbance during classes.

Student toilets

- Although the final location of the bathroom block is different from the programming design, it has adequate numbers of toilets and the location of this block makes access easy and least disruptive.
- The design of the toilets has minimized the issue of toilet malfunction.
- Although the design called for showers and dressing areas, these amenities were left out in the final construction and there is agreement that they should be in there to best serve the students.

Library/technology

Everyone agrees the library is large enough to provide the flexibility for simultaneous multi-function during any given time period. They also agree that there needs to be computers for the students to learn the technology required to advance into today's world.

Figure 33 (Left), 34 (Right). Photos of the Classrooms

Figure 35 (Left), 36 (Right). Photos of Circulation

Figure 37 & 38. Photos of the Student Toilets

Figure 47. Photo showing Outside Plumbing Connections at Student Toilets

Community development and outreach

Although this space was programmed for the known functions at the time, the program requirements have changed and will need to be revised for future school projects. This part of the facility was not built, but two of the unused classrooms in the front of the school have been converted to offices and community meeting rooms.

School hall

What was programmed was a covered multi-purpose space to be used for school assemblies, cooking and lunch, and recreation to minimize environmental issues and conflicts. This is still what everyone feels they need. They agreed that there also needed to be more outside hand washing locations.

Outdoor classrooms

These were built as programmed. In interviews we heard conflicting evaluations of these spaces. FLMI administrative staff felt they were not needed or used. The school interviews showed us strong support for these areas and their ongoing use. Like a covered School Hall, there was consensus that a covered outdoor classroom with a walkway would allow year round use.

Site and building security

The site security developed was the traditional type used at all schools- an 8' high cmu wall with barbed wire on time and a manned security gate monitoring and checking all vehicles and individuals entering the compound. This was not addressed in the programming process with the understanding that the existing system worked well

Systems including plumbing, electrical and drainage

- It is evident in the walk-through that the ideas of alternative fixture use integrated into the design have made a difference in the overall function and maintenance of the facility.
- To conserve water, we suggest installing gutters and downspouts and rain collecting barrels to collect water to be used in the toilet flushing activity and to minimize adverse drainage conditions during the rainy season. These had not been install yet and all were calling for this system to be installed.

Figure 39 (Left), 40 (Right). Photos of Site for Future Development and Community Outreach Space

Figure 41 (Left), 42 (Right).

Figure 43 (Left), 44 (Right). Photos of the Open School Hall Area

Figure 45. Photos of the Outdoor Classrooms Hall Area

10. CONCLUSION

This is the second of three school projects where the programming was developed using the participatory action research methodologies described in the paper. It is the only one completed using this novel programming methodology so far. We feel confident from this initial post- occupancy evaluation of the two school projects that the project developed using our multidisciplinary project team and these methodologies were more successful in meeting and supporting the educational needs of all involved. Our program identified not only the spaces needed but also integrated the organizational requirement into the

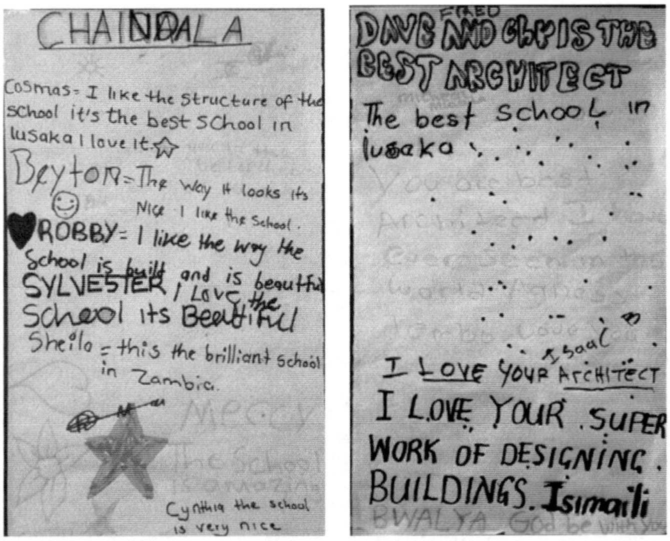

Figure 46. Photos of the Chainda Student Comments When Asked What They Thought of Their School

design, a process that has made the Chainda school a benchmark for school designs in Lusaka. Now, beyond Family Legacy schools, other government and private schools are viewing the success of this school as a model. These comments written by the students we interviewed support the comment made by the teachers and administrators about how pleased they were with this facility and the way it supports the educational system and programs.

These comments were written by the 6th year pupils at Chainda, all of whom had attended other schools before this opened last year. Their comments reflect the joy of the teachers and administrators we talked to. We did not use quantitative assessments to measure the success of the project, but rather employed qualitative methods that brought the community together. The process used in engaging the students, teachers, and administrators, at a personal and community level, truly identified community needs through photovoice, focus group, and charrette exercises. The design programming outcome does show a more successful building outcome.

REFERENCES

E. Cherry, and J. Petronis. (2016). Architectural Programming. The Whole Building Design Guide- a program of the National Institute of Building Sciences, Washington, DC.

G. Mwakikagile, (2010). Zambia: Life in an African Country. New Africa Press. p. 159. ISBN 9789987160112. Archived from the original on 2 May 2018.

C. Wang, and M. Burris. (1997). "Photovoice: Concept, Methodology, and Use for Participatory Needs Assessment". Health Education & Behavior. 24 (3): 369–387.

C. Nykiforuk, H. Vallianatos and L. Nieuwendyk. (2011-01-01). "Photovoice as a Method for Revealing Community Perceptions of the Built and Social Environment".

RW Strack, C. Magill, and McDonagh, K. (2004). "Engaging youth through Photovoice" (PDF). Health Promotion Practice. 5: 49–58.

C. Wang. (2003). Chapter Nine: "Using Photovoice as a Participatory Assessment and Issue Selection Tool" (PDF). Archived from the original (PDF) on 2003.

E. Colucci. (2007). "Focus groups can be fun": The use of activity-oriented questions in focus group discussions". Qualitative Health Research. 17 (10): 1422–1433. Archived from the original on 2018-05-09.

T. Greenbaum. (2000). "Moderating Focus Groups." Thousand Oaks, California: Sage Publications, Inc. ISBN 0-7619-2044-7.

H. Sanoff, C. Pasalar, and M. Hashas, (1977). School Building Assessment Methods. School of Architecture, College of Design, North Carolina State University with support from the National Clearinghouse for Educational Facilities.

Architecture 4970 and 5960 (2015). "Community-Based Participatory Research," A report to complete requirements for a service-learning course from the College of Architecture, University of Oklahoma in the Summer of 2015 for Family Legacy Ministries International. Lusaka, Zambia.

Architecture 4970 and 5960 (2015). "Re-envisioning Family Legacy Primary Schools: A Design Process." A report to complete requirements for a service-learning course from the College of Architecture, University of Oklahoma in the Summer of 2015 for Family Legacy Ministries International. Lusaka, Zambia

E7 - Transform Processes and Outcomes for Equity

Historic Preservation Survey Methodology for Equitable Outcomes

Tera Williams (University of Washington)

1. ABSTRACT

As the built environment continues to evolve, what we decide to save is just as important as what we build next. When planning for the future, it is very important that the cultural resources and historic landmarks for marginalized populations are preserved. When part of town gets upzoned, the assets of marginalized communities are often lost to the process of development, erasing the history and social cohesion of community in the margins.

This case study analyzes ways to make the historic preservation process more equitable through survey methodology. It looked specifically at the current policy on a national and local level in Washington state's King County, reviewed survey and inventory methodology, and suggested improvements in the historic preservation survey and inventory process moving forward. Using the case studies research approach to conducted interviews, archival research, and procedural comparison insured internal validity in the overall investigation and the resulting embedded case studies.

The investigation determined that current policy is being interpreted in a way that unnecessarily limits the assets that can be preserved, local organizations need better foundational plans to implement the current standards to their fullest potential, the current inventory method is inconsistent and lacks validity across surveys, and the assets being detected are a manifestation of internalized bias.

The implications of this work are that the field ultimately would benefit most from being more diverse and having a higher level of public input.

1.1. Keywords

Equity, Historic Preservation, Public Engagement, Planning, Community Design

2. INTRODUCTION

For the past three years, a historic preservation coalition called Beyond Integrity has researched sites, pinpointed inequities, and explored new ways to protect historic properties. Beyond Integrity is interested in "elevating equity in preservation standards and practices" and "looking beyond architectural integrity when deciding what places we save" ("Beyond Integrity | 4Culture," n.d.). The coalition is housed under 4Culture, King County's cultural work funding and support agency. The aims of this case study were developed to build upon the work of previous Beyond Integrity internships. Those internships focused on reviewing previous preservation efforts within King County, analyzing those efforts to see whether they were equitably executed as is, looked at properties that were not designated after nomination or had not yet been nominated, revised and added extra information on the cultural significance of resources to designation reports, and analyzed alternative preservation tools (Aguila, n.d.). The focus of this investigation was to create an in-depth case study exploring survey and inventory methodologies with the aim of making recommendations for the improved identification of culturally significant properties for marginalized communities.

3. BACKGROUND

Historic resource surveys are the initial step to figuring out which assets have significance. The first step to doing a historic survey is developing a course of action. This plan determines what area will be studied and what the focus will be. The actual survey process is made up of "archival research and development of a historic context, field survey, and recording of information" with each step of the process feeding into the others (City of San Diego, 2018). A historic context is a narrative history of the survey area. This document is crucial to the planning process for the field survey. Field surveys involve going around and collecting information on resources based on their exteriors. They can be either reconnaissance or intensive level survey. Reconnaissance level surveys are exploratory surveys of "a broad group of historic resources," sometimes as a precursor to an intensive survey (Department of Archaeology & Historic Preservation, n.d.). Intensive surveys look more closely at resources and include more research time. They build off of a reconnaissance survey to develop a report on the resource (Department of Archaeology & Historic Preservation, n.d.). Historic resource surveys are a vital part of the preservation and identification process.

Historic preservation work means different things to different groups of people. Most can agree that it is about preserving historically significant elements of a city so that those elements can carry on and continue to benefit future generations. Significant elements and their benefits can manifest in many different ways and often, that is where things become confusing. Within our current national guidelines, significance is determined by analyzing a resources' location, design, setting, materials, workmanship, feeling, and association to define its significance. Historic preservation most frequently focuses on physical structures and the way that those structures are perceived to of been maintained. After significance is arrived upon, the physical features of the building are analyzed for the way that they push forward that significance; judged on their visibility on the building; compared to other properties; and from that point, the asset is analyzed to determine if the categories of integrity that most align with the significance and valued physical features are there. Integrity within historic preservation is the ability of an asset being reviewed to retain its identity.

The standards for determining significance and integrity are colored by America's dominant culture's history and values, but culture is very subjective and open to interpretation.

When American culture is referred to as "mainstream," it must be acknowledged that this nation has a mainstream culture, but there are ethnic communities with special concerns and issues in addition to the national culture. Typically, mainstream American culture is understood to mean the culture and embodied values of the dominant, majority population: White and predominantly of Western European heritage (Holland-Moore, 2016).

Culture's subjectivity makes it very difficult to apply mainstream American standards to the cultural resources valued by marginalized populations. The way historic preservation is currently set up allows for "'implicit' or 'unconscious biases' that are beyond our control and activate involuntarily" to impact the way that every aspect of the preservation process takes place (Holland-Moore, 2016). Unconscious bias operates in ten different ways overall with three of the ten manifesting in the preservation process: diagnosis bias, or our unconscious urge to go with our first impression of something; value attribution, or the impulse to assign something qualities based on that first impression; and priming effect, the tendency to answer something based on presupposition springing out of past situations. Implicit bias theory argues that: the inherent biases of the majority population would favor places and things particular to the history of their cultural group. In the case of historic preservation, the majority of its practitioners and decision-makers are White, reflecting the majority population. Those inherent biases are manifested through the determination as historically significant of those properties perceived as important or related to the majority population (Holland-Moore, 2016). Unconscious bias manifests in both the standards for designation and the selection of resources; that has had a lasting impact on what ends up being considered historically significant. People that do not belong to marginalized groups often lack the tools and understanding to identify important cultural resources. If a marginalized group's history is not preserved, it will not be available to be studied and utilized moving forward (Holland-Moore, 2016).

The importance of cultural resources in marginalized communities centers around intangible qualities and as such, it is difficult to recognize them through traditional survey methods. Working to capture intangible qualities in survey and inventory methodologies would prevent some of these resources from being overlooked. Some examples of spaces that can hold intangible cultural significance across marginalized groups are:

- Points of origin
- Routes of migration
- Places of experience
- Places of suffering and struggle
- Places of achievement
- Places of interaction
- Spiritual places
- Milestones of international relations
- Place of education and of presentation (Kaufman, 2004)

For marginalized groups, space is what you make of it; the memories made in the space are what gives it its value, not necessarily its architectural significance or integrity. Many of these spaces that hold value in marginalized communities are not considered architecturally significant. The nature of these environments is to continually change to meet the needs of the community it serves (Holland-Moore, 2016). In these communities, modifications are a manifestation of spaces being well used and those same changes exclude resources from being designated.

4. RESEARCH OBJECTIVES

Determine a clear methodology to figure out what was missed in past King County and Seattle survey and inventory processes.

- How do we take the cultural knowledge, experience, history, and hope for the future and reflect that in the assets that are preserved?
- What is a "high level" of cultural significance?

- How do survey and inventory methodology need to change to find resources of high significance for marginalized communities?

5. METHODS

The purpose of this case study was to analyze current survey and identification methodology to prevent community assets from slipping through the cracks of the current preservation methodology. To do that, I conducted research; created a case study protocol; interviewed people partaking in historic preservation efforts in Seattle and King County; did two reconnaissance level surveys to determine the current state of previously surveyed properties; and created embedded case studies out of two properties with a high level of cultural significance. My case study protocol guided my research and interviews. It was reviewed by Beyond Integrity for merit before undertaking interviews. The sites were picked based on my interviews with King County and Seattle preservationists. I did supplementary online, library, and archival research unpacking preservation and what it would mean for it to be equitable. I chose to create embedded case studies so that my overall case study would feel complete without having to research each property in the survey. It also enabled me to dive deeper into the individual properties. I created a case study database in the form of a bullet journal and documented all of my observations, interview notes, and avenues for further exploration within it. I supplemented my interviews and observations with additional research and presented my preliminary findings to Beyond Integrity and the preservationists I interviewed to increase internal validity.

6. RESULTS

The purpose of this case study was to find alternative methods of identifying and preserving the assets and narratives important to marginalized communities. Preserving assets that represent the rich diversity through the investigation of local survey techniques in our communities is very important to the continued vitality of our diverse spatial fabric. The methodologies reviewed from the King County and Seattle surveys represent different ways of approaching survey work. Reviewing the body of research congruently with these surveys revealed interesting new recommendations for preservation.

Statewide, cultural resources with high cultural significance in marginalized communities are being overlooked (DAHP, 2014). For marginalized communities, history is a vital part of connecting with heritage. They are already working hard to preserve their legacy, but that work is not typically happening through official preservation channels (Holland-Moore, 2016; Kaufman, 2004). If these community assets and their accompanying narratives are not preserved, they will not be available for future generations to research or enjoy (Holland-Moore, 2016). It is important that preservationists from non-marginalized backgrounds work to understand the factors that make cultural assets important to marginalized communities and how that relates to integrity.

The importance of history as a mode of understanding heritage emerged with surprising strength from the Assessment. Surprising, at least, to a white researcher, perhaps because for many members of the majority history has become a sort of luxury good, a pleasant if vaguely defined part of heritage that can be pretty much taken for granted. For many respondents, on the contrary, history can never be taken for granted. It requires active definition and constant attention, because it shapes identity and describes relationships with the majority culture that, in turn, define life in crucial ways. Evidence of achievement must be unearthed, underlined, spotlit. Memories of discrimination and suffering must be maintained. And sometimes evidence of existence – of simple presence within the larger story – must be discovered and defended. For much of history lies lost, forgotten, or buried, and before it can become part of heritage it has to be rediscovered...

History, then, is a crucial part of heritage. It has little in common with the genteel appreciation of the "finer things" that the word heritage frequently connotes, or with the "souvenir history" that Puerto Rican poet Martin Espada derides: the superficial and usually congratulatory commemoration of symbolic highlights in American history. It does not paint the past as "simpler times." It is instead a relentless struggle to discover, uncover, rediscover, and recover facts about the national past that have been swept from public consciousness either because they are uncomfortable or because the evidence is ephemeral (Kaufman, 2004).

For this case study, I chose to review two very different survey methodologies. The Seattle case study is on an area survey of Downtown Seattle, while the King County case study came out of a survey specifically focused on the cultural assets remaining in White River Valley from the Japanese Americans that populated the area for 50 years before being interned. The Downtown survey looked at a general area while the King County survey specifically analyzed the built environment with respect to culture. King County typically does area surveys rather than culturally specific surveys. The thought process behind selecting their culturally-specific survey is firstly, to have a wider range of material to discuss and secondly, to provide suggestions for improvement of culturally-specific survey methods. The different methodologies come together to inform the resulting recommendations in their own unique ways.

7.1. Seattle Survey Context

The 2007 Downtown Historic Resources Survey and Inventory was funded by Seattle's City Council to preemptively survey and inventory historic properties in Downtown Seattle preceding an upzone of the area. An upzone increases the development potential for an area and can result in the loss of cultural resources to development. The survey aimed to analyze all historic buildings in the downtown area, largely prioritizing architecturally significant buildings. Through this survey process, many buildings were designated as landmarks.

The 2007 survey was a multiple phase survey. In the reconnaissance level research and development phase, surveyors analyzed the survey area to locate the 127 buildings in the area constructed before 1966. The properties were grouped into categories based on their architectural significance, as perceived by surveyors. 22 of the buildings were determined to not actually have any historic features or significance at all; 23 of them were not especially changed, but because of their "scale and/or physical character" appeared to be devoid of significance; 52 were well preserved and were considered potentially historic; and 30 were intact and had significance (Krafft & Krafft Architecture/CRM, 2007). The 84 properties deemed most significant by surveyors and department staff were analyzed in the intensive survey phase. For these properties, landmark nominations were completed. Most of the remaining properties had limited inventory forms filled out.

A large problem with the survey was the way results were distributed. The City of Seattle's historic preservation system as a whole is in need of an overhaul to be more cohesive between city departments and to be more transparent for the public. The lack of transparency around the survey report initially made it difficult to understand the methodology used. When the survey was completed, these were translated into a new category system for sorting historic buildings by the Seattle Department of Neighborhoods staff. In this system, Category 1 buildings were all buildings though to be the most significant, Category 2 buildings were buildings that were less significant, but still eligible, Category 3 buildings were buildings that were worthy of inclusion in the historic resources inventory, but not eligible to be designated as a city landmark, and Category 4 buildings were considered "so altered that they would not qualify as Seattle landmarks" (Department of Neighborhoods, 2007). The survey took place without any public announcement and as such, had little to no public input. After the survey was completed, notices of what category properties were placed into were sent out to property owners. In reality, it is impossible to definitively rule a cultural resource out as a Seattle landmark without doing an intensive level survey. Seattle Municipal Code specifies that:

An object, site or improvement which is more than twenty-five (25) years old may be designated for preservation as a landmark site or landmark if it has significant character, interest or value as part of the development, heritage or cultural characteristics of the City, state, or nation, if it has integrity or the ability to convey its significance, and if it falls into one (1) of the following categories:

- It is the location of, or is associated in a significant way with, an historic event with a significant effect upon the community, City, state, or nation; or
- It is associated in a significant way with the life of a person important in the history of the City, state, or nation; or
- It is associated in a significant way with a significant aspect of the cultural, political, or economic heritage of the community, City, state or nation; or
- It embodies the distinctive visible characteristics of an architectural style, or period, or of a method of construction; or
- It is an outstanding work of a designer or builder; or
- Because of its prominence of spatial location, contrasts of siting, age, or scale, it is an easily identifiable visual feature of its neighborhood or the City and contributes to the distinctive quality or identity of such neighborhood or the City (Municipal Code Corporation, 2019).

The grammatical formatting of the code leaves its interpretation a bit clouded, but it appears that the integrity of the asset itself is not actually a requirement for designation, as is rather more of an alternative within being able to convey its significance. With the assets of marginalized communities more often lacking architectural integrity, making blanket statements surrounding designation and architectural integrity ignores assets that are cultural in nature.

7.2. Seattle Case Study

Located at 2018 1st Avenue, the Vogue Hotel is a three-story brick building that was constructed in 1908. It was originally constructed as a hotel. The building has survived many different uses. The Vogue Hotel was "surveyed" in the 2007 Downtown Seattle Survey. Though the Vogue Hotel does not seem historically significant at

Figure 1. The Vogue Building (2019). Photo by the author.

first glance, a deeper look at it shows that sometimes, digging under the surface to see what a building means to the community is necessary to see what resources should be preserved.

Vogue Hotel is a Queen Anne style late victorian with a rectangular plan and a flat roof. Its standout features are bay windows going from the second to the third floor, brick construction, and cornice details. It previously featured a large fire escape from the roof with balconies on the second and third floors. The side of the building was formerly used for concert promotion before the construction of a building next door. There was a fire between the buildings which potentially damaged what remained of the mural in 2007 (Bruscas, 2007).

While the Vogue Hotel was originally a hotel, it has cycled through many uses since. After its stint as a hotel was over, it became a brothel. The downstairs space became Johnny's Handlebar, a leather bar catering to gay men. In 1979, Johnny's Handlebar closed and became WREX ("Vanishing Seattle - Posts," n.d.). WREX served the same crowd but showcased punk music. WREX started booking live music in the very early 1980s. The first show was a performance by The Fastbacks with Duff McKagen (of Guns N Roses fame) on the drums. WREX began booking bands three nights a week and played music videos with "video wallpaper" throughout the week and between acts. As WREX grew, it became a space that "was just as much a 'gay' venue as a 'punk' venue that welcomed anyone with a love for the alternative scene" (Stevens & White, 2017). WREX closed on March 18, 1982, citing cash-flow problems. The club reopened a year later, rebranded as The Vogue. The Vogue primarily played dance music, but still occasionally booked live acts and was "one of the first venues to host bands that would one day be known as 'grunge'" (Stevens & White, 2017). Nirvana played what is largely considered their first Seattle show in the Vogue on April 24, 1988 (Humphrey, 2016). The Vogue closed moved out of the Vogue Hotel building and closed in 2007 (Scanlon, 2007). Now, the building houses VAIN, a hair salon "committed to encouraging personal self-expression and challenging conventional understandings of beauty." VAIN opened in 1996 and "has built a reputation for delivering cutting-edge style in a comfortable and empowering atmosphere" ("VAIN our story," n.d.). The upper floors serve as artist studios.

The Vogue Hotel has a rich social history within alternative communities. It has changed to both meet the needs of Belltown and create a space for cultivating counter-culture communities for decades. It was first documented as a "beer parlor" in 1934 when the staff was caught selling liquor over the counter, a violation of liquor ordinances of the time ("7 Arrested By Vice Squad," 1934). The building was originally low-cost rooms on a short term basis with the downstairs retail section activated as a counter-culture gathering hub. The building in and of itself is significant because of its value as a third space for people that are marginalized from the dominant culture. VAIN is a salon where people can get their hair done in ways that do not subscribe to the dominant norms of respectability. As such, VAIN often serves as a third space for mothers and aging weirdos who belong to the counterculture and are struggling to define who they are in adulthood (Ariel, 2013; "How to dye your armpit hair | Offbeat Home & Life," 2014). Before VAIN, Johnny's Handlebar provided a home away from home for gay men in the leather scene. WREX was meant to invoke a sense of chaos. It was decorated with car and airplane parts meant to invoke a wreck. WREX created a synergistic environment where people who did not fit within the dominant culture were free to be themselves. The Vogue continued in the work of WREX and out of that, fostered and encouraged the transformative change of the music world as a whole by supporting the developing grunge movement. The Vogue building is largely remembered for the role it played in the development of grunge, but it is only because of its history as a counter-culture space that it could get to that point. Nirvana played there six times and fans still come to VAIN to visit the site (Garbacik, n.d.; Scanlon, 2008). Mudhoney played their first two shows at The Vogue in 1988 ("Mudhoney: Tourbook, 1988," n.d.). The use of the space has imbued it with cultural meaning for many over the course of its lifespan.

The Vogue was not considered historically significant in the 2007 survey because it was not constructed to stand out or do anything out of the ordinary. While the 2007 survey had nothing at all to say about the Vogue, later surveys indicated that:

This building has been significantly altered and no longer retains integrity of feeling, association and materials, so it does not warrant inclusion in the NRHP.

Constructed as a small hotel in 1908, this building was one of many small hotels along First Avenue that catered to laborers, seamen and other low-income residents. The building's bay windows and bracketed cornice indicate its early date. However, it has had extensive window and storefront alterations ("PROPERTY: #99138 Ridpath Hotel," 2009).

We can already ascertain from this description that the hotel served a low-income population representative of the early use of the neighborhood. The structure's nature as a counterculture hub inherently indicates that it should change and transform as a manifestation of its purpose. It is worth arguing that changes to the traditional form are what create and maintain the integrity of a counterculture space for the population they serve (Desmond, McDonagh, & O'Donohoe, 2000). Counterculture refers to a coherent system of norms and values that not only differ from those of the dominant system (where this and nothing else is the case we speak of subcultures) but also comprise at least one norm or value that calls for commitment to cultural change, that is to a transformation of the dominant system of norms and values (Burnier & Roomsch-Katholieke Universiteit (Nijmegen, 1975). If counterculture itself requires a dedication to cultural change, it follows that that same dedication would manifest in authentic counterculture spaces.

In 2005, the City of Seattle Office of Film and Music created a snapshot map of Seattle's music legacy in the downtown area. This map pointed out places considered important to Seattle's music history. The map highlighted Belltown and The Vogue in particular, stating:

In the shadow of the Space Needle, Belltown evolved during the go-go 1990s from an edgy industrial area where underground arts flourished to a mecca for Gen X bohemians and dot-com entrepreneurs. It's now a highly walkable hub of boutiques, eateries and nightspots...

The Vogue: 2018 1st Ave. Nirvana played its first Seattle show in this former crucible for the Seattle sound, opening for Blood Circus in April 1988. The original venue now houses the rock and roll hair salon Vain (City of Seattle Office of Film and Music, 2005).

If there were a way to more openly communicate between government organizations and collaborate on collective goals, oversights such as this one would not happen. In addition, the Vogue Hotel has two addresses and four historical names. In the parcel viewer, one must search for the current address of VAIN in order for it to pop up, but it pops up referencing the secondary address. On WISAARD the property is referenced as Ridpath Hotel, in the 2007 survey it is referred to as the Riopath, and in archival research, it is called the Preston Hotel or Hotel Preston.

The narrative of the building has persisted because of do-it-yourself approaches to public history. A 2003 performance installation by Diana Falchuk named Artifact aimed to reconfigure the systems and authorities that turn material culture into official history via the speculative origins of a 100-year-old decorative floor rediscovered underneath the sidewalk in front of Vain in 2001" (Falchuk, 2003). Though it will be on hiatus until the close of 2020, there is currently a walking tour called Seattle Grunge Redux: The Walk That Rocks Through Seattle's Music History led by Eric E. Magnuson. The tour, which visits the Vogue, is said to "Explore Seattle's grunge era (from the mid-1980s through the late 1990s)" (Magnuson, 2019). Eric says that "many clubs, bars and record shops have been around for decades and are part of their identity for people, eating out and then watching a band - that's just part of Seattle"(Tiedemann, 2017). The building itself also works to speak towards its history with interior finishes exposing the original bones of the building and a large mural on the parking lot side that references its historic context and current role as a counter-culture space.

7.3. King County Survey Context

The survey selected for King County was the

Figure 2. Mural on the side of the Vogue Building (2019). Photo by the author.

Japanese-American Legacies in the White River Valley Historic Context Statement and Inventory from 1997. This survey was prepared for the King County Landmarks and Heritage Program by Mildred Tanner Andrews to pay homage to the rich Japanese-American community that lived in the White River Valley before being interned in 1942. To conduct this survey, Ms. Andrews accessed archival materials, did field work, and conducted oral history interviews with Japanese-Americans whose families were in the White River Valley region before being displaced and sent to internment camps. The age of the survey somewhat limited the amount of original research remaining. The original methodology included fieldwork, inventory forms, an addendum to an existing inventory report to adequately reflect the cultural ties of the property, a master index of the primary properties researched, and recommendations for the next steps in the process. The survey was informed primarily by interviews and site visits with Japanese-American residents whose families lived in the region before the internment in WWII. The historic context statement focused on cultural narrative and values and while the resources investigated were places and moments of importance for the community. The recommendations section of the report appears to be largely undeployed.

The Japanese-American Legacies in the White River Valley Historic Context Statement and Inventory was not a typical survey in that it focused more on cultural significance than anything else. There was a heavy focus on the Japanese-American agricultural history in the area. They started with dairy farms, then transitioned to vegetable and berry farming. Japanese farmers grew all kinds of vegetables, berries, and flowers and "by the beginning of World War I, Japanese farmers from the White River Valley and other areas occupied 70%" of the market stalls at Pike's Place Market (Andrews, 1997). In the 1920s, Japanese farmers in the area started growing hothouse rhubarb in the winter. The crop was only grown in 3 parts of North America at the time, Ontario, Michigan, and Washington. Farmers used local, Japanese-owned packing companies to ship their produce. "In 1941, the Vegetable Shipping Industry, Western Washington District, reported that Japanese organizations shipped "80% +" of locally grown major commodities, namely lettuce, peas, cauliflower and celery" (Andrews, 1997). In 1941, there were 285 farms in the area. 28 of them were owned and 257 were leased. There were 1,141 people involved in the farms. 456 of them were immigrants and 685 of them were citizens (Andrews, 1997).

The Japanese-American Legacies in the White River Valley Historic Context Statement and Inventory's focus on culture is of great value to this research, but the lack of focus on the preservation or promotion of the properties themselves are an oversight. Overall, the Japanese-American Legacies research is not accessible enough to the public. The properties inventoried do not have any of the inventory information stored in Washington Information System for Architectural and Archaeological Records Data (WISAARD). The goal of this research presumably was to tie the story to remaining resources. In not promoting that narrative, the effort fails to truly meet that goal. King County normally does more traditional surveys and so it was understandably harder for the organization to move forward with the preservation or promotion of these resources. The code for historic preservation in King County is as follows:

A) A historic resource may be designated as a King County landmark if it is more than forty years old or, in the case of a landmark district, contains resources that are more than forty years old, and possesses integrity of location, design, setting, materials, workmanship, feeling, or association, or any combination of the foregoing aspects of integrity, sufficient to convey its historic character, and:

- Is associated with events that have made a significant contribution to the broad patterns of national, state or local history.
- Is associated with the lives of persons significant in national, state or local history.
- Embodies the distinctive characteristics of a type, period, style or method of design or construction, or that represents a significant and distinguishable entity whose components may lack individual distinction.
- Has yielded, or may be likely to yield, information important in prehistory or history.
- Is an outstanding work of a designer or builder who has made a substantial contribution to the art.

B) An historic resource may be designated a community landmark because it is an easily identifiable visual feature of a neighborhood or the county and contributes to the distinctive quality or identity of such neighborhood or county or because of its association with significant historical events or historic themes, association with important or prominent persons in the community or county or recognition by local citizens for substantial contribution to the neighborhood or community. An improvement or site qualifying for designation solely by virtue of satisfying criteria set out in this section shall be designated a community landmark and shall not be subject to K.C.C. 20.62.080.

C) Cemeteries, birthplaces or graves of historical figures, properties owned by religious institutions or used for religious purposes, structures that have been moved from their original locations, reconstructed historic buildings, properties primarily commemorative in nature and properties that have achieved significance within the past forty years shall not be considered eligible for designation.

Figure 3. Current Shed (2019). Photo by the author.

However, such a property shall be eligible for designation if they are:

- An integral part of districts that meet the criteria set out in subsection A. of this section or if it is:
- A religious property deriving primary significance from architectural or artistic distinction or historical importance;
- A building or structure removed from its original location but that is significant primarily for its architectural value, or which is the surviving structure most importantly associated with a historic person or event;
- A birthplace, grave or residence of a historical figure of outstanding importance if there is no other appropriate site or building directly associated with his or her productive life;
- A cemetery that derives its primary significance from graves of persons of transcendent importance, from age, from distinctive design features or from association with historic events;
- A reconstructed building when accurately executed in a suitable environment and presented in a dignified manner or as part of a restoration master plan, and when no other building or structure with the same association has survived;
- A property commemorative in intent if design, age, tradition or symbolic value has invested it with its own historical significance; or
- A property achieving significance within the past forty years if it is of exceptional importance. (Ord. 17635 § 2, 2013: Ord. 10474 § 4, 1992: Ord. 4828 § 4, 1980) (King County, 2016).

This code specifies that the property needs to have integrity in addition to overall historic significance, but does not actually define integrity with any rigidity. That flexibility allows for different interpretations of code.

7.4. King County Case Study

The Yasumura's Shed, doing business as Sumner Packing Co., was run by I. Yasumura and his sons to sell, process, package, and transport his produce and that of other Japanese farmers. Located at 204 H St NW, the site was identified as a cultural resource in the 1997 Japanese-American Legacies in the White River Valley Historic Context Statement and Inventory. The original shed burned down in 1950 and the current shed was rebuilt to serve the same purpose in 1957 ("Break Ground Thursday for Bulbs and 'Barb," 1957, "Fire Destroys Rhubarb Shed at Sumner," 1950). For some, it can be hard to see the historic significance in a cultural resource that has burned down.

This code specifies that the property needs to have integrity in addition to overall historic significance, but does not actually define integrity with any rigidity. That flexibility allows for different interpretations of code.

7.4. King County Case Study

The Yasumura's Shed, doing business as Sumner Packing Co., was run by I. Yasumura and his sons to sell, process, package, and transport his produce and that of other Japanese farmers. Located at 204 H St NW, the site was identified as a cultural resource in the 1997 Japanese-American Legacies in the White River Valley Historic Context Statement and Inventory. The original shed burned down in 1950 and the current shed was rebuilt to serve the same purpose in 1957 ("Break Ground Thursday for Bulbs and 'Barb," 1957, "Fire Destroys Rhubarb Shed at Sumner," 1950). For some, it can be hard to see the historic significance in a cultural resource that has burned down.

The Japanese-American Legacies in the White River Valley Historic Context Statement and Inventory ended up identifying the Yasumura Shed through interviews with Japanese-American families that lived in the Auburn area before and after internment. The Yasumura family packed produce and shipped it out by train, making the shed's location next to the train tracks a vital element of the business. On May 22nd, 1942, though, the shed was used as a make-shift train stop for Japanese-American's to board evacuation trains to Pinedale Assembly Center, outside of Fresno, CA, before being sent to internment camps that they would occupy on a longer-term basis. While the previous survey identified the shed as a significant resource for Japanese-Americans in the White River Valley, I would argue that the landscape itself and the route that they traveled to Pinedale Assembly Center represent the feeling the resource is intended to invoke more.

The original shed was owned by the Union Pacific and Milwaukee Railroads and rented out to the Yasumura family for the operation of Sumner Packing Co (Andrews, 1997). After the Yasumura family evacuated, the shed was rented out to the Sumner Rhubarb Growers Association ("Fire Destroys Rhubarb Shed at Sumner," 1950; Parker, 1944). In 1950, an oil stove explosion ignited the shed, causing irreparable damage. Though the Yasumuras themselves had previously been targeted by arsonists in 1938, the shed fire appeared to be accidental and unrelated ("Epidemic of Blazes Laid to Firebug," 1938, "Firebug Blamed for Rural Fires," 1938, "Fire Destroys Rhubarb Shed at Sumner," 1950; Staff Corespondent, 1938). In 1957, a new, slightly bigger shed was constructed on the original site to meet the needs of the Sumner Rhubarb Growers Association and that is the shed currently on the site ("Break Ground Thursday for Bulbs and 'Barb," 1957). The shed is a concrete block warehouse with a gabled, corrugated tin roof. It speaks to the agricultural history of the area. The agricultural heritage of the area for both Japanese-American and non-Japanese-American residents is very important, especially when taking into consideration the connection to Sumner, which considers itself the rhubarb/rhubarb pie capital of the world (Peterson, 2019). Though there were other packing companies of more notoriety in the area, the layers of connection that this one holds make it especially significant.

The shed is currently being used as warehouse space and surplus parking for Seattle Automotive Distributing, Inc. while the tracks around the site are largely owned by Puget Sound Energy (King County Department of Assessments, 2019; Seattle Automotive Distributing, n.d.). It retains its desolate industrial/warehouse feel. Next to the warehouse, the old Milwaukee tracks remain, though the Interurban tracks have been converted into the Interurban trail. The site itself maintains integrity. The site itself is representative of a harrowing event and marks the dismantling of 50 years of work to build the White River Valley community. I view the site as a lens through which to view the route that Japanese-Americans traveled from White River Valley and the symbolic representation of that journey. Japanese immigrants were taken first. On Thursday, May 21st and Friday, May 22nd of 1942, the people leaving were American citizens (Flewelling, 2002, p. 231). They met at the Yasumura's shed. 540 people left on Thursday with another 458 leaving on Friday (Flewelling, 2002, p. 233). The train trip to the Pinedale Assembly Center outside of Fresno was 36 hours long (Flewelling, 2002, p. 231). Shuji Kimura described the initial feelings of departure as follows:

We didn't feel so bad leaving with all the excitement of leaving. But soon when six p.m. came and the train began to move, and we saw old Mr. Ballard waving his hat at us, his coat collar turned up against the rain, mother began to cry. I couldn't see through my tears either. I saw the Main Street Crossing -- there were more people waving. The train began to go faster and the berry rows, the rhubarb, the lettuce fields, the pea fields, began to slip past our window like a panorama. My throat hurt, but I couldn't take my eyes from the familiar fields and pastures slipping so quickly away (Flewelling, 2002, p. 234).

Pinedale Assembly Center was formerly part of the Sugar Pine Lumber Company. It was a temporary site used to hold people until they could be sent to relocation centers. Most people were sent to either the Poston or Tule Lake relocation center (Japanese Americans in World War II, n.d.).

8. CONCLUSIONS

This research process was guided throughout by the foundational research questions, designed to examine the methodologies being used for inequities. Through those questions, four goals were set for inquiry. I examined the survey and inventory process, conducted embedded case study research, and channeled that research into recommendations for methodology changes moving forward. This inquiry is the first step, but the hard work lies ahead.

The first goal was to select previous surveys and develop a methodology for selecting resources of high cultural importance from within those surveys. Within the Seattle survey, I was specifically looking at resources that had been deemed ineligible for designation in the 2007 downtown area survey but still hold meaning to the community. Within the King County survey, per the request of the department, I was looking for a resource that potentially created a means for promoting all of the resources found through the Japanese-American Heritage survey process. To select the resources to focus on, I looked for places that spoke to the intangible values listed in the National Parks Service's Cultural Heritage Needs Assessment. The Vogue building is a place of experience, achievement, and interaction. The Yasumura's shed and the shed site are a route of migration, place of experience, place of suffering and struggle, place of interaction, and (negative) milestone of international relations (Kaufman, 2004). The two resources were deeply impactful, but their history has not been amplified through official channels.

The second goal was to, through the process of analyzing those resources myself, figure out the kinds of steps that needed to be taken to ensure that the resources being preserved are not only reflecting one cultural story. As it turns out, the process as a whole had no tangible way of valuing culture in and of itself. The inventory process was set up to survey for things related to integrity. Integrity had not been defined on a local level and was being interpreted to mean architectural integrity rather than the integrity of identity. A broader interpretation of local and national guidelines could resolve most of the equity problems in the selection and evaluation of historic resources. In addition, restructuring the archival research stage of the survey process to include public engagement and organizational collaboration will lead to more representative results. Overall, the process needs to be framed as more of a creative problem-solving process than a rigid protocol.

My third goal was to figure out a more refined metric for a high level of cultural significance. I came to the conclusion that resources with a high level of cultural significance had to speak to an intangible quality of some sort within their associated community and would also likely have the integrity of location, workmanship, feeling, and association in order to be highly significant. There is an argument to be made that any design changes would be because of cultural significance, making those a manifestation of use. Displacement could potentially impact the integrity of setting, but the integrity of location would speak to the history of the area. Centering culture in the analysis of the resource is the most important aspect of the process. A high level of cultural significance is integrity with cultural significance.

The fourth and last goal was to change methodology so resources of high cultural significance are initially recognized in the survey and inventory process. Foundationally, preservation plans would have to center equity and create a work plan on achieving it, heavily featuring public engagement and transparency. Staff and board members would need to be retrained on evaluation criteria with a renewed focus on the importance of context statements, interpreting those statements for significance, expanding our interpretation of integrity, and creative means of designation. Creating organizational change in this way is not going to be easy, but the results will be worth it.

Making sure that the history important to marginalized groups is preserved and accessible is vital moving forward. Preservation policy is significantly more flexible than it is given credit for. Reviewing resources for significance and integrity does not work out of context. Changing the foundational way that we engage in historic preservation to center equitable and just outcomes benefit everyone.

9. DISCUSSION

As things stand, resources that represent the heritage of marginalized communities are not being captured by the current survey and inventory methodology and even when they are, they often do not qualify as landmarks. Many of the metrics to determine if a resource is worthy of designation, such as integrity, come down to judgment calls made by the investigator. If, instead, integrity was expanded to be a more dynamic concept and a metric for measuring the integrity of a resource overall, rather than just the architectural integrity was created (Zancheti & Loretto, 2015). Foundationally, surveys need to be designed to be reliable and valid in order for historic preservation research itself to have integrity.

- **Reliability** refers to the consistency of a measure. Psychologists consider three types of consistency: over time (test-retest reliability), across items (internal consistency), and across different researchers (inter-rater reliability).
- **Validity** is the extent to which the scores from a measure represent the variable they are intended to. But how do researchers make this judgment? We have already considered one factor that they take into account—reliability. When a measure has good test-retest reliability and internal consistency, researchers should be more confident that the scores represent what they are supposed to.
- **Face validity** is the extent to which a measurement method appears "on its face" to measure the construct of interest.
- **Content validity** is the extent to which a measure "covers" the construct of interest. For example, if a researcher conceptually defines test anxiety as involving both sympathetic nervous system activation (leading to nervous feelings) and negative thoughts, then his measure of test anxiety should include items about both nervous feelings and negative thoughts.
- **Criterion validity** is the extent to which people's scores on a measure are correlated with other variables (known as criteria) that one would expect them to be correlated with.
- **Discriminant validity**, on the other hand, is the extent to which scores on a measure are not correlated with measures of variables that are conceptually distinct (Price, Jhangiani, & Chiang, 2015).

Essentially, survey methodology needs to ensure that the information it is reporting holds up over time, is consistent between resources being analyzed, is consistent across different resources, makes sure all of the cultural resources of note are adequately documented, that the standard cultural assets are being held to correlates with the values of the community, and that values that are not supposed to have a correlation with survey outcomes are not excessively represented in the survey results. All current historic preservation surveys currently have face validity. Making sure that validity goes deeper and reliably assesses cultural resources is a foundational goal for where survey and inventory methodology should go next.

9.1. Practical Implications

While there are general guidelines for how to conduct a survey, much of the survey process is left up to personal interpretation. A different survey procedure or goal drastically impacts its outcomes. The two surveys investigated here had very different methodologies. Moving forward, the survey and inventory process has to change to accommodate resources outside of the dominant culture.

In the Downtown Seattle Survey, the intersection of the building age and the architectural and historical significance were analyzed. Their methodology did not leave room for culturally significant resources in the area that lacked architecturally significance. The narrative surrounding the Downtown Survey was very diluted with no public access to the actual final report. The rating system was created after the survey was completed and inaccurately stated the intended results of the survey. The lack of transparency throughout the entire process and continued lack of transparency after its reveal along with the focus only on architectural significance represent the distinct power imbalances in this survey.

The Japanese-American Survey analyzed resources remaining from the Japanese-American community in White River Valley. The survey went the complete opposite direction of the Seattle survey and centered their survey work around interviews with Japanese-Americans that lived in the area before and after the internment. The narrative was centered around describing the culture of White River Valley for Japanese-Americans in the 50 years leading up to 1942's internment. The internment, redistribution of previously Japanese-American owned or operated assets, lack of traditional "integrity" in the properties, and the difficulties in generating awareness for this history represent some of the power constructs present in preservation the process.

The current survey methodology needs to be expanded to preserve assets of high cultural significance that are not able to meet the current interpretation of the standards of integrity. Speaking practically, it would make more sense to reach out to the community and historical research/preservation organizations that focus on the area being surveyed as the first step of archival research. Outside of that, a simple key to finding cultural resources is to look at the historic context statement and unpack how the changes to a survey area over time are currently represented in the built environment. It is important to be able to find these resources in a typical survey because otherwise, the inequity of the results will indicate a lack of methodological validity. Surveys of specific populations and their resources are important, but those hold more value in the preservation of narrative and heritage and would be better served through a more collaborative project. If historic preservation cannot preserve the assets of marginalized communities, what is it saving? If the work being done is not accessible to the public, why are we doing it?

9.2. Limitations

There are many limiting factors in applying this research. The primary barriers are the lack of funding and the lack of time required to restructure the process. Survey change will require change around the way current policy is interpreted, changes to interdepartmental information sharing and transparency with the public, and funding to get it all done. For obvious reasons, that kind of change can prove to be very challenging, especially with the level of collaboration required to create those changes. Finding a way to revamp the process with the goal of consistent validity. Consistently equitable outcomes are vital to the process moving forward, but it is not going to be easy.

9.3. Recommendations

Changes to the way the historic preservation process is being implemented, both big and small, can make a huge change in its outcomes. Survey and inventory methodologies need to be revised to allow for the detection of more diverse and representative cultural resources and to better represent the full capabilities of the policies currently in place. Developing a preservation plan that includes equity as a goal is the most tangible and transparent way to build the foundation of preservation justice. Overall, the policy as it stands is not inherently inequitable, but its implementation on a local, state, and national level has and continues to be ineffective for populations outside of dominant American culture. Methodologies for survey and inventory need to be developed that ensure equitable outcomes. The process needs to be overhauled to exclude bias as much as possible, resulting in survey and inventory reliability and validity. The standards that determine whether or not a resource holds significance need to align with the values of the community. Equitable representation of cultural assets is a requirement for survey validity.
Context statements are a cornerstone of the survey and inventory process. In the National Registry's qualification process, context statements play a vital role in the determination of significance.

The significance of a historic property can be judged and explained only when it is evaluated within its historic context. Historic contexts are those patterns or trends in history by which a specific occurrence, property, or site is understood and its meaning (and ultimately its significance) within history or prehistory is made clear...
In order to decide whether a property is significant within its historic context, the following five things must be determined:

- The facet of prehistory or history of the local area, State, or the nation that the property represents;

- Whether that facet of prehistory or history is significant;
- Whether it is a type of property that has relevance and importance in illustrating the historic context;
- How the property illustrates that history; and finally
- Whether the property possesses the physical features necessary to convey the aspect of prehistory or history with which it is associated (Quaide, 2001).

The goal of the historic context is to create a scale against which to measure significance (Quaide, 2001). Looking at the results of the public input; forming those along with foundational knowledge, regional research, and anything available on HistoryLink to unpack the way the area has changed over time into a context statement; and then moving forward with a survey to figure out where and how the changes, use, and the culture of the area are represented in the built environment. Building infrastructure for engagement and dedicating time to interfacing with the public could potentially crowdsource its way into preservation justice. Interfacing with the public makes it easier to craft historic context statements that truly reflect the intangible values of a community (NATIONAL TRUST FOR HISTORIC PRESERVATION, 2018). Public outreach is a vital part of equitable historic preservation, but its value is often overlooked. Reaching out to the community and historic research/preservation organizations that focus on the survey area should be the first step of archival research in the survey process. Moving forward, it would be most beneficial to streamline the public outreach process and integrate it with public engagement. The Austin Historical Survey Wiki is essentially an area survey for the architectural character that then allows for the public to input their own information in regard to the significance of the properties surveyed (The University of Texas at Austin School of Architecture, 2012). Other areas have created walking tour apps, done cultural mapping workshops to map traditional cultural properties for tribes with ArcGIS, created do it yourself tool kits for residents to move forward with preservation efforts on their own, created walking tours in Google Maps, and utilized virtual reality to simulate landmarks that were no longer there (Ciecko, 2018; City of Pasadena, 2019; NATIONAL TRUST FOR HISTORIC PRESERVATION, 2018; Peter, 2013; Philadelphia Historic Preservation Task Force, 2019). These types of interventions may currently seem out of reach, but a few of these projects were done by their local universities for low or no cost to the municipality and most were grant-funded.

Equity-Centered Community Design, used as a framework for this study, is a great framework to use in designing public outreach and engagement moving forward. It dismantles the structures around engaging the public and can help to create community meetings that are within people's window of tolerance, or do not trigger the fight, flight, freeze, or fix impulse, for the marginalized people in attendance (Williams, 2018). It is very important to make the meetings relaxed and try to have people actually doing work in the community plan them. Work to share about existing resources, explain the process informally to the community, make sure that your definition of "community" is very thoughtful and inclusive of groups that are often excluded from these discussions (marginalized populations, children, the elderly, low-income, unhoused, people who speak English as a second language, LGBTQIA+, etc), eliminate jargon from all handouts and presentations, serve food, play music, and maximize the opportunity.

Throughout the recommendations, I will be referencing the National Register Criteria for Evaluation. Those criteria state that:

The quality of significance in American history, architecture, archeology, engineering, and culture is present in districts, sites, buildings, structures, and objects that possess integrity of location, design, setting, materials, workmanship, feeling, and association, and: A. That are associated with events that have made a significant contribution to the broad patterns of our history; or B. That are associated with the lives of significant persons in our past; or C. That embody the distinctive characteristics of a type, period, or method of construction, or that represent the work of a master, or that possess high artistic values, or that represent a significant and distinguishable entity whose components may lack individual distinction; or D. That have yielded or may be likely to yield, information important in history or prehistory (NATIONAL TRUST FOR HISTORIC PRESERVATION, 2018).

On a base level, these criteria inform preservation criteria throughout the country. On a local and national level, historic preservation professionals largely only survey for criterion C, when criteria A, B, and D are also viable pathways to designation (NATIONAL TRUST FOR HISTORIC PRESERVATION, 2018). Rewriting the inventory form to ask about events that have happened at an asset, people they may be associated with, and whether or not they have the potential to reveal more information specifically would encourage designation through the other criterion. Making this change would also require a redefinition of integrity to mean retained identity rather than architectural integrity.

Integrity is usually interpreted to mean architectural integrity, but is not defined on a local level and on a national level is significantly broader than the way it has previously been interpreted. Nationally, integrity grows out of significance and manifests in "location, design, setting, materials, workmanship, feeling, and association" (Quaide, 2001). The way that integrity is currently being assessed locally is largely just by analyzing the architectural features of the building. The National Register, on the other hand, has a thorough methodology for assessing integrity. The National Register bases integrity on the significance of a property and goes through the steps of defining which physical features are vital to that significance, determining whether or not those "features are visible enough to convey their significance," determine if the property needs to be compared to similar properties, and determining which of the seven "aspects of integrity are particularly vital to the property being nominated and if they are present" (Quaide, 2001). They define integrity as the resource retaining the "identity for which it is significant" (Quaide, 2001).

Changes made by marginalized groups should be seen as a manifestation of use and a sign of integrity in properties significant for their association with those groups. The fact that these changes are seen instead as detraction is little more than a manifestation of the respectability politics within the field, making the uniformity of resources an unspoken goal. For assets of marginalized communities, then, it would make sense to evaluate the changes in materials or design as a manifestation of workmanship. Often, these changes are viewed as a compromise of their integrity, but culturally, that kind of integrity is not valued in the margins of traditional American culture.

The current inventory form does not actually allow for resources to be interpreted in this manner. A revision would be necessary to successfully implement the current standards for more equitable outcomes. It would be worthwhile to include the National Register's list of all of the areas of significance on inventory forms. It would also make sense to include traditional cultural properties on the same form. Resources that do not fall into the other categories could then go into consideration for designation as traditional cultural properties. Some additional considerations include adding a note about including all known names for a property on inventory form, including primary and secondary addresses, and providing links to the property in other city or county databases (NATIONAL TRUST FOR HISTORIC PRESERVATION, 2018). Having a standard set of consistent criteria to use across survey work would help to create validity across the data collected. No survey should ever state that properties are ineligible for designation. Only the landmarks board can make that determination. This example of survey criteria classifies properties without putting forth the claim that they are ineligible for designation:

- Rank I Properties deemed immediately eligible for Local Landmark status or nomination to the National Register because of their essential historic importance and retention of architectural integrity. 75 properties were classified as Rank I.
- Rank II Properties that retained a high level of integrity and played a crucial role in the community's overall development. 690 properties were classified as Rank II.
- Rank III Properties that did not themselves retain a high level of architectural integrity but were deemed culturally significant or likely eligible for designation as part of an historic district. Approximately 2800 properties were classified as Rank III.
- Unranked Properties that featured limited integrity, were built in the recent-past, and/or exhibited potentially significant features but could not be determined eligible for historic designation. 1500 unranked properties were included in the HRI (Convergence Architecture, 2018).

The most important parts of making sure survey and inventory methodology have more equitable outcomes is really more about the foundational system that the preservation is happening in. As such, it is important that King County and Seattle have historic preservation plans that list public outreach and engagement, equity, transparency, and accountability as goals.

The City of Portland's Comprehensive Plan includes a bold policy for advancing a more inclusive historic preservation program:

- Policy 4.53. Preservation Equity. Expand historic resources inventories, regulations and programs to encourage historic preservation in areas and in communities that have not benefited from past historic preservation efforts, especially in areas with high concentrations of under-served and/or underrepresented people.
- Portland's existing inventory of historic resources, those included in the 1984 HRI and those that have been voluntarily designated since, are heavily weighted toward the central city and architectural landmarks associated with Euro-American Portlanders. An updated historic resource inventory with a new public database and mapping application could more comprehensively tell Portland's story and provide plentiful opportunities for public education into the future (Convergence Architecture, 2018).

If Seattle and King County made preservation

equity a goal in the same way that Portland has, changes to the preservation policy would follow. Some of the tools that should likely be used to implement those goals include:

- Increased landmark inventory functionality
- Increased awareness around the use of easements and tax incentives for property owners; a redeveloped inventory form
- Staff and landmark board training on equitable use of criteria
- In the case of Seattle, increased interdepartmental communication and a thorough troubleshooting of the current communication pitfalls
- Increasing the diversity of preservation staff to decrease implicit bias in the survey results
- Updating the existing inventories to include narratives and resources of marginalized populations that were not captured in previous inventories.

REFERENCES

Arrested By Vice Squad. (1934, October 14). Seattle Post Intelligencer, p. 60.

Andrews, M. T. (1997). Japanese-American Legacies in the White River Valley: Historic Context Statement and Inventory. Retrieved from King County Landmarks and Heritage Program website: https://play.google.com/store/books/details?id=oiLKGwAACAAJ

Ariel. (2013, March 4). Ok yes fine, I got a haircut: on ditching your signature style | Offbeat Home & Life. Retrieved August 2, 2019, from Offbeat Home & Life website: https://offbeathome.com/haircut/

Break Ground Thursday for Bulbs and 'Barb. (1957, July 14). THE NEWS TRIBUNE, p. 31.

Bruscas, A. (2007, May 3). Old Vogue mural goes up in flames. Retrieved August 2, 2019, from Seattle@ Nite website: https://blog.seattlepi.com/night/2007/05/03/old-vogue-mural-goes-up-in-flames/

Burnier, A., & Roomsch-Katholieke Universiteit (Nijmegen, N. (1975). Science Between Culture and Counter-culture. Dekker and Van de Vegt.

Ciecko, B. (2018, December 18). Using Technology to Enhance Historical Narratives. Retrieved August 20, 2019, from Saving Places website: https://forum.savingplaces.org/blogs/special-contributor/2018/12/18/using-technology-to-enhance-historical-narratives

City of Pasadena. (2019). Historic Preservation. Retrieved August 20, 2019, from City of Pasadena - Planning & Community Development website: https://ww5.cityofpasadena.net/planning/planning-division/design-and-historic-preservation/historic-preservation/

City of San Diego. (2018, November). Historic Contexts and Surveys. Retrieved June 26, 2019, from City of San Diego Official Website website: https://www.sandiego.gov/planning/programs/historicpreservationplanning/contexts

City of Seattle Office of Film and Music. (2005). Seattle Music Map. Retrieved from https://www.seattle.gov/Documents/Departments/FilmAndMusic/SeattleMusicMapComplete.pdf

Convergence Architecture. (2018). Updating the Historic Resource Inventory Actionable recommendations for documenting historic and cultural resources to advance a more inclusive, diverse, and accessible historic preservation program for Portland, Oregon. Retrieved from Portland Bureau of Planning and Sustainability website: https://www.portlandoregon.gov/bps/article/675769

DAHP. (2014). Getting the Future Right: Washington State Historic Preservation Plan 2014-2019. https://doi.org/10.1038/s41557-018-0199-0

Department of Archaeology & Historic Preservation. (n.d.). Survey Levels. Retrieved June 26, 2019, from Washington State Department of Archaeology & Historic Preservation (DAHP) website: https://dahp.wa.gov/historic-preservation/historic-buildings/historic-building-survey-and-inventory/survey-levels

Department of Neighborhoods. (2007). Historic Preservation. Retrieved July 22, 2019, from Department of Neighborhoods website: http://web.archive.org/web/20081201221626/http:/www.seattle.gov/Neighborhoods/preservation/survey_and_inventory_07_results.htm

Desmond, J., McDonagh, P., & O'Donohoe, S. (2000). Counter-culture and consumer society. Consumption Markets & Culture, 4(3), 241–279.

Epidemic of Blazes Laid to Firebug. (1938, September 5). Seattle Post-Intelligencer, p. 2.

Falchuk, D. (2003, December 13). Artifact. Retrieved August 2, 2019, from Diana Falchuk website: http://www.dianafalchuk.com/work/artifact/

Firebug Blamed for Rural Fires. (1938, September 5). Seattle Post Intelligencer, p. 1.

Fire Destroys Rhubarb Shed at Sumner. (1950, January 27). THE TACOMA NEWS TRIBUNE, p. 10.

Flewelling, S. (2002). Shirakawa: Stories from a Pacific Northwest Japanese American Community. University of Washington Press.

Garbacik, 11:06 Jaimee. (n.d.). RESIDENTS' PODIUM FOR SEATTLE LEGACY SPACES. Retrieved from http://www.seattleghosts.com/wp-content/

uploads/2016/10/RESIDENTS-PODIUM-TRANSCRIPTION-EDITED.pdf

Holland-Moore, L. (2016). ETHNIC MINORITY HERITAGE VALUES AND U.S. HISTORIC PRESERVATION SIGNIFICANCE POLICY (Master of Arts in Historic Preservation, Goucher College; R. Wagner, R. Donaldson, & de T. P. Tiller, Eds.). Retrieved from https://mdsoar.org/bitstream/handle/11603/3817/MAHPthesis_Holland-Moore.pdf?sequence=6&isAllowed=y

How to dye your armpit hair | Offbeat Home & Life. (2014, October 3). Retrieved August 2, 2019, from Offbeat Home & Life website: https://offbeathome.com/how-to-dye-your-armpits/

Humphrey, C. (2016). Loser: The real Seattle music story. MISCmedia.

Japanese Americans in World War II. (n.d.). Retrieved from https://upload.wikimedia.org/wikipedia/commons/a/a0/Japanese_Americans_in_World_War_II%2C_a_National_Historic_Landmark_theme_study.pdf

Kaufman, N. (2004). CULTURAL HERITAGE NEEDS ASSESSMENT: PHASE I. Retrieved from National Park Service website: https://www.nps.gov/subjects/tellingallamericansstories/upload/PhaseIReport.pdf

King County. (2016). Landmarks Ordinance 10474 (Version 20.62 PROTECTION AND PRESERVATION OF LANDMARKS, LANDMARK SITES AND DISTRICTS). Retrieved from https://www.kingcounty.gov/~/media/services/home-property/historic-preservation/documents/resources/KCC_20_62.ashx?la=en

Krafft & Krafft Architecture/CRM. (2007). Historic Property Survey Report - Downtown Commercial Core - Seattle WA. City of Seattle Department of Neighborhoods Historic Preservation Program.

Magnuson, E. (2019, June). Seattle Grunge Redux. Retrieved August 2, 2019, from Seattle Grunge Redux website: https://www.seattlegrungeredux.com/

Mudhoney: Tourbook, 1988. (n.d.). Retrieved July 29, 2019, from https://www.ocf.berkeley.edu/~ptn/mudhoney/tourbook/1988.html

Municipal Code Corporation. (2019). Seattle Municipal Code (Version July 9, 2019). Retrieved from City of Seattle website: https://library.municode.com/wa/seattle/codes/municipal_code?nodeId=TIT25ENPRHIPR_CH25.12LAPR_SUBCHAPTER_IIILAPRBO

NATIONAL TRUST FOR HISTORIC PRESERVATION. (2018, May 17). HISTORIC PRESERVATION & BUILDING REUSE: BEST PRACTICES RESEARCH [Letter to PHILADELPHIA HISTORIC PRESERVATION TASK FORCE]. Retrieved from http://planphilly.com/uploads/media_items/national-trust-for-historic-preservation-best-practices.original.pdf

Parker, T. (1944). Yasamura family. Mr. Jobu Yasamura manages a cooperative store in Brooklyn, New York. Back home in Auburn, Washington, he and his father and only brother owned and operated a produce packing and shipping house. At the time of evacuation, Jobu and his family, wife and two children, Seiichi, 11, and Ailyn, 7, were sent to the Minidoka Relocation Center. At the center Mr. Yasamura was Asst. Mgr. of the center co-op store. In January of 1943, he brought his family to New York, where they found a comfortable home in Upper Manhattan near Riverside Drive. Mr. Yasamura is an issei, though he came to the United States when he was four years old. Retrieved from https://oac.cdlib.org/ark:/13030/ft9h4nb5hd/?docId=ft9h4nb5hd&order=2&brand=oac4&layout=printable-details

Peter, L. (2013, July 22). Free Milwaukee Soldiers Home Walking Tour App Launches This Week. Retrieved August 20, 2019, from Patch website: https://patch.com/wisconsin/waukesha/free-milwaukee-soldiers-home-walking-tour-app-launches-this-week_7f11ac76

Peterson, J. (2019, July 23). The rhubarb pie capital of the world gears up for annual homage to tart vegetable. Retrieved August 1, 2019, from thenewstribune website: https://www.thenewstribune.com/news/local/community/puyallup-herald/article232863932.html

Philadelphia Historic Preservation Task Force. (2019). Neighborhood Toolkit. Retrieved August 20, 2019, from Philadelphia Historic Preservation Task Force website: https://www.phlpreservation.org/copy-of-resources-3

Price, P. C., Jhangiani, R. S., & Chiang, I.-C. A. (2015, October 13). Reliability and Validity of Measurement – Research Methods in Psychology. Retrieved July 13, 2019, from https://opentextbc.ca/researchmethods/chapter/reliability-and-validity-of-measurement/

PROPERTY: #99138 Ridpath Hotel. (2009, September 22). Retrieved August 2, 2019, from WISAARD v2.1.3 website: https://fortress.wa.gov/dahp/wisaardp3/

Quaide, R. A. (2001, November 28). National Register of Historic Places Bulletin (NRB 15). Retrieved August 20, 2019, from National Register of Historic Places website: http://www.nps.gov/history/nr/

publications/bulletins/nrb15/index.htm
Scanlon, T. (2007, June 29). The Vogue nightclub is closing. Retrieved August 27, 2019, from The Seattle Times website: https://www.seattletimes.com/entertainment/the-vogue-nightclub-is-closing/
Scanlon, T. (2008, January 31). R.I.P.: Dead nightclubs of Seattle. Retrieved July 29, 2019, from The Seattle Times website: https://www.seattletimes.com/entertainment/rip-dead-nightclubs-of-seattle/
Staff Corespondent. (1938, September 11). Firebug's Trail Hot. THE TACOMA NEWS TRIBUNE, p. 1.
Stevens, J., & White, D. R. (2017, July 28). WREX. Retrieved July 23, 2019, from Jive Time Records website: https://jivetimerecords.com/northwest/wrex/
The University of Texas at Austin School of Architecture.

E8 - Transform Processes and Outcomes for Equity: Cultural Aspects

Just US: Equity-Centered Revitalization in Traditionally BLACK Neighborhoods

Tera Williams (University of Washington)

1. ABSTRACT

This theory review analyzes existing foundational research on the intersections of equity, revitalization/ displacement, and culture to develop a new, more equitable, model of revitalization. We compiled sources to create a new model for equitable revitalization, tested and refined the model, presented a case study at EDRA, and finalized the method in this form. The model centers meetings led by changemakers from within the community to diffuse power imbalances that prevent equitable revitalization in the existing community development structure.

The revitalization of traditionally Black communities looks at land as a commodity rather than a foundational piece to the story of its inhabitants. This revitalization often works to bring in outsiders and subsidize their visions rather than supporting and amplifying existing parts of the community's social structure. This process ultimately erodes the social fabric of a neighborhood and drives up the rent, displacing people from their homes and removing the communal aspects of living in the city from the neighborhood.

Revitalizing a neighborhood equitably based on the cultural values and narratives surrounding the community will improve the health of that community without displacing long-time citizens. Revitalization should be about the overall health of the existing community members.

1.1. Keywords

Equity, Revitalization, Culture, Justice, Community

2. METHODS

The purpose of this theory review was to construct a new model of revitalization that facilitates non-violent communication between changemakers and city government. Also called participants in practice, changemakers are people on the ground in the community trying to improve their community through grassroots organizing efforts (McElroy, 2016). Changemakers understand the existing community narrative, history, and the forces that their community has been fighting against; they are the ones who know what is going on on a grassroots level within their communities and are the most qualified people to create change. These people have the necessary information to figure out what a community needs most, but we as a society are not empowering them to feel like they can make a difference. They are the ones who could best utilize the revitalization process outlined here, if only they could find a way to bring that process to the planning table. My hope is that I can utilize this project to find a way to empower marginalized communities in the revitalization process.

To do this review, I initially worked by identifying the main contributing factors to my model. The factors I identified were equity, revitalization/displacement, and culture. I then worked to find sources that overlapped those concepts. After months of searching for a theoretical model for equitable revitalization that overlapped those categories, I decided to compile my own. I completed a draft of the initial model, presented a case study created with that model at the Environmental Design Research Association, and revised and further finalized the model. My model of equitable revitalization centers around holding community meetings led by changemakers. The goal for these meetings is to facilitate conveying the communities opinions to local government without activating fight, flight, or freeze in community members. My model incorporates Marshall Ganz' theory of Public Narrative, Narrative theory, and Creative Reaction Lab's Equity-Centered Community Design process with references to Craig Wilkins Aesthetics of Equity, placekeeping and placemaking, Asset Based Community Design, and Regenerative Design. It centers actively problem-solving with the community in revitalization efforts work. To further center an intersectional and multidisciplinary audience and make the work as approachable as possible, the writing style used in this report is intentionally less formal.

3. INTRODUCTION

3.1. Background

Throughout the United States, discriminatory

personal beliefs rooted in anti-blackness strongly influenced the design of our communities, resulting in segregation and ethnic enclaves within cities. Anti-blackness is a form of anti-indigeneity specifically targeting Black people and stems out of anti-darkness. It is a precursor to racism (Brown, 2018). Anti-darkness was, at some point, intertwined with white supremacy, the belief system that white people are superior to all other races, to create anti-indigeneity and justify the subjugation of other races of people. While Black people were negatively impacted by anti-indigeneity and are indigenous to the continent of Africa, they are rarely thought of as such and are frequently discriminated against, even by other communities of color. Within American culture, racism was reverse engineered to justify the heavily prevalent sentiments of white supremacy and create cognitive dissonance about the inhuman treatment of enslaved Africans.

While structurally racist policies, such as Jim Crow and Redlining, have since been abolished, sentiments of anti-blackness still persist as part of the American cultural toolkit. Even as a black or non-black person of color, it takes continual work to unpack the internalized racism you have consumed (Brown, 2018). Moving forward, it is impossible to look at the way that marginalized people interact with the built environment within America without first acknowledging the fact that it was either not built for them or, in the case of newer development, built for them by people that think they are inherently inferior.

While marginalized populations often suffer from discrimination and limited opportunity in society as a whole, ethnic enclaves serve as a break from that. They are there so that members have a comfortable space to exist. The Architecture of Happiness discussed the idea of community as an extension of the home -- a place that a member of that community can feel comfortable to be themselves (De Botton, 2008). Enclaves function as a buffer zone between the harsh reality of the outside world and home.

Having that space where they are free to speak their own language/dialect, can support their local economy, and have strong social connections, defined here as social cohesion, is a large indication of their health and mental well being (Bruhn, 2009; Chuang, Chuang, & Yang, 2013; National Center for Environmental Health, 2009). Ethnic enclaves provide an opportunity to start businesses specifically targeted toward the community they are in (Ruef & Grigoryeva, 2017). W.E.B. Du Bois spoke on this phenomenon when he said, "it is [the] density of Negro population in the main that gives the Negro businessman his best chance" (Du Bois, 1899). While historically very important to marginalized communities in the United States, these enclaves are often spaces that cities considered an afterthought. Enclaves have traditionally been "located close to an urban core, in areas of mixed land use, old housing, poor services, and low or decreasing socioeconomic status," from which members try to "spatially assimilate" elsewhere, or leave the ethnically defined space to join accultural society once they have earned enough money (Massey, 2001; Smith, Pride, & Schmitt-Sands, 2017). And to an extent, leaving makes sense. In American, the zip code you live in has a strong correlation with how old you will be when you die (Dwyer-Lindgren et al., 2017). Ethnic enclaves are likely to contain vulnerable populations because of the old housing stock, lower incomes, and the disadvantages that ethnic minorities face in society as a whole.

3.2. Problem

With increasing urban population growth, ethnic enclaves are being revitalized. A revitalization movement is an "organized attempt to create a more satisfying culture" (The Editors of Encyclopædia Britannica, 2011). Revitalization in marginalized communities frequently centers the feelings of outsiders about the community to create a more satisfying culture for outsiders rather than existing community members. When outsiders revitalize a neighborhood, they often do not focus on preserving the existing pillars of the community because they are unaware of them and instead focus on attracting further financial opportunities from the outside. Revitalizing based solely on economic growth displaces people and erases their presence from the landscape.

This process is often referred to as gentrification. From the term's origination, gentrification has always meant "remaking a neighborhood for new incomers and to the detriment of current residents" (Moskowitz, 2017, p. 32). Gentrification happens in distinct and predictable phases. First, individuals move to an area and tell their friends about it. Next, people see the changes occurring and start buying real estate in the area to be a part of the change. It is at this stage that displacement starts to begin. The third stage is when the newcomers start to hold "more prominent roles" in the neighborhood and promote the neighborhood to people outside of the original community. At this stage, banks start to lend more within neighborhoods where loans were previously inaccessible and police presence increases to make sure the newcomers "feel safe." Stage four is when the gentrification is complete and the neighborhood starts to become exclusive and wealthy. The people that originally began the gentrification process, the "pioneers" start to be displaced. Some places have even surpassed stage four and moved into a fifth stage in which only the rich elite can afford to live in previously affordable neighborhoods. Before these stages, though, is a stage zero in which local governments decide to allow gentrification to happen. Popular rhetoric scapegoats gentrification onto developers, white people, or some combination of the two, but the government puts all of the pieces in play to create landscapes that are unequal.

In every gentrifying city—that is, in every city where there is a combination of new coffee shops and condos, hipsters, and families struggling to hang on—you can usually trace the start of that change not to a few pioneering citysteaders but to a combination of federal, local, and state policies that favor the creation of wealth over the creation of community. Usually those policies come in the form of the deregulation and privatization of urban services: transportation, education, and especially housing. By the time the hipsters arrive, the political and economic forces that paved the way for them have been at work for years (Moskowitz, 2017, p. 26).

After Ronald Reagan drastically cut government spending in the 1980s, cities had to figure out how to fund themselves. Cities with "basic social safety nets" and "not enough income" were unable to borrow money to support themselves, so they then started functioning more like corporations, prioritizing making money over the welfare of the people and utilizing gentrification as a tactic to generate funding after decades of federal budget cuts (Moskowitz, 2017).

Gentrification is not the result of "individual or institutional" actors, but rather as "a logical consequence of a system in which real estate is viewed as an unrestrained commodity" (Moskowitz, 2017, p. 121). If cities value making money above the needs of marginalized populations, gentrification is a natural progression. Rosa Luxemburg theorized that in a landscape with excess capital, cities would end up being used to "absorb" those funds. At that time, Luxemburg meant things like "grand architecture, monuments, parks, and beautiful streetscapes as ways to attract the rich and beef up a city's tax base," but in communities that have been systemically neglected for generations, that ends up manifesting as "coffee shops, streetcars, and art galleries" (Moskowitz, 2017, p. 121). Gentrification is a governmental revitalization tactic.

In the United States, whatever land is the least profitable ends up housing the poor. Most other industrialized nations have realized that a free real estate market will never meet the needs of the poor and as such have set aside land or created regulation to make sure the needs of everyone are factored into the planning equation. Mass displacement only happens in countries that have limited policies on housing regulation. Within the US, the selection process is largely an automatic decision of putting the people with the least resources into neglected or toxic landscapes. Urban life is often an existence fraught with unequal access to opportunity. Looking at the spatiality of social issues yields different results than looking at the issues from a standpoint of access without regard to distance. In order for gentrification to exist at all, the landscape of the United States had to be inherently inequitable. Gentrification cannot happen without some land having a lower cost than other land. This spatial inequity within itself creates the landscape of gentrification.

Spatial Justice is very interrelated with social justice in that social issues cannot exist without space and space exists to be social. Social equity theory is basically a judgment of whether the distribution of resources is fair to all parties involved. Spatial equity is social equity applied to geography. Historically, things have not been fair on a social level and social inequity was made spatial through redlining.

While "gentrification is often presented as a sort of corrective to the suburbs" it is instead a continuation of the suburbanization experiment that acts a new form of segregation because only newcomers are granted subsidies and the geographies that are being constructed are not for the existing populations (2017, p. 102). Gentrification is being subsidized in a similar fashion to the way the suburbs were fifty years ago. Then, "subsidies came in the form of billions spent on highways and suburban housing," but in the current day subsidies are coming "in the form of billions used on tax breaks for stadiums and condos, for renovations of storefronts and homes, for streetcars and bike lanes" (Moskowitz, 2017, p. 102). The suburbs "were built to reinvigorate capital" (Moskowitz, 2017, p. 128). Now that poor people are being displaced into suburbs or rural areas, they are not getting the same level of community or city services that they are acclimated to. The sidewalk life within cities gives their citizens the safety and community that people in the suburbs pay for. The suburbs replaced collectivism with consumerism and gentrification is the transplantation of that consumerism to an urban environment. One has to wonder how the suburbs will function when they are primarily filled with the poor, considering that suburbs are still heavily subsidized (Moskowitz, 2017).

Often, this is referred to as gentrification, but calling it gentrification centers the actions of outsiders rather than the impact on the community. Within this work, I discuss mass displacement, erasure, and economic disempowerment rather than gentrification to center the experience of marginalized communities. Gentrification is a revitalization strategy, not some strange unknown cultural phenomena; in order to equitably revitalize, the focus needs to be on equitably revitalizing. Creating a more satisfying culture should be about providing the things within a community that its members need to thrive in their neighborhood. Revitalizing a neighborhood equitably based on the cultural values and narratives surrounding the community will improve the health of that community without displacing long-time citizens.

3.3. Motivation

Space is created culturally and manifests as both a noun and a verb. The way the construct of space is defined makes space a noun and "the desirability of space is fundamentally and intimately formed by race" because of how space presents as a verb, or how we use the space (2007, p. 22).

Danger remains largely, if not solely, a condition of color and the evidence of its effect is inarguable: African Americans rarely enjoy the benefits of space, as they are historically and routinely rendered "visually transparent, peripheral, part of the landscape, ready to be moved, cleared and discarded for spatial use and improvement" or to remove "danger."

[S]pace is an arena where the black male is thought of as the bogeyman, a volatile predator, ready to pounce on anyone caught slipping. The idea that he might be concerned about his safety is regarded as a ridiculous notion. Yet, it is often he who is unable to pause and relax without physical, psychological, and other non-verbal harassment when almost any movement is viewed with suspicion (2007, p. 23).

Arthur Symes said that,"[t]here are no architectural problems, per se, they are part of society as a whole," and with that in mind, it is interesting that we often think of our spaces as neutral and design them without first unpacking our own biases (Wilkins, 2007, p. 23). If society as a whole struggles with white supremacy, why then would we assume that we do not apply that same lens to our spaces? Understanding then that people have very different perceptions of space based on the foundational narrative of their culture is a significant portion of what this project entails.

The dominant culture of America generally views space as an inert thing that just exists, whereas in marginalized communities, space is defined by how it is used. Marginalized people are still a part of society, they are just interpreting things through a different lens and with a different narrative template. It's not necessarily that they live by different rules, it is just that they interpret space differently. The thing about existing in the margins is that sometimes, there is overflow. People that live in the center either get forced out of their position and feel entitled to space in a marginalized community or decide to intentionally relocate themselves further from the center into the margin.

In a model where space is defined by its use, being critical of vital aspects of the community means that one should not inhabit that space. For marginalized communities that have made something from nothing, space is the community. Spaces in the margin were created through the community. People living in the margins have had to fight for the right to exist. The spaces that they live in are often the spaces they were relegated to and they have built them up brick by brick into something that feels like home. This different understanding means that often, people are not aware of the space they take up in moving to a marginalized community. People feel entitled to space because they pay for it, not understanding the foundational community narrative surrounding that space and how it will impact the community to invite a different market into their space. Marginalized communities rely on space to maintain their social fabric and that is foreign to people that do not see a connection between the spatial and the social.

4. MODEL

Marshall Ganz' theory of Public Narrative is foundational in the way the revitalization process should be reimagined. Public Narrative rests on the idea that "discontented, but compliant" communities can bring about change through community-led "organized collective action to challenge the status quo, as opposed to the occasional outburst of resentment" (Ganz, 2011). Such social movements are necessary for the revitalization process because marginalized communities cannot rely on established systems to work for them and instead have to work to translate their values into action on their own. Ganz calls that translation process Public Narrative. Foundationally, a Public Narrative cannot be created without a narrative template or understanding of the community being served (Stapleton & Wilson, 2017).

Narrative is not a concept that is at all foreign to most members of marginalized communities. Within marginalized communities, the idea of narrative is largely understood. If you ask someone about their story and they tell you where they are from, you know a lot about them already. That is all you need to know whether or not the two of you have a common narrative and the level on which you will be able to relate. A narrative template is not a story that you are building, but a template that everything else about a community depends upon for validity. Without that template, the community would not truly have a collective identity. Wertsch's concept of "collective memory" is the basis for a narrative template that then interfaces with the "cultural tool kit" to create a portion of a community member's identity within that narrative. In a way, we could think of the collective memory and cultural toolkit parts of a narrative template as the subconscious slant within a community member's own telling of their story. It manifests in things like word choice, tone, and focus without the author actively trying to portray any feelings one way or another.

The Public Narrative process identifies three distinct challenges in community organizing work-- hands, head, and heart or action, strategy, and narrative, but primarily creates a framework to share narrative. Public Narrative centers action in values. It uses those values to both figure out which actions are worth pursuing and garner support for action. Within the Public Narrative framework, strategy and narrative work together to create action. It is easy to identify with the character in a story and that allows the listener to experience the feelings of the issue instead of static debate points. Public Narrative works to connect three different narratives to create one cohesive Story of Action. It connects the story of the individual, the story of the collective, and the reason action needs to happen now.

Within Public Narrative, Ganz subscribes to Jerome Bruner's argument that people interpret all information either analytically or narratively (Bruner, 2009). In other words, we analyze the information we are absorbing and then try to make a judgment about that information based on what we already know. Ganz states that "when we consider action in the face of uncertainty, we have to ask ourselves three questions: why must we act, how can we act, and what must we learn to do" (Ganz, 2011). The narrative is what answers our why, critical analytic thinking answers our how, and the conclusions of critical analysis lead us to our what.

Through the Public Narrative process, value-based feelings are turned into stories. The underpinnings of that process are understanding our values, emotions, how we experience them, and how they intertwine. Our values are the compass for our feelings. Making choices that should be based on values without being in touch with one's feelings, positive or negative, is extremely difficult. Different types of feelings can either stop or encourage an action (Ganz, 2011).

Three emotional systems for these types of reactions are fight, flight, or freeze; surveillance; and disposition. Fight, flight, or freeze is an emotional system that initiates the responses that people have when they feel threatened. The surveillance system compares base knowledge of how things are supposed to happen with what is actually happening. As such, anxiety is often a tool of this system preventing people from functioning primarily out of habit. Finally, the dispositional system is basically a range "from depression to enthusiasm, or from despair to hope" (Ganz, 2011; Marcus, 2002). The systems all overlap and work together. Depending on what our disposition is when our surveillance is triggered, we will either end up in fight, flight, or freeze; think critically; take action; or somewhere in the middle. Understanding how these systems interrelate and impact emotional outcome makes it possible to counter negative reactions and encourage more people to act. Public meetings within the revitalization process can activate these emotional systems in one way or another. Working to structure these types of conversation differently can result in an increased comfort zone for participants.

A Public Story, like any other story, consists of a plot, character, moral, and setting. A plot is "a challenge, a choice, and an outcome" (Ganz, 2011). It only teaches us something if it has a relatable character. Having a character within the story that they can connect with helps the listener to empathetically experience the story. The moral of a story is about making a point. A plot and a character mean nothing without a point to the story. The setting of a story anchors it in space. In work rooted in a specific location, the setting becomes a character in the story of its own right. A firm understanding of the public story is imperative to the construction of the public narrative.

The individual narrative tells the story of the values that shape you. Referred to by Ganz as the "Story of Self," an individual narrative designed to encourage others to take action centers around "choice points." A choice point is like a smaller story within your overarching story. It marks a moment when you "faced a challenge, made a choice, experienced an outcome, and learned a moral" (Ganz, 2011). The selection of specific choice points indicates what values are important to you. Telling these stories is important because they communicate with the memories of listeners. Our individual narratives are our stories and those stories define our identities.

Collective identity, called the "Story of Us" by Ganz, focuses on points of intersection for members of a specific group of people. It "brings forward the values that move us as a community" and shows how the storyteller integrates into the collective story. Community narrative is a part of self-identity, not a persona that someone puts on. Culture is largely a collection of stories about collective identities (Ganz, 2011).

Creating a story of collective identity requires determining who gets to belong to that identity and what values are the most important for that group (Ganz, 2011). This work looks primarily at spatial identity. Things that define spatial identity will include common experiences and landmarks, terminology, and can even define base understandings of broader concepts. "The 'where' of our sensory experiences in the world have a profound influence on our ability to create individual and collective identities -- to become, know, and name who we are" and that becomes a sort of "spatial membership" in which one is defined based on the space they occupy (Wilkins, 2007, p. 7). People from outside of the collective identity often want to center their voices in conversations about identities they are not a part of. This conversation is usually not very well informed and steeped in stereotypes about the area and its inhabitants. There may be a kernel of truth here, but the outsider narrative is generally ill-informed and should not be used to define the story of the collective identity. Figuring out who belongs to the collective identity determines whose voices hold the most weight in conversations centering community issues.

Community members interpret and present the community narrative in a variety of ways because of the way that the surveillance and disposition systems function for different people. Some people internalize criticism or praise from the outside. This can impact their judgment in day to day life and ultimately even their perceptions of themselves. As put forth by Eminem, a participant in practice who very much so identifies based on his spatial background, the viewpoint for these people ends up becoming, "I am whatever you say I am. If I wasn't, then why would I say I am?" ("Eminem – The Way I Am," 2000). Some people internalize an undying loyalty to their community in spite of outside criticism. This often manifests as bravado, but can also function as a protective instinct. Some people are not internalizing anything external, but just taking it for what it is. This internal dialogue for community members determines how they present or perform the narrative and can be different on any given day.

Ganz' "Story of Now" is the call to action within the Public Narrative. It sets out to articulate what a community is up against, lay out the choices available, and provide context for what the best possible outcome could be. The Story of Now connects the Story of Us to time. It inspires us to act now through an articulated threat paired with hope. It is a vision of the future built off of a reflection of the past. "In a story of now, we are the protagonists and it is our choices that shape the outcome" (Ganz, 2011).

While the Public Narrative framework addresses the heart aspect of community organizing, it neglects the "head," or critical thinking aspect of the process that leads to action. Within this project, Creative Reaction Lab's Equity-Centered Community Design process fills that role. Equity-Centered Community Design is, "a unique creative problem-solving process based on equity, humility-building, integrating history and healing practices, addressing power dynamics, and co-creating with the community. This design process focuses on a community's culture and needs so that they can gain tools to dismantle systemic oppression and create a future with equity for all" (Creative Reaction Lab, 2018a). Equity-Centered Community Design is an extension of the Equity-Centered Design Framework created in part by Stanford's The d.school. That framework appears to have initially been designed to be applied in the classroom but is very flexible in its thinking (Stanford d.school, 2016a, 2016b). The d.school puts a high value on figuring out who you are so that the design de-centers the individual designer (Stanford d.school, 2016a). Coming from that, Equity-Centered Community Design applies it specifically to the design of equitable communities for marginalized people.

Creative Reaction Lab outlines the Equity-Centered Community Design process as "inviting diverse co-creators, building humility and empathy, history and healing, acknowledging and dismantling power constructs, defining and assessing the topic/community needs, ideating approaches, rapid prototyping, and testing and learning" (Creative Reaction Lab, 2018b). The Equity-Centered Community Design Field Guide builds upon that to explain what each section is, why it is necessary, tips on how to move forward with that step in the process, a sample activity for each section, and a space to reflect on the section. Equity-Centered Community Design is a reimagining of traditional design thinking to further fit the needs of a community.

Bringing different people to the table is very important when doing equity work. Inviting a variety of voices into the conversation creates a deeper knowledge base and a greater level of understanding when trying to solve problems. ECCD argues that "we especially need to include the individuals and communities affected, also known as living experts, who are often excluded from design and decision-making processes" (Creative Reaction Lab, 2018b). In bringing an array of voices to the conversation, ECCD requires "diversity, inclusion, equality, and equity" as a prerequisite (Creative Reaction Lab, 2018b). The guide recommends not confining people to the roles you brought them in to play. Having people there that represent differing interests is about more than checking boxes, it is about making sure everyone's voice is heard.

Without humility and empathy, it is hard to see things from other people's point of view. In this context, "humility is recognizing the influence of your own biases and perspectives when trying to understand another's emotions, thoughts, experiences, and actions" and "empathy is the active attempt to understand another person's perspective by imagining how you would feel, think, or act if put in their situation." The combination of the two "help us understand others while also acknowledging our biases" (Creative Reaction Lab, 2018b). To facilitate that, the field guide suggests getting comfortable with being uncomfortable, acknowledging the limits to your knowledge base, and letting go of assumptions.

The history and healing section is about relearning the cultural heritage of a community without a lens of oppression. To do this, people have to question the messages they have been fed. The field guide suggests you ask, "'Who wrote this narrative?' 'What was its purpose?' 'In what time context did this occur?'"(Creative Reaction Lab, 2018b). Reclaiming this history and restoring the narrative that has been erased is healing for a community.

Power constructs, "invisible delineations of power that are created and maintained by larger systems such as government, media, education, and private industries," hold marginalized communities back (Creative Reaction Lab, 2018b). Much like privilege, everyone benefits from power dynamics in one way or another. It is important to look at the power dynamics in a project. Within the revitalization process, power dynamics traditionally play a large role in preventing marginalized communities from being heard. This can result in communities not feeling like their voices will be taken into account in final decisions and not seeing why they should bother to make suggestions. Creating a space that is free from those types of situations takes work, but ultimately allows for participants to be significantly more comfortable sharing. Defining and assessing the community needs is the first step toward creating a plan for moving forward. Working to understand a community makes it easier to figure out why those needs are not being met (Creative Reaction Lab, 2018b). To do this, ECCD includes community members throughout the entire process, rather than just getting a say in the beginning. In this stage of the ECCD process, the community needs stem from the information used to develop the Story of Us. At this point, a community can pull directly from the Story of Us to better outline what needs are and are not being met.

The ideating process is about brainstorming potential solutions to the problems identified by the community. This step is what, "moves the process from theory, research, and competency building to the development of interventions, initiatives, and other approaches" (Creative Reaction Lab, 2018b). This process is just about getting ideas out as a group. Going into it, it is important to let go of any ideas about what things have to look like and go with the flow. Refining needs to happen later.

Rapid prototyping is a stage of the design process that allows for testing of what works and what does not by creating 2D and 3D representations of potential solutions. These representations, also known as prototypes, can help to think through a design, explain what someone is thinking to others, and allows for designers to make their mistakes early before they cannot be corrected (Creative Reaction Lab, 2018b). Rapid prototyping is about making multiple drafts of an idea and improving upon versions as you go.

Testing and learning come after the prototyping phase. Eventually, a prototype of the idea is made for scrutinization and testing by community members. The testing and learning process is there to improve the idea and elevate it to a level where it can best fit the needs of the community. The field guide suggests bringing in the community to test the idea but to also include other members of the community for additional perspective. "Since Equity-Centered Community Design is about designing with, not for, it's important that there are clear, consistent channels for all stakeholders -- especially those directly impacted by the designs and decisions at hand -- to share feedback" (Creative Reaction Lab, 2018b). Testing and learning help to refine ideas.

Equity-Centered Community Design frames the critical thinking and community involvement components of the revitalization process completely and allows for the marriage of foundation theories to further frame guiding principles. Asset Based Community Development is rooted in the idea that you can take what you have in the community and build from there rather than looking at the negatives present. Asset Based Community Development is basically like providing a framework for ECCD specifically focusing on the revitalization process. Regenerative Design is framed in three ideas that I feel are very important to the ECCD process. The theory of Regenerative Design strives for "a radical change for the better; creation of a new spirit; returning energy to the source" (Östlund, 2017). Within this work, Asset Based Community Design factors in as Placekeeping and Regenerative Design factors in through Placemaking and overall planning efforts. Combining different theories to provide different approaches to the problem-solving component of revitalization creates a richer Equity-Centered Community Design process.

5. CONCLUSIONS

The development of this methodology led to the realization that community buy-in is a large part of the process. Developing a public narrative leads to public buy-in and makes it easier to talk about a project's intended outcomes. Moving through the ECCD process with firmly grounded intentions enables the revitalization process to have community buy in and the process itself allows for equity to be centered in the community engagement segment of the revitalization process. The lack of research on equitable revitalization methodology in and of itself shows that more work is needed in the field.

America was built on a foundation of trauma, tainting the soil through its colonization. We are growing and building on tainted soil. As time passes, we move further from the origin of the trauma, but the only thing we have to draw from is contaminated soil. We are poisoned by our surroundings and as such, even when trying to improve things, all we have to offer has been tainted. No matter how hard something tries to grow, it stems from toxicity (Brown, 2018). Healthy growth requires that the trauma is addressed.

Contaminated soil needs to be decontaminated to reach its full potential. Soil decontamination is often handled through a complex human-led process. Traditionally, contaminated soil is removed or diluted. Removing it treats the soil as the problem rather than handling the contamination. It shuffles the contaminated soil elsewhere instead of removing the actual contamination and healing the soil (Kaller, 2014). The way that soil decontamination is handled often ends up instead being site decontamination. Often seen as a side effect, displacement is actually a choice that is made.

Soil can be decontaminated naturally through a process called bioremediation. Soil can be purified by certain plants in a process called phytoremediation or mushrooms through mycoremediation (Belik, 2013; Kaller, 2014). Different plants and mushrooms can clean up different types of toxins without being poisoned by them. Bioremediation starts with a site getting examined to see what toxins are present. A plan is then developed to use the appropriate plants to detoxify the area (Kaller, 2014).

The displacement of soil has many overlaps with the displacement of people. Within both contaminated sites and urban neighborhoods, displacement is utilized as a treatment strategy. In fact, displacement shifts things that the system sees as undesirable to locations that have less value. Black people within the built landscape are viewed as a problem rather than a population that has been poisoned by foundational inequalities. If we instead, work together to create a more equitable process, displacement can be largely avoided. Looking at these communities as soil, examining their foundational contaminants, and then figuring out who can come in and work together as a team to fix that contamination can solve the inequality without displacing the communities. In Tupac's posthumously released spoken word album, The Rose That Grew from Concrete, he states that, "you wouldn't ask why the rose that grew from the concrete had damaged petals. On the contrary, we would all celebrate its tenacity. We would all love its will to reach the sun. Well, we are the rose, this is the concrete, and these are my damaged petals. Don't ask me why, thank God nigga, ask me how" (Serpick, 2000; Shakur & Giovanni, 2000). Tupac was referring to his own tenacity in rising above his situation, but what if we did ask the people how? Just listening to the way that communities work to survive would likely inspire a different kind of design. Understanding that different communities interact and interpret the built environment in different ways should be a vital part of the revitalization process. Community needs to be framed as a place to feel safe and grow together. Equitable revitalization should be a means of building communal wealth and assets for the traditional residents of the area.

REFERENCES

Belik, V. (2013, August 9). Nature's way of cleaning contaminated soils - Yukon News. Retrieved October 17, 2018, from Yukon News website: https://www.yukon-news.com/letters-opinions/natures-way-of-cleaning-contaminated-soils/

Brown, R. (2018, November). Healing from Internalized Racism. Presented at the Healing from Internalized Racism: A Workshop for Womxn of Color, Seattle.

Bruhn, J. (2009). The Concept of Social Cohesion. In J. Bruhn (Ed.), The Group Effect: Social Cohesion and Health Outcomes (pp. 31–48). Boston, MA: Springer US.

Bruner, J. S. (2009). Actual Minds, Possible Worlds. Harvard University Press.

Chuang, Y.-C., Chuang, K.-Y., & Yang, T.-H. (2013). Social cohesion matters in health. International Journal for Equity in Health, 12, 87.Creative Reaction Lab. (2018a). About Equity-Centered Community Design. Retrieved February 19, 2018, from Creative Reaction Lab website: http://www.creativereactionlab.com/eccd/

Creative Reaction Lab. (2018b). Equity-Centered Community Design Field Guide. Creative Reaction Lab.

De Botton, A. (2008). The Architecture of Happiness.

Knopf Doubleday Publishing Group.
Du Bois, W. E. B. (1899). The Negro in business. Atlanta University.
Dwyer-Lindgren, L., Bertozzi-Villa, A., Stubbs, R. W., Morozoff, C., Mackenbach, J. P., van Lenthe, F. J., ... Murray, C. J. L. (2017). Inequalities in Life Expectancy Among US Counties, 1980 to 2014: Temporal Trends and Key Drivers. JAMA Internal Medicine, 177(7), 1003–1011.
Eminem – The Way I Am. (2000, September 7). Retrieved February 28, 2018, from Genius website: https://genius.com/Eminem-the-way-i-am-lyrics
Ganz, M. (2011). Public Narrative, Collective Action, and Power. In Accountability through Public Opinion (pp. 273–289). The World Bank.
Kaller, B. (2014, August 11). Using plants to clean contaminated soil - Resilience. Retrieved October 17, 2018, from Resilience website: https://www.resilience.org/stories/2014-08-11/using-plants-to-clean-contaminated-soil/
Marcus, G. E. (2002). The Sentimental citizen. University Park. PA: Pennsylvania State University Press.
Massey, D. S. (2001). Residential segregation and neighborhood conditions in US metropolitan areas. America Becoming: Racial Trends and Their Consequences, 1(1), 391–434.
Moskowitz, P. (2017). How to Kill a City. Special Markets Department at Perseus Books, 2300 Chestnut Street, Suite 200, Philadelphia, PA 19103,: Nation Books.
National Center for Environmental Health. (2009, October 15). Health Effects of Gentrification. Retrieved November 19, 2018, from CDC Healthy Places website: https://www.cdc.gov/healthyplaces/healthtopics/gentrification.htm
Östlund, S. L. (2017). Regenerative Place-Making (LICENTIATE OF PHILOSOPHY, Chalmers University of Technology; J.-H. Kain, Ed.). Retrieved from https://www.mistraurbanfutures.org/sites/mistraurbanfutures.org/files/regenerative-placemaking-comp.pdf
Ruef, M., & Grigoryeva, A. (2017, March). Jim Crow, Ethnic Economies, and Status Attainment: Occupational Mobility among U.S. Blacks, 1880-1940 *. Retrieved from https://pdfs.semanticscholar.org/ad/2a4806e5327db0ee7dd57295a0445a840cf6.pdf
Serpick, E. (2000, November 24). Tupac Shakur: The Rose That Grew From Concrete, Volume I. Retrieved November 20, 2018, from Entertainment Weekly website: https://ew.com/article/2000/11/24/tupac-shakur-rose-grew-concrete-volume-i/
Shakur, T., & Giovanni, N. (2000). The Rose That Grew from Concrete. In The Rose That Grew from Concrete V. 1 [CD]. Retrieved from https://genius.com/Nikki-giovanni-the-rose-that-grew-from-concrete-lyrics
Smith, R. J., Pride, T. T., & Schmitt-Sands, C. E. (2017). Does spatial assimilation lead to reproduction of gentrification in the global city? Journal of Urban Affairs, 39(6), 745–763.
Stanford d.school. (2016a). K12 Lab Equity-Centered Design Deck MASTER. Retrieved February 27, 2018, from Equity-Centered Design Framework website: https://docs.google.com/presentation/d/1S-7fZojfgGs3M3T110vaXZFztRvjmMdkCjJ4UiIQ5i0/edit
Stanford d.school. (2016b, May). Equity-Centered Design Framework. Retrieved February 27, 2018, from Stanford d.school website: https://dschool.stanford.edu/resources/equity-centered-design-framework
Stapleton, K., & Wilson, J. (2017). Telling the story: Meaning making in a community narrative. Journal of Pragmatics, 108, 60–80.
Wilkins, C. L. (2007). The aesthetics of equity. University of Minnesota Press.
Yisak, I. (2017, October 24). An Incomplete List of Resources for the Equity-Centered Designer. Retrieved February 27, 2018, from Medium website: https://medium.com/equal-space/an-incomplete-list-of-resources-for-the-equity-centered-designer-4f57b410e606

E9 - Transform Processes and Outcomes for Equity

The Analysis of Bias with Visual Illustrations

Li Han (Virginia Commonwealth University-Qatar)

1. ABSTRACT

This paper proposes a methodology to examine confirmation bias through a data analysis illustrated with graphs. The focus is to understand bias through color-coded graphs and the interrelationships and associations of the textual content in order to reveal any implicit bias. The goal is to provide a graphic analysis method which is rooted in social science analysis but which at the same time is visually direct in order to examine the question of bias.

1.1. Keywords
Implicit Bias, Confirmation Bias, Graph Illustration

2. INTRODUCTION

Biases are developed through a faulty cognition process. In the 1970s, psychologists began to demonstrate that human beings make predictable errors in reasoning. These errors result in people arriving at unjustifiable conclusions and engaging in irrational behavior (LES, 2019). One of the reasons is that our cognitive system is often depended on intuitive heuristics (Kahneman, 2013). However, certain heuristics have been proven to lead to systematic errors. These systematic errors are called biases.

There are explicit biases and implicit biases. Current research focuses more on implicit bias due to its nature often being unnoticeable and unprovable. Implicit bias is never an easy concept to be grasped. "The notion of implicit bias can be puzzling. The idea is that people harbor mental associations based on race, gender, and other social categories that may lead to discrimination without intent, or possibly even awareness" (Payne, Vuletich, & Lundberg, 2017, p. 233). Scholarship on systemic prejudice has highlighted that women and minorities will be systematically disadvantaged regardless of the attitudes of individuals (Payne, Vuletich, & Lundberg, 2017, p. 233).

Currently, the most widely used method for measuring attitudes in the behavioral and social sciences is the Implicit Association Test (IAT) (Siegel, Dougherty, & Huber, 2012). The IAT test was first introduced in 1998. Everyone can take the IAT test through a webpage hosted by Harvard University ("Take a Test," n.d.). The topics cover gender, age, religion, race and more. However, one limitation of IAT is that it requires subjects to volunteer their participation.

Text analysis is a vehicle that can offer opportunities to interpret meaning. One online tool is called InfraNodus. As its website states, though, "Currently there are no tools that can measure how biased a text is. Various text mining APIs categorize texts based on its content and sentiment, but there are no instruments that can measure the level of inclination towards a certain point of view in text" (Paranyushkin, 2018). InfraNodus' computer-generated graphs still need human interpretation and their effectiveness in detecting bias is questionable. In addition, current research for implicit bias does not offer a good methodological framework for visual analysis on bias based on the meaning of existing documents.

To complicate matters, there are also many forms of implicit bias. Many psychologists have proposed that human reasoning is subject to positive confirmation bias. The term confirmation bias was first introduced by English cognitive psychologist Peter Wason in 1960 (Gale & Ball, 2002). This is a tendency, when testing an existing belief, to search for evidence which could confirm that belief, rather than for evidence which could disconfirm it (Jones & Sugden, 2001). In addition to searching for confirming evidence, Findley and Scott suggested that confirmation bias also includes the tendency to judge information supporting one's beliefs as more important than ignoring disconfirming information (Koriat, Lichtenstein, & Fischhoff, 1980; Bogaard, Meijer, Vrij, Broers, & Merckelbach, 2014).

3. RESEARCH OBJECTIVES

In educational settings, many reviews are conducted at various levels. Unfortunately, not all the reviews are objective. Some reviewers might be biased. However, the subject who has been reviewed—whether a student, a faculty member, a department or the school—often has no tool to challenge the result intellectually even when it was unfair. The following graphs may reveal some cognition errors that can happen during any review process. This allows the one who has been reviewed to spot biases if there are any.

This method is designed to map biases with visual illustrations. It can be particularly helpful for designers and artists who are visually oriented to understand content that might be biased. The kind of implicit biases that are at the center of this research are confirmation biases.

4. CASE STUDY

4.1. Data Mining Methodology

There are two stages to this data-based analysis of value judgements. The first stage is data collection. The second stage is data processing and interpretation. Both faulty data and a biased process can lead to inaccurate judgements.

In the case studies here the author has classified value judgements into five categories—Positive, Negative, Neutral, Factually Incorrect and Impermissible Criteria.

It is at the information processing stage that there is the potential for confirmation bias to be introduced. During that processing, some data will be filtered out because these representations are by necessity simplified abstractions of the real objects. There is a danger that these simplifications are inadequate to capture important qualities (Sculley & Pasanek, 2008). One common methodology employed in data mining is to use a 'bag of words' model, in which a document is represented by a histogram, with a unique bin for each word that contains a score related to the number of times that word appears in the document. This representation has been extremely successful in the text classification domain (Mitchell, 1997).

Building on this technique, this analysis uses lines of text rather than individual words. Facts and organizational materials without additional value judgement are left without a coded color.

Where confirming data is found, the links between the content are illustrated in Black. Where disconfirming data is found, the links between the content are illustrated in Red. Confirmation bias can come from the elimination of disconfirming data, and confirmation bias can also come from the amplification of confirming data. If both happen at the same time, the chance for confirmation bias is high.

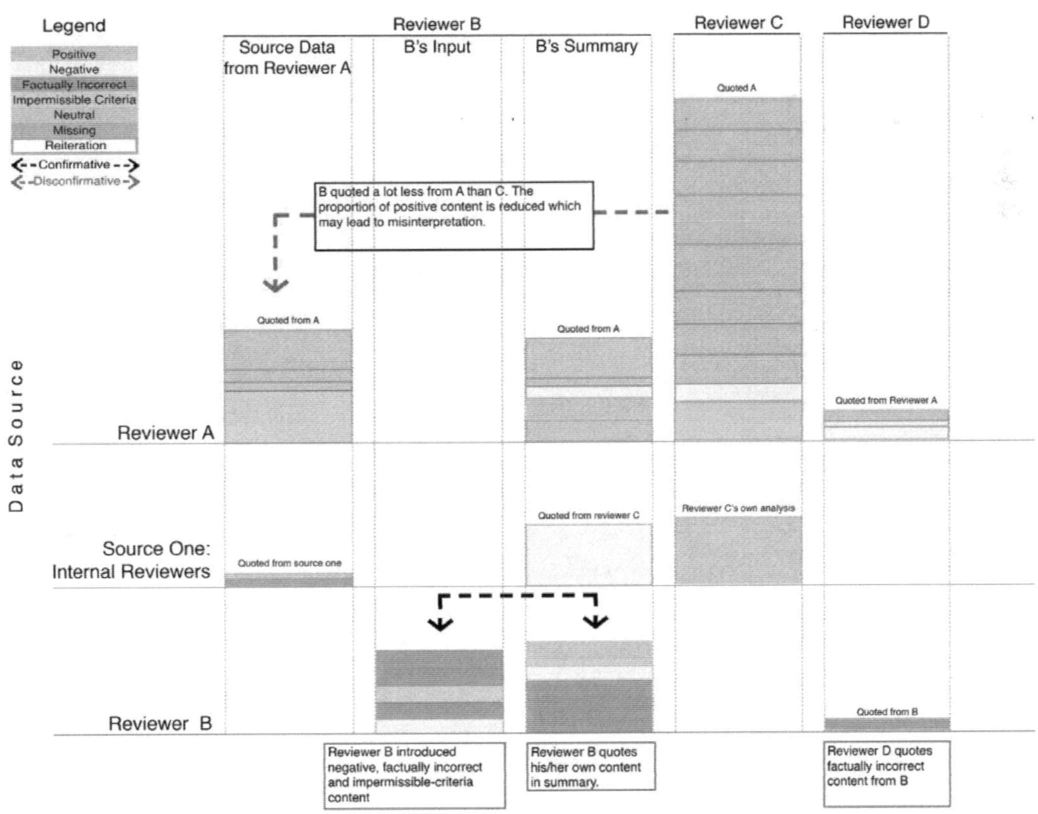

Figure 1. Reviewer B Introduces Negative, Factually Incorrect and Impermissible-Criteria Content

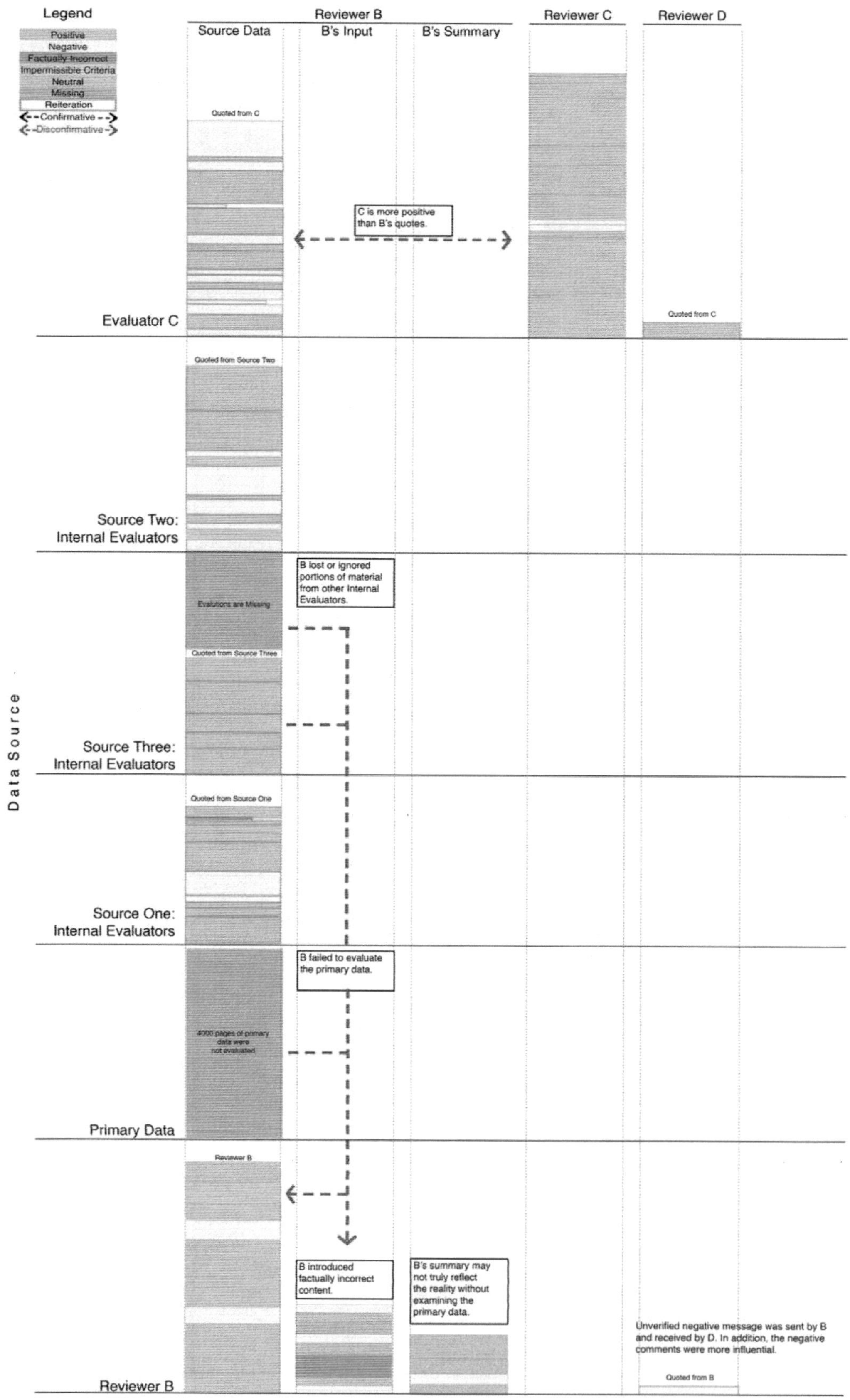

Figure 2. Reviewer B Fails to Evaluate the Primary Data

Three case studies are illustrated with graphs. All three cases are performance reviews conducted by the same group of reviewers. They are labeled as reviewer A, B, C, and D. Reviewer A is an independent external reviewer. A's review is the first in the process and often is quoted by B, C and D. B, C and D are internal reviewers. First B and C reviewed independently. Sometimes B quoted from C. However, C has not quoted anything from B. D then received the results from A, B and C and then concluded the review process. A, B and C examine the same data. D has not directly reviewed the data. D's review was based on the results collected from A, B and C. However, in all three reviews, they have concluded differently. The question is which one is closer to truth.

4.2. Case One

Looking at Figure 1, the raw data are a mixture of positive, negative, neutral, impermissible-criteria, and factually incorrect information.

Figure 1 shows discrepancies between reviewer A and reviewer B. First, B quoted from A, who are mostly positive. Then B injected their own review, which contains factually incorrect information, negative information and impermissible criteria. B quoted less from A compared to C who also quoted A. Possibly there is a confirmation bias causing the disconfirming information to be filtered out. It is possible that the faulty material introduced by B diluted the integrity A's content. What can be said with certainty is that B's summary confirmed its own "findings," which have been shown to be untrue. Those same "findings" were quoted by D, because D would have had no reason to doubt their truth.

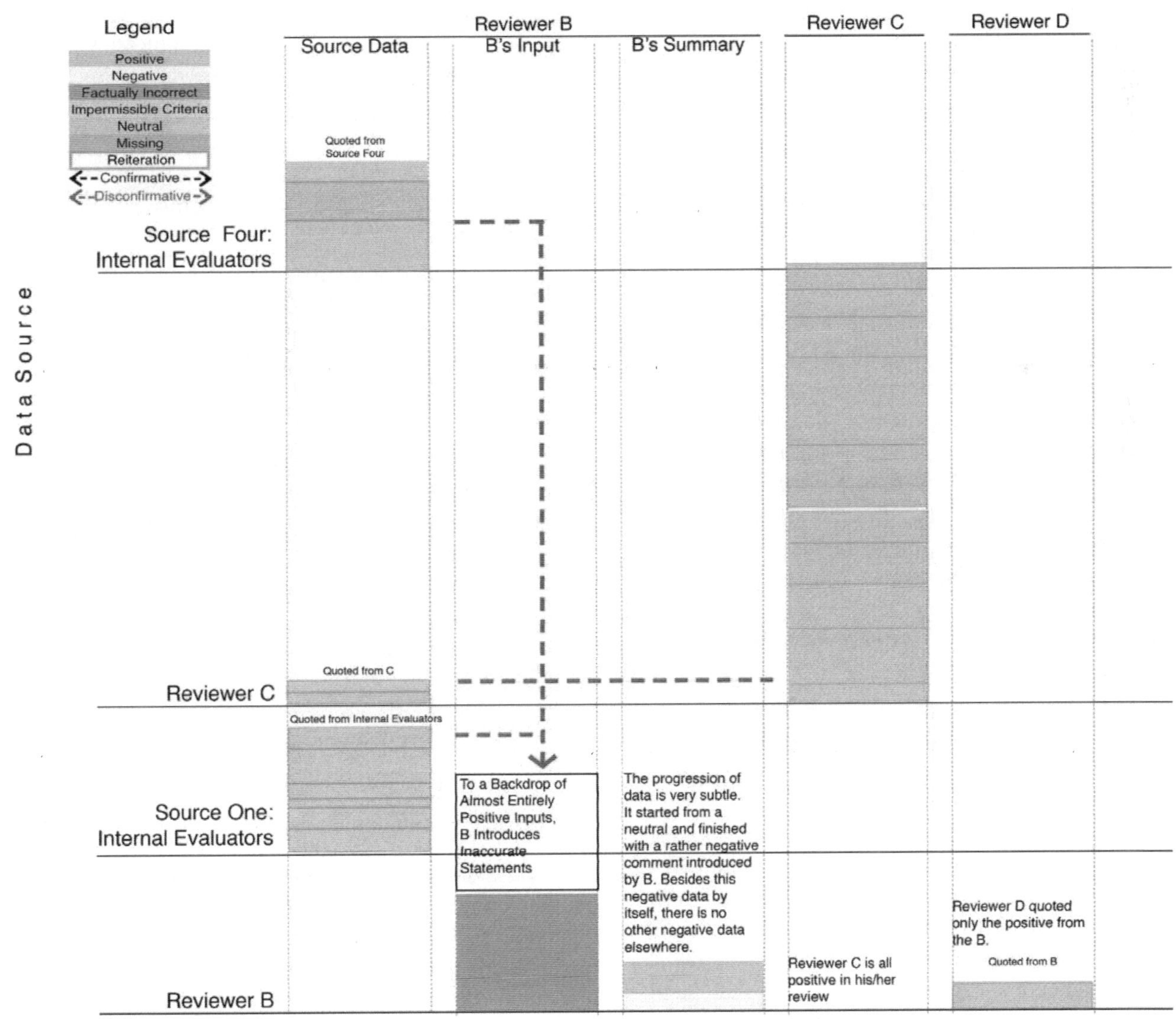

Figure 3. B Introduces Inaccurate Statements

4.3. Case Two

Looking at Figure 2, B has not reviewed all the content that was required to be reviewed. Therefore, the findings of B are questionable. Using the same raw data, the negative information quoted by B is more than all the rest of reviewers. C's report is much more positive.

4.4. Case Three

Looking at Figure 3 shows that the raw inputs about are entirely positive. Reviewer B then introduced factually incorrect statements. There is no indication that B reviewed the actual work which is required.

5. CONCLUSIONS

Implicit bias has not been eliminated in the workplace. It only has taken a more disguised form. It resides in our cognition system. However, it is difficult to be scientifically proved and recognized at institutional and organizational levels.

Often a visually oriented person is at the disadvantage in building an argument against a sea of documents. To unfold many layers of information from various sources, sometimes conflicting with each other, can be quite daunting. Because many people like to function in the domain of visual representation and intuition, there is a need to create ways to communicate visually and directly.

Based on the content analysis, reviewer B may have taken an initial negative position which led he/ she to systematically undermine the positive data which represents the majority of the raw data collected from various sources, introducing some faulty data and nurturing the negative data before drawing their conclusion.

Research shows that negative information tends to influence evaluations more strongly than comparably extreme positive information (Ito, Larsen, Smith, & Cacioppo, 1998). Therefore, unless someone is consciously fighting to avoid bias, it resides in all of us. In any analysis there is some noise—noise that obscures the melody. When we listen to a piece of music we normally ignore the noise but listen to the music itself. However, when someone believes the noise is more significant than the music itself, they will find confirmations that the noise represents the music. Whether we filter out the music or the noise is a judgment that everyone has to make.

REFERENCES

Bogaard, G., Meijer, E. H., Vrij, A., Broers, N. J., & Merckelbach, H. (2014). Contextual bias in verbal credibility assessment: criteria-based content analysis, reality monitoring and scientific content analysis. Applied Cognitive Psychology, 28(1), 79-90. https://doi.org/10.1002/acp.2959

Chamorro-Premuzic, T. (2013, August 22). Why Do So Many Incompetent Men Become Leaders? Harvard Business Review. Retrieved May 12, 2019 from: https://hbr.org/2013/08/why-do-so-many-incompetent-men

Gale, M., & Ball, L. L. (2002). Does Positivity Bias Explain Patterns of Performance on Wason's 2-4-6 Task? Proceedings of the Annual Meeting of the Cognitive Science Society, 24. Retrieved from https://escholarship.org/uc/item/5w46f77h

Ito, T. A., Larsen, J. T., Smith, N. K., & Cacioppo, J. T. (1998). Negative information weighs more heavily on the brain: The negativity bias in evaluative categorizations. Journal of Personality and Social Psychology, 75(4), 887-900. http://dx.doi.org/10.1037/0022-3514.75.4.887

Kahneman, D. (2013). Thinking, Fast and Slow (1st edition). New York: Farrar, Straus and Giroux.

Koriat, A., Lichtenstein, S., & Fischhoff, B. (1980). Reasons for confidence. Journal of Experimental Psychology: Human Learning and Memory, 6(2), 107–118. https://doi.org/10.1037/0278-7393.6.2.107

LES (The London School of Economics and Political Science). (2019). MBA Essentials: Heuristics, Biases, and Decision Making. London: LSE.

Paranyushkin, D. (2018). Measuring Discourse Bias Using Text Network Analysis. Towards Data Science.

Retrieved May 12, 2019 from: https://towardsdatascience.com/measuring-discourse-bias-using-text-network-analysis- 9f251be5f6f3

Payne, B. K., Vuletich, H. A., & Lundberg, K. B. (2017). The Bias of Crowds: How Implicit Bias Bridges Personal and Systemic Prejudice. Psychological Inquiry, 28(4), 233–248. https://doi.org/10.1080/1047840X.2017.1335568

Sculley, D., & Pasanek, B.M. (2008). Meaning and mining: the impact of implicit assumptions in data mining for the humanities. Literary and Linguistic Computing, 23(1), 409-424.

Siegel, E. F., Dougherty, M. R., & Huber, D. E. (2012). Manipulating the role of cognitive control while taking the implicit association test. Journal of Experimental Social Psychology, 48(5), 1057–1068. https://doi.org/10.1016/j.jesp.2012.04.011

Take a Test. (n.d.). Retrieved May 12, 2019 from: https://implicit.harvard.edu/implicit/takeatest.html

edra51

Tempe, Arizona

Abstracts - Individual Presentations

A1 - Transform Design for Resilience & Sustainability: Planning, Design & Behavior I

Cooling Potentials of Rooftop Shade: A Promising Strategy for Reducing Urban Induced Heating in Kuwait's Residential Neighborhoods

Saud R AlKhaled (Kuwait University); Paul Coseo (Arizona State University); Ariane Middel (Arizona State University); Anthony Brazel (Arizona State University); Chingwen Cheng (Arizona State University)

Urban induced heating is a pressing problem confronting cities. The situation is further exacerbated in hot urban deserts (HUDs) due to already high temperatures, aridity and projected climate change (McCarthy, Best, & Betts, 2010; Pal & Eltahir, 2015). In parched, water-stressed, and radiation-abundant HUDs, applications of commonly proposed heat mitigation strategies are controversial. Radiative cooling through increased urban albedo and transpirative cooling using vegetation-based strategies result in hydroclimatic sensitivity (Georgescu, Morefield, Bierwagen, & Weaver, 2014) and water-temperature trade-offs (Gober et al., 2012) respectively. Shade through urban form is increasingly seen as an approach that cities can leverage to reduce cooling energy demands (Futcher, Mills, & Emmanuel, 2018; Krüger, Pearlmutter, & Rasia, 2010) and increase the use of outdoor spaces (Middel, Selover, Hagen, & Chhetri, 2016). Yet the power of shading urban surfaces by urban form is generally perceived as an early stage planning strategy rather than a retrofitting design mitigation strategy due to the difficulties in altering the physical arrangement of existing urban form. However, shaded rooftops as an additive architectural element provides an opportunity to achieve a pseudo effect of deeper urban canyons with higher aspect ratios (H/W) while facilitating ventilation and limiting nocturnal heat entrapment. As a case study, the microclimate model ENVI-met 4.4, combined with field measurements, was used to simulate the effect of rooftop shade-sails for a mid-summer day (June 21) in an existing Kuwaiti neighborhood. Near ground thermal effects and spatiotemporal variability from obstructed solar radiation and shade patterns across the neighborhood are analyzed and presented. Preliminary results suggest that street level mean radiant temperature and air temperatures may be reduced by as much as 27K and 1.5K respectively. The study demonstrates how applications of rooftop shade-sails can have significant near ground cooling effects and thus provide an alternative to paradoxical mitigation strategies in HUDs.

Making the Case for Synergistic Performance Measurement across Landscape Architecture, Planning, and Urban Design

Charles Walter Ware (University of California Los Angeles)

Currently trending project design performance measures are silo'ed. This presentation argues that when performance measures are integrated, synergistic effects happen. The performance measurement trajectory is explored as it transitions from independent measurement toward an integrated approach. Design disciplines are focused on assessing performance with metrics classified by three separate pillars of the classic sustainable development model: economic, environmental and social. As measurement "silos," these autonomous elements sometimes even occur in competition with one another, e.g. on a broad level, environmental health versus economic development. Independently these realms do not realize the full potential for achieving sustainable development.

In order to investigate this transition, primary certification and rating systems, such as LEED are analyzed. Project performance across a range of scales is also examined. Limitations and opportunities associated with existing measurement approaches are identified. Looking forward, potential interdependencies between environmental, social, financial and artistic performance realms are suggested. Preliminary synergistic principles and protocols are gleaned.

Analysis findings show a transition from independent measurement silos to more aggregated and integrated models. Furthermore, a transition to models with comprehensively organized, cross-category benefits is evident. The trajectory suggests a move toward truly synergistic potential, whereby interdependent relationships between topics are yielded. The theory suggests that sum total performance can exceed the sum of its independent parts.

Having reviewed the current performance measurement trajectory, we advocate a synergistic approach. Such as shift is expected to merge creative and performative design thinking to address persisting global sustainability issues. This project will set a preliminary foundation to move toward synergistic rating programs. It will be challenging to achieve this without mutual cooperation between multi-disciplinary designers, government agencies, academics, and communities. This research highlights the role of landscape architects in facilitating a more robust approach for sustainability assessment and establishing cooperative relationships among stakeholders.

Does School of Design Care about the Environment?

Louise Manfredi (Syracuse University); Seyeon Lee (Syracuse University)

Large format printing and prototype fabrication are a vital facet of design education, yet waste generated from repeated design iterations is unavoidable. These projects contribute to landfill waste because our design community are faced with unclear or absent reuse or disposal guidelines. As McDonough and Braungart (2010) state, lack of sustainability awareness is a big contributor to excessive waste, and in order to have a long-term impact on sustainability issues, a combination of teaching (integration sustainable design strategies in and outside the classroom), including experiential learning, (intentionally creating awareness by engagement) and lifelong education (action in communities) is required (Oyelola, 2012).

This paper explores systematic strategies implemented at the School of Design, Syracuse University, to use materials more sustainably and responsibly in the design studios. First, we describe findings from waste bin experiments, in addition to surveys and interviews with students to understand their attitudes and behaviors toward sustainable material reuse, reduction, and recycling. Second, we outline the process to develop material exchange centers where students share and reuse materials, and our strategies to deliver waste management guidelines and information.

Results from the interviews with students (n=32) showed that most students were unclear about the waste management systems used at the university and in the county of Onondaga. While all students welcomed the material exchange initiative, they were hesitant to utilize them across design programs. This effort began a sustained conversation between faculty, staff and students about the excessive landfill waste generated by studio projects, with ideas being generated about how to make this new system self-sustaining.

This research will be beneficial to anyone attempting to raise awareness and understanding of institutional waste management strategies at university level.

A3 - Transform Design for Resilience & Sustainability: Planning, Design & Behavior III

A Sociological Approach to Creating an Ecologically Sustainable Built Environment

Lubomir Popov (Bowling Green State University); Franklin Goza (University of Wisconsin Whitewater)

The construction industry and built environment in the U.S. account for about 12% to 20% (according to different sources and methodologies) of the country's ecological footprint, including greenhouse gas emissions and energy use, as well as a number of other sustainability issues. The sustainable environment problem has engaged the entire society. The design professions are at the front line, working on issues like saving energy and resources, renewable materials, reclaiming and recycling materials, controlling environmentally damaging emissions, and so forth. Regarding sustainability, designers have construed, codified, and institutionalized a problem area and have developed productive and effective solutions. However, architects are constrained to a large extent by the strategies and requirements of the clients, even when design firms try to educate their clients.

The objective of this paper is to discuss an alternative approach to environmental sustainability based on developing and promoting sustainable lifestyles. This is a theoretical paper. Our methodology is based on social construction of technology, sociology of sociotechnical systems, and sociology of consumption. We propose a radical change of user and client's values and attitudes, leading to client requirements and architectural programs that will inform sustainable design solutions. In this context, sustainability will be produced by environment-friendly social actions and lifestyles that involve the appropriate volumes and types of consumption. This approach fosters the idea of codifying sustainability issues as a social problem rather than a design problem. From this perspective, clients and designers will have a much wider area for problem definition and design solutions, ranging from modifying human values and behavior, to sustainable lifestyles, and traditional sustainable design methods and techniques. In this presentation we will also sketch the outlines of a research program on implementing life styles and models of consumption that lead to social and environmental sustainability.

Proposal for an Off-Grid Mobile Architecture Prototype for Disaster Relief

Shota Tajima (Chiba University); Satoko Nasu (Chiba University)

This study aims to develop an off-grid mobile architecture prototype for use in disaster relief.
In recent years, climate change has led to an increase in the number of natural disasters. For example, power was cut to 930,000 households due to a massive typhoon that hit Chiba Prefecture, Japan in September 2019. The outage lasted up to 2 weeks in some areas. To respond to such situations, it is necessary to implement a community continuity plan to minimize long-term damage to local communities and facilitate a quick recovery.
Mobile architecture (e.g., trailers and container units) offers housing solutions that help victims maintain resilience after a disaster. However, there are few practical examples of off-grid mobile architecture that can be used in the event of a power outage. We previously presented such a solution at the Solar Decathlon 2014. In addition, we categorized available data on off-grid mobile architecture.

Based on our previous research, we propose a conceptual prototype of off-grid mobile architecture for disaster relief with the following features: 1) It comprises units that can be transported and assembled anywhere in the world. It can accommodate any community size by combining multiple units. It can be temporarily extended using a tent, allowing disaster victims to stay for several days. 2) It incorporates the infrastructure necessary for community maintenance, including solar power generation, storage battery, inverter, toilet, and basic air conditioner. According to our trial calculations, it has a power generation capacity of about 1.29 kW, which can be increased by expanding the solar panels as needed.
We aim to soon realize the proposed prototype while taking into account to the needs of victims and local governments.

Speaking Their Language: Are Environmental Communications Aligned to Occupant Dispositions in Undergraduate Residence Halls?

Erin Miller Hamilton (Texas Tech University); Apoorva P Rane (Texas Tech University)

Despite the growing prevalence of sustainability initiatives among university campuses broadly (Watson, Hegtvedt, Johnson, Parris, & Subramanyam, 2017), there still remains relatively little published research on promoting environmentally responsible behaviors (ERBs) in these settings. Past studies focused almost exclusively on explicit behavior change interventions (e.g., Karp, McCauley, & Byrne, 2016) that often do not consider the college students' existing inclinations to participate in ERBs. In a study of students living in four green and two non-green residence halls, this study explores evolving trends of the ERBs of undergraduate students (n=575) from the Fall to Spring semesters. This study identifies the dominant psychological factors that support students' ERBs, and critically examines how environmental communications within residence halls frame ERBs according to a variety of values and motivations.

Linear mixed-effects regression analyses revealed a diverse array of personal dimensions related to students' self-reported energy, water, materials, and travel behaviors. A thematic analysis of photographs of existing signage within the residence halls reveals a disconnect between the values and motivations frames that are commonly used to induce ERBs and those that significantly drive undergraduate ERBs. Results will be discussed in light of implications for those seeking to adapt the physical and informational environments of green buildings to better support the existing values and motivations of undergraduate students.

Natural Hazard Governance and Sustainable Communities in Sundarbans Delta (India) – A Case Study of Tropical Cyclone Aila, 2009

Indrajit Pal (Asian Institute of Technology); Tuhin Ghosh (Jadavpur University); Sylvia Szabo (Asian Institute of Technology)

Economies in developing countries are impacted most from the natural as well as anthropogenic hazards. Over the last decade, natural disasters have caused more than 780,000 fatalities and destroyed physical properties worth a minimum of $960 billion. Sundarbans, Indian, was declared as a 'world heritage site' in 1994 and as a 'biosphere reserve' in 1989 (Small and Nicholls 2003). By virtue of its proximity to the Bay of Bengal, the land is very much vulnerable to the tropical cyclones and tidal disturbances during the monsoon (Pal and Shaw 2018). The major part of the Sundarbans area is facing the problem of waterlogging because of its basin-like island geography. In 2009, Cyclone Aila devastated the Sundarbans delta and impacted more than 3500 km of very old earthen embankments. Over 500,000 houses are damaged either fully or partially, and a similar amount of cropped area is affected. Over 60,000 people rescued have been accommodated in government relief camps. The population affected in the state was more than five million.

The present exploratory research examines the effectiveness of the Indian policy pertaining to the disaster risk governance and management in achieving effective mitigation and response to a disaster thereby minimising the casualties and losses to the community caused by the cyclone Aila in 2009. Study also analysed how the implementation of the policy enabled the government of the Indian state of West Bengal to minimize the casualties and damage to property. The implementation gaps at the various levels also explore that natural hazards management efforts by one branch of government could be undermined by the socio-economic and socio-political drivers. The study finds that establishing cyclone early warning systems and building cyclone shelters will not be helpful and effective if the natural coastal systems are degraded for commercial shrimp farming and depletion of mangrove forests.

A4 - Transform Design for Resilience & Sustainability: Planning, Design & Behavior IV

City-University Partnerships for Thermally-Resilient Communities

Paul Coseo (Arizona State University); Braden Kay (City of Tempe, Arizona); Ariane Middel (Arizona State University); Jennifer Vanos (Arizona State University); David Hondula (Arizona State University); Zoe Hamstead (University at Buffalo); Emmanuel Frimpong Boamah (University at Buffalo); Nicholas Rajkovich (University at Buffalo); Saud R AlKhaled (Kuwait University); Debarati Majumdar Narayan (The Pew Charitable Trusts)

The escalating socio-ecological challenge of thermal extremes – prolonged periods of heat and cold – have significant impacts on human health and urban ecosystems. In fact, these extremes claim more lives in the U.S. than all other weather-related events (Berko, Ingram, Saha, & Park, 2014). . Researchers, practitioners, and public officials must navigate this growing urban challenge, as events become more frequent. City officials have responded with policies to reduce effects of extreme heat (Larsen, 2015; Stone, 2012). Yet, cold-related mortality is greater than or equals that from extreme heat (Berko, Ingram, Saha, & Park, 2014). Few cities have collaboratively addressed thermal vulnerability in a holistic and systematic way. In fall 2017, researchers from two universities partnered with local governments in Tempe, AZ and Buffalo, NY to codesign a city-university participatory action research agenda. Funding from the Health Impact Project, a collaboration of the Robert Wood Johnson Foundation and The Pew Charitable Trusts, supported advancing that agenda, with the goal of crafting a healthy and equitable infrastructure policy for Tempe, using a multi-pronged approach to capture thermal experiences associated with different types of infrastructure. The team conducted a (1) citywide heat and health survey to capture resident perspectives, along with a deeper examination of one Tempe neighborhood. The (2) microclimate and (3) participatory heat assessments for a select number of parks, playspaces, multiuse paths, and sidewalks were designed to test the effectiveness of a living labs-policy approach for climate policy and actions. Data from these activities informed a (4) climate action workshop with the aim to co-produce policy and design solutions. This presentation will discuss the results of this approach with a focus on barriers, opportunities, and lessons to creating healthy, equitable climate policies. We will also discuss next steps for piloting this approach in Buffalo.

Ecotone Planning: A Gateway for Resilience and Preservation

Phillip Zawarus (University of Nevada Las Vegas)

The 7J Ranch, recently acquired by The Nature Conservancy (TNC), requires a sustainable master plan that invests in the current and long-term resilience and sustainability of its diverse wildlife ecology, sensitive water resources, and cultural impact on residents and visitors. Located in Beatty, Nevada. The project serves as a gateway to northern Nevada and Death Valley, attracting a wide variety of regional and international visitors. In addition to the preservation of natural systems and ecology, the plan aims to incorporate eco-tourism, research laboratories, artist residence, sustainable infrastructure, and public outreach (Beck, 2013). Due to the dichotomy between the natural and the built environment in rural location, the key goals for this master plan are to balance the interfaces between intensive and passive programming.

The rich and lush ecology of flora and fauna of the site is supported by the Amargosa River which teeters between the Mojave Desert and Great Basin ecoregions. As climate changes begins to affect global weather patterns, the ecological composition may shift at various temporal intervals leading to a ripple effect on the other systems supported by this ecotone. Springs and aquifers sustain this healthy ecology with year-round surface and subsurface drainage for fish, amphibians, mammals, and birds. Due to this sensitive ecology revolving around its precious water resource a strategic plan was developed to protect these rare resources from being over-used or contaminated by inappropriate land use practices.
Students from the University of Nevada Las Vegas (UNLV) analyzed the ecology, wildlife, and history of the area

to create interpretive signage throughout the project site for educational purposes to visitors, residents, and stakeholders. This phase of the project framed the programmatic development of the site's master plan to preserve sensitive microhabitats and cultural artifacts while introducing renewable energy (Czechowski, Hauck, Hausladen, 2015), agriculture, recreation, and trail networks.

Englewood Chicago Is Rising: Volunteers in Service to America - Design for Resilience and Sustainability!

Douglas A. Williams (University of Illinois Urbana-Champaign)

At the precipice of Americas' 1960s Civil Rights Movement, where foreign shores were ubiquitous for services missions in developing nations' favelas, shanties, and trench towns, President Lyndon Baines Johnson created AmeriCorps – Volunteers in Service to America. For over 60 years, V.I.S.T.A. has been an equitable domestic resource champion for strengthening neighbors linked to hoods, for their self-sufficient sustainable investment to reestablishing neighborhoods of belonging. With growing disparities of class and race, reminiscent of spatial geographies of the past, many well intended programs maintain paternal dependency. Recently, one of the largest numbers of V.I.S.T.A.s in a Chicago neighborhood has been building numerous non-profits' capacities to turn the tides of climate change. What role do architects have in serving expert citizens for the positive transformation of their environment?

This ethnographic case study is a part of a longitudinal evaluation of the Englewood Chicago AmeriCorps V.I.S.T.A. program, sharing local people's place-keeping stories in the inner-city. These findings support social capital theory: when socioeconomic crisis persist, communities come together (Putnam, 2000), using the built environment to foster a more sustainable and healthy community (Fullilove, 2016) and social-cultural relationships of claiming space (Allen, 2001; Finney, 2014; Glave, 2010) are values of invested in communities.

ChangingClimateChangeBeliefs: Does It Make a Difference?

Robert W Marans (University of Michigan); Noah Webster (University of Michigan); Edmundo Roberto Milipillan (University of Michigan); John Callewaert (University of Michigan)

Despite denial by the current U.S. administration, national data indicate that an increasing number of Americans believe that climate change is happening. At the same time, the data reveal a frustration with policy makers in their willingness and ability to address the issue. Recent polls indicate that not only is there a growing awareness that climate change is happening but that more and more Americans are concerned about its impact on society and them personally. Similarly, a recent poll of teenagers shows that nearly 6 in 10 are afraid of climate change and that 1 in 4 are taking action. Data from 5 years of student surveys at the University of Michigan also show a growing concern about climate change. But the question is the extent to which students are willing to take action to address the issue.

Following an overview of University of Michigan's Sustainability Cultural Indicators Program (SCIP) and the types of survey data collected, the paper discusses changing student beliefs about climate change over the 5 year period. Changes in the degree of certainty that climate change is happening, the belief that climate change is caused by human activity, and the relative importance of climate change to each student personally are reviewed. Findings are then compared with findings from a national sample of the U. S. population. Next, models showing how these changing views impact student engagement environmental activities on campus and off-campus and pro-environmental behaviors are presented. Pro-environmental behaviors include eating meat, mode of travel to and from campus, and efforts to conserve energy and reduce waste. Both cross sectional and panel data are used to test the models. Finally, the implications for training future environmental activists and educating the public are discussed.

A5 - Transform Design for Resilience & Sustainability: Planning, Design & Behavior V

Sunblock: Exploring the Potential of a Distributed, Decentralized District Energy System

Jonathan Bean (University of Arizona)

Envision a desert neighborhood that runs on a closed loop of shared water instead of fossil fuel, and where existing underutilized equipment at a public school keeps 75 disadvantaged residents' homes comfortable at summer's peak. As demonstrated by the three finalist student teams in the 2019 Solar Decathlon Design Challenge, this vision can be achieved with technologies that are available today. The system, called SunBlock, is an affordable, effective, and sustainable way to harvest and share abundant solar energy that addresses several of the most challenging aspects of carbon reduction. SunBlock turns energy from an individual cost to a shared asset, making entire neighborhoods net-positive with an innovative system that pairs photovoltaic panels, super-efficient heat pump technologies, and distributed thermal storage, benefiting the environment while creating more opportunity for people. SunBlock uses existing technologies to transform abundant solar energy into a shared resource available around the clock. First, photovoltaic panels transform solar energy into carbon-neutral electricity. Second, this is then used to power an ultra-efficient air-to-water heat pump, which uses one unit of electricity to move as many as nine units of thermal energy. The result is heated or cooled water, which is stored in underground tanks to condition individual buildings during the day and at night -- no need to burn coal to run AC compressors. Third, the tanks are interconnected in the SunBlock district loop to share thermal energy between neighborhood buildings. SunBlock significantly reduces refrigerant use, which according to the Drawdown project has single greatest potential to reduce the effects of climate change. SunBlock also reduces grid stress. Furthermore, SunBlock addresses two common challenges in Western cities: car dependence and water shortage. SunBlock tanks can store rainwater, while existing backyard utility easements provide a route for the district loop and the potential for new pedestrian and biking paths.

Vitu: The Linkages between Indigenous Crafting, Sustainable Livelihood, and Concepts of Wellbeing in the Context of Social Entrepreneurship

John Hiroomi Takamura (Arizona State University)

As more indigenous peoples begin to take on the challenges of sustainable community development for themselves, more research into understanding the linkages between indigenous crafting (Tice, 1995, Swain 1993, Cruz-Torres 2008, Little 2004), sustainable livelihood (DfID 1999, ILO 2010), and indigenous wellbeing (Wills 2007, Walters & Takamura 2015, Ngata et al 2012, Corntassel 2003, Gudynas 2011, Watts 2013, Basso 1996) becomes critical. Additionally, such research is needed to study how social entrepreneurship affects indigenous wellbeing in terms of the social, cultural and economic factors specific to poverty alleviation. This research aims to understand how indigenous crafting impacts capital (specifically social cultural, and spiritual capital) and how these impacts affect indigenous wellbeing in terms of spirituality and cultural identity in order to eventually create a native or indigenous tribally owned and operated fair trade fashion product brand venture in Mexico for indigenous Mazahua communities. This fair-trade fashion product brand venture called Vitu™ will function as a social enterprise to help alleviate Mazahua poverty as well as to positively impact Mazahua society in terms of Mazahua language/culture preservation, and personal wellbeing. Through artisan interviews and a visual ethnography of artisan life, the goal of this research and fieldwork is based on answering the following general research query: How might social entrepreneurship (specifically indigenous entrepreneurship) positively affect human wellbeing (specifically indigenous wellbeing)? Answering this question is essential in determining social entrepreneurship's impact on Base of the Pyramid (BoP) community development within the context of poverty alleviation. In summary, this research is critical in laying the foundation of understanding for future design & art interventions like indigenous branded social enterprises in the alleviation of poverty at the economic Base Of the Pyramid (BOP) so that "no harm" will be done to the fragile indigenous communities willing to engage in social entrepreneurial activities to improve their livelihoods.

Biomimicry to Transition Infrastructure Design Principles Towards Resiliency

Alysha Helmrich (Arizona State Univeristy); Samuel Markolf (Arizona State Univeristy); Samantha Hayes (Griffith University); Mikhail Chester (Arizona State Univeristy); Cheryl Desha (Griffith University); Nancy Grimm (Arizona State Univeristy)

As the world becomes increasingly complex and uncertain, the systems that support the institutions and services that maintain a functioning society must be resilient to expected and unexpected disturbances such as climate change or disruptive technology. Today, infrastructure design is commonly approached from a fail-safe method that designs infrastructure to an acceptable level of risk (Ayyub, 2018), and the infrastructure is considered to be in a 'fail' state if it is not providing its intended service. By not allowing for failure in design, infrastructure managers can only optimize the performance of a system based on what is known and may not consider the consequences when the unknown (re: unexpected failure) occurs (Kim, Chester, Eisenberg, & Redman, 2019). Additionally, due to complexity and uncertainty, these unexpected failures become more frequent, leading to advocacy for resilient infrastructure. Resiliency is commonly defined by engineers as robustness, but it can be generally defined as the ability for an infrastructure system to not only withstand predicted disturbances but continue to function beyond its design envelope (Woods, 2015). Current engineering design principles need to integrate these ideas of resiliency, and this leads to question: how can infrastructure managers design infrastructure to be resilient? In order to make this transition, engineers can examine nature's life's principles. Nature has long withstood disturbances within and beyond its designed capacity and can serve as a model for engineers. This work compares current design principles to overarching themes identified in resilient infrastructure literature and utilizes life's principles to explore transitions to resiliency. Current engineering design principles do not exhibit an integration of life's principles, which are more closely mirrored by resiliency principles. This study illustrates a case study of how life's principles in resiliency can be integrated into existing design practices.

Unintended Consequences and Trade-Offs of Urban Heat Mitigation Strategies: A Perspective Analysis

Florian Arwed Schneider (Arizona State University)

Urban heat mitigation strategies such as increasing tree and vegetation cover, implementing green and cool roofs, cool pavements, and climate-sensitive urban form intend to cool the built environment, mitigate the urban heat island effect (Taleghani, 2018; Wang, Berardi, & Akbari, 2016), and reduce negative health outcomes. However, those interventions are no silver bullets without trade-offs. Scientific results on interactions with the environment and people can be unclear, unstated, hidden, incomplete, or unknown, potentially leading to unintended negative consequences that could outweigh the intentional heat mitigation benefit. Incomplete research may lead to malformation and biased evaluations of the strategy by policy- and decision-makers, which could inadvertently create unhealthier, less sustainable, and less livable urban spaces. Thus, it is essential to find and investigate the urban heat mitigation trade-offs as well as their representation within the scientific and practitioners perspective.

Three perspectives on the interventions have been reviewed for consequences: scientific heat mitigation perspective, scientific perspective different than urban heat mitigation, and the practitioner's perspective. The three-fold review yields consequences, feedback on consequences' representation within the perspectives, and whether they are intended or unintended. A disconnect between the knowledge needs of practitioners and the knowledge provision of heat mitigation science has been found. Most adverse consequences have received less attention in both the scientific and practitioner perspective. A clear distinction between an intended and unintended consequence is not possible since it depends on the perspective used. This dichotomy highlights the need for research with respect to a significance threshold that could justify neglecting or considering the assessed consequences. The additional knowledge would allow researchers to better and more holistically inform action. Informed action and design are key to ensure benefits over adverse impacts and to maximize the potential for those benefits (van den Berg, Maas, Verheij, & Groenewegen, 2010).

A6 - Transform Design for Resilience & Sustainability: Planning, Design & Behavior VI

Strategic Collaborations and Partnerships to Accelerate Climate Resilience at the Local Level

Debarati Majumdar Narayan (The Pew Charitable Trusts)

As communities across the United States feel the impacts of climate change, those that are healthy are more likely to be resilient. Beyond individual health, communities with strong social networks, and equitable access to healthy housing, food, jobs, quality air and water, transportation, and other amenities, are healthier. According to the recent National Climate Assessment (USGCRP, 2018), direct and indirect impacts of climate change on these social, economic, environmental, and political factors, can threaten community wellbeing, safety, and quality of life, while exacerbating existing inequities. As local leaders in the U.S. committed to acting on climate changeare exploring data-driven approaches to climate policies (Climate Mayors, 2018), decision-making around those policies present an opportunity to consider the socio-economic and environmental drivers of health. While some local jurisdictions are investing in co-beneficial solutions such as developing pedestrian masterplans to both reduce emissions and promote physical activity, other resource-constrained jurisdictions could benefit from strategic partnerships and collaborations to act more quickly on resilience strategies.

In 2019, the Health Impact Project, a collaboration of the Robert Wood Johnson Foundation and The Pew Charitable Trusts, supported four cities, Baltimore, Minneapolis, Princeton, and Tempe, in implementing strategies at the intersection of climate change and health to improve communities' preparedness and resiliency. The cities translated research related to resilience strategies into action, to generate replicable model policies/programs for other cities, by bringing health and equity evidence to decision-making, and working collaboratively across city agencies. This presentation will provide an overview of how individual city efforts contributed to resilience and showcase how a cross-sector and cross-jurisdiction peer-to-peer learning model supports accelerating the development of healthy, equitable climate policies and programs. Each city will provide examples of challenges, wins, and potential recipes for success in implementing strategies that address climate change and health at the local level.

Toward Smart Sustainable Flood Management Under Climate Change: Learning from Sustainable Blue-Green Infrastructure Design in Ancient Cities

Marzieh Rezaei Ghaleh (Imam Khomeini International University)

Cities around the world are facing challenges in preparing for, mitigating, and adapting to climate change impacts. Information and communication technology (ICT) plays a critical role in climate change adaptation and sustainable flood management (e.g., Flood Resilience, Smart Water Management). Smart cities use ICT to implement their Smart systems, including Internet of Things, sensors, geospatial technology, artificial intelligence, and block chain, to collect and deliver information and knowledge, improving quality of daily life and projecting to the future (Dameri, 2017; Choudhary, 2018). In spite of these technological revolutions, lessons from ancient civilizations can guide future cities. For example, historical cities in Iran, with the nine thousand-year-old civilization, has implemented a specific green-blue infrastructure system that not only manages the seasonal floods and runoff but also produces food and reduces the urban heat island effect. This system has played an essential role in supplying groundwater resources, mitigating the water shortage, supporting and protecting the ecological health of water resources, and providing healthy food. This system has served multiple benefits in environmental, social, and economic sustainability in arid and semi-arid regions for thousands of years and is still working effectively. By learning from the ancient wisdom of historical knowledge of sustainability in ancient civilization in Iran and the implementation of blue-green infrastructure in the process of flood management, this research aims to analyze opportunities and challenges in urban flood management in a sustainable and resilient manner, through the use of ICT. The lessons learned from this study can be applied to other arid and semi-arid regions around the world, including the Phoenix metro area in Arizona, USA. This research can serve as a catalyst for designing smart and sustainable future cities while achieving UN sustainable development goals of combating climate change and its impacts.

Loose-Fit Urbanism: Policy and Regulation from Pattern to Performance

Neal LaMontagne (University of California Los Angeles)

Planning and design of complex urban environments in an age of increasing uncertainty and aspiration requires models that balance active engagement with the dynamic forces shaping the city with an openness to invention and adaptation. Generative models such as pattern languages provide some clues of how to strike that balance - identifying key qualities and relationships without precluding inventive solutions or learning from past experiences. However, their 'loose-fit' nature present challenges to standard modes of planning practice and design regulation.

This paper builds on a case study of over 40 years of planning and urban design governance in Vancouver (BC, Canada) from an early application of a pattern language in the experimental neighbourhood of False Creek South to the evolution of a discretionary regulation regime that employs an evolving principle of urban design performance. The paper will demonstrate how a 'loose-fit' model allowed for emergence of creative solutions but also assisted in the socialization and adaptation of rules. Lessons from practice, and an extensive series of interviews with planning, design, development and political actors, provide insights and recent cases from beyond Canada employing a pattern approach are considered.

Lessons learned are contextualized in a broader literature on governance, generative methods, and complex urbanism and applications to other complex policy challenges, including sustainability, are examined.

A8 - Transform Design for Resilience & Sustainability

Unbuilding - the Decommissioning of the Built Environment

Ursula Emery McClure (Perkins & Will)

The global ramifications of sea-level rise are much greater than merely changing coastlines. The rising seas will not only require populations to retreat but cause ecosystems to shift, storm-proofing to change, constructed landscapes to go to ruin and/or require remediation, and coastal buildings to be addressed. These future conditions demand a rethinking of how one builds, environmentally, structurally, systematically, and architecturally. Simultaneously, they demand us to consider how we unbuild environmentally, structurally, systematically, and architecturally. Couple these watery futures with the unbuilding industries' current position as one of the world's largest solid waste producers with their byproduct, construction debris, accounting for half of all solid waste generated every year and where to put it safely becomes a paramount concern.

This abstract will present research developing alternative and more sustainable decommissioning strategies/policies for the coastal built environment. Currently, most published policy, regulation, and methods of decommissioning are based on the unsustainable practice of breaking down buildings for landfills or the more sustainable practice of preparing a construction for adaptive reuse. In the context of the built environment and the sea, however, what does adaptive reuse mean? What methods of decommissioning does one employ for coastal unbuilding? The methodology for this research is twofold. The first phase consists of case study analyses of current decommissioning strategies (grounded buildings, ships, offshore-oil production, brown fields, etc) and the collation of known hazards the built environment has on the sea's health and welfare. The research findings synthesize the data for developing responsible procedural methodologies for this new category of decommissioning. The second phase of this research develops a procedural methodology in the form of a policy statement/set of guidelines stylistically modeled after the FEMA P-55 Coastal Construction Manual and synthesized in a policy statement such as the Policy on Decommissioning of DOE Facilities under CERCLA.

Addressing Human and Ecological Health in the Urbanizing Amazon through Interventions in the Built Environment

Hope Freije (University of Washington); Rhiannon Neuville (University of Washington); Rebecca Bachman (University of Washington)

Urbanized during the late 19th century, Iquitos is the largest city in the Peruvian Amazon. Despite immense biodiversity surrounding it, Iquitos is reported as having a mere 1.2 m2 of green space per capita, far from the WHO's recommendation of 9-12m2/capita (Zucchetti and Carranza, 2018, p. 77). This deficit of urban vegetation and permeable green spaces contributes to human and ecological health issues. For example, an overwhelmed stormwater system leads to standing water, which acts as a breeding ground for mosquitoes, such as Aedes aegypti, the vector for the likes of dengue, yellow fever, chikungunya virus, and Zika (Alarcón, 2016, p. 14).

In partnership with TRACTION Labs, Peruvian-led, interdisciplinary design research nonprofit, our team of three landscape architecture graduate students spent two months in Iquitos investigating opportunities and strategies for increasing the quantity and quality of green space. Through multidisciplinary literature review and interviews with Iquitos' biologists, architects, and urban planners, we identified a potential to increase green space in "right-of-ways" (spaces between sidewalks and roads), yet a dearth of data quantifying it. Whereas costly municipal plans to increase urban green space in Iquitos have repeatedly been delayed or abandoned, individuals can create green space in the right-of-ways. By quantifying existing right-of-way area, we aimed to create a compelling argument for governing bodies to incentivize citizen action.

The preliminary results of our study found approximately 393,000m2 of right-of-way in the district of Iquitos, 93% of which is unplanted. Through this data we can communicate the potential to increase human and ecological health in Iquitos by improving stormwater conditions, reducing the heat-island effect and adding more habitat for pollinators within right-of-ways. Further, we can propose designs that celebrate how people in Iquitos are already using these spaces, and imagine a citizen-created streetscape that celebrate both nature and city in the urban Amazon.

Redesigning Our Streets to Support Micromobility

Beth Ferguson (University of California Davis)

This research project presents an overview of creative cities that have embraced micromobility and bicycle infrastructure to reduce congestion and carbon emissions. As cities plan for the challenges of the 21st century, they must incorporate new forms of multifaceted sustainable transportation services into their climate resiliency plan. Rethinking on a massive scale is required to change global energy consumption within the next 10 years to meet the goals set by the Paris Agreement. [2]

Could micro-mobility services and flexible infrastructure help cities reshape the land they have previously dedicated to the automobile and reduce carbon emissions? Transportation leaders project that future mobility options will focus on multi-model services and not private vehicles. Mobility innovators are launching shared e-bike services that connect with public transit and bicycle infrastructure. "In the United States, federal law allows an e-bike to be regulated as a bicycle when it uses less than 750 watts of power, has functional pedals and a maximum speed of less than 20 miles per hour." [1] Ride-hailing services like Uber and Lyft are investing in urban e-bike share systems because they see 2-5 mile trips being done faster and cheaper on two wheels. Expanding micro-mobility services are proving to be cheaper and more convenient than car ownership for a range of demographics. What role should cities play in supporting the micro-mobility regulation processes to help reshape their sustainable transportation infrastructure? Case studies will include e-bike cargo programs in NYC, superblocks in Spain, and e-bike share networks in California.

A9 - Transform Design for Resilience & Sustainability: Health / Active Living by Design / Gerontology

Rethinking Depression in Long-Term Care Facilities from an Environmental Perspective

Marwa Abdelmonem (Virginia Commonwealth University-Qatar); Kristi Gaines (Texas Tech University); Debajyoti Pati (Texas Tech University)

Transitioning into a Long-Term Care Facility (LTCF) has been associated with immense stress. A number of studies show that newly admitted residents of LTCF experience different levels of depression during their transition (Wright, 2014). Despite the adverse complications of depression (e.g. social isolation, suicide tendency, mortality, etc.), it is usually overlooked and considered part of the aging process (Carpenter & Huang, 2011). Environmental causes of depression are inadequately addressed in the Environment-Behavior research. This study proposes a conceptual framework that links depression to place attachment and the psychological processes during transition (Manzo & Wright, 2014). We examined the proposed framework using a grounded theory approach. This qualitative methodology consisted of two main parts: (1) self-report questionnaire (Geriatric Depression Scale) followed by semi-structured interviews, and (2) a follow-up questionnaire. The study was conducted with 11 participants in a nursing home in St. Louis, MO.

The study yielded five themes of findings: (1) variations in transitional periods among participants were linked to their prior experiences in changing homes, (2) the ease of transition depends on the resident's level of mobility, (3) types of previous residences (temporary versus permanent) contribute to the degree of attachment and transition duration, (4) access to nature provides residents with a restorative experience that facilitates transition, and (5) passive observation provides a restorative experience to residents who used to be city-dwellers. In conclusion, this paper establishes an index for further development of both the line of inquiry, depression during the transition into LTCFs, as well as the methodology to better define the physical attributes of the built-environments that can alleviate depression in LTCFs.

Understanding Patient, Care Partners and Care Provider Perspectives on Key Home and Community Environment Challenges Encountered By Older Adults Undergoing Joint Replacement Surgery

Anjali Joseph (Clemson University); Sahar Mihandoust (Clemson University); Lisa Hoskins (Clemson University)

While the majority of older adults prefer to recover in their home post-surgery, most homes in the U.S. are not designed to support the needs of older adults (Gillespie et al., 2012). Older adults undergoing joint replacement surgeries will likely experience significant challenges navigating their environment post-surgery and are at a higher risk for falls (Matsumoto et al., 2012). Home interior and exterior design may pose as movement barriers and create risk for falls. Home modifications would prevent falls and would help elderly age in place (Granbom et al., 2019).

The purpose of this study was a) to understand the specific home and community physical environment challenges encountered by older adults undergoing joint replacement surgery and b) to understand acceptance for technology-based solutions that may support proactive home assessment. This study consisted of a series of semi-structured interviews and focus groups. Semi-structured interviews were conducted with 20 patient-care partner dyads before and after joint replacement surgery and focus groups were conducted with involved clinicians. Data collection focused on: a) Challenges with home environment for joint replacement surgery patients, b) Processes in place to proactively address home modification, c) likelihood of using technology-based home assessment tools, and d) Potential to integrate the tool into existing healthcare information systems. Recorded data was transcribed and themes were extracted after content analysis.

Patients-care partners identified entrances, stairs,

and bathrooms as areas with major moving challenges after surgery. Although patients and care providers agreed on the importance of conducting proactive home assessment and modification prior to surgery, not many patients modified their home prior to surgery due to lack of knowledge or services. Technology-based home assessment tools were appealing mostly to the younger patients- ease of use, proper graphics, and quick correspondence were the desired tool features. Clinicians had concerns about liability and patient privacy.

Active Design: Postwar Tract Housing and the Hidden Benefits of Natural Movement

Alexis Holcombe (Virginia Commonwealth University)

In the country's 50 largest metro areas, adjacent suburbs comprise 79% of the total population (Urban Land Institute, 2016). Suburban dwellers are also younger (75% of 25- to 34-year-olds) and more racially and ethnically diverse (76% of the U.S. minority population) than one might expect. Active Design promotes the inclusion of "natural movement" into the built environment. Most Active Design research focuses on public buildings and adjacent exterior environments in the urban context. Yet U.S. adults spend 68.7% of their time in a residence (Klepeis, 2001). Research into "Blue Zones" has examined five areas of the world where austerity and hardship create environments that encourage people's bodies to work harder. By contrast, modern conveniences lower resiliency and can lead to significant negative health impacts (Buettner, 2015).

Case studies will analyze the layouts of three 1960's era split level homes in suburban Virginia. It will be shown that these designs can optimize movement by maximizing the path of travel in plan and section. Virginia was a leading center for experimental residential design in postwar tract housing. Understanding the hidden benefits of living more actively today in 1960s era tract housing of suburban Virginia would therefore have important implications for many Americans, as the designs created by these local experiments eventually spread to suburbs nationwide.

With their characteristic stair runs as the defining feature, a important benefit of split level design contrasts with one of the key goals of Universal Design: creating barrier-free one-level living for a rapidly ageing U.S. population. The challenge, therefore, is to convince Americans that living in a home with easy-to-navigate levels can actually lead to a longer, stronger and healthier life span. Research suggests that we must "deconvenience" our homes so that we move naturally throughout the day.

Confidence Levels of Members of a TRANS-Disciplinary TEAM of Professionals before and after Viewing a Design Solution for Aging in PLACE Modifications Communicated with Architectural Visualization Technology.

Denise Rae McAllister (Purdue University)

Providing home modifications to empower aging individuals to live independently in their own homes involves the participation of a trans-disciplinary team of professionals (Tofle, 2015). Making up an ideal team are health care professionals, construction professionals and design professionals working together or independently to provide an assessment of a person's needs, to devise a plan outlining changes to the home environment and then to communicate that solution to the client or patient (Lockwood, 2017). Communicating the proposed design solution to end users is done in a variety of ways, many of which do not successfully convince them to implement the recommended modifications. Utilizing architectural visualization technology in the form of a digital walk through can increase the confidence in the recommended modifications by those viewing the walk through (Weakley, 2017). The walk through in the study was created by exporting a 3D model created with building information modeling software into gaming software. Perceptions of the attractiveness, construct-ability, ease of use and safety of typical home modifications for aging in place were queried, both before and after viewing the digital walk through. Study participants were divided into five groups, aging individuals, friends and family members of aging individuals, construction professionals, health care professionals and design professionals. Results reveal differences in the impact the use of architectural visualization had on members of the five groups as well as differences in perceptions of the effectiveness of modifications most frequently recommended.

A10 - Transform Design for Resilience & Sustainability

Designed Landscapes, Urban Climate, and Mental Health: A Commentary on Approaches in Urban Planning and Design for Addressing Urban Mental Wellbeing

Peter Crank (Arizona State University); Paul Coseo (Arizona State University)

An understudied intersection of urban landscapes and mental health is a focus on urban climate triggers of mental disorders. The design professions influence urban landscapes and thus microclimates through policy, projects, and discourse (Childers et al., 2015). Researchers have shown that these constructed landscapes create benign or harsh corresponding microclimate conditions, which have significant impacts on mental health (Kuo and Sullivan, 2001). Thus, designers play a large role in creating better microclimates to support community mental health.

A review of literature in landscape architecture and planning over the last 150 years also finds a connection between mental health and urban climate in arguments for greening and low density development. This undertone appears in several prominent calls for parks and green space to address mental health symptoms (specifically schizophrenia) that result from negative microclimate conditions (Howard, 1946). These designers made their recommendations for the most part based on anecdotal evidence, observations, personal experience, and unproven theories.

This paper is a commentary that focuses on examining the urban design, urban climate, and mental health nexus through a case study in the city of Phoenix about the importance of the design of urban environments for urban climate and mental health. We discuss the current breadth of knowledge at the intersection of design, climate, and mental health. Then we discuss best approaches to addressing the stressors for schizophrenic crises and episodes through heat, sunlight exposure, and nature contact; how the design of urban landscape help/hurt the schizophrenic population in Phoenix; and the strategies that maximize thermal comfort also aiding in reducing the environmental exposure to schizophrenia crises stressors. Finally, we take a neighborhood in Phoenix as a case study in how this perspective can be applied to an existing neighborhood redesign to improve holistic wellbeing for residents.

Visualizing Change Scenarios for Flood Buyout Properties in Conway, South Carolina

Lucy Rummler (Penn State University); Charles Andy Cole (Penn State University); Kenneth Tamminga (Penn State University)

Residents of Conway, South Carolina endured destructive flooding from successive hurricanes in recent years. The City of Conway acquired 66 parcels of impacted residential and commercial floodplain property through the Federal Emergency Management Agency (FEMA) Hazard Mitigation Grant Program, and city leaders hope to use the land for flood protection, ecological protection, and community benefits (Hymann & Williams). The complexity of costs and benefits and how these change over time sets up a risk of inaction towards the city's goals using these parcels ("Floodplain Buyouts", 2017). In this presentation, we develop change scenarios with three levels of resource intensity for three prototypical locations that include flood buyout properties. The change scenario levels are no action, moderate action, and aggressive action. To determine the most suitable land use change for moderate and aggressive change scenarios, we use geospatial analyses and the restoration action decision tree developed by Miller and Bestelmeyer (2016). We visualize the future impact of each change scenario to critical systems including flood safety, transportation, ecological protection, city spending, development, and city image. The visualizations are designed to be informative to residents, business owners, and city leaders so that they may proceed with community-led action. Lessons learned from addressing the issues facing Conway may be more widely applied, as these issues are representative of dozens of coastal plain communities in the South Atlantic United States.

Promote Salutogenic Design through Evidence-Based Design Research

Zhe Wang (Henan University); Huiling Meng (Henan University)

Evidence-based design research (EBDR) supports design decision making. Promoting humanity and addressing user needs, salutogenic design is advocated in China. Based on case studies in China and the USA, this paper investigated the ways of conducting salutogenic design through EBDR.

Methods of EBDR were analyzed using a study of cancer care environments as an example. This study had three phases (pre-design research, research-informed design, and post-occupancy evaluation/POE) and used both qualitative and quantitative methods to clarify user needs and satisfaction. During the pre-design, a literature review on design precedents and credited findings were conducted; qualitative data analyses clarified patient needs during infusion treatment. To address these needs, specific design strategies were developed and applied in the design of a new infusion center in Louisiana. During design, infusion patients and their families were invited to review design and support decision making. After the new center was occupied for two years, post-occupancy data was collected through on-site observation and survey. Statistical analyses of the data clarified significant design strategies. Suitability of these EBDR methods for use in China was investigated using a study of senior-living environments. Pre-design research was conducted and required for acquiring project approvals. Most of the research data are descriptive and in need of object measures to support design decision making. Methods of literature review and qualitative and quantitative data collection focusing on user needs were applied. Associated with fast economic developments, the processes of design run fast in China. Designers have limited direct collaboration with the users. Applied design strategies are recorded and POE is scheduled upon the completion of project. POE is generally suggested to projects invested by the government but rarely conducted due to tight schedules. Similarities and differences of the experience using EBDR to promote salutogenic design in these studies will be discussed.

B1 - Transform Education for Our Common Future: Education I

The Role of Basic Design Studio in Design-Based Education

Ozlem Demir (Amasya University); Ahmet Baskan (Amasya University); Merve Ozkaynak (Amasya University)

This research aims to address two vital issues in design education in Turkey: 1) prior knowledge of the candidate about the program, and 2) their intention to study design-based programs. University entrance examination in Turkey is administered once a year and contains questions on various disciplines. The Urban Design and Landscape Architecture (UD&LA) program requires a lower score compared to other fields. Furthermore, there is no other examination specific to design-based programs to assess candidates and examine their prospects. Those who decided to study design-based programs such as architecture, landscape architecture, interior design, and industrial design are not required to have prior design-education.

The purpose of this research is to understand the role of basic-design studios in the first-year design-education in design-based programs. The research was conducted as a pilot study at one of the public universities in Turkey. A semi-structured interview was utilized as a data collection method. The intention was to evaluate students' knowledge before and after and the primary intention of the first-year students in the UD&LA program. 41-students volunteered to participate in the interview. The primary goal of interviews is to identify and characterize various factors that can be an indicator of design education for those who do not intentionally join the program. The study further aims to outline obstacles that reduce students' learning skills and using basic-design principles in their work correctly. This research incorporates observations of experts in diverse design disciplines on students' progress.

The results of before and after interviews indicate that there are significant differences in knowledge of design and willingness to study the program. Most students did not seem to have a good understanding of the program prior to joining. The findings suggest various strategies like group-work, 1/1 scale-project to engage pieces, encouraging students to explain their design-work with the basic-design principles.

Bringing out Their Best: Applications of the Reasonable Person Model for the Gen Z Design Studio

Erin Miller Hamilton (Texas Tech University); Michelle Pearson (Texas Tech University)

It has been ten years since the Kaplans challenged environmental psychologists to discover "the ways in which the environment can bring out the best in people" (Kaplan & Kaplan, 2009). Since then, scholars and practitioners from multiple disciplines have explored the applications of the Reasonable Person Model (RPM) for helping people to engage in behaviors such as "sharing knowledge, giving and receiving help, earning and showing respect, and working to achieve common goals" – in short, to be their best (Basu & Kaplan, 2015, p. 1). While previous scholarship has explored the implications of RPM for classroom teaching, this work examines the utility of this model specifically within the context of the Gen Z design studio. In environmental design academia today, the majority of students belong to Gen Z. The experiences of this generation are unique because they are the first generation to never have experienced life without the internet (Turner, 2015); and therefore, most Gen Z students expect and know only instant access to information and communication.

How do these life experiences affect the interior design studio and what are some strategies to help students be successful? This presentation will include a review of the relevant literature including learning styles, social interaction, and generational characteristics. Key findings indicate that growing up in a technology-saturated environment has shaped the way Gen Z students manage their interactions with technology, approach information-seeking, and navigate social relationships. We will discuss concrete strategies for creating a supportive educational environment for the Gen Z design student. We will conclude with a discussion of an emerging empirical study exploring the effectiveness of implementing these strategies in the studio. The implications of this research are significant for educators to best serve their students by understanding that despite some generational differences, we all share several universal information-processing needs.

Teaching Social Factors in Design Realms: Examples from Landscape Architecture

Patsy Eubanks Owens (University of California, Davis); Yiwei Huang (University of California Davis); Jayoung Koo (University of Kentucky)

Instruction relating to the social and psychological impact of design has been integrated into landscape architecture curricula for approximately four decades (Francis, 1982). Design studios and lecture courses charge students with learning and applying information related to how people mentally and physically interact with the landscape. Environmental psychologists, architects, sociologists, and others generate much of the research shared with students (Zeisel, 1984). That research explores how the physical environment influences the individual's perceptions, satisfaction, and use of place, and how the incorporation of these findings, as well as methods for understanding specific communities, has been inconsistent in landscape architecture courses and curricula.

This paper examines the evolution of integrating learning objectives related to social factors into landscape architecture coursework with a focus on the exercises and assignments used to convey this aspect of people-place understanding and design generation in the classroom setting. Many education exercises, such as the environmental autobiography (see Marcus, 1992), have remained effective and consistent over the years while others warrant reflection and refinement to meet the current population's needs, advances in technologies, and other shifts in the people-environment relationship. For example, early methods of recording behavior with sketchbook and pencil are being replaced with digital tools (see, for example, Loebach & Gilliland, 2016; Cox, Loebach, & Little, 2018; and/or cite other articles presenting behavior mapping with ArcGIS Collector or other forms of technology)

An overview of design-relevant social factors topics will be presented and potential lecture outlines, references, in-class exercises, and assignments discussed. Exercises appropriate for classroom instruction will be critiqued. In addition, curricular approaches for studio projects incorporating social factors research and education will be discussed. Lastly, trends and predictions for future teaching methods will be elaborated on along with suggestions for future courses, materials, and methods of social factors education in landscape architecture.

Using Discomfort in Design Education

Olivier Vallerand (Arizona State University)

Building on feminist and queer critiques of architecture and design education and practice, educators have sought to imagine modes of teaching that embrace a queer and feminist ethics in both content and methods. Through interviews with educators in North America, Europe, and Australia, I have identified feminist- and queer-informed strategies for design pedagogy: disrupting traditional methods and encouraging productive failure, supporting students, engaging with activism, and embracing radical visibility. The strategies focus on bridging a gap between different groups, on helping designers to acknowledge the limits of their designs, and on building relations through layered opportunities. However, many educators interviewed have also identified strong resistance from both students and other faculty. Borrowing from a pedagogy of discomfort framework (Boler, 1999; Zembylas & McGlynn, 2012), this paper suggests how these resistances can be productively used to challenge students' understanding of architecture and design – and the professional context – but also how they can be used to create relations with other disciplines and to challenge institutional barriers.

B2 - Transform Education for Our Common Future: Education II

The University As a Public Sector Practice

Cathi Lynn Ho Schar (University of Hawaii)

Among all of the services that governments provide—social, cultural, political, economic, and ecological expertise and analysis, for example—design thinking and design services are typically absent. Yet, there is an emerging movement in design-thinking practices toward government integration (Bason, 2018), that presents lessons for environmental design research, pedagogy, and practice. Back in 2002, the Danish government established MindLab, an innovation unit within the ministries of Business and Growth, Employment, Children, and Education. Other similar units have emerged worldwide. In the U.S., this integration is reflected in for-profit companies like IDEO, non-profits like Bloomberg Philanthropies, and in-house governmental operations like LAB OPM and the Boston Office of Urban Mechanics. This presentation identifies a new space for public sector environmental research and design, through a survey of existing and emerging practices, including the University of Hawai'i Community Design Center (UHCDC), a newly established research, planning, and design resource that began as a public-interest practice, and evolved into a university-wide public sector practice, based on three years of close partnership with the Hawai'i state government. In 2017, the state legislature appropriated $1.8M to five state agencies to initiate projects with the center, which has grown into a platform connecting over 20 faculty members across 6 departments, 10 full time staff, 70 part time student interns, and hundreds of students in academic courses enged on projects for 12 state agencies. This presentation will provide an overview of the center and details related to two projects for the Office of Planning and Department of Public Safety that highlight the university's ability to act as a multi-disciplinary research and design hub for the state. The work reflects the center's shift to to scopes of work that jointly rethink systems, services, and environments as a way of advancing the public good and serving the public sector.

Utilizing Theories As a Design Tool in Interior Design Studios

Tina Patel (Kent State University); Sara Marie Angne Alfaro (University of Missouri-Columbia)

Theories are not intended to be prescriptive, but projected to inform the thought processes and design, activate questions, inform hypothesis, and assist in the development of thinking tools (Tilley 2006, 10). Engagement with theory is therefore crucial as it allows designers to reflect, articulate and engage in discourse within their own and other disciplines with intellectual rigor. It is essential for designers to recognize the theory that they already have, although they may not recognize it as such, in terms of concepts of form, perception, and creativity, is an excellent launching point. An understanding of these concept-based methodologies along with critical thinking about the creative design process, is one part of the knowledgeable designer and informative educator.

Susan Close reveals that there is a logical connection between theory and practice in interior design that involves content, context and narrative. Pertinent issues such as space, place, globalization, gender identity, branding, migration, performativity and privacy that are significant areas of theoretical study are all transferable to the more revisionist study of interior design being constructed in the twenty first century (Chalmers and Close 2007, 79).

This presentation focuses on how two interior design programs have embedded theories from environmental psychology, social sciences and humanities in the research and programming phase. Findings reveal how students embraced these theories to build hypothesis for the design issue and these theories later informed guidelines to cultivate a more human-centered design response. Reflection statements from the students capture the value of theory in interior design education to assist directly in their work by improving their position and power. Conclusively, the authors acknowledge the role of theory is to cultivate, develop and support the advancement of interior design as a discipline through interaction, mediation, and discourse.

Urban Planning and Environmental Design Educational Experiences Towards Sustainability

Ana Karinna Hidalgo (University of Regina)

The fields of urban planning and environmental design urge to incorporate theory, methods and discussions about sustainability, climate change and resilience along the learning process. Students should not only learn how to solve problems and develop great buildings and spaces but also to relate their knowledge to current global issues.

This presentation includes three educational experiences from the courses of planning theory, architecture design and environmental design. These courses are aimed at understanding how environmental design, architecture and urban planning can provide with a healthy environment that respects nature, culture, Indigenous knowledge and topography towards the creation of sustainable and resilient communities. During an academic term students developed comprehensive design projects that integrate the many concepts and theories needed with a strong emphasis on the SDGs, Indigenous knowledge, natural landscape (topography) and well-being.

The first project, Urban Streets, pushed students to observe, use, live and analyze from a different perspective streets in Metro Vancouver. They were asked to critically analyze aspects related to transportation, sense of place, physical and mental health and design elements in order to redesign them towards a comprehensive approach to its function. The second project, First People's Graduate House, challenged architecture design students to place a real project located in British Columbia in different provinces in Canada (Saskatchewan, Ontario and Alberta). They incorporated concepts of passive house, sustainability, site analysis, vernacular materials and Indigenous territories. The diversity of the outcomes also reflected the diversity of culture, natural landscapes and Indigenous knowledge around Canada. The third project, Eco-Museum / Eco-Housing was aimed at developing an energy efficient habitat in a shipping container that considers sustainability, endemic vegetation, LEED/ BREEAM and biophilic design in order to create a healthy and sustainable environment. Students were asked to solve both the interior design and outdoor space (parking lot).

Hybrid Pedagogy: Transfer of Knowledge to Teach Landscape Vision Planning and Design in Studio

Yujia Wang (University of Nebraska Lincoln)

Landscape architects are playing an increasingly visible role in the deepening integration of knowledge of various disciplines to effectively define the future of communities. However, landscape architecture education tends to focus on the design of space and places, which is better suited to define spatial and programmatic aspects on a site-and-scope-specific basis, than the projective aspects of defining the future itself. For the latter, public landscape has proven to play a key role in providing a coherent urban structure, framing and energizing developments, and addressing urban risks. The aim of this studio setup is to equip the students with the capacity to operate in this context.

The design of this hybrid pedagogy is for an integrated approach to a) develop the mindset and skills for projective thinking and design, by transfer of knowledge from urban and regional planning, taking precedents from practices already established, and well-structured; and b) continue to reinforce the ability to define public spaces with character and scale, using traditionally available toolsets in landscape architecture education. The two are reconciled on a project scale where the impact of both show with adequate definition and a continuation in full spectrum of design can be achieved.

With inputs from colleagues at Cardiff University School of Geography and Planning, several well-established concepts and processes are identified, adapted and introduced to the studio. This process consists of data-based analysis, scenario planning that identifies key capacities, defining vision and objectives, and making spatial strategies that synthesize the conceptual framework and define it in structure of space. The studio chooses its context in a local community in Omaha, within which there is active interest to think about the future. This allows the engagement with stakeholders to be an important part of the studio, reinforcing the university's mission as a land grant institution.

B3 - Transform Education for Our Common Future: Education III

Design Criteria of Various Types of Study Spaces for Millennials in a University Residence Hall

Sharmin Kader (TreanorHL)

Today, new types of learning environments are emerging in college-residence hall facilities to accommodate new pedagogical approaches and the new generation of students. The newly designed residence halls are designed with various types of study areas. For example, students will have access to three to four different types of study spaces in a hall. This research study is based on two newly constructed residence hall in Texas: case-A is with700 beds, and case-B with 500 beds. The residents of these two halls have access to three different types of study spaces, excluding their bedrooms: floor study rooms, floor lounge areas, and a central community learning center at the first-floor level. These spaces provide a wide range of elements to foster student learning in new and creative ways by offering study options with various types of comfortable furnishings to encourage creativity and support peer-learning. The community learning center includes a PC lab with advanced software and printing options, multiple workstations and reading areas for individual study and collaborative work, and instruction rooms with multi-media. Though the idea of providing various options of study spaces, there has not been an in-depth examination of the design criteria of these spaces in a residence hall. To develop evidence-based guidelines for design, this study has considered a mixed method of data collection: a) online student survey; and b) interviews with staff and students. The study measured the utilization rate of these spaces by the student, reasons behind the preferences or selections, benefits of having these options, and identified what design concepts worked and what didn't from the results. A list of design criteria was developed based on student feedback, and these findings carry a great significance for architects and university housing professionals for future design and renovation.

Evaluating a Workshop for Interdisciplinarity in Environmental Design Research

Jinoh Park (North Carolina State University); Byungsoo Kim (North Carolina State University); Charleston Yi (North Carolina State University); Sharon Joines (North Carolina State University)

Transforming society sets what an opportunity education has to offer. As more complex problems arise in modern society, diverse perspectives and approaches are required to make decisions. To meet the needs, design research education emphasizes implementing interdisciplinary research culture at the institution level (Davis, 2015) and at the student level (Park, Kim, Lee, Hands, & Rider, 2019). This research team has developed a workshop to begin collaborative interdisciplinary research in design at the graduate student level (Kim,Park,Rider,&Joines, 2020). This research aims to evaluate the workshop frameworks developed in a previous study, whether it is serviceable to derive common research topics and goals from student research teams with different backgrounds in environmental design.

This research conducts an exploratory case study by organizing a workshop with graduate students in environmental design research. The research gathered participants through an environmental design research association. The researchers organized the groups of three/four participants in various backgrounds. The groups will be asked to participate in the workshop including 1) workshop preparation, 2) team discussion, 3) article sharing, and 4) brainstorming for seven days. All the activities will be reported by each group, and a survey will be conducted after the workshop to collect data of; the groups' research themes and goals; and their comments/suggestions about how to improve the workshop in the future. Finally, a three-month follow-up survey will be conducted to understand how many workshop participants groups continue research based on the predetermined theme and goal and their progress. The results of this research will present the impact of the interdisciplinary design research framework in the previous study (Park,Kim,Lee,Hands,&Rider, 2019) to design research in the environmental design field in a workshop format. This research will present the benefits, limitations, and the next steps of the workshop.

Collaboration Skills in Co-Designing for Complex Problems

Mauricio Mejia (Arizona State University); Yumeng Xie (Arizona State University)

Contemporary design education has inherited a master-apprenticeship, in which instructors are expert designers that are not able to explain the principles that guide their actions (Frascara, 2007, p. 63). The curricular approach in design is mainly the studio pedagogy, in which students engage in a flexible environment to experiment and create with instructors that provide mentor feedback rather than authority direction (Sawyer, 2017, pp. 109–110). The traditional studio pedagogy has shown strengths and provided unique skills for making; however, students often lack integrative knowledge and abilities within or across disciplines (O'Brien & Soibelman, 2003). Since practices such as environmental design are increasingly complex, integrative, and participatory, designers need to learn skills from multidisciplinary resources (Gericke, Qureshi, & Blessing, 2013) and be able to collaborate with other disciplines, communities, governments, and stakeholders (Moreno & Villalba, 2018, p.48; Gibbons et al. 1994, p. 28).

In this presentation, we will review current research of transdisciplinary collaboration in design education. And we will discuss and provide more insights on four collaboration skills for design students: translational communication, facilitation management, iterative prototyping, and transdisciplinary empathy. Translational communication consists of the ability of designers to communicate well among different research and practice cultures that have different methods, worldviews, and use of language. Designers skills include exchanging discipline specific terms/language relevant to the project and discussing teamwork expectations. Facilitation management is the ability to lead collaborations and bring together diverse stakeholders with equal active and continuous involvement. Iterative prototyping consists of rapid, rough, and cyclical prototyping, so transdisciplinary teams can learn and discuss based on tangible ideas and solutions. And transdisciplinary empathy is a capability to accept other members' concerns based on their practice experiences. We will also propose a general framework of how to provide design students a better environment to learn transdisciplinary collaboration skills.

Cultivating Inclusive Campus Spaces for Diverse Students: A Higher Education Case-Study Using a Multi-Methods Approach

Kevin Bonnell (Wilfrid Laurier University); Simon Coulombe (Wilfrid Laurier University)

The physical environment is often seen by the public and decision-makers as having a relatively negligible impact on people's lives, but it can be inherently oppressive for marginalized people, as spaces are often designed using dominant visions and knowledge (Laws, 1994). On university campuses, which are increasingly diverse in terms of students' identities, it's important to consider how the physical environment can convey messages of racism, discrimination and trauma, and alternatively, how it can promote diverse students' resilience and wellbeing. It is critical to center the research and design processes around student voices to help explore how the campus design can challenge or support marginalized students (Dutton & Grant, 1991). This presentation describes a case study in Ontario, Canada, using student-engaged research methods to explore how campus spaces influence students' wellbeing. We use a multi-methodological approach integrating three methods: walking interviews, photovoice and narratives. Using multiple methods offers "multiple angles" through which we can understand the nuances of students' experiences. These methods have been used in the past to understand the experience of individuals within various environments, but to our knowledge, no study has integrated these to comprehensively explore the experiences of university students. This research also incorporates the voices of campus stakeholders by working with a Research Advisory Committee, including students and staff that identify as or support marginalized students. The results highlight students' stories of resilience and wellbeing in relation to campus spaces, contributing to empirically derived recommendations to create inclusive and wellbeing-promoting campus spaces. Students' stories are presented to emphasise the importance for higher education professionals, designers and architects to collaborate with marginalized students to create campus spaces that truly address their diverse needs.

B4 - Transform Education for Our Common Future: Interior Design

An Intercollegiate Collaboration: Examining the Charrette As a Mode for Vertical Studio Engagement

Stephanie Sickler (Florida State University); Amanda Gale (University of North Carolina Greensboro); Travis Hicks (University of North Carolina Greensboro)

Community-based partnerships have been widely accepted as favorable pedagogy for student engagement (Conway, Amel, & Gerwien, 2009). While most partnerships include sustained collaboration between stakeholders and students, this project examines a short-term engagement, facilitated through an intercollegiate design charrette, proposing it as an alternative model for engaged scholarship. The intense and collaborative nature of a charrette format provides an immersive experience, closing the gap presented by a shortened engagement period, thereby offering similar outcomes to the traditional model.

A case study will be presented, which details a charrette engaging design students from two universities with a community partner where persons with disabilities live and work together on a sustainable farm. The vertical format of the charrette engaged students from second through fourth year levels in their respective programs, integrating students at a variety of skill levels. Students were presented with a design challenge prior to the charrette, and completed initial research and programming prior to meeting. Though short in duration, the project conformed in every way to Zollinger, Guerin, Hadjiyanni, and Martin's model of service learning in the applied arts (2009) as the charrette required participants to (1) relate to course objectives, (2) apply course knowledge, (3) connect to the community, and (4) reflect on learning.

Students engaged in this short term collaboration gained a tremendous amount of experience as a result of the collaboration not only with the community partner, but also with students from a different university and different levels in their design studies. Reflection papers captured learning outcomes and students' perceptions of the charrette. This multi-faceted design challenge removed an extended design development phase but replaced it with rapid fire prototyping from a diverse group of design thinkers. The result was an alternative and ultimately achievable model for student engagement offered by charrette experiences.

The Power of Pattern: Utilizing Historical Textiles to Inspire Design for Health and Well-Being

Stephanie Sickler (Florida State University); Kari Fisher Culberson (Kari Fisher Designs)

Instrumental in returning the focus of the textiles industry to human-centered design, William Morris believed that the industrialization of textiles caused a dehumanization of the industry by removing the innate connection with nature and humans ("V&A Introducing William Morris"). His passion for the anti-industrial movement expressed through pattern has been echoed in other voices such as Alexander's A Pattern Language (1977) where authors asserted that pattern must positively influence happiness and tranquility. Frontczak et al. (2011) determined that comfort of furnishing, colors and textures were indicators of occupant satisfaction for indoor environmental quality. However, finish material studies in design coursework often overlook the importance of pattern outside of aesthetics. This project challenged students to embrace the power of pattern and natural elements to influence health and well-being for occupants. Students have shown increased emotional recognition, empathy, and awareness of multiple perspectives when taught to critically analyze and recognize patterns in art (Shapiro, Rucker, & Beck, 2006). Students in a materials course were tasked with designing an original hand-blocked textile then apply it to furniture in a home where residents planned to age in place. Students were given historical Morris patterns to analyze and reinterpret into original design aimed at supporting the needs of the end user by influencing tranquility and well-being through pattern.

Final designs reflected influences from Morris while indicating a shift toward contemporary styles well-suited for application in an environment designed with well-being in mind. Designs presented a new aesthetic for aging in place. This reflection back toward textile heritage is relevant as nature and health are being connected in industry, social media, and society in general. This project is timely, connects students to industry as well as history, and deserves a deeper look at our connections to the world around us and the history of our industry.

Critical Settings: Interior Territories of Action and Impact

Rana Kamal Abudayyeh (University of Tennessee)

Interior spaces carry the code of everyday life and formulate the backdrop of spatial memories. They play an integral role in foregrounding identity (Upton 2002, 719). The resilience of such native patterns of occupancy offers a common denominator in the midst of changing typologies of dwelling, community, and culture. Interiority is uniquely positioned to respond to arising demands and complexities of social and environmental challenges around the world. Yet, interior design has often contended with many misconceptions with regards to its social outreach and impact. Even interior pedagogy has veered away from complex narratives; preparing students for the pragmatics of interior spaces, but not their inherent implication and visceral reading.

With this agenda, a series of upper-level interior architecture studios worked to foster connections to the community and promote issues of civic responsibility. They involved collaboration among a diverse team of faculty, design professionals, and community organizations while building on industry partnerships. Within this robust collaborative framework, students develop skills necessary for transitioning into the profession. More importantly, such collaborations invite reciprocal exchanges between the academy and the community. They inform an active feedback loop where interior design assumes integral roles within public interest and wellbeing. Our main design explorations focused on the application of innovative interior strategies to spaces of impact. The studios began with conceptual studies of the sensorial effects of materials and textures and the resultant reading of space (Kolarevic & Klinger 2013, p. 10). Students applied these explorations to the design of socially engaged programs such as a youth outreach center, and a special needs school. Through this structure, the courses questioned how diverse perceptual modalities generate new potentials for learning environments. Students worked directly with local communities and were involved in formulating interior solutions for settings that can deliver significant influence towards positive social change.

Spatial Configuration and Informal Learning-Related Behaviors in the University Student Union: A Comparative Case Study of Two Multi-Use Learning Environments in Higher Education

Renae N Mantooth (North Carolina State University); Celen Pasalar (North Carolina State University)

There has been over a 10% growth in campus built space in the last decade and almost half of university funds spent on the built environment are for informal learning spaces (Sightlines, 2016). Informal learning is both an activity (Rogoff, Callanan, Gutiérrez, & Erickson, 2016) and space typology (Morieson, Murray, Wilson, Bronwyn, & Karli, 2018). As campuses have evolved with innovative technologies and diverse perspectives, the student union has adapted as a place for learning, belonging, and community (Rullman & Harrington, 2014). Prior research on informal learning spaces in higher education is within the context of the academic library (Walton & Matthews, 2018) and less is known about the relationship between the spatial configuration of student unions and student behaviors that support the overall experience of undergraduate students. The conceptual model that frames this study integrates both Bandura's (1989) Social Cognitive Theory and Stokols' (1995) Social Ecological Model to capture the multifarious variables that contribute to the relationship between the physical environment and the learning process. Within a comparative case study design, we examined the relationship between spatial implications and informal learning-related behaviors. A spatial analysis was conducted on Talley Student Union at NC State University and Gatton Student Center at the University of Kentucky with computational Space Syntax techniques by classifying how accessible and how visible (integrated) specific zones of informal spaces are within each student union. Low, medium, and highly integrated spaces within each building were identified to purposefully select locations for systematic behavior observations. Observations were conducted to understand how various spaces support informal learning-related behaviors. Findings suggest that patterns of informal learning-related behaviors are influenced by spatial configuration, furniture layout, and circulation patterns. This research informs designers and administrators about what spaces are needed, how spaces are utilized, and why students choose them on university campuses.

B5 - Transform Education for Our Common Future: Education IV / Cyberspace & Digital Environments

Activist Phenomenologies: Game Engines As Design Research into Diverse Embodiment

Matthew Seibert (University of Virginia)

In response to regular comparison of the built environment to the procedural complexity of a computer--open to hacking, reprogramming, and optimizing--Shannon Mattern aptly calls for "new models for thinking about cities that do not compute. ...urban epistemologies that embrace memory and history... recogniz[ing] spatial intelligence as sensory and experiential" (2017). Embracing these insights, this presentation counter-intuitively posits an acutely computational medium and representational tool for investigating the experience and production of the built environment: the game engine. As the software traditionally used to develop videogames--combining the capacities of 3d modeling, physics simulation, and effects generation--it enables intentioned world-building of navigable, interactive environments. This creative act of spatio-temporal making, deconstruction, and re-composition is a powerful view into one's unique bodily way of knowing the world, one that can be iteratively studied and shared.

Building upon the theorization of embodied experience following French phenomenologist Maurice Merleau-Ponty's body-schema (2012), cultural geographer John Wylie's definition of landscape as the materialities and sensibilities with which we see (2006), and media scholar Brendan Keogh's play of bodies bridging across physical and virtual realms (2018), a new design research heuristics is presented through example projects. This is a new computational tool open to engaging an individual's (or community's) spatial intelligence as sensory and experiential--as embodied and lived--in a new, diverse, and collaborative dialogue where the open worlds and alternative spatial readings produced by game engines constitute a new form of landscape representation. Virtual worlds thus crafted subtend a novel, prismatic palette of cybernetic phenomenologies, "a way to unveil the particularities of different embodiments" and their informing of "what bodies, institutions, and architectures can do" (Norwood, 2018). In a context of global crisis around climate change, social inequality, and toxic nationalism, this heuristic methodology of experimentation without optimal solution holds great promise to affect change through design activism.

Co-Creating with Community: Participatory Design in Environmental Design Education

Yumeng Xie (Arizona State University); Chingwen Cheng (Arizona State University); Paul Coseo (Arizona State University); Mauricio Mejia (Arizona State University)

In conventional design practice, professional designers work for clients rather than working with stakeholders. Design practices are becoming more complex, requiring multidisciplinary collaboration, and demanding the participation of communities (e.g. Moreno & Villalba, 2018; Gericke, Qureshi, & Blessing, 2013). Lacking engagement with stakeholders will result in dissatisfaction from direct users and may result in unwanted outcomes. Participatory design (PD) provides an effective approach to involve stakeholders in the design process.

This research provides an overview of both challenges and potentials of PD in design education and uses a Master of Landscape Architecture (MLA) studio course at Arizona State University to illustrate the effects of PD. In this course, MLA students collaborated with Mo'ili'ili community members in Honolulu, Hawai'i, to co-design Old Stadium Park redevelopment in three phases of the PD process – 1) field study, 2) community workshop (people 50+), and 3) pop-up design conversations (people 100+). Students' learning process was enriched by incorporating the PD approach, which posed challenges and generated benefits. Major challenges in teaching include: (a) trust-building with stakeholders, (b) time management, (c) uneven access to centers of knowledge, (d) flexibility, and (e) data analysis skills. Major benefits for students include: (a) increased motivation and passion about designing with communities, (b) understanding of equitable design process, (c) increased awareness of their responsibilities and the critical social consequences of their work, and (d) enhanced identification of certain genres that are more useful for design decisions like emotions drive place attachment in the community. This study suggests that it is time for design educators to strengthen future designer's PD skills for a more equitable and sustainable design approach that solves complex real-world problems. The studio pedagogy and research framework is not only applicable for all design disciplines education, but also contributing to building social capital in impacted communities.

Transforming Design Education: Measuring Acceptance of Urban Designs through Physio-Psychometric Tools.

Luis A Mejia Puig (Oklahoma State University); Md. Azizul Islam (Oklahoma State University); Maral Esmaeili (Oklahoma State University); Tilanka Chandrasekera (Oklahoma State University)

Urban design opens the opportunity to use of virtual reality (VR) as a tool to enhance the design process [1][2]. Through VR, future urban design developments can be researched, assessed and evaluated. This type of analysis permits design proposals to cater to stakeholders of city planning and administrator's needs.

Previous research has shown how urban designs, influenced by pedestrian features generate higher levels of acceptance from the viewer [3]. Additionally, pedestrian paths in urban environments are the outcome of visibility attributes and cognitive demand [2]. Cognitive Load Theory (CLT) defines cognitive demand and currently plays a critical part in the relationship between urban environments and users. Three elements structure CLT: Intrinsic cognitive load, extraneous cognitive load and germane cognitive load [4]. The study conducted by Noland et al. [3] used eye tracking in combination with visual preference surveys as tools for better understand user´s acceptance; nonetheless, their research did not experiment with immersive environments. This study proposed the use of functional near-infrared spectroscopy (fNIR) to measure cognitive load in correlation with visual preference surveys as a measure of design's acceptance in immersive virtual reality environments.

A walk-by through two different urban areas was documented using a 360-degree video recorder. Thirty participants were recruited and randomly assigned to one of the groups. Participants were asked to view a walk-by video through a VR headset. An fNIR device was attached to the participants' scalp to account for cognitive load in addition to a visual preference survey for correlation.

The collected data was analyzed comparing the calculated cognitive load and the outcomes of the visual preference surveys. The analysis revealed how different designed attributes affected the cognitive load of the users, and influenced their acceptance of the urban space. The results of this study provide implications for future urban design decisions.

A Multifaced Collaboration: Impacts on Pedagogy

Cherif M. Amor (VCU-Q); Sherin Karawiya (ACADEMIA)

Design and allied disciplines face complex pedagogic challenges relative to the implementation of collaborative endeavors. While a growing body of collaborative efforts have been initiated, a substantive deficiency of collaboration and collaboration mechanisms is detected in design and allied disciplines (Amor, 2008; McCoy, 2012; Webb & Miller, 2006). This deficiency, in addition to the compliance with accreditation standards, advisory boards recommendations, and university strategic plans compelled the development of an interior design course that brought together four bodies:
1) an academic institution, 2) a design firm/real estate, 3) a governmental agency, and 4) an art exhibition organization.

The purpose of this line of inquiry was 1) to explore the feasibility of this multifaceted collaboration based on a real-life design project, 2) to examine the impact of collaboration on the generation of design processes and products, and 3) to investigate the implications of collaboration on subsequent design studios.

To do so, a junior studio course, including 21 students, offered during the fall semester, served as the setting for this inquiry. A design firm provided the programmatic data for this project (site, plans, program), an incentive represented in a design competition that culminates in prestigious awards, and served as client/critic for this project. The governmental agency facilitated the application and vetting of sustainability principles. The art exhibition institution offered a venue for students to curate and exhibit their work. Data were analyzed using two approaches: 1) analysis and evaluation of three different design phases, and 2) open coding (Strauss & Corbin, 1990), which consisted of breaking down, conceptualizing, and reconstructing data in new ways. Corroborating previous research, one significant finding from this study underscored the comprehensive learning and application of design concepts as influenced by different intervening entities. Additional findings, pros and cons, will be shared with the conference attendees for feedback.

Developing a Learning-Teaching Strategic MODEL for Facilitating Creativity in Architectural Design Studio Based on Personal Construct Psychology

Ahmed Agiel (University of New South Wales)

Designing is an active learning process of knowledge to construct perception skills as well as reasoning that are cognizant, imaginative, intuitive, creative, selective and intellectual. Architectural education should embody the design knowledge and the associated tools for practically applying this knowledge to specific issues and construing personal meanings in design. Researches suggest that the conventional methods of teaching design depend on the instructor-student relations, repetitive exercises or systematic knowledge of an applicable software. Architectural design educators, highlight that there is an uncertainty in the content of delivered design knowledge and lack of documented theory, definite methods or tools related to design pedagogy. Integrating the theories of 'constructivism' and 'personality', the authors postulate the applicability of Personal Construct Psychology (PCP) (Kelly, 1955) and its analytical components as a means to facilitate creativity in architectural design. We hypothesize that in comparison to the conventional design approach, the PCP based design approach aids creativity by enabling the learners to construe their own meanings and conceptions to meet the needs of the local image of architecture. An experimental pilot study was conducted to evaluate the design creativity achieved on the performance of a convenient sample of architecture students in a design process at United Arab Emirates University, UAE. A specific conceptual model framework comprising of four design phases were systematically combined to develop the PCP based design model. Results substantiate that the design outputs based on PCP enhanced and facilitated creativity enabling the students to construe concepts from their conceptions, establish relations between the concepts and inherent knowledge and generate inferences based on their reasoning. Instructor serves as a facilitator. The findings of the study identify a new conceptual theory for the transformation of conventional design studio to an open constructivist environment based on 'constructivist concept learning'.

B6 - Transform Education for Our Common Future: Children, Youth & Environment / Transform Processes and Outcomes for Equity: International Housing Research

Collective Agency and Decision-Making Processes in Post-Disaster Housing Recovery: A Case Study from Southwestern Uruguay

Silvina Lopez Barrera (Mississippi State University); Diego Thompson (Mississippi State University)

The natural and built environments of rural communities of southwestern Uruguay have recently experienced significant transformations due to climate change and intensification of agriculture. Community or collective agency is the capacity to make changes at local level through collective actions of individuals, groups, and/or institutions, being a critical aspect of how communities respond to disturbances. Based on semi-structured interviews with key informants, participant observation, and digital spatial analysis (between 2003 and 2018), this article studies community governance for adaptation to recent environmental challenges and natural disaster events in southwestern Uruguay. Using the case study of the small-town Dolores, this article explores collective agency and decision-making processes in post-disaster housing recovery, including temporary emergency housing and housing reconstruction. This study reveals housing complexities of short-term post-disaster relief as a resilient strategy but with long-term sustainable challenges and critical implications. Findings show the main challenge of post-disaster housing reconstruction was to include vulnerable populations in decision making processes. Lack of inclusion in decision making processes resulted in increasing risks from the re-implementation of vulnerable construction materials and systems, which mitigated pressing needs but did not solve long-term housing needs of those affected by extreme weather events. Our discussion focus on how these adaptive responses could increase exposure of vulnerable communities under similar natural phenomena. When collective mobilization and collaborative efforts emerge to solve phenomena such as extreme weather events, it is critically important to include affected populations in decision-making processes and explore the characteristics and nature of the adaptive solutions which could represent future challenges for communities facing natural phenomena such as climate change. Recommendations center on collective agency as a key component for small rural communities to reclaim and improve natural and built environments while strengthening collective decision-making in post-disaster housing relief process.

Is the Promise of ' Better Homes ' Delivered? : Assessing Impacts of Slum Redevelopment Policy in India

Uchita Vaid (Cornell University)

This paper examines the assumption that redevelopment of slum communities, results in improved housing quality, by conducting a longitudinal study of an in-situ redevelopment policy in Ahmedabad, India. Under an in-situ redevelopment policy, low-rise apartment-style housing is built on the same site where slum settlements were originally located. I conduct a pre- and post-redevelopment assessment of housing quality in a settlement that underwent redevelopment, along with a wait-listed control group. This quasi-experimental study design helps to conduct a "place-based person-centered" evaluation by adding a physical environment lens to existing evaluation lenses. Existing evaluations of redevelopment policies focus largely on socio-economic consequences1 or implementation processes2. A missing element in these evaluations is an examination of the physical environment effects of redevelopment policies which affect residents' health and well-being. This focus on physical environment is critical because such policies uproot people from their existing life under assumptions of providing improved living conditions. This paper aims to bring voices of inhabitants to forefront, in the evaluation process.

I use a mixed-methods protocol that employs standardized observer-based housing quality assessment tool to identify specific aspects of housing quality that improve/deteriorate. I also conduct in-depth interviews and focus groups to understand the changes in housing quality post-redevelopment. This paper uses mixed-effects models to examine housing quality in a settlement, pre- and post-redevelopment (2.5 years after), in comparison to a wait-listed slum settlement that did not undergo redevelopment. It further compares the housing quality of a resettled community at two time periods, 2.5 years post-redevelopment and 8.5 years post-redevelopment to explores changes in housing quality in the medium-term. Our analyses reveal an improvement in housing conditions, in the short term, but settlements show higher levels of deterioration in the mid-term, especially in structural quality and maintenance. The paper further discusses the policy implications of these findings.

Nature-Based Learning Environments: A Sensory Approach for Inclusion

Kristi Gaines (Texas Tech University); Charles Klein (Texas Tech University); Malinda Colwell (Texas Tech University); Angela Bourne (Fanshawe College); Michelle Pearson (Texas Tech University); Huili Wang (Texas Tech University)

The purpose of this study was to address the importance of sensory input within the outdoor learning environment as a child development strategy. Typically, people receive information about the surrounding environment through their senses collectively (sensory integration). However, sensory processing disorder may occur when sensory signals do not integrate to provide appropriate responses (Gaines, Bourne, Pearson, and Kleibrink, 2016). As a result, the environment may cause a child to feel confused or irritated. As with all sensory symptoms, severity may vary, and both hyper- and hypo-sensitivities may be present.

Best practice design indicators for indoor (Gaines, Bergan, and Curry, 2014) and outdoor (Cosco, Moore, and Smith, 2014) learning environments were initially evaluated. The research addressed in this investigation builds upon these previous studies. The result is the development of inclusive evidence-based design recommendations for children of all abilities.

Sensory Integration (SI) theory provided the theoretical framework for this study. SI refers to the use of sensory information that helps an individual interact with the environment. A mixed methods approach was utilized to gather data including 1) focus group, 2) interviews, 3) observations and 4) surveys. A total of over 600 subjects participated. The target population was individuals with sensory integration disorder and their caregivers.

The findings show that individuals with sensory processing disorder view their environment differently than the general population. The data gathered was analyzed and coded to reflect six sensory categories: sight, touch, hearing, taste, smell, and motion (includes proprioception and vestibular senses). The result was the development of a sensory approach for design for outdoor learning environments. The research also showed that all children benefited from the integration of the inclusive sensory design principles. This presentation will illustrate each principle and explain the ways child development is promoted through their use in outdoor learning spaces.

Metacognitive Regulation Practices Evidenced By Adolescents in Residential Informal Learning Environments.

Amy M Frisz-Conlon (University of Missouri)

Through advances in technology, learners have been liberated from the restrictions of formal learning environments. With this liberation, comes challenge and risk to regulate learning in a range of diverse environments. The purpose of this qualitative, ethnographic research study was to understand the relationship of spatial preference, metacognition and technology across adolescents' learning environments.

Dabbah and Kitsantas developed a framework involving self-regulated learning (SRL) in Personal Learning Environments (PLE's) (2011). Pintrich et al. (2000) indicate that there is no perfect measure of metacognition. Tobias and Everson (2009) discuss the two principal ways used to study metacognition — observing students' performance on cognitively complex tasks and using self-report inventories. This research study focuses on self-report inventories predominantly through photovoice blogs, in addition to interviews, observations and cognitive maps created by a cohort of culturally diverse adolescents over the period of one year. These teens revealed coveted details of their gestalt residential learning environments. Themes that surfaced as a result of the data include topics of evolution and affordances, collaboration and isolation, comfort and safety, and technology as cognitive conduit and inhibitor. Results are explained through the elaboration of a new Learning Environment Taxonomy, based on Garrison's Community of Inquiry and the Revised Bloom's Taxonomy. Conclusions feature guidelines for designers, educational staff, parents and students relevant to the residential environment.

B8 - Transform Education for Our Common Future: Cultural Aspects / Sustainable Planning, Design & Behavior

Using the Geonarrative to Surface the Cultural Landscape of Singapore: A Spatial Visualization of in/Formal Spatial Practice

Jessica Diehl (National University of Singapore)

After independence in 1965, Singapore endeavored to develop quickly and has now achieved status as a global city-state. But a result is that Singaporeans are left without a strong sense of who they are or where they came from. This paper is based on a master's level landscape architecture community design studio to understand, capture, and re-envision local culture and heritage to develop new landscape design strategies that enhance the place-ness of Singapore for the future. The concept of the cultural landscape, coined by Carl Ortwin Sauer in 1925, is both simple and complex; it is any landscape that is understood as a manifestation of both tangible and intangible attributes and processes over time in relation to humans (Longstreth, 2008). Using a geonarrative approach to interpret the cultural landscape of Singapore, this study is an investigation of where and how the landscape reflects and produces local identity.

A geonarrative, with roots in critical cartography, integrates GIS with ethnography and grounded theory by combining qualitative GIS, narrative analysis, 3D GIS-based time-geographic methods, and computer-aided qualitative data analysis (Kwan & Ding, 2008). Students focused on Tanjong Pagar and Jalan Besar districts, comprised of a diverse group of neighborhoods and residents. The sites are rich with ecological, social, and historic diversity, new growth is planned, and users are a mix of multi-generational Singaporeans, new migrants, expats, and tourists. Surveys, observation, and mapping techniques were employed to understand travel habits, activities, favorite places, desire for new functions, walkability, food culture, and places of memory. Findings were compared to the macroscale context (land use, infrastructure, and historic maps). The geonarrative was a useful method for surfacing local texture in a global fabric. Overlaying formal and informal spatial practice provided a complex and nuanced understanding of the current state of Singapore identity.

The Mapping Resilience Project

Tasoulla Hadjiyanni (University of Minnesota)

Projections show that by 2040, almost half of Minnesota's population will be people of color and much of that growth is expected to come from international immigrants. As the cultural landscape of our cities and communities diversifies, so must the ways of knowing and engaging communities that design educators employ. This paper shares The Mapping Resilience Project, a collection of digital stories created by students in the course Design and Globalization.

The course uses digital storytelling as a tool for community engagement, one that expands the interior design historical canon in ways that increase understanding of the role design plays in the production of difference, inequality, and marginalization. In teams of four, students use interviews with community members, historical archives, music, sound, script writing, and powerful imagery to tell the story of places and of the people who work and live there--from Hmong-owned hair salons to Polish and Mexican markets.

Students benefit from dialogues with others, attaining leadership skills by learning to be critical of their own perceptions, assumptions, and reactions. Indeed, direct and indirect positive interactions between members of different racial, ethnic, and sexual orientation backgrounds have been noted to alleviate intergroup anxiety, reduce bias, and promote more positive intergroup attitudes and expectations (Broockman & Kalla, 2016; Godsil, Tropp, Goff, & powell, 2014; Zúñiga, Williams, & Berger, 2005). The instructor can also use the films to assess students' comprehension of the complexity surrounding place-making: Do students move beyond stereotyping? Do they capture contradictions and tensions? Lastly, digital stories can be shared with the organizations and individuals featured as well as the general public. As each year's stories are added to The Mapping Resilience Project, a repository of lived experiences is created that can reveal knowledge and insights that could remain hidden through conventional pedagogical approaches.

Reclamation to Re-Creation: Creating Site Furniture Prototypes with Plastic Bottles

Qing Lana Luo (Oklahoma State University)

Reusing materials has been widely recognized as a means of sustainable practice. This study was to demonstrate a pedagogical model of design education that uses reclaimed materials as design elements to create site furnishing prototypes with functionality and aesthetics.

This study developed a pedagogical framework in a landscape architecture course for a constructive learning environment with four steps: (1) facilitate collaborative design processes for idea generation, (2) recognize the versatility of visual art principles to arrive at aesthetically designed objects with intended functions, (3) apply the sustainability principles of reusing materials for newly created objects, (4) evaluate the finished design products as to their function and usage. The study began with the students being assigned to design ten recycling bins with plastic bottles as the key materials. In the design process, each group of students went through brainstorming preliminary concepts to create a recycling bin prototype. All teams documented their design processes with sketches, materials, and reflections as the supporting design process while debating on the concepts and inspirations of sustainability. Students applied the visual art principles to arrive at the design solution and assembling of materials. A follow-up survey was conducted to evaluate the application of visual art principles, form design, function effectiveness, and sustainable messages delivered. The data were analyzed with means, standard deviation, and t-test.

The results were presented in data tables and preference distribution curves, and indicated that students were able to inventively use scrap materials to create functional and aesthetic recycling bin prototypes. While consumption of plastics is increasing yearly, students learned about the harm caused to the environment, and proactively conceived creative and sustainable solutions to reduce waste going into the landfill. The study also examined the multi-benefits provided by reusing materials have an effect on health, social responsibility, and urban design.

Unlocking Nature's Secrets: A Nature-Based Tool for Sustainable Visual Communication Design

Michelle Fehler (Arizona State University); Sara El Sayed (Arizona State University)

Biomimicry is a discipline that seeks sustainable solutions to human problems in nature (Benyus, 1997). The emerging discipline has the potential to become a leading approach for sustainable design (DePauw, 2014). Biomimicry provides for a transdisciplinary experience by exposing students to the sciences such as biology and chemistry. Drawing inspiration from the world around us, however, comes with many challenges for designers (Rowland, 2017).

This research explores the impact of a tool that was based on the Life's Principles (LP's) (cc Biomimicry 3.8). These LP's make up a framework that describes how all living organisms thrive and create conditions conducive to life and therefore provides a reference to how to design sustainably. Biomimicry has been used as a tool by designers to reach sustainability (Faludi and Agogino, 2018), but no tools have been designed specifically for visual communication designers. For this tool, the LP's were translated into sets of questions that provide a score of "life-friendliness" of a design idea. Each question has 3 possible answers: "met", "partially met" and "not met". The tool then lists possible areas to explore to improve the overall score and prompt ideas for sustainable solutions. The goal of this tool is to allow designers to tap into 3.8 billion years of research and development with minimal time investment.

During the Spring 2020 semester, 45 students of the Visual Communication Design program at the Design School at ASU will explore the impact the tool has on their capstone projects. This study will engage qualitative and quantitative research into student reflections, evaluations through five essential key sustainability competencies (Wiek, et.al, 2011), as well as expert critiques of final design projects. The research will specifically investigate if the LP-based tool can bring about fresh ideas and improve the sustainability of a VCD design solution.

Sustainable Communities in a Warming World: Developing Interdisciplinary Design Studios to Address Climate Change and Urban Communities.

Robert Williams (University of Massachusetts Amherst); Carey Clouse (University of Massachusetts Amherst)

To address the myriad complex issues facing cities in the era of climate change, design educators should embrace innovative, interdisciplinary pedagogical models for educating future leaders in architecture and related design disciplines. The design studio is not only an appropriate place to test this type of learning, but when comprised of different types of students also offers opportunities for collaborative exchange that can be carried into professional practice.

Our departments have been testing this type of interdisciplinary studio course model for nearly a decade. A joint course between the landscape architecture and architecture departments brings together graduate students from both disciplines to work collaboratively on designing sustainable communities.

The current iteration of the studio – "Housing and Community in the Post-Industrial City" – charged teams of students to design mixed-use housing communities on a brownfield site in a post-industrial, flood-prone neighborhood in Providence, RI. Students were asked to frame sustainability in terms of cultural, economic, and environmental constraints. They confronted questions of climate justice and environmental equity through readings and discussions. We also challenged the students to examine the different impacts of climate change on urban sites and design new typologies that could be generalized beyond this particular site and studio.

This presentation will share the model, the work, and the opportunities of this studio design. We will discuss the pedagogical objectives of the studio, the approach to building interdisciplinary teams, and the specific structuring of the course. Then, we present the range of student work produced in the studio. Finally, we suggest opportunities and limitations of this course format to distill more general principles that could guide the development of future design studios in other institutions. Ultimately, our goal is to outline an approach to interdisciplinary design studios that address complex urban issues in the face of climate change.

B9 - Transform Education for Our Common Future: Phenomenology / Work Environments

Value-Based Education: A Case for Biased Pedagogies and Practices in Detroit

Claudia Bernasconi (University of Detroit Mercy); Virginia Stanard (University of Detroit Mercy); Ceara O'Leary (Detroit Collaborative Design Center)

Objectivity and neutrality in education have long been questioned (e.g. Freire 1970/2000). Furthermore, while evidence-based models of education and practice attempt to ground learning in "objective" data, the adoption of such models in disciplines in the social domain produce knowledge deficits, efficacy and application deficits, and issues of reduction of complexity and normativity, as social processes cannot be framed in a deterministic way and require value judgement (Biesta, 2010).

In order to promote value-based education for students in the Architecture and Community Development programs, the School of Architecture at University of Detroit Mercy has implemented a curricular structure that includes four learning opportunities: an undergraduate community-based Public Interest Design studio that introduces students to community engagement principles and practices; a graduate Master of Architecture service-learning elective, through which students engage with K-12 students in Detroit; a graduate Master of Community Development curriculum, including a comprehensive capstone project that addresses real world issues in partnership with local community organizations; and the Detroit Collaborative Design Center, which employs students through experiential internship programs.

Through these programs, students engage in learning processes that enable them to develop a critical consciousness of social justice and to design and implement critical interventions within local contexts. Students participate in cycles of thought, action and reflection, and strengthen their ability to understand social contexts and broad patterns, i.e their sociological imagination (Astin et al. 2000).

In order to assess long-term impacts and benefits of these learning opportunities, in-depth interviews with alumni from each of the four learning opportunities have been conducted. Pedagogical themes that emerged from the analysis of interviews will be discussed with focus on perceptions of the learning experience, personal outcomes, the role of designers, design decision-making and project drivers, and graduates' own roles as citizens and their future career plans.

Workplace Manager Background: Proposal for an Interdisciplinary Course

Chiara Tagliaro (Politecnico di Milano); Sofia Patera (Politecnico di Milano); Margherita Davoli (Politecnico di Milano); Stefano Ribes (Chalmers University of Technology)

Nowadays, the professional figure of "workplace manager" is achieving greater relevance due to changes affecting the nature of contemporary ways of working (Mitchell-Ketzes, 2003). Despite the emergent market demand, this role is relatively young and its job description remains largely undefined in between many fields of knowledge (Appel-Meulenbroek, Clippard, & Pfnür, 2018). Today's workplace complexity asks for people able to satisfy promptly a variety of user needs through an interdisciplinary approach (World Economic Forum, 2018).

We undertook a large investigation of academic classes and professional courses in EU to understand the extent to which structured education programs are supporting the background of the future workplace managers. Through the comparison of about 500 academic classes in real estate and 20 professional courses, the research put together a critical framework of the current educational offer. This, gained support through a customized scraper software, which has validated the empirical results. Afterwards, data were cross-examined thanks to interviews with workplace professionals.

We found that few programmes focus on workplace management, therefore educational courses cannot fully meet the market needs, yet. This lack is partially filled with professional courses, but they only consider some aspects of workplace.

Consequently, our investigations led us to propose an innovative academic course, offering an interdisciplinary approach to workplace management education. This aims to meet the needs of the professional world and job market demand with academic training.

Transformation of Profession and Discipline: Towards a Pedagogical Model for Healthcare and Wellness Design

Saif Haq (Texas Tech University)

Healthcare design vis-a-vis contemporary attitudes such as responsible design decision-making, stakeholder satisfaction, transparent investments, response to social and technical changes, evidence-based design etc. seem to epitomize the founding principles of the Environment Design Research Association. Fifty years after EDRA's founding, we now discuss healthcare design as being both research and design, including both creativity and science, and as an investment for desired outcomes. As such, we expect design practitioners to use both sides of their brain. So, what are some appropriate teaching models for such an area of architecture and design? While there has been a few schools in the United States with a long tradition of healthcare design pedagogy, we also have a rise in other schools that see the need in this focus area.

This presentation is about transformative education practice relating to healthcare design. It will begin with a review of:

1. Requirements of relevant professional and scholarly organizations such as the American College of Healthcare Architects, Academy of Architecture for Health, Center for Health Design (EDAC certification), and Architecture + Health Educators group
2. Pedagogical models that are practiced in some well-known architecture schools.

Then it will report the results of a questionnaire survey sent to professional healthcare architects and related scholars (n=25).

Based on the three described above a preliminary structure of a pedagogy for healthcare facilities design will be outlined for university education that might also allow specific strengths of academic departments to flourish. Since this is an ongoing project, audience feedback will be recorded to contribute to a full-length paper in the near future.

C1 - Transform in Transitions: Health / Gerontology

Impact of Visibility on Healthcare Workers Hand Hygiene Behavior: A Comparative Case Study of Two Nursing Units in an Academic Medical Center

Intisar Ameen Tyne (University of Kansas); Hui Cai (University of Kansas)

Nosocomial infection is the leading causes of death in the United States (Joseph, 2006). Most nosocomial infections transfer through the hands of healthcare professionals (Joseph, 2006). Therefore, hand-hygiene is one of the most important factors to reduce nosocomial infections (Lankford et al., 2003). Several studies have identified the importance of physical environment on improving hand-hygiene behavior. However, limited research has provided the empirical evidence between visibilities of hand sanitizer dispenses on hand-hygiene compliance. Additionally, there is a lack of an objective tool to analyze the visibility of HHSs location.

The paper aims to empirically evaluate if visual cues of HHSs have any impact on healthcare professionals' hand-hygiene compliance through a cross-sectional comparative study in a newly constructed nursing unit (unit-A), with an existing one (unit-B), located in the same academic medical center. Unit-A is rectangular shaped, bended in middle, containing 44 HHS. Unit-B is square shaped with 17 HHS.

The study applied quantitative research methods including visibility analysis and observation. Visibility analysis was conducted using space syntax to provide quantitative measures of visibility of each HHS location. The observation was conducted based on 10-hour behavior mapping and 10-hour behavior tracking of nurses during 1-week in each unit.

The results showed that the visibility of HHSs are significantly higher in unit-B than in unit-A ($P<0.001, t(59)=4.727$). 70.5% HHSs of unit-B have integration value over 7, whereas only 31.8% HHSs of unit-A have the value over 7. Both the BM and BT data showed that higher average frequency of hand-hygiene activity occurred in unit-B compare to unit-A (4.41% vs 1.41% and 18.26% vs 9.7%). When compared the frequency of hand-hygiene behaviors between the two units using BT and BM, the differences are statistically significant ($p<.001, t(17.678)-2.807$ and $p<.001, t(17.863)=3.811$). The result demonstrated that higher visibility of HHSs might lead to higher hand-hygiene compliance.

Supporting Human Behavior and Needs within Physical Environments: A Concept Analysis of Environmental Congruence.

Becky Gill (Arizona State University)

Person-environment (P-E) fit is a process theory that posits well-being and positive behaviors result from the compatibility between individuals or groups and their environment (Caplan, 1987). Environmental congruence is a concept within this theory that has not been clearly defined or widely referenced in literature, and its application to physical environments has only been mentioned in three studies (Dendaas, 2011; Heerwagon et al., 1995; & Kaplan, 1983).

The study's purpose was to provide a cohesive and comprehensive understanding of P-E fit theory, discover the role of environmental congruence within this theory, create a parsimonious definition of the concept, develop a common language relevant to physical environments, and determine means for investigating and measuring the concept. Peer-reviewed journal articles focused on environmental congruence and the P-E fit theory were selected, reviewed, and analyzed. Walker and Avant's (2011) eight-step approach to concept analysis was utilized since it is a widely accepted, easy to understand, and systematic approach. Data was collected by way of a systematic literature review (SLR), analyzed using qualitative thematic coding techniques, and organized within a literature review matrix.

Criteria was established to recognize and evaluate the concept environmental congruence as it applies to physical environments through the use of defining attributes, antecedents, consequences, and empirical referents. A notable finding was the concept's focus on the design or redesign of the environment to deliver positive outcomes. Environmental congruence is defined as the positive state resulting from attributes of an environment being designed or redesigned to support the functional and psychosocial needs, values, goals, abilities, and resources of an individual or group within a given setting and/or situational event.

A concept analysis of environmental congruence extends the knowledge base within research and professional practice for design disciplines by merging information under a unifying umbrella.

Design Practitioners' Attitude Toward Research Application in Healthcare Environments Design

Sara Bayramzadeh (Kent State University)

Design of healthcare facilities is one of the most demanding and complex project types in the architecture industry. Therefore, there has been a growing attention to research to evaluate healthcare facilities design in relation to healthcare outcomes, such as patient safety, patient experience, and efficiency. However, design practice tends to rely on previous design experience in developing design solutions as opposed to research. Research can facilitate evaluation and creation of innovative design solutions to reduce the risk of infection, reduce staff and patient stress. Integrating research and design has been a challenge for the design industry due to lack of scholarly access to research results and a lack of understanding of applied research. In other words, any translation of research into design is lost. This study aims to reveal design practitioners' (1) attitudes toward research application, (2) perception of research role in design success, and (3) experience with access to research resources.

The study used a web-based survey tool to investigate design practitioner's attitude toward incorporation of research during design processes. A snowball sampling method was adopted to recruit participants. The survey was sent to architecture and design firms and professional organizations nationwide. The survey invitation asked design practitioners to share the survey link with their peers. To ensure reliability for the questionnaire, the survey questions were vetted by two design practitioners with research experience.

This project is ongoing. Hurdles in accessing research and understanding technical language will be explored. The results are expected to demonstrate practitioners' willingness to adopt research in their design process. Based on findings, solutions to help design practitioners adopt and apply research in their work will be suggested. The study is expected to pave the way to integrating research and practice in meaningful ways and create a base for future research in this area.

Examining Home Modifications for Aging-in-Place and Older Adults' Attitudes Towards Gerontechnologies for Fall Prevention, Detection, and Monitoring

Gabriela Pereira (Illinois State University); Mihyun Kang (Oklahoma State University)

Home modifications are adaptations to living environments intended to increase ease of use, safety, security, and independence. Older adults have reported home modifications as one of the ways to get a safer place and live with more independence and autonomy (Hwang et al., 2011). Among different levels of home modifications, home gerontechnologies promise to facilitate aging-in-place by being pro-active in preventing and detecting falls, as well as on monitoring older adults activities at home environments to rapidly respond in case of a fall (Chen & Chan, 2014; Hawley-Hague et al., 2014).

The purpose of this study was to examine the occurrence and reasons behind home modifications among older adults living in southwestern USA and their attitudes towards home gerontechnologies for falls. A self-administered survey was conducted with 331 older adults. Descriptive and Inferential Statistics through linear regression was used to analyze data. Results show that while most participants were willing to age-in-place in their own residences (83%), the majority has not considered any home modifications so far (67%). Those who have considered or done any home modification in their home environments indicated safety and living independently the main reasons for such. Results of the linear regressions suggest that home modifications consideration do not predict attitudes towards fall prevention, detection, and monitoring. However, home modifications implementation appeared to predict attitudes towards home gerontechnologies for fall prevention and fall detection. Older adults who are proactive into improving their home environments for aging-in-place tend to be more inclined to embrace fall prevention and fall detection home gerontechnologies as they see necessary for aging-in-place.

C2 - Transform in Transitions

Refugee Camps As Behavior Settings: The Case of Gaza Refugee Camp in Jordan

Zaid Awamleh (Izmir University of Economics); Deniz Hasirci (Izmir University of Economics)

This research study is about understanding how the built environment shapes behavior in protracted refugee camps. Often permanently occupied, the "temporary" mentality in designing such spaces may promote inappropriate shelter solutions.

Studies of protracted refugee camps and behavior settings of families in private residences are limited, and also not included in Roger Barker's survey (Barker, 1968), the founder of behavior settings theory (Wicker, 1988). The research is carried out in Gaza refugee camp in Jerash, Jordan for a period of three years. Along with observation and questionnaire methods, behavior settings survey was used as a primary research tool with a contribution from the researcher.

Comparative analyses were produced in different time periods of the research within the same behavior settings regarding the architectural components of the residential environment, and the patterns of behavior using behavior setting survey and behavioral mapping instrument. Findings showed that occupants greatly restrict their behaviors to conform to the settings they occupy, while settings demonstrated the tendency to be redefined and serve for multifunctional purposes. Meanwhile, with the absence of substitutes, refugees had to remain at unsatisfying settings, and the settings had to sustain occupants with incompatible behaviors since their patterns of behavior, unlike other settings, are constantly under excessive control and disciplinary power by outsider authorities.

The results of this research have also indicated a psychological dimension of the behavior setting, which had not been adequately considered at the previous applications of the behavior settings survey.

The contributions presented in this research along with the behavior settings survey provided a potential mechanism to compose behavioral changes or to detect problems that threaten the existence of the setting or any of its essential components. These findings contribute to existing literature and could be utilized to improve the living standards of refugees all over the world.

Pleasure Vs. Well-Being in Environmental Design.

Phillip Gunn Mead (University of Idaho)

Since 2006, four books have boldly claimed environmental design's ability to increase our happiness or well-being: The Architecture of Happiness, Happy City, The Blue Zones of Happiness, and A Place to be Happy. Only one is written by an architect and none by academic theorists or environmental psychologists. None meaningfully reference key concepts found in today's Positive Psychology movement or classic philosophical writings on the subject such as in Aristotle's Nicomachean Ethics or Heidegger's Building, Dwelling, Thinking. What is missing from design writings on well-being is a more rigorous academic perspective drawn from established writings on the subject. This proposal questions environmental design's aims to improve our quality of life in relation to what is current in today's broader academic discussions on the subject of well-being/happiness. In particular, this proposal will focus on the relationship between pleasure and well-being/happiness in environmental design. Here, there appears to be conditions when pleasure and happiness can be at odds, and conditions where they complement. In environmental design there appears to be an abundance of academic writings and aesthetic design principles that focus more directly on the concept of pleasure and elements that attract our attention. Lacking are writings that more overtly focus on the architectural conditions that lead to well-being and happiness. Positive psychology's founder Martin Selegman's five conditions for Well-Being (PERMA...Positive Emotions, Engagement, Relationships, Meaning and Achievement) can provide clarity on how pleasure and happiness can conflict and overlap in environmental design.

Vacant Housing: A Non-Gentrifying Neighborhood Revitalization Strategy for East Chattanooga

Alex Hunt (University of Tennessee at Chattanooga); Eun Young Kim (University of Tennessee at Chattanooga)

Vacant housing has been linked with decreased property value and blamed for attracting criminal activity (Kelly, 2004). Above average numbers of vacant housing in an area can be an indicator of neighborhood decline and impending gentrification (Bates, 2013). There is an above average concentration of vacant housing in East Chattanooga as compared to the averages in Hamilton County, in Tennessee, and in the United States. Since the problems associated with vacant housing increase with the number of vacant houses, it is in East Chattanooga's best interest to bring these properties back into productive use and ultimately revitalize their neighborhoods. This study examined the vacant housing issue in East Chattanooga with the goal of identifying ways to utilize them in a non-gentrifying neighborhood revitalization strategy. Existing data on East Chattanooga as well as resident participation in the form of in-depth interviews were used to evaluate the problem and identify solutions. A purposive sampling method was used to recruit seven East Chattanooga Leaders as participants for the study. Content analysis was used for the interviews. The findings revealed a need for more owner occupied housing, a desire for social impact investors, mixed feelings about short term vacation rentals, and a demand for lenders who serve low income populations. This study will serve as a resource for neighborhood leaders, and a call to action about how vacant housing in their community should be prioritized, and what tools they can use to do the work.

Hyperactive Sites of Memory in Post-Yugoslav Context

Erika Lindsay (University of Detroit Mercy)

This research explores shifting signification of iconic sites of memory within the post-conflict landscape of former Yugoslavia through comparative analysis of fieldwork case studies, uncovering circumstances which contribute to the iconic construction, destruction and resultant reconstruction of iconicity. Partisan memorials of former Yugoslavia continue to patina as these Second World War memorials find reincarnation in their ruinous state.

Many sites have been de-badged, stripped of their commemorative meaning through removal of all identifying elements. Layer by layer these monuments are re-inscribed with contemporary meaning as graffiti, attempts at removal, and wartime scars shape new narratives. Transformed through iconoclastic acts, these monuments wield more power in their ruined state. These monuments are survivors, their contemporary condition sitting in stark juxtaposition to original, intended meaning. A recontextualization of these sites through fieldwork and analysis situates the monuments in their current context as multivalent signifiers of memory. These hyperactive sites of memory continue to accumulate traces of the present while they write history for the future, demonstrating the power of memorial elasticity.

C3 - Transform in Transitions

Collaborating for Sustainable Persuasive Technologies

Eli Typhina (North Carolina State University)

Participants in this presentation will explore how inclusion or exclusion of human and non-human actors early in technology development dramatically impacts the interface developed and, ultimately, the types of sustainable behavior prompted or not prompted. To exemplify the impact collaboration makes, the presenter will compare the creation process of two mobile phone applications that both aimed to prompt sustainable forestry practices. The development process of one app followed the status-quo, developer centered method, which included the developer and several representatives from two forestry organizations. The second app followed the actor diagramming and tracing method, which incorporates concepts from participatory design, actor network theory, and deep ecology to prompt collaboration across human actors and increase knowledge of non-human actors' roles. In the case study discussed, the developer engaged in collaborative exploration of sustainable forestry practices with 16 diverse human actors, from farmers to forest rangers to corporate loggers. The diagramming method also required the developer to go beyond human actors by immersing herself with the non-human actors of sustainable forestry, such as invasive beetles and 4-wheelers.

Through vivid stories and images of the actor diagramming and tracing method in application, participants will learn how to increase collaboration while developing sustainable persuasive technologies. Participants will also learn how to transition the insights gleaned using the diagramming and tracing method into designs by watching how the developer in the study transformed sustainable forestry insights into prototype designs and mobile app features. The presentation will have a hands-on component where participants can play with the method by developing an initial diagram for a sustainable technology of their interest. Finally, participants will walk away with concrete examples of how collaboration during the technology development process dramatically changes the types of sustainable behavior prompted and the outcomes of that behavior on society and our environment.

Rosetta Stone for Research Informed Practice

Heather Burpee (University of Washington)

Rosetta Stone for Research Informed Practice is a collective research project launched in 2016 by the University of Washington's Integrated Design Lab (UW IDL) Partnership Initiative, a partnership between the UW IDL and 13 of Seattle's leading architecture firms. The Partnership is a formalized mechanism for practitioners to coalesce around a collective research vision, creating a body of work that benefits all participants. The Rosetta Stone project aims to bridge research and practice around topics of High Performance Design (HPD). Economic, environmental, and social benefits of HPD are well documented, but researchers and practitioners must coordinate their efforts across disciplines in order to develop solutions that translate research into action to simultaneously combat climate change and promote health and equity in the built environment. The project has developed a growing collection of primary research, abstracts, and high-level content that is currently available to Partnership Initiative members. Translation of primary research articles occurs through the lens of value cases (health, satisfaction, perception, etc) and design elements (daylight, biophilia, air quality, etc) and is disseminated to firms to implement on built projects or more broadly within their practices. Feedback suggests that the multiple levels of content—from generalized talking points to in-depth research briefs—address various motivations and entry points for decision-makers. We anticipated that the tool would help firms fill in gaps in applied research knowledge; through beta testing we have learned that it has multiple modes of implementation, both expected and unanticipated. We will present how to overcome the challenge of content development and delivery on a larger scale, and future steps for developing a digital tool for greater accessibility and a pilot program to test the Rosetta Stone framework with a wider audience.

Assessing User Experience in Built Environment Using Extended Reality Technology.

Zahid Islam (University of North Texas)

This study investigates how extended reality(XR) based spatial user experience(UX) evaluation can benefit the user-centered design(UCD) approach in the interior design studio. Research has shown that design plays a crucial role in enhancing human experience. The technological advancement and its recent shift from the Information-Age to the Experience-Age enable designers to implement XR-based UX assessments to transform the designer-centered design process to inclusive, integrated and user-centered design process. XR is a relatively new platform that incorporates characteristics of VR(Virtual-Reality), AR(Augmented-Reality) and MR(Mixed-Reality). UX focuses on the interactions between the users, the context of use and the products. XR based interactive virtual environments, and spatial prototypes provide designers with greater control of the design process by UX measures from the end-users at various levels for informed decision-making while resulting in higher ecological validity.

This study simulated multiple interactive interior environments using XR technology. The methods used to evaluate XR based UX through subjective self-reports, and their potential use for user-centered design solutions, are identified. This study also highlighted the potential application of XR-based spatial prototypes to overcome several limitations regarding the assessment of UX in the design process. The significance and applicability of Extended Reality to measure User Experience during the design process to achieve user-oriented design solutions are discussed. Findings of this study also demonstrated an XR-based framework to evaluate UX relevant to the built environment.

Several studies have investigated building performance through post-occupancy evaluation; however, research on measuring UX for creative and inclusive design is scarce. Spatial UX evaluation may benefit from an XR-based methodology (spatial prototyping) within the context of human-space interaction (experience). The quantifiable impact of XR-based spatial-UX has the potential to offer a systematic approach to producing a differentiated experience focused on the users of the physical space.

C5 - Transform in Transitions

Public Rental Housing Development in Zhengzhou, China: An Effective Effort to Create an Inclusive Built Environment?

Weican Zuo (University of Michigan)

Since 2010, Public Rental Housing (PRH) has become a critical part of the social welfare system in Zhengzhou, China, providing decent homes for its growing population. To this end, the Zhengzhou municipal government has required that new housing developments include at least 10% PRH units. More importantly, this policy - similar to Western inclusionary housing policy - aims to construct an inclusive built environment. In order to assess the implementation of PRH and examine whether PRH has helped foster spatial and social inclusion, the author conducted research on PRH development in Zhengzhou through intensive fieldwork in the summer of 2019. The first stage of the research focused on PRH on an urban scale; this entailed interviews with officials and the collection of data from the Zhengzhou Housing Security and Management Bureau, with the aim of outlining Zhengzhou's PRH development, in comparison with other public housing programs regarding policy, implementation, and distribution. The second stage of the research investigated PRH on a neighborhood scale. Nine neighborhoods in Zhengzhou were examined and compared: three with concentrations of public housing; and six commodity housing neighborhoods containing a certain amount of public housing units as required. In each neighborhood, the author interviewed staff who work for the property management enterprise, and four to five inhabitants (40 in total). Through investigating the built environment of these neighborhoods and in-depth interviews, this part of the research revealed the public housing development on a human scale. The outcomes of the research indicated that compared with other public housing programs in Zhengzhou, PRH development has effectively reduced residential segregation on an urban scale. However, the goal of fostering spatial and social inclusion has not been achieved by PRH on a neighborhood scale. To reach this goal, the author offers constructive suggestions on PRH policy development and architecture design.

On the Architecture(s) of Dissent, Dissonance and Disobedience

A. Sameh El Kharbawy (California State University)

This paper examines the agency of architecture as an instrument for expression, empowerment, activism, and advocacy in conflict areas. It reveals how architecture can effect (social, political) change; how it can function an instrument for activism, mobilization, and protest.
By way of example, I look at radical (dissident, disobedient) architecture in the "new" (post-Spring) Middle East, in order to reveal how it contributed to individual agency and empowerment; how it created spaces for activism and dissent in political uprisings; how it offered a space for dialogue and collaboration between groups in conflict; how it communicated stories across borders and raised awareness about a region that has been troubled and terribly misruled for decades.

This study draws from extended field research as well as from a vast archive of literature on social-political movements in the "new" Middle East. During several visits to the region (from 2011 to 2019) I witnessed vast changes to its cities and communities, and documented them in paintings, photography, and video. Some of the resulting images have been difficult (and difficult to explain), and yet, this visual tapestry quickly revealed a unique spectacle of a new identity; a "new" Middle East of which I was a witness and perhaps, to some extent, even architect.

In writing about dissident, disobedient architecture, I maintain a critical consciousness,; a vigilance in employing those instruments of historical, humanistic and cultural construction from which architecture's history has often been victim. Focusing on the telling detail rather than detailed telling, my aim in this paper is not to bring any new truths to bear on architecture, but to bring the realities of the New Middle East to bear on our (more realistically, my) understanding of architecture's political and social agency, and responsibility.

Beyond Icons: Embracing Diversity in Photographic Representation of Public Landscapes.

Anne C Godfrey (State University of New York)

Every successful public space is supported by diverse groups of people coming together for a common purpose. This diversity is wide: design team, client, community stakeholders, government agencies, fund raisers, donors, champions, consultants, subcontractors, and specialist work together to bring a project to fruition. During the entire life of a project photographs are made by individuals across these groups. Photography is the most important and accessible means through which to communicate design success. Yet the diversity of these voices, through the diverse photography they make, is rarely formally collected or documented as a means to tell the whole story of a project. There is much to learn, yet the opportunity is lost, those voices fade, and the details disappear.

Currently a monographic approach to the photographic documentation of public design work is the common mode. This narrative type portrays a much narrower story, usually that of the design firm. Monographs rely upon built-work photography, represent a more select spectrum of contributors, and often rely upon single iconographic photographs (Ward 2001). Thus, many parts of the story of public place making told through photographs are never documented and archived for future study. We lose critical knowledge and lessons when only a narrow set of photographic narratives survives the life of a project.

The presenter will examine and articulate sources from which a diverse set of photographs may be collected. This is a means to explore how a more complete photographic history of public projects may be collected and shared. This is in pursuit of accumulating photographic histories representing the widely diverse people and narratives that make successful public spaces.

Broken Corners : An Architect's Participation

Gregory Crichlow (University of Kansas)

The establishment of Chocolate Spokes Bicycle Studio exemplified an architect using the medium of a bicycle shop to address the inequities of the marginalized urban core of Five Points Denver. The divergent role looked at design engagement to create spacial relevance by occupying a disenfranchised retail corner. The "broken corner" presents the potential for transcendent and intuitive interaction between two opposing scales of social capital with understanding the sequence of entry, transaction and exit. The resurrection of the exterior was not a complete overhaul but a series of small gestures of vibrant, inviting color with the removal of "protection bars" to express a level of trust. An architect's ability to spatially conceptualize future existences, established a motive for the corner as an investment in the community with a space that is sensitively choreographed to encourage constructive discourse.

In defining Chocolate Spokes, the architect presented the community the tangibles of bicycle service, bicycle design and bicycle manufacturing in the same space, and inviting the next generation to learn through mentorship. "There is nothing about bicycle technology that lends itself to race and class division."1 Chocolate Spokes presented the bicycle as a tool for accessibility both in the urban fabric and interpersonal evolution. "In certain quarters bicycles have come to be viewed as "expressways for gentrification".2 As the neighborhood becomes increasingly recognized and the much-needed infrastructural improvements are initiated, lessons have been learned through an iterative process of conversations between municipalities and community members. With each lesson Chocolate Spoke's impact on how the Five Points community now views the corner as a catalyst for economic, public and environmental health can stand as an example for how architects can directly engage and understand the potential for creating prototypes of equalizing entities that occupy other broken corners.

D2- Transform Places with People and the Environment: Interior Design I

Does the Physical Setting of an Exhibition Affect Audiences' Understanding of Narratives?

Kyoungmee Kate Byun (Northern Arizona University); Lesley Maradik Harris (University of Louisville)

Not only do museum exhibitions have to allow for contemplation of the exhibited archives or objects, but they also have to communicate to audiences the exhibition's educational purpose. Therefore, many scholars have researched museum exhibition design, especially on how spatial layout aids in effective communication to audiences. However, there is a lack of research on human perceptions related to the interrelation between physical settings and narratives in an exhibition. Since audiences can freely move around in exhibition environments, they can be affected by the physical settings to become emotionally, intellectually, and physically implicated into the narrative. Therefore, physical factors must be considered in exhibition design as a part of the narrative environment. The purpose of this study is to examine how physical factors of narrative environments affect human perception to facilitate communication based on four spatial narrative theories. Narrative environments are intentionally designed in a gallery at an urban mid-western university based on the narratives created by qualitative photovoice analysis from a project titled "Understanding HIV Risk and Resilience among Adolescents who have been Orphaned by HIV/AIDS in Hai Phong, Vietnam." The gallery is designed to examine different physical factors (e.g., form, scale, and light). A defined group of 30 participants (total 6 groups) will experience the exhibition respectively to explore their perceptions on the effectiveness of interactions between physical settings and narratives. All participants will complete a Likert scale survey and have semi-structured interviews for data collection. Findings from this study will contribute to exhibition designs for curators and museum designers for effective communication of narratives.

Transforming a Campus Library from Book Boxes into Students' Learning Commons

Jae Hwa Lee (Iowa State University)

University libraries are in a dramatic and indicative period of flux. Instead of having spaces for stacks, libraries need to provide more diverse spaces to meet students' academic needs today. How Generation Z and Millennials approach information is distinctly different than prior generations, prioritizing digital access, speed, and the ability to multi-task. Moreover, studies point out that the preference for individual or group work is a key factor in determining what type of spaces students prefer to study in (DeClercq & Cranz, 2014; Oliveira, 2016), which may also fluctuate throughout the day and the semester. In order to respond to changing expectations in their academic library environments, this research identified how students actually use the current library spaces on campus and what they really want within a wide-open seating area that is designated as a group-study zone. As a case study approach, this current study employed multiple methods: 20 behavior mapping of 429 seats, 443 user online survey, and individual interviews with 16 users at one public university in the Midwest, U.S. The behavior mapping and the online survey data suggested that more students using this area were for study and research in individual with their personal laptops than for group working or socializing. For this reason, more than 70% of them prefer to choose sitting with more privacy for their individual studying. In addition, the users reported issues with the noise level, a variety of furniture options, wayfinding, crowdedness, and a lack of outlets during their personal interviews. The findings of this research provided an understanding of the full spectrum of the current user's needs and behaviors in the library environments, to support the different levels of privacy and different types of studying. The design implications to draw multiple options were also discussed.

Evaluating the Impacts of Integrating Visual Access to Nature in Nurses' Work Environments and Deep Interiors

Apoorva P Rane (Texas Tech University); Andrew Johnston (Providence St. Joseph Health); Debajyoti Pati (Texas Tech University)

The problem of nurse burnout is still persistent (Maslach & Jackson, 1981; Ross, 2010; Vahey et al., 2004), often resulting in medication administration errors, compromised quality of care, and seemingly unempathetic or irritable behavior (Garrett, 2008; Keers et al., 2013; Sorenson, et al., 2017; Trinkoff et al., 2007). This ultimately manifests in reduced patient safety and satisfaction (Lockley et al., 2007; Rogers, 2008). Although previous studies have implemented multiple interventions for stress reduction such as high-quality break areas, napping strategies, downward substitution of work, and centralized and decentralized nurse stations; none of them have been sufficiently effective (Duffield et al., 2011; Halm, 2018; Potash et al.,2014).

Nurses spend more than 95% of their time in their work environments (Hendrich et al., 2008) and avoid restorative breaks (Nejati, 2016). Therefore, this quasi-experimental research investigates stress reduction in nurses following the introduction of biophilic features in their workspaces. In this study, the researcher investigates reduction in stress before and after the installation of Sky Factory Ceiling tiles™ in adjacent corridors around the two nurses' stations in the Emergency Department; through cognitive, psychological and physiological outcomes. The sample population consists of dayshift nurses at the ED of a hospital located in the Southwestern United States. The outcomes include Acute Stress (AS), Reaction Time (RT), absenteeism, Heart Rate Variability (HRV), Heart Rate (HR), Blood Pressure (BP), and participant feedback. This three-step longitudinal project is currently at the third data collection stage. Preliminary findings of the initial two waves of data collection indicate that 50% of the participants show a significant reduction in AS. Moreover, more than 60% participants show significant improvement in the RT and incorrect to correct response ratio. A comprehensive report of the dependent variables (AS, RT, absenteeism, and feedback of the participants) will be discussed in this presentation.

Analyzing Changes in Awareness of Nature in Fifth Grade Children

Amanda Gale (University of North Carolina Greensboro); Anna Marshall-Baker (University of North Carolina Greensboro)

"Biophilia" or "love of nature" was first developed by Fromm (1973) and later expanded by Wilson (1984) who acknowledged an "innate" connection with nature. Because experience with natural conditions can foster cognitive abilities and act as a gateway to developing knowledge of natural systems (Maller, 2009), we hypothesized that introducing biophilic changes to a classroom environment would positively affect children's awareness of nature. Biophilic attributes derived from Kellert (2008) were installed into a fifth-grade classroom. The exploratory study employed a pre- and post-test research design with a convenience sample of 38, fifth-grade elementary school children in the suburb of Charlotte, NC. Following a pre-test to determine baseline responses, the treatment variables -- biophilic attributes that included potted plants, a large fish tank with colorful plants, and bird feeders outside the classroom windows -- were installed within the classroom and a post-test administered after four weeks. The pre- and post-tests consisted of the same writing assignment in which the children described any meaningful activity or event. Content analysis was employed on the writing assignments using 16 pre-determined biophilic attributes. The kind of attributes were documented as well as the frequency of their occurrence in the pre- and post-tests. Results indicate that water, animals, and views and vistas were the most commonly represented attributes found within the students' writing. Color was the least represented attribute within the writing assignment. Further, students displayed an increased awareness of animals and views and vistas from the intervention; however, a decrease was revealed regarding water, air, and natural materials. The length of time between the intervention, the amount of biophilic attributes installed, and the age of the children are factors that need further investigation. The study will be presented in detail and include recommendations for future work involving installation of biophilic attributes into interior space.

D3- Transform Places with People and the Environment: Interior Design II

Perception of Priorities and Strategies for Active Shooter Preparedness in University Classrooms

Kristy Kellom (Michigan State University); Linda Nubani (Michigan State University)

Educational facilities have experienced 20% of the active shooter incidents in the U.S. between 2000 and 2018 (U.S. Department of Justice and Federal Bureau of Investigation, 2019), prompting schools and universities to increasingly respond with emergency planning, response, and training methods (Ford and Frei, 2016; Roman, 2018). The purpose of the research is to identify architectural characteristics of a university classroom that enable or hinder Run, Hide, Fight trained response against an active shooter threat. Data was collected during two focus groups to compare the perception of classroom users to a group of practitioners and first responders regarding perceived priorities for prevention and response strategies against active shooter events. Focus groups incorporated three phases including 1) training, 2) classroom simulations within a 360-degree visualization lab, and 3) strategic thinking exercises and discussions. Before and following each phase of the focus group, participants were surveyed on their perceived levels of fear and preparedness against active shooter threat. Within eight classroom spaces of various instructional environment types, participants identified physical attributes that enabled or hindered Run, Hide, Fight responses. From these attributes, five categories emerged from which participants ranked categories of highest priority from a classroom user perspective, and categories policy makers should prioritize. Means of ingress and egress were considered the highest priority architectural category, focusing on the occupants' ability to evacuate a space (Run) quickly and without barrier as utmost importance. This finding supports current life safety regulations requiring corridors to be well illuminated and barrier-free. However, a significant difference was found regarding categories of perceived importance among the user group compared to the professional group. The results of this study improve baseline knowledge of safe and secure design practices against active shooter impact and support recommendations to existing life safety policies through adaptation in design.

A Reflection of Wellbeing in Higher Education: First Year Interior Architecture Students

Amanda Gale (University of North Carolina Greensboro)

Students studying design and architecture spend more time in their classrooms, typically studio settings, than other majors (Syrkett, 2017). Mental health issues, including suicidal thoughts, are increasing, these are linked to in part to isolation and lack of support (Kafka, 2019). Therefore, the design and presence of student lounges near studios is important. As a result of this trend, it is pertinent to examine the relationship between the interior environment and wellbeing of college students. Specifically, the study sought to determine which features of the built environment influence students' perceptions of wellbeing within student lounge. Students studying in applied programs such as design have more hours in their university setting while in class or working on coursework outside of class. A comprehensive search of literature from healthcare, nutritional science, occupational psychology, and design revealed nine interior features that positively affect wellbeing. These include: views of nature; opportunities for social interaction and physical activities; presence of plants; access to daylight, drinking water, and healthy foods; and use of color and natural materials (Al-Amari & Al-Khamees, 2015; Lipscom & Rollings, 2017). Thirty-six first-year design students completed the online survey. Participants were shown a series of images and asked to select the area on the image that most reflected wellbeing and to rate the degree to which wellbeing was represented within the image.

Heat map survey questions allowed respondents to select parts of the images that indicated elements reflecting wellbeing within the lounge. The number of clicks, as well as a color-coded heat map images were generated identifying perceived features influencing wellbeing. Respondents reported access to daylight and drinking water as the most important features concerning their wellbeing but found all features at a minimum of very important to their wellbeing. The images most representative of wellbeing encompassed all nine features.

D4- Transform Places with People and the Environment: Sustainable Planning, Design & Behavior I

Public Values Related to Federal Lands Management: The Case of the BLM's Western Oregon Plan Revision

Stacey Schulte (University of Colorado)

This research analyzes public comments made in response to a controversial regional land planning process extending over the last three decades in Oregon. We used comments submitted to the Bureau of Land Management (BLM) through the Western Oregon Plan Revision (WOPR) process to assess the complexity of anthropocentric values related to preferences for public lands management. This process received the most public comments of any NEPA process at the time. While people often express ecocentric values in surveys, they express primarily anthropocentric values in the public comment process (Bengston, Xu, & Fan, 2001; Manning, Valliere, & Minteer. 1999; Proctor, 1998). People use a variety of arguments and values to justify forest preservation but almost all reflect a human-oriented value system. Findings show that natural resource values expressed in public comments are different from those expressed in other forums. This may more accurately reflect values which runs counter to the notion that natural resource values are shifting to more intrinsic values. It may also reflect the belief that the human-oriented benefits of conservation are compelling argument to decision-makers. This assessment contributes to the understanding of public values regarding management of natural landscapes under federal ownership which may be particularly useful in more localized decision making with stakeholders.

Thermal Comfort in Outdoor Environment: Assessing the Impact of Psychological Adaptation Factors in the Experience of Heat

Mohsen Garshasby (Arizona State University); Chingwen Cheng (Arizona State University)

Increasing air temperature in urban environments has been a prevalent issue over the past few decades, especially as it directly affects thermal comfort as a sensible measure for individuals. Given the reliance on outdoor spaces for thermal comfort, architects and planners need to not only find better ways to consider thermal comforts in outdoor spaces but to obtain more knowledge on the users' experiences for evidence-based design practices to create more sustainable places. Built upon the adaptive thermal model in the past two decades, psychological adaptation- as the individual's ability to psychologically adapt with the thermal environment- measures were identified as important factors that influence perceived thermal comfort. Nikolopoulou, Baker, & Steemers (2001) argue for the presence of a psychological aspect when thermal comfort within outdoor spaces are concerned. Nikolopoulou et al. (2003) introduce several factors namely naturalness, expectations, experience (short-/long-term), time of exposure, perceived control and environmental stimulation. However, it is still unclear to what extent and how each of these factors impacts overall thermal comfort.

This study aims to address whether and to what extent familiarity, as one of the psychological factors, influence the perceived thermal comfort of individuals. Therefore, a mixed method approach was utilized to examine the impact of familiarity within different groups of students through studies in four time periods; February 2018 (n=110), June 2018 (n=35), October 2018 (n=130), and November 2018 (n=190). Data was collected through surveys as well as walking interviews and analyzed with SPSS. The results confirmed that there is a significant correlation between familiarity and reported levels of comfort in outdoor environments. Understanding the impact of these psychological factors on the perception of heat would be valuable for architects and planners by recognizing the individuals' experience and ultimately enriching the design opportunities.

D5- Transform Places with People and the Environment: Sustainable Planning, Design & Behavior II

Transformation Revisited: Phoenix Sonoran Preserve Twenty Years after the Master Planning Process

Joseph Michael Ewan (Arizona State University)

In 1999, the City of Phoenix adopted the Sonoran Preserve Master Plan (SPMP) (Burke and Ewan, 1999). The SPMP won a National ASLA Presidents Award in 2000 and has been used as an exemplar in ecological landscape planning (Steiner, 2008), specifically for its incorporation of ecological principles into the urban planning process (Ewan, Fish Ewan & Burke, 2004). In the twenty years since adoption of the SPMP, voters approved funding towards the acquisition of 21,500 acres (8,700 hectares) of land identified in the plan for preservation and publicly accessible desert open space. Phoenix also weathered the economic downturn of 2008 and experienced increases in urban heat and shifts in local climatic conditions. To date, the City of Phoenix has acquired 9,600 acres (3,885 hectares), and developed three trailhead locations and 36 miles (58 kilometers) of trails. In order to understand the role of the SPMP and its impact on providing natural open space for urban dwellers over time, it is helpful to revisit the plan in concert with ground-truthing its implementation through case study analysis. "Case study analysis is an effective way for landscape architecture to advance and mature as a profession, providing a promising tool for the profession to train students, develop a research base, and improves practice" (Francis, 2001, 2019). Utilizing Mark Francis's Case Study Method for Landscape Architecture, this presentation provides insight into the human dynamics of urban open space planning and implementation, using the Phoenix Sonoran Preserve as a case study subject, to reveal ways physical planning can better serve human environments.

Would the Cognition of Relationship between Walking Distance (How Far in Miles) and Time (How Long in Time) Make People Walk More? a Case Study in Springfield, Massachusetts

Yanhua Lu (North Carolina State University)

As the urban core of western Massachusetts, the city of Springfield grew fast in the early 20th century due to the industrial development, but then declined rapidly after World War II. In order to bring this legacy city back to life, the city of Springfield together with the Massachusetts Department of Public Health initiated a pedestrian wayfinding project in 2015, to encourage people to walk more, hence reactivating the streetscape, revitalizing the city, and improving people's well-being.

In April 2016, 54 temporary wayfinding elements showing walking time and distance to nearby points of interest were installed in downtown Springfield. Both a pre-installation (n=110) and a post-installation (n=90) interview were done to understand if the cognition of walking time and distance would encourage more walking. Random samples were selected, including residents, workers, and visitors in downtown Springfield. The interview asked participants how much time in minutes, and how much distance in miles they would be willing to walk. It also asked if the signage had made any influence on participants' travel behavior.

The results showed that 36% of the participants misread the amount of time it takes to walk for a certain distance, for example, numbers of participants thought a half-a-mile walking needs over 20 minutes, which demotivated their willingness to walk. Further, 63% of the participants think the number of pedestrians increased after the signs were installed. And 82% of the participants, who noticed the signs walked more because of the information from the signs.

The data reveals that a clear cognition of how long a person can walk in a certain amount of time can help influence one's travel behavior. And the signs with information of walking distance and time can influence travel behavior, as long as it is noticeable.

A Useful Strategy for Enhancing Pedestrian Comfort with Focus on Mitigating Canopy-Heat Islands

Samain Sabrin (University of Alabama Birmingham); Maryam Karimi (University of Alabama Birmingham)

In the car-centric cities, planners are trying to shift paradigm by encouraging people to walk more. Over the last few decades, extreme heat events have surged in high density city-centers due to the Urban Heat Island (UHI) effect which makes walking more unbearable for pedestrians. Therefore, city-planners' effort to convert into walkable cities is failing in terms of changing people's mobility choice. One of the strategies for reducing heat islands is to increase green cover, which is also constrained by a canopy's height and width ratio. Our study shows that this ratio plays an important role in determining the initial comfort which improves by a certain degree with additional trees. During the hot hours, pedestrian comfort varies in the upper and lower sides of an east-west (E-W) oriented street, which can govern the tree planting decisions. We measured the impact of built environment (different canopy ratios) and vegetation on the thermal comfort using human biometeorological evaluations (mean radiant temperature, Tmrt; physiological equivalent temperature, PET) in E-W streets. We utilized the RayMan model (capable of calculating short and long wave radiation fluxes affecting the human body) to estimate biometeorological parameters in the urban canopy-level incorporating buildings, streets and trees. Few studies employing Rayman model have established the connection between pedestrian-level comfort and canopy ratio in the lower and upper side of an E-W road segment. For the hottest days of major summer months in Philadelphia, we estimated shading period, Tmrt and PET along the most populated places of a sidewalk at an average human height to be compared in different canopy ratios. The study identifies the best canopy ratios in terms of maintaining pedestrian comfort. Urban canopies with the candidate ratios can be further intervened with a combination of tree canopy, shading and evaporative cooling structures to enhance pedestrian use.

The Rustbelt Herbarium-Understanding the Processes Shaping Landscapes in Legacy Cities.

Mark Lindquist (University of Michigan); Daniel Phillips (University of Michigan)

The Rustbelt Herbarium is an evolving, multi-city celebration of spontaneous urban vegetation. The project employs the collective power of citizen science to reframe the conversation (and challenge negative stereotypes) surrounding "weeds" in cities. Inspired by Emerson's assertion that "A weed is a plant whose virtues have not yet been discovered", we believe that touching, identifying, and collaborating with these forces holds immense potential for urban sustainability challenges. The current project aims to conduct deeper research into the context of the so-called "legacy cities" of southeast Michigan, with a particular focus on the city of Detroit. This involves several qualitative and quantitative lenses, including 1) Fieldwork to measure in-situ biodiversity within neglected landscapes throughout the city 2) perception studies which assess the perceived biodiversity and value of these landscapes amongst multiple groups of local stakeholders, and 3) direct engagement with the individuals and organizations charged with maintaining the overgrown landscapes in the city of Detroit. We present our initial findings, which suggest gaps between perceived and measured biodiversity, and many complex socio-ecological factors which continue to limit broader acceptance of novel ecosystems as a viable landscape typology within cities. The research also identifies a new landscape typology, currently emerging in Detroit—the open space community park–and it's potential role in experimenting with new modes of "intended wildness" (Hwang and Yue 2019).

D7- Transform Places with People and the Environment: Work Environments

Exploring the Relationship between Thermal Comfort and Productivity in an Office Environment in Raleigh, North Carolina

Helia Taheri (North Carolina State University); Kristen A Ambrose (RATIO Architects); Traci Rose Rider (North Carolina State University); Sarah Wood (RATIO Architects)

The United Nations stated that by 2030 the world population who live in the urban area will increase to 60 percent (United Nations, 2017). Most of these people are working in office buildings (ASHRAE, 1993). Given that 90 percent of the operation cost of businesses are spent on staff, including salaries and benefits (World Green Building Council, 2014), anything impacting the productivity of staff influences the financial status of the organization. As such, the environment should be one of the major concerns for any organization. The indoor environment such as the office layout, thermal condition, acoustics, lighting level, etc. has a relationship with employees' productivity (Frontczak et al., 2012; Mawson, 2002; Roelofsen, 2002; van der Voordt, 2004; Al Horr et al., 2016). In this paper, we conducted a study on the relationship between thermal comfort/conditions and the impact of environment on employees' productivity in an office building in Raleigh, North Carolina. The methods in this study are survey and environmental monitoring. Dry bulb temperature and relative humidity were measured in several locations of the office between August 16th to September 6th, 2019. Also, we randomized the distribution of surveys among participants 12 times during these three weeks. The results showed that the temperature is not well distributed, and most people far from the windows and in the conference rooms feel colder than people close to windows. Also, they showed that thermal comfort has a relationship with the employees' perception about the impact of the environment on their productivity in space, which means that participants who feel colder think that the environment interferes with their productivity. The next step in this study is to suggest some changes to improve thermal comfort, and have post occupancy evaluation to see how these changes impact on the sensor data and survey data.

Acoustical Challenges of the Open-Plan Office and the Vagaries of Implementing Remedial Policies

Anne Herndon (Lake|Flato Architects); Estefania Barajas (Lake|Flato Architects)

As workplace configurations have densified, research has demonstrated a decrease in workers' acoustical satisfaction, bringing into question the relationship between open-plan offices and the ability to perform focused work (Haapakangas et al., 2018). Recent design trends like activity-based working attempt in part to remedy acoustical challenges by introducing flexibility of work environment.

Prompted by an architecture firm's impending renovation, this study began as an attempt to understand in what ways their existing open-plan configuration corresponded with negative effects on focus, primarily using one-on-one interviews. It unexpectedly became an exercise in the politics of applied research within a workplace setting. In the interviews, participating workers were asked, among other things, to rate the extent to which a variety of environmental factors affected their abilities to perform focused work over the preceding week. The findings corroborate existing research on open-plan offices that generally suggests acoustical factors as primary detractors (Banbury & Berry, 2005; Haapakangas et al., 2018; Haka et al., 2009).

Results and recommendations for the upcoming renovation were publicly presented to the firm, along with policy-based suggestions executable within existing spaces. However, recognizing that the 33% interview participation rate did not constitute majority representation, a series of tools were created including guided group conversations and follow-up surveys tailored to feedback from these discussions to further develop potential policy interventions. The aim was to hear as many voices as possible, regardless of individuals' willingness to publicly express opinions.

Since implementation of two follow-up practices,

feedback has been mostly positive, ranging from relief at having new quiet havens from "disruptive" studio environments to frustration at the behavioral changes required. Despite consensus on many points, there exists a certain resistance to change. We will continue to facilitate the feedback and policy re-adjustment loop, seeking to ensure changes are generated by those whom they affect.

A Case Study on Employee Wellbeing; How Does the Built Environment Influence Behavior in the Workplace?

Laura Frederick (Arizona State University)

This research study seeks to understand how workplace wellbeing is affected by the interior environmental settings of the open office, and how the built work environment influences employee behaviors. Employee wellbeing is a top concern for many organizations, as it has been linked to job performance and organizational commitment. Research also suggests that overall wellbeing is important to employees. Organizations are significantly investing into upgrading workplace environments, and there is a need for a clear understanding of how those improvements truly impact employee wellbeing. Current workplace research reveals that the open-office floorplan accounts for more than 70% of office layouts in the U.S. and most commonly used for the benefits of collaboration and efficiency (Gallup, 2017). However, the open office layout ranks poorly in current employee wellbeing studies with a number of office environment stressors such as noise, distractions, and privacy concerns are noted in to impact employee wellbeing (Bodin-Danielsson Et al 2008). The knowledge work performed in office environments require high amounts of cognitive tasks and combined with filtering distractions in the workplace can increase strains caused by common office stressors, thereby impacting employee wellbeing (Bridger et al 2011). This study examines common stressors from the open office environment and compare employee perceptions of their work environment before and after renovations, as well as observations and behavioral mapping that recorded how the built environment influences the behaviors of the occupants. This study was conducted in a facility currently in a multi-phase renovation of the interior office space which presented the opportunity to survey employees before and after the renovation to view contrasts if any in employee wellbeing. The findings will be presented to help understand how the designed interior workplace impacts the wellbeing of the users and ultimately overall organizational health.

D10 - Transform Places with People and the Environment: Children, Youth & Environment I

Redesigning Schools: Facilitating Children's Involvement

Rokhshid Ghaziani (De Montfort University)

Children are skillful communicators capable to express themselves in many different ways. As experts of their own experiences it is important to include them in decisions regarding their own spaces. A review of literature on school design reveals there is a gulf between children's experiences and designers' methodology. Although children were involved in several projects in the UK, including 'The School I'd Like' (Burke and Grosvenor, 2003), 'Joinedupdesignforschools' (Sorrell, 2005), 'The Young Design Programme' (The Sorrell Foundation, 2006 and 2007) and Creating Excellent Primary Schools (CABE, 2010), their involvement in the design or re-design of schools have not been explored meaningfully for years. School environment affects children's health, performance and emotions. The effective design of school buildings makes these places more pleasant and more functional; and increases the quality of children's spatial experience. Children's views are important and need to be more effectively integrated in the school design projects. However, there are challenges for academics, designers and policy makers in determining which methods are appropriate for listening to children's views and ensuring their effective participation. This paper explores the views and expectations of children regarding their school environments and aims to provide an overview for involvement of children in redesigning their schools. The empirical study was undertaken by 210 children (8-11 years old). The children re-designed three types of indoor spaces in four Primary schools in UK.

Qualitative and quantitative research methods were used, including questionnaires, drawings and model making to gather and analyse children's views. This paper presents the findings that emerged from children's perspective. The initial findings highlight the importance of involving children in the school design process that could then inform the decision-making processes of architects and designers. The findings would have implications for school design practice, demonstrating how research can be embedded in primary schools.

"Our Journey to School" Behaviors and Social Representations Among Children Attending Public Schools in Zapopan, Mexico

Juan Torres (Université de Montréal); Alexander Grigorievich Nizhelski Álvarez (Université de Montréal); C. Pettersen Jean-Christophe (Université de Montréal); Danielle Labbé (Université de Montréal)

Commuting between home and school is an important part of children's daily mobility. This is also true in disadvantaged neighborhoods in the Global South, where urban environments can raise important challenges for children in terms of transportation and access to education. How these trips are made is determined by multiple factors (Mitra 2013), related to children (capacities, attitudes, beliefs, etc.) and their environment (built, social, familial, etc.). How do children commute between home and school in disadvantaged urban neighborhoods in countries like Mexico? What are the views of these children on the conditions of their journeys?

This communication presents preliminary results from a study conducted in 2018-2019, in collaboration with the municipality of Zapopan, Mexico. The study included a survey with children (n= 223) from 6th grade of public elementary schools and 1st grade of public secondary schools in 5 neighborhoods in Zapopan, at the periphery of Guadalajara, Mexico. The study also included socio-spatial portraits of the five neighborhoods, drawing on secondary sources.

Preliminary results show differences in trip lengths and transportation patterns among children related to school level (elementary or secondary school) and neighborhood but also depending on travel direction (school-bound or homebound), and gender. Children from elementary schools had a more positive view of their daily trips to school than those from secondary. Among the transportation modes, our participants perceived public transit as the less pleasant and less safe option. This presentation will include recommendations for further research, urban planning, and street design.

The Relationship between Outdoor Play Settings and the Level of Children's Social Development from an Ecological Approach

Reem Hassan Bagais (Texas Tech University); Sharran Parkinson (Texas Tech University)

Socialization, an important factor in children's lives, has a great impact on emotions, decisions, behaviors, and interpersonal interactions. Absence of social skills negatively affects cognitive performance and executive functioning. The environments that children are exposed to have an important role in these social developmental skills (i.e., environmental awareness, cooperation, interaction and communication, and confidence and self-esteem). This study is conducted from both an analytical and theoretical point of view using both Affordance Theory and Social Learning Theory. According to the Social Learning Theory, the environment is one of the main factors that supports these skills. The Affordances Theory then elaborates on how the environment may further influence children's social behaviors.

A literature review was undertaken using keywords to define interrelated categories. Previous literature guides the design considerations for the social affordances used in the study. This paper analyzes the affordance characteristics of different outdoor play settings related to social skills and it provides a guideline and suggests design solutions offered for each affordance. These design solutions support four social skills revealed in this study: 1) environmental awareness, 2) cooperation, 3) interaction and communication, and 4) confidence and self-esteem. This study explores certain play settings characteristics offered by the physical outdoor environments that are important for encouraging and inspiring children's social play. These design features can help designers originate different solutions that can be applied to neighborhoods available to the community.

D11 - Transform Places with People and the Environment: Children, Youth & Environment II

Innovative Learning Spaces As a Transformative Tool of Communities: An Experimental Design Exercise

Nouran M El Begermy (The American University in Cairo); Hisham S Gabr (The American University in Cairo)

The city of RHO is a commune of the Metropolitan city of Milan, famous for its Fiera Milan SpA, which is one of the largest trade fairs in the world. The fair is connected to the Milan city center via a metro line and a suburban train that passes through RHO, splitting the city in half, resulting in fragmented districts with their own urban fabric. A proposed school rehabilitation experimental design project aims to develop the community by creating magnets that attract community members to "cross over to the other side" and reconnect with the heritage of RHO. and remaining agricultural landscapes. The paper focuses on the impact of the proposed school design on the inhabitants of RHO, especially the children. Today, the school's teaching space is no longer just the classroom. The design experiments with new teaching methods and the notion of an Open School and promotes new learning spaces in and outside of the school to engage with the community and the heritage of RHO. and allows for activities that are oriented towards the well-being of children as self-realization process. The role of the Open School is to be a recognizable place with a specific identity, a place for social exchange and enrichment. The redesign of the common open spaces emphasizes the sense of belonging to the children, to the teachers, to the parents, and to the school. The idea is to involve the community in a pedagogical way where every member has the advantage of making a pedagogical contribution to the school. The approach promotes giving identity to abandoned and anonymous spaces within the school and to highlight and utilize the community of RHO's sense of belonging to the school with a goal to transform and redefine the relationship between the school and the children and RHO.

What Physical Design Features May Reduce Aggressive Behavior Among Residents of Juvenile Detention Centers?

Elnaz Nahirafee (Texas Tech University); Erin Miller Hamilton (Texas Tech University)

Research shows that aggression in juvenile detention centers (JDC) is one of the most controversial and frequent issues, that in many cases leads to hospitalization, suicide, homicide, sexual assault and many other unsecured and life-threatening situations. 17,600 acts of suicidal behavior and 24,200 incidents of the juvenile on juvenile reported in a year in the United States. Evidence was investigated through a systematic literature review based on keywords such as juvenile detention, youth aggressive behavior, juvenile suicide, juvenile offense, etc. Most of the articles, documented in this literature review, reported peer-reviewed empirical studies.

To examine the embedded components for aggression among residents in these settings, the importation and deprivation model has been adopted. The importation model explores aggressive behavior from the perspective of resident's characteristics and from the deprivation model point of view, aggression can happen due to settling in detention facilities with loss of freedom and services. Accordingly, the risk factors of aggressive behavior among juveniles in JDCs can be the environmental factors that trigger stress, anxiety, loss of autonomy, repression, and fear. This study aims to identify these physical design features.

The most important physical design factors that have an impact on aggressive behavior are territory, crowding, and control. The results of this study have been used to design a JDC to reduce sense of crowding in public spaces, create balance between residents' sense of control and restriction by staff; promote surveillance in private, semi-private and transitional spaces; promote sense of territory and privacy in semi-private and private spaces, and promote safety in restrooms and bathrooms. The design phase excludes the design of spaces that are not serving for juvenile residents, like kitchen, staff rooms, visitors' entrance, storage.

Biophilia in Winter: Sensory Promenades in Primary Schools

Melanie Watchman (Université Laval)

Experiencing nature in primary school can contribute to child development and well-being (Kahn & Kellert, 2002). Principles of biophilic design are growingly becoming used by architects and designers to foster a connection with nature in the built environment. Natural elements such as daylight can be objectively documented and measured with instrumental equipment. However biophilia also relates to the experience of space and peoples' response. Few tools currently help architects and designers to assess the subjective biophilic dimensions of buildings during site visits. This research explores the documentation of the experiential aspects of biophilia in learning environments during an architectural promenade. The biophilic ambiance rose is developed to assess abiotic natural elements such as sunlight, daylight, wind, air, and snow, based on tools that document sensory perceptions and biophilic design strategies (e.g. Demers, Potvin, & Dubois, 2009; McGee & Marshall-Baker, 2015). A group of researchers completed these biophilic ambiance roses during three types of promenades between inside and outside school spaces. Site visits took place during winter in two primary schools in Quebec City, Canada; a narrow school building on a mineral site and a school with a thick floor plan on a site with abundant trees. The biophilic ambiance roses describe harmonious or abrupt transitions perceived between school spaces. The absence of transition spaces between inside and outside generate an intense contrast both visually and thermally that were identified as a source of dissatisfaction. The presence of snow in the schoolyards also influences the degree of naturalness perceived within the buildings. The study highlights the potential of the biophilic ambiance rose to discuss the experiential dimensions of biophilic architecture. It further reveals the importance of seasonality when assessing and designing biophilic buildings in cold climates.

D13 - Transform Places with People and the Environment: Children, Youth & Environment III / Participation / Communication

Getting Closer - Transforming Places with People through Games

Anna Szilagyi-Nagy (Eberhard Karls Universität Tübingen); Broković Dodig Marta (Union Nikola Tesla University)

In order to transform places with people, there is a need for methods and tools that can successfully engage various stakeholders and empower typically neglected groups such as children and youth or older people. Games as teaching, research and design tools in architecture and urban planning have the power to bring people from all walks of life, with different educational backgrounds and abilities, together (Brandt 2011). Such games might be analogue or dialogue, could come from various game genres (role plays, card games or board games); and have various purposes. Through the game play participants can communicate, negotiate, explore, co-design, evaluate and reach consensus on a myriad of issues related to our built environment (BE) (Sanoff 2000). Some games aim to build consensus or strategy to solve specific BE problem, i.e. urban ecology, climate change mitigation, heritage protection, while the others support sharing and formulating opinions about various BE typologies, i.e. residential, work, educational, healing, urban, etc. (Tóth and Szilágyi-Nagy 2019)

Authors argue that games are efficient tools for engaging various groups in transforming many different BE typologies. This paper introduces inspiring game examples suitable to be applied when making, implementing and maintaining decision about various past and future BE issues. Game examples are going to be systematically analysed through a matrix taking into consideration the BE typologies, purpose of the game, potential players, game genre, planning phase in which the game was implemented. The paper will provide a much-needed cross sectional view on the use of games as tools to transform BE with people, identify challenges to their current use and develop recommendations for the future development of the games.

Space and Society: The Contemporary Role of Media in Public Space

Melanie Duffey (IDEC)

Literature has established that public space is a physical place, where social interactions occur, and that the physical and the social environment work in a reciprocal relationship and shape social interactions. The goal of this research was to better understand public space users' behaviors in public spaces with free access to Wi-Fi. One will find that there is a dynamic relationship between the urban experience and media as it relates to users' social interactions within public space with the on set of device usage or interfaces. According to de Souza e Silva and Frith (2012) interfaces should be viewed as "systems that enable people to filter, control, and manage their relationships with the spaces and people around them" (de Souza e Silva & Frith, 2012, p. 5). Information communication technologies are interfaces introduced to the public realm, and their impact on the human experience is still not entirely understood.

This study identifies 7 public space case studies that examine users experiences with public space today. The central finding is that Wi-Fi, something that is seemingly everywhere yet invisible, has affected societal perceptions of place and is one contributing factor to why some users visit a particular public space. However, while Wi-Fi is important to public space users, users still perceive that physical attributes of a space are more important than less visible amenities, such as Wi-Fi. One outcome, was the creation of 'user profiles' based on the observational data collected in this study with a follow up survey. User profiles also illustrate how device users in public space use wireless communications and other low-tech interfaces as a tool to negotiate the social environment present in public space. Findings will be presented applying the same principles of social interaction, as defined by Gehl (1987), but within the context of mobile media.

A Comparison of Identified Neighborhood Areas in an Interval of Eleven Years

Mohammed Zakiul Islam (Bangladesh University of Engineering and Technology)

This is a study of comparing neighborhood boundaries identified by a group of 26 respondents in an interval of eleven years. In total 109 respondents identified their respective neighborhood area in 2007. Out of these 109 respondents, 26 respondents who are still living in the same neighborhood (and those who could be traced back) agreed to participate in 2018. Respondents again mapped their respective neighborhood boundary following the same method that was followed in 2007. These identified neighborhood mappings of 2007 and 2018 were studied.

Three methods were followed to collect information: Questionnaire survey, Neighborhood area mapping using satellite image, and In-depth interview. Questionnaire survey was used to collect the factual and behavioural information, Neighborhood area mapping using the Google. Earth's satellite image was used to identify the neighborhood area and in-depth interview was used to understand the respondent's interpretation for the changes of the identified neighborhood boundary. Out of 26 respondents, the neighborhood areas have increased of 19 (79%). On average, the neighborhood area has increased three times in size when compared with 2007. No correlation was found between the size of neighborhood areas of 2007 and 2018.

The following findings of the study may lead to future research questions. 1) Busy vehicle roads, boundary walls and friend's house have been a major factors for determining the neighborhood boundary. 2) No relationship was established between independent mobility and the size of the neighborhood. 3) The neighborhood is not necessarily a continuous area but it can be segmented too. 4) New urban project influences the neighborhood area 5) Identified neighborhood area remained similar when changes in the built environment is minimum.

D14 - Transform Places with People and the Environment: Gerontology

Living in Darkness: Lighting Issues in a Senior Housing

Daejin Kim (Iowa State University); Carl Chang (Iowa State University); Jennifer Margrett (Iowa State University)

In the normal aging process, more than 30% of older adults have vision impairment which has a negative effect on their ability to stay in their home and community independently and safely as long as possible. It is common that smaller pupil size and lens changes lead to reduced ocular light transmission and increased glare, which can become increasingly problematic in daily life. Research demonstrates that older people need more light to achieve the same level of visual acuity as younger people, and to safely perform routine daily activities (Boyce, 2014). A general rule of thumb is that older adults need 3-4 times more light and 2.5 times the contrast compared to a 20-year-old to perform the same tasks (Bieniek, Frei, & Rousselet, 2013; Owsley, 2011). Adding to the difficulty, older adults frequently have inappropriate lighting equipment which cannot provide proper illumination and compensate for eye impairment (Kim & Portillo, 2018). The purpose of this research was to assess lighting conditions in senior housing and examine older adults' perception of the lighting. This research included visits to 50 independent housings in two large retirement communities in the Midwest. The research team probed types of lighting (e.g., ambient lighting, lamps, night light) and measured multiple illuminances of the lighting systems in different rooms (e.g., living room, bathroom, bedroom). Residents were asked to describe the adequacy of light levels in each location and their specific tasks and needs related to lighting. The research findings provide empirical evidence regarding lighting preferences, needs, and challenges in senior housing. Discussion will focus on why older adults are satisfied or dissatisfied with their lighting conditions and methods to address lighting issues using ambient intelligence technology.

A Social Psychological Perspective on the Role of Community Spaces in the Wellbeing of Older Adults.

Polly Fong (University of Queensland); Catherine Haslam (University of Queensland); Tegan Cruwys (Australian National University); S. Alexander Haslam (University of Queensland)

We live in an aging society, and the vast majority of community-dwelling older adults' in countries such as the U.S. and Australia wish to 'age-in-place' or remain living independently in their own homes for as long as possible. Whether our cities' infrastructure is ready to cope with the challenges of an aging, car-dependent, suburban population is a question that policymakers must address urgently. Relatedly, urban sociologists claim that the social infrastructure, such as public libraries and neighbourhood centers in our cities are vital to community wellbeing (Oldenburg, 1989; Klinenberg, 2018), and yet research shows that only a third of seniors, who are at risk of social isolation, make use of such spaces. This paper presents the case for a novel social psychological approach to examine when and why older people engage with community spaces and how they support wellbeing from their usage. Using the case-study of a successful suburban neighbourhood recreational club, we draw on focus group interviews with club members (N=31) to explore the nature of their participation and meanings tied to belonging, place and identity. The results of our thematic analysis suggest that (1) spatial and social context of community centers can both enable and restrict participation, (2) community spaces support valued social identities, which are those that provide a positive personal and collective sense of self and (3) enactment of valued social identities in community spaces facilitate a sense of wellbeing. The implications of our findings for strategic planning of social infrastructure for implementing initiatives for healthy aging will be discussed.

The Analysis of Community Environment Design for Self-Helping Aged People in China

Chen He (China Architecture Design & Research Group)

With the aging of society and the promotion of pension policies, the elderly in China are increasingly following the"aging in community" living model. The Chinese community, as the fundamental social living unit, is not only the social support system, but it also maintains the interrelationships and social networks. Therefore, the physical and non-physical environment for a community has a crucial role in meeting the physiological and psychological needs of the elderly. However, according to recent research, the existing mainstream design is not always suited for self-helping aged people, causing problems of safety, accessibility, and interactivity. This research will explore the relationship between the self-helping aged people and the built environment in the community. It is structured as follows: The first section will analyze the cultural ideology and historical significance of the communities environments in the development of urbanization, and further, it will explore the influence of technological developments and changes in social values; By researching typical communities, the second section will analyze the activity status and spatial distribution characteristics of self-helping aged people within the community and identify any features and problems within their living environment. Also, it will examine the evolving needs (physical, medical and mental) of self-helping aged people; Then, by doing empirical case study under environment and behavior studies platform in Beijing, the third section will summarize the status of the relationship between the variety of needs of self-helping aged people and the built environment. Finally, it will analyze and integrate the collected information and propose a system of ergonomic parameterization, spatial typologies, and implementation strategies for self-helping and the aging-friendly environment in the community.

D16 - Transform Places with People and the Environment: Cyberspace & Digital Environments

Understanding Art Museum Usage Patterns By Utilizing Social Media Data

Dingyi LIU (Beijing Jiaotong University); Zhongzhong Zeng (Beijing Jiaotong University); Bo Zhang (Oklahoma State University)

Social media has become a major platform for people to share their feelings and opinions. It was also employed for assessing the quality of building usage. By reading text reviews from social media users, such as TripAdvisor and Google Place, researchers can identify usage patterns for different built environments. Compared with traditional data collection methods such as questionnaires, interviews, and observations, etc., the abundant, diverse, self-posted, and ever-growing social media data provides researchers with much more data to study the relationship between people and environments. This research aims to take advantage of social media to explore art museum usage, with higher richness and complexity. In this process, a new POE method using social media data is exemplified. The researchers use Kimbell Art Museum in Texas and Philbrook Museum of Art in Oklahoma as two case studies. Employing Python, 1,570 and 1,239 TripAdvisor reviews from 2010 to 2019 are mined. These 150,000 words were categorized, into 50 activity themes, 34 object/scene themes, and 27 experience themes. Five research questions are answered as follow, (1) What objects and scenes caught visitors' attention; (2) How did visitors feel about their museum visits; (3) How did museum programs affect visitors' feeling; (4) What were the different visit dynamics between these two museums with different architectural features and exhibit features; (5) What are the temporal patterns of the museum visits dynamics. Gephi is used to visualize the relationships between museums and experiences. This study reveals the gap between architectural design intentions and visits attention, the importance of outdoor space for a museum, the seasonal popularity of museum programs, and the importance of exhibit versus architectural space. This study will help to inform future art museum planning, design, and management. The new POE method using social media data could also be applied in other architectural types.

Understanding Usage and Perception of Three Memorials from Social Media Data

Bo Zhang (Oklahoma State University); Yiwen Zheng (Shaoxing University)

Designed memorials in public space are a special type of built environments, which not only accommodate outdoor activities but also endow places with meanings. The nature of memorials provokes design creativity and societal attentions. Yet, if we ask, how are the memorials used and perceived by the public? rarely any empirical research has been conducted to reveal the activities, emotions, and thoughts associated with memorials. The main reason could be that the conventional data collection methods (e.g. survey and structured observation) are not readily applicable to reveal the deep and instant memorial visit experiences. Thus, it is always unclear if the symbolisms claimed by the memorial designers are realized.

This study takes the advantage of the availability of social media data to explore the usage and perception of designed memorials. We deem the supplies of huge, efficient, self-posted, and ever-growing data have great potentials to reveal how designed memorials are viewed, interacted, perceived, and understood. The 9/11 Memorial in New York City, the Vietnam Memorial in Washington DC, and the Oklahoma City National Memorial are selected as three case studies. 47813, 1135, and 5827 posts from Googleplace and 92430, 11614, and 4,855 posts from Tripadvisor are mined respectively for each memorial. Photos and texts from the posts are coded into three themes respectively, including: (1) activities, (2) objects/scenes, and (3) experiences/facial expressions. The study conducted four separate analyses, such as: (1) popularity of different areas; (2) human gestures and activities in different areas; (3) human facial expressions in different areas; and (4) human experiences and emotions in different areas.

The results yield new dynamics about memorial space usage and experience, present the social capital of designed memorials, and reveal environmental effects of different design elements, interestingly, some of which don't align with designers' initial design ideations.

Virtual versus Real Usability Evaluation of Healthcare Furniture in an Outpatient Surgical Waiting Room

Roxana Jafarifiroozabadi (Clemson University); Rutali Joshi (Clemson University); Anjali Joseph (Clemson University)

Over the past decades, virtual reality (VR) has tremendously improved prototyping process and decreased time and costs associated with design projects (Ingrassia & Cappello, 2009). In the field of healthcare design, several studies have used VR to obtain user feedback and identify the optimal design in healthcare environments such as operating suites (Wingler et al., 2019). Nevertheless, the reliability of this methodology has not been thoroughly investigated and it is unclear what design features can be similarly assessed in VR and real environments.

To address the above-mentioned gap, an IRB-approved observational study was conducted utilizing VR and scenario-based simulations. A total of 92 participants were prompted to complete 4 scenarios with varying tasks that a care partner would most likely engage in during wait time in a virtual outpatient waiting area. Participants were asked to select a seat in each scenario and were then exposed to the actual furniture they selected during the VR experience. An online survey comprised of 15 questions was administered to the participants to evaluate the real furniture based on physical comfort, flexibility, and aesthetics. Semi-structured interviews were conducted to obtain users' feedback regarding similarities and differences between the virtual and real furniture.

Survey results depicted participants' rankings of design features associated with the selected furniture. Content analysis of the interviews indicated that features including furniture comfort, color, texture, dimension, flexibility, and back angle were perceived differently while furniture shape and configuration were perceived similarly in experiencing the virtual versus real furniture. In some cases, differences in perception of such features influenced how participants evaluated seat efficacy in supporting the scenario tasks.

This study contributes to understanding the limitations associated with VR incorporation in design process, and therefore informs future developments required to enhance the prototyping process as well as users' experience in virtual environments.

Impact of Biophilic Elements on Airport Anxiety: A Pilot Study on Perception of Time and Crowding in an Immersive Virtual Airport

Nicholas He (Cornell University); So-Yeon Yoon (Cornell University)

Airports, especially TSA security checkpoints, are well-known for their stressful atmosphere that negatively impacts both passengers and employees. Biophilic experiences with elements such as nature scenes, plants, and trees are known to reduce stress and anxiety by involuntary attention and mental restoration (Kaplan, 1995). More recently concluded, multi-sensory experiences of nature are even more effective than nature scenes alone (Sona, Dietl & Steidle, 2019). The impacts of biophilic design have been studied mostly in the context of workplace. Research on the effects of biophilic elements in more extreme environments consisting of loud noise and crowding has been scarce.

This pilot study investigates whether and to what extent biophilic elements have the same recovery and restoration effects on an individual's health in a crowded environment. For this study, two scenarios of a typical airport were created in immersive high-fidelity virtual environments; one with biophilic elements and one without). Eighteen subjects participated in the lab experiment. Each participant was invited and asked to wear an HTC VIVE head-mounted display and an HR wristband to interact with the two scenarios in a random and balanced order. Questionnaires pertaining to a stated choice task, positive-negative emotional responses, crowding perception, and time perception were conducted immediately after the experiment. The participant then viewed two additional conditions of the same two environments with different crowding levels in random order and completed the questionnaire for choice modeling to examine the extent of biophilic preferences. The benefit of biophilic design in reducing airport anxiety and the process will be presented with findings from survey and heart rate data analyses.

D17 - Transform Places with People and the Environment: Health I

Can Home Layout Influence Navigation of Early and Middle Stage Alzheimer's Patients?

Mahnaz Ensafi (Texas Tech University); Debajyoti Pati (Texas Tech University)

This study examines the orientation challenges of patients with Alzheimer's Disease (AD). It focuses on the types of memory affected by AD and analyzes related theories to provide design solutions to decrease wayfinding issues. A literature review and analysis was conducted between September 2018 and May 2019 by searching Elsevier, JSTOR, and ProQuest. The inclusion criteria included articles published in English between 1960 and 2019 with a focus on behavioral and cognitive issues of AD and related theories.

Changes in cognitive function affecting memory and behavior are the most common symptoms of AD. Memory, executive functioning, and attention are the main cognitive domains reduced in AD patients, while navigation and orientation complications are among the first behavioral symptoms. In the moderate stage of AD, wandering occurs continuing with anxiety and aggression. Therefore, recognition theories, in addition to different types of memories were explored to have a better understanding of the extent to which they are influenced by AD. Furthermore, the theory of affordances, legibility, and imageability were considered in relation to their impact on the behavior of Alzheimer's patients to be able to provide design solutions. People in the early and middle stages of AD typically live in their own homes and prefer to stay there as long as possible. The statistics indicate that even greater numbers of older adults will face AD in the future requiring greater costs for the government and healthcare industry. However, resultant costs of care can be reduced by modifying the physical environments. Flooring, materials, finishes, colors, cues, number of walls, furniture's orientation, lighting, distance and relationship between different spaces, type of openings, and area of each environment were considered based on the theories to address the navigation issues. The proposed design solutions were used to create a readable, understandable, walk-able, inescapable, and recognizable environment.

The Impact of Infusion Center Layout on Staff Collaboration

Kara Freihoefer (HGA); Nancy Doyle (HGA); Amanda Simpson (St. Elizabeth Healthcare); Mahshid Jalalianhosseini (University of Wisconsin-Milwaukee)

The National Cancer Institute approximates 1.8 million people in the U.S will be diagnosed with cancer in 2019. Outpatient infusion centers have become important settings in which cancer patients receive life-saving chemotherapy treatments.

Staff collaboration and teamwork are essential elements in the delivery of quality care in any healthcare setting and are especially important in the infusion center environment where care teams prepare, deliver, and administer powerful drug solutions. Xyrichis and Lowton (2008) identified physical proximity of team members as an important factor contributing to successful teamwork. Little is known about how the designed environment may influence collaboration and ultimately the delivery of care in infusion centers. The goal of this research was to investigate the effects of two different infusion center layouts—open bay and semi-private bay designs—on nurse-to-nurse and nurse-to-pharmacy collaboration.

A mixed-method approach comprised of shadowing and surveying was used in the study. Shadowing took place in both infusion center designs on two separate occasions. Nurse activities and locations were recorded for a total of 34 hours. Processes and workflows related to medication delivery in each unit also were assessed through shadowing. Survey questions gauged staff satisfaction with collaboration, as well as satisfaction with medication delivery.

Results of the study revealed interesting differences in nurse collaboration in the open bay and semi-private designs. Additionally, differences in nurse collaboration with pharmacists and the process of medication preparation and delivery also were observed. The implications for future infusion center design and the benefits of embedding pharmacists within infusion centers will be discussed in the presentation. Specifically, plans for a new infusion center, with fully enclosed infusion bays, decentralized charting stations, and dedicated pharmacy space in each infusion pod will be described, as well as plans for a post-occupancy evaluation to collect comparison data in the new center.

Post Occupancy Evaluation of Community Health Facilities in Terms of Staff Spaces and Clinician Interactions

Shermineh Afsary (The University of Kansas); Mahbub Rashid (The University of Kansas)

In the USA, Community Health Centers (CHCs) can provide affordable primary healthcare service to all patients including the uninsured, low-income ones, or those living in remote areas. Today, many CHCs have old buildings, which are not designed well. They need to be evaluated and renovated ensuring better physical environment for high quality primary healthcare service. Previous studies show that better staff spaces can provide comfort and increase performance of staff; and frequent staff-staff interactions can reduce errors and increase collaboration in healthcare settings (Rashid & Boyle, 2014) (S.Zadeh & Shepley, 2012). Most of them has focused on hospitals and less on CHCs. Therefore, the main purpose of this research is to evaluate the physical design features of CHCs in terms of staff spaces and staff-staff interactions. It aims to understand how the design features of staff spaces can enhance staff-staff interaction, and how to create a more effective tool to evaluate and suggest design solutions for improved staff spaces and staff-staff interaction. The case studies for this study are the Douglas County Community Health Facility in Lawrence, Kansas and the Samuel U. Rodgers CHC in Kansas City, Missouri. In these case studies, staff spaces and clinician interactions are evaluated by data collection from 40 staff members in a questionnaire survey. The survey uses two modules of a Post Occupancy Evaluation (POE) tool developed by the Center for Health Design (CHD, 2017). These modules are related to design features of staff spaces and clinician interactions. Based on the findings of the statistical analyses of the survey, the CHD POE tool will be revised to improve the POE of CHCs; new indexes of design will be developed to evaluate staff spaces and clinician interactions in CHCs; and innovative design solutions will be proposed to enhance clinician interactions and to improve staff workspaces.

D18 - Transform Places with People and the Environment: Health II

Domesticity on Display: A Study of the Objects That Trace Our Daily Experiences

Alyssa Kuhns (University of Tennessee)

Our homes serve not only as a space for living but as a record of our daily lives. Benjamin (1986) states "To live means to leave traces", and we leave these traces through the proximity, amount, and arrangement of our personal belongings (p. 155). It is through these belongings or objects that we document the cultural and behavioral impacts within our homes. Despite the ability of our objects to document our activity, we reject signs of life and notions of imperfection within our domestic interiors. Modern forms of media including home good catalogs, Instagram, and Pinterest present the domestic interior as an "idealized phenomenon used to encourage the consumption of goods" (Sparke, 2008, p. 35). Cultural pressures encourage us to eliminate traces of living from our domestic spaces. However, a highly curated domestic environment is often unobtainable for most of the population, whether based on financial or practical limitations.

To understand these conflicting states of living and display within domestic interiors, this research examines private residences through observational research methods and photo documentation. Traveling to over 20 private residences throughout the Midwest and Northeast United States, I documented major spaces within domestic interiors using consistent one-point perspective views as a means for comparison. These images, arranged into a matrix, organizing residence and room type, served as a dataset to analyze collections of objects within domestic spaces and understand their impact on spatial reading and behavior.

This presentation investigates parallels and contrasts between the perception and function of objects within domestic interiors. It argues that there is opportunity for new design explorations based on a method of studying traces left behind by objects in domestic settings rather than an aesthetic ideal. It is the aggregate and arrangement of objects, whether curated or naturally occurring, that define how we experience domestic space.

Significant Factors of the Environments in Senior-Living Facilities – Review of the Literature

Zhe Wang (Henan University)

The oldest old (80+) is the faster growing group in older population. China, America, and India are the countries with the largest populations of the oldest old. Across country and culture, the needs of the oldest old for institutionalized elderly care are similarly increasing. Aim: Promoting environmental support for elderly care in senior-living facilities, this study aimed at confirming the existence of relevant research evidence. Significant factors were reviewed and the direction for future research was identified.

Focusing on empirical research of senior-living environments, a literature review was conducted to summarize credited findings published in English and Chinese in recent 30 years. Applicable settings range from independent living to nursing care. A review-based analysis of peer-reviewed English/Chinese-language literature was completed to identify both qualitative and quantitative findings focusing on the environmental correlates of residents' health, well-being, quality of life, and satisfaction in the facilities. Literature reviews and unpublished works were not included. Fifty-nine studies (51 in English and eight in Chinese) were included.

The environmental correlates were clarified in terms of theoretical subject (homelike environment, social interaction, access to nature, and active living) and of spatial design (site planning, building design, and key interior spaces including residential rooms, common places, and amenities). A research agenda was created to summarize previous research and identify gaps. Design and research guidelines were noted in relation to site, building, and interior environments.

Considerable opportunities exist for use of environment to promote elderly care in senior-living facilities. Environmental factors at the site, building, and interior levels should be carefully considered if building new senior-living facilities or renovating existing ones. Multiple theoretical subjects may be applied. More research is required to link studies for a better understanding of local needs and new technology for elderly care.

The Analysis of Visibility and Circulation in the Community Health Clinic, Using Space Syntax: A Case Study in Baltimore

Shermineh Afsary (University of Kansas); Mohammad Gharipour (Morgan State University)

Using Space Syntax Analysis (SSA), this research aims to find out how layout configuration in a Community Health Clinic (CHC) affects the visibility and circulation. CHCs are the health centers providing affordable primary and preventive healthcare service to all patients including the uninsured, low-income ones (NACHC, 2015; Taylor, 2004). The majority of CHCs in the USA deal with budget issues. It has resulted in aging architecture suffering from architectural issues and affecting their performance. In addition, most of the research in the USA have focused on the architecture of hospitals and less on CHCs. This research contributes to a better understanding of visibility and circulation in health clinics, especially in CHCs to find out how to improve it and recommend affordable design solutions. Spatial layout design for efficient visibility and circulation is one of the most complex subject in healthcare design. SSA theory and method allows the quantification of layouts to produce independent variables to analyze visibility (Haq, 2012), circulation and its impact on social behaviors. The visual attributes are connectivity, integration and the visual step depth from key spaces analyzed by SSA. The correlation between local connectivity with global integration defined as "intelligibility" shows that the spatial configuration is understandable and predictable for the user's movement. The research setting is Shepherd's CHC, in the city of Baltimore, USA. The study on circulation and visibility reveals significant shortcomings regarding the layout and organization of spaces. Findings on visibility demonstrate that visual connection, integration and step depth inter-between the waiting areas, nursing stations, doctor's office and exam rooms are not proper. Findings on circulation and intelligibility of this clinic (R-value <0.7) shows that this layout is not understandable enough for the users. The research is concluded with affordable design solutions in layout configuration to improve the efficiency of spaces.

D19 - Transform Places with People and the Environment: Health III

Post-Occupancy Evaluation of a Therapeutic Garden That Helps People Heal

Terri Zborowsky (HGA); Emanouil Spassov (HGA)

The objective of this research study was to conduct a post-occupancy evaluation of a Rehabilitation Therapeutic Garden where there is a paucity of research. Since Ulrich's study (1984), that revealed how a view to nature impacted patient care outcomes and his 1997 theory that access to nature reduces stress in patients, family and staff members, many others have studied similar relationships. Cooper-Marcus and Barnes (1995) used a variety of methods to evaluate four hospital gardens and found that access to the gardens provided restoration from stress and an improved mood. Davis (2011) conducted a POE on a rooftop hospital garden used primarily for physical therapy and found that administrative staff perceived high accessibility while patients and staff reported low accessibility, among other results. This current study setting, a rooftop garden, was built directly adjacent to a newly constructed inpatient rehabilitation unit. It was designed based on therapeutic goals for the rehabilitation of both inpatients and outpatients who use the garden under direct supervision of their therapists. Research methods included Sachs' (2016) therapeutic landscape evaluation tool and staff interviews. This study received IRB approval. Data were analyzed using the Lincoln & Guba's(1985) methods of naturalistic inquiry. The garden succeeded in achieving its intended goals in a number of ways: three broad categories of findings will be shared at the presentation. The research results demonstrate the value of landscape architecture to clients by identifying the enduring impact of therapeutic gardens on patients, family and staff members. This study will add original insight to the growing knowledge base of therapeutic landscapes, as well as fill gaps in literature and current knowledge.

Transitions in the Emergency Department: Does the Built Environment Impact Interruptions during Physician Handovers?

Rutali Joshi (Clemson University); Anjali Joseph (Clemson University); Ronald Pirrallo (Prisma Health); Michelle Ossmann (Herman Miller)

End-of-shift handovers between physicians in the emergency department (ED) are crucial episodes of patient-related information exchange achieved through collaboration and communication. Communication breakdowns are known to occur in fast paced and dynamic ED settings due to frequent interruptions that contribute to medical errors. Approximately 11% of these medical errors are attributed to problems during handovers. While some interruptions may be important to functioning of the ED, others may be unnecessary, especially during handovers. Despite the impact interruption can have on work processes, no study has examined them in the context of cognitively demanding and crucial tasks like handovers. To date, healthcare research in the ED has focused primarily on the content and process of handovers, and little on the impact of the larger system comprised of the organizational policies as well as built environment, tools and technology on interruptions.

The objective of this study is to understand the relationship between the ED built environment and interruptions faced by physicians during handovers. 3 ED pods with varying physician workspace configurations – (a) open plan layout, (b) enclosed physician workspace, and (c) semi-enclosed physician workspace were compared. Unnecessary and critical interruptions were captured in real time during 60 handovers using an observation tool. Physicians perception of interruptions, collaboration and impact of built environment on these behaviors were garnered using a survey. In addition, space syntax was used as an analytical tool to measure mutual visibility to and from the handover location and visibility to artefacts central to handovers like status boards.

This correlational study reported differences in interruptions and perception of collaboration during handovers between the three physician workstation configuration with specific design and syntactic characteristics. This session will present findings to support design decisions that may help mitigate unnecessary interruptions and support only critical interruptions within the ED during physician handovers.

Global Birth Environment Design Network: Developing an International Collaborative Birth Environment Design Resource.

Davis Harte (Boston Architectural College); Doreen Balabanoff (OCAD University); Nicoletta Setola (Università di Firenze); Elizabeth Newnham (Griffith University); Alessia Macchi (Università di Firenze)

Globally, women request respectful childbirth experiences (Newburn, Singh, 2003). Birthspace design qualities can influence psychological/physiological states and birth processes (Hodnett et al., 2012). Research suggests the environment is implicated in increased medical intervention, which impacts healthcare costs (Miller et al., 2016). Designers' access to birth design evidence is inadequate, while knowledge is expanding. Our project aims to synthesize this knowledge, assess gaps, and develop a digital knowledge mobilization resource.

How might a design resource best provide knowledge to inform decision-making and meaningful approaches for salutogenic (supporting health and well-being) birth environments? This question drives our inquiry into participatory methods useful in achieving the desired design outcome – an accessible, comprehensive, richly layered and clearly organized resource for designers.

Through online collaboration, we seek to define objectives and outcomes, including, thusfar, the following: 1) Literature review: Comprehensive and peer reviewed meta-review of literature. 2) Guidelines and evaluation tools review: Identify resources, approaches, gaps in criteria, and need for classification, critical analysis, organization for user needs. 3) Workshop(s): Utilizing a co-design approach called service design, identify, invite and work with diverse users to inform and inspire architects and designers to fulfill the research aims. Service design is an interdisciplinary approach with an outside-in, iterative process that facilitates human-centered services. 4) Digital resource: Develop a knowledge mobilization resource – an organized set of resources easily translated into practice. To our knowledge there isn't currently such a resource. 5) Planning process: Document process and methods as valuable to research methodologies for bringing evidence, best practices and ongoing knowledge development to service providers in practical applications Salutogenic birth design is essential to provide better design for women, newborns, and subsequently – community health. This research will amplify the birth environment as a focus for designers, to enhance contributions for more optimal and satisfying birth experiences.

Towards Optimizing the Spatial Arrangement of Patient Rooms for Minimum Nurse Travel Distance in a Hospital Unit

Jisun Lee (Cornell University); Hyunsoo Lee (Yonsei University); So-Yeon Yoon (Cornell University)

One of the critical considerations for healthcare design providers in planning a hospital nursing unit is the link between the patient rooms and nurse stations. The goal of this study is to provide a novel approach to evaluate spatial layout options in the decision-making process towards optimizing the arrangement of patient rooms for minimum nurse travel distance. With increasing demands on medical care services, hospital inpatient units tend to consist of mixed patient room arrangement of single/double-bed and multi-bed rooms in a nursing unit. This influences the nurse-to-patient assignment and often causes an unbalanced workload of nurses. Improving the environment to support a well-balanced workload distribution and reduce walking distance is necessary for the nurse's work efficiency and well-being (Ulrich et al., 2008; Hendrich et al., 2008).

An agent-based simulation was conducted to model nurses' movement with virtual agents representing the nurses in three inpatient units. Grasshopper plug-in "PedSim" is used to generate nurses' walking trails in Rhino modeling spaces and the distances were automatically calculated for comparison. PedSim is based on the social force model (Helbing & Molnar, 1995) with agents driven by multiple forces, namely a target force, person repulsion, and obstacle repulsion for basic movement mechanism (https://www.food4rhino.

com/app/pedsim). A nurse's regular traveling cycle for medication to the assigned patient rooms was included in the simulation.

The spatial features that impact on reducing nurse travel distance were discussed, and revised layouts were suggested and evaluated for a before-and-after study. Patient room arrangements that supported contiguous patient assignments and even walking distance among nurses were discussed. This study contributes to extending the application area of agent-based simulation to an evaluation framework for spatial configuration analysis as an easy-to-access method. This approach can be further applied for healthcare environmental design solutions in the outpatient as well as inpatient units.

D20 - Transform Places with People and the Environment: Health IV

Confirming the Recreational Attractiveness of Adaptive Heritage Reuse

Juo-Ying Shen (National Chung Hsing University); Hung-Ming Tu (National Chung Hsing University)

Although adaptive reuse is a tool to attract travelers and preserve heritage effectively, some cases of adaptive heritage reuse (AHR) have failed to attract travelers. Past qualitative interviews had conceptualized six factors making AHR attractive from the recreational viewpoint, including heritage and activity of reuse environment, natural environments, regional environments, self-growth, health benefits, and social benefits. The purpose of this study was to use quantitative factor analysis to confirm the recreational attractiveness toward AHR, which can be applied as landscape planning principles of sustainable AHR. The initial pool of 58 scale items capturing recreational attractiveness toward AHR was generated from the results of past interviews. All items were assessed using the question 'If you want to visit a heritage site, please evaluate the importance of the following characteristics', and were scored on a 7-point scale. This study included two stages. In the first stage, initial data were collected from 13th to 14th March 2019 at Calligraphy Greenway, a major recreational space with several heritage sites located in Taiwan's Taichung city. A total of 332 valid questionnaires were collected. Exploratory factor analysis was used to explore the dimensions of recreational attractiveness towards AHR. Six dimensions and 24 items of recreational attractiveness towards AHR were extracted: heritage of reuse environment, activity of reuse environment, natural and openness, regional environments, obtaining recreational benefits, and pursuing long-term values.

In the second stage of the study, the structure of the above six dimensions and 24 items were confirmed by confirmatory factor analysis. A total of 193 valid questionnaires were collected from 8th to 9th June 2019 at the same site. The results showed that the structure of the six dimensions were confirmed. These six dimensions should therefore be considered as landscape planning principles of sustainable adaptive heritage reuse to preserve heritage effectively.

Exploring Relationships between Human Comfort and Health in Buildings: Establishing Opportunities for Future Research

Helia Taheri (North Carolina State University); Traci Rose Rider (North Carolina State University)

People spend 90% of their time indoors (Leech et al., 2000), illustrating the possible impact of the built environment on human comfort and health (Olesen, 2005). According to ASHRAE Guideline 10P (2014) there are four comfort conditions which contribute to a comfortable environment for humans: thermal, visual, indoor air quality, and acoustics. Each condition has different measurable factors. For example, thermal comfort factors include dry bulb temperature, radiant temperature, relative humidity, air velocity, etc. In addition to human comfort, health and well-being are also affected by the built environment and its elements. Several studies show that there is a relationship between each of the human comfort factors and human health in the built environment (Clausen et al., 2002; Derby M.M., 2017; Hedge et al., 2014; Ormandy and Ezratty, 2012; Stoops, 2004). From WHO, health is defined as "a state of complete physical, mental and social well-being and not merely the absence of disease or infirmity" (WHO, 1946; Olesen, 2005). This paper synthesises the literature addressing the relationship between human comfort parameters and human health in the built environment to understand existing gaps in which research is needed. Through this exploration, a foundation is created for future studies at the intersection of human comfort, health and well-being in the built environment, and suggests future studies for sustainable buildings.

The Impact of Daylight and Views on Patients with Heart Disease in Cardiac ICUs

Roxana Jafarifiroozabadi (Clemson University); Andrea Franks (AnMed Health); Anjali Joseph (Clemson University)

Being diagnosed with heart disease can lead to symptoms such as depression, anxiety, sleep disruption, and delirium which can cause prolonged hospital stays among patients (Cho, Min, Hur, & Lee, 2013). Evidence suggests that presence of daylight and window views can influence these symptoms positively (Walch et al., 2005; Ulrich, 1984). However, no studies to date have differentiated the impact of daylight from window views on patients with heart disease or investigated which factor can best support patient recovery in cardiac intensive care units (CICUs).

This study investigates the impact of access to daylight and window views on patients' length of stay in a CICU through evaluating factors including anxiety, depression, sleep disruption, and delirium.

An IRB approved two-phase study is conducted in a CICU comprised of rooms of the same size with three unique types of daylighting and viewing accommodations including rooms with daylight and no access to window views, windowless rooms, and rooms with daylight and window views. The first phase of the study entails analyzing retrospective patient data form 2011 to 2019. The second phase is a prospective study with a quasi-experimental design in which 36 anticipated patients will be recruited through purposive sampling. The study has already been pilot tested with four patients. Data regarding depression/anxiety, sleep quality, and lighting are collected through questionnaires, actigraphy, and light data loggers respectively. Other data regarding delirium and patients' length of stay are obtained from medical records after discharge. The difference between these factors across the rooms will be assessed through a series of t-tests.

Findings from this study will delineate similarities and differences between health outcomes such as length of stay among patients assigned to CICU rooms with different types of daylighting and viewing accommodations, and thus will inform the design of future CICUs.

The Role of Micro-Restorative Experience in Indoor Spaces in Restoring Attention

Hossein Saedi (Iowa State University)

The purpose of this study was to explore the potential impact of micro-interactions with natural green elements on attention restoration, as a component of health and well-being.

The world's urban population is expected to increase to 68% by 2050 (UN DESA, 2018). This population needs accommodations. One way of responding to this demand is high-density developments. Densification usually means no front or back yards and limited access to a natural environment (Beer et al. 2003; Swanwick et al. 2003, Wang et al., 2019). Living in these new dense cities brings a new lifestyle with it that limits the citizens` interactions with green environments.

Interaction with natural environments provides people with numerous health benefits (Baycan et al., 2009) such as promoting cognitive functioning and sustained attention (Keniger et al., 2013). So it is important to maximize the benefits that citizens can receive during their limited time that remains for their interaction with natural elements.

Attention Restoration Theory indicates that sustained attention is one of the critical factors for successful cognitive functioning. This experimental study tests the potential for natural green element micro-interactions to increase the level of sustained attention. This study compared the results from a Sustained Attention to Response Task (SART) for 35 residents of a high-rise residential building. Participants were randomly assigned to experience one of two versions of a building`s threshold modeled in a 3D virtual reality environment. One model included natural green elements and the other did not. Participants completed the SART twice. Once before experiencing the threshold and once after. The results indicate that those who interacted with natural green elements in the building threshold obtained demonstrably higher SART score and expressed less cognitive errors. This research outcome supports the potential positive effect of micro-interaction with natural green elements on health and well-being.

D21 - Transform Places with People and the Environment: Health V

Transformation of Public Space with Displaced Citizens

Aaron Brakke (University of Illinois)

Architectural histories have presented Latin American development through descriptions of uneven socio-politico circumstances categorized as 'formal' and 'informal'. In the Colombian city of Bogota, the periphery has become an area where tensions between sovereign power and bio-power are clearly evidenced. To further complicate the context of our work, approximately 10% of the inhabitants of these neighborhoods include people who have been displaced due to the armed conflict. To address this problem, the Colombian government allotted 583 million dollars in 2013 for the construction of low-income housing to help alleviate the stress of this tremendous influx of urban residents. Unfortunately, this infrastructure does not account for the complexities of the heterogeneous demographic that has been displaced from rural areas of the country. This paper will discuss the scope of several projects of "vivienda interes social" built in Bogota and assess both the positive aspects of these projects as well as their shortcomings. To counter these issues, we have been working with our students to create "insurgent architectural projects" with Participatory Action Research established by Orlando Fals Borda (Borda, 1999, Borja, 2014 and Harvey, 2008). One project will be described and discussed in detail. This project involved a design process to create new imaginaries for a space that had been used as a waste site for leftover construction materials. We worked with residents to develop and articulate the imaginaries into a vision for a vibrant space that was designed and co-created with them. The transformation of this leftover space into a park was a complex undertaking that mitigated tangible social and political tensions. This paper will include what was learned from this PARS process and what has been observed in a post occupancy evaluation.

How Do Streetscapes Communicate Walkability?: Understanding Non-Expert Perceptions through Photovoice and Physiological Sensing

Chester Harvey (University of California Berkeley)

An important role of streetscape design is to communicate the appropriateness, or as Lynch (1981) describes it, the "fitness," of certain uses along streets. If walkability can be communicated through streetscapes, it might be used by designers and planners to promote walking behavior, even along streets with substantial vehicle traffic (Jacobs, Macdonald, & Rofé, 2002). However, design experts may understand walkability in streetscape design differently than the non-expert public. Limited evidence about how expert theories of streetscape walkability coordinate with non-expert perceptions makes it difficult to incorporate streetscapes into planning analyses or projects aimed at encouraging walking. In this presentation, I report initial findings from research funded by EDRA's Brill Grant in Urban Communication. I use a mixed-methods, quasi-experimental field study to investigate whether non-experts identify streetscape walkability in ways that align with the expert-identified qualities described by Ewing and Handy (2009): spatial enclosure, human scale dimensions, transparent boundaries, visual complexity, and imageable features. Study participants traverse a walking route in Oakland, California that varies substantially in terms of these qualities. Participants' perceptions of walkability, and underlying perceptions such as stress, safety, interest, and boredom, are measured through a variant of the photovoice method. As they traverse the route, participants photograph scenes or features that they perceive as influencing walkability and debrief these photographs in a follow-up interview. Additionally, the study investigates how physiological indicators of stress, recorded by a watch-like sensor worn by the participant while walking the route, may identify subconscious perceptions. Photovoice and stress sensor data are compared to Ewing and Handy's streetscape qualities, assessed at regular intervals along the route based on their expert-defined criteria. By evaluating whether participant responses coordinate with these qualities, the study examines the efficacy of contemporary streetscape-based walkability theory for understanding public perceptions of walkability and walking behavior.

Identifying the Theories in Healthcare Design Research and Challenges in Theory Development, Use and Evaluation

Suining Ding (Purdue University)

The theory is an ambiguous term that has been given different definitions that are not universally agreed upon. Although the theory has long been thought to play a central role in research for the development of knowledge, the definition of theory has been problematic. In healthcare design research, the primary focus is to create knowledge and use scientific research evidence to inform design decisions to achieve the best possible outcomes for patients and caregivers. Therefore, what are the relevant theories for healthcare design research? What are the challenges for theory development, use, and evaluation? The findings of this study provide answers to these two questions.

The purpose of this study is to critically review theories relevant to healthcare design research and identify the challenges for theory development, use and evaluation so that design researchers can benefit from the outcomes of this study. This research took a qualitative approach with grounded theory methodology. The method used for this research is a literature review. Findings show that the following theories can be used in healthcare design research as scholars often reference these theories in different terms: explanatory theory, positive theory, normative theory, procedural theory, and substantive theory. The positive theory is described as analytic and predictive, while the normative theory as creative. Based on the literature surveyed, challenges in theory development, use, and evaluation were identified. Given the nature of the healthcare design process that fulfills with creativity and intuition, the challenge of theory development is that the theories have been strongly normative and weakly analytic. Theory development not only should involve describing the world, but also should involve an explanation. The challenges for theory use and evaluation are how to interact with scientific research evidence and how to implement the research evidence in design solutions.

The Impact of Physical Distance to a Healthcare Facility on Heart Disease Outcomes in Anderson County, South Carolina

Roxana Jafarifiroozabadi (Clemson University); Patricia Carbajales-Dale (Clemson University); Andrea Franks (AnMed Health)

Heart disease is the leading cause of death in the United States and a major health concern in the state of South Carolina (American Heart Association, 2017). Heart disease outcomes are highly dependent upon prompt access to emergency medical treatment. Evidence suggests that the majority of adverse health outcomes pertaining to heart disease occur within one hour after the symptom onset (Yamashita & Kunkel, 2010). Any delays in treatment can increase the likelihood of adverse outcomes such as mortality among patients (Fang et al., 2008). Although early identification of the symptoms and immediate seek of medical help is crucial to mitigate heart disease outcomes, it is postulated that physical distance from a healthcare provider can also impact these outcomes among patients. This IRB-approved study investigated how physical distance to a hospital in Anderson County, S.C. impacted heart disease outcomes among the patient population in proximity of the facility. Patients' addresses from 2011 to 2019 (n = 22,500) were obtained from the hospital's electronic medical records and geocoding techniques were utilized to map where patients resided. Using Network Analyst, the travel time from each data point to the hospital was determined. Spatial statistical analysis was employed to determine the relationship between proximity to the hospital and heart disease outcomes including mortality and length of stay. Further analysis was conducted to examine associations among variables including patients' demographics, medical history, socio-economic factors, heart disease outcomes, and hospital distance in this county.

Clarifying the relationship between distance to a healthcare provider and heart disease outcomes enhances our understanding of health-related burdens that the built environment can potentially impose on us.

Such disparities can be circumvented by paying closer attention to the service area that each healthcare facility can cover to further discover the need for areas that fall outside the service area.

Measuring the Brain Responses to the Environment: A Literature Review on fNIRS, fMRI, and EEG Studies

Homa Pesarakli (Texas A&M University); Zhipeng Lu (Texas A&M University)

Importance and Goal. The physical environment has a profound impact on human cognitive activities, behavior, and health. Different tools have been developed to record human physical and mental responses to environmental features objectively. Neuroimaging—including EEG, fMRI, fNIRS, PET—is one of them that has the potential to provide reliable measure. Yet, little is known regarding how these technologies can be applied appropriately to studies on human behavior and physical environment relationship. This review aimed to evaluate the three primary neuroimaging technologies that have been implemented in environment and behavior research: electroencephalography (EEG), functional near-infrared spectroscopy (fNIRS), and functional magnetic resonance imaging (fMRI).

The review started by identifying empirical studies, via google scholar, that used targeted neuroimaging technologies (EEG, fMRI, and fNIRS). Then, we screened the articles by reviewing the full text and assessed the research quality in terms of research design and the reliability of the measurement tools. Selected studies were further analyzed regarding their methodologies, related brain regions/waves/signals, outcome variables, type of stimuli (independent variables), confounding variables, and pros and cons of the measurements if any.

The results showed that each of fMRI, fNIRS, and EEG has specific attributes that make them suitable for different areas of application. The main differences between these technologies were that first, each of them measures different variables that are related to neural activities. Therefore, the information that is provided by each of these technics can be inferred differently Second, both fNIRS and EEG monitor regional neural activities while fMRI monitors whole brain activities. Third, each of them is proper to measure neural activities in response to specific types of environmental stimuli due to their size and enforced pose that they cause. The conclusion provides recommendations for future research studies that combine neuroscience with human-environment research.

D24 - Transform Places with People and the Environment: Cities & Globalization / Residential Environments

Post Military Housing for Differently-Abled Military Veterans

Eric Weber (University of Nevada Las Vegas); Dak Kopec (University of Nevada Las Vegas)

Military life differs from civilian life. New military personnel undergoes 6-13 weeks of 'boot camp, which helps the enlistee learn how to function within a hierarchy — boot camp clusters men and women into barracks. According to Military.com, there is a lengthy and comprehensive list of potentially disqualifying health conditions that limit those who make it to boot camp. Disqualifying conditions ensures base-line abilities among all enlistees.

Four years of duty and experience converge, leavening many soldiers with a health condition, which is a factor for roughly 49,933 homeless veterans (Olenick, Flowers, and Diaz, 2015). To address this issue, we propose a reintegration camp that mirrors the 6-week boot camp model. The camp would provide diverse small housing stock designed to accommodate physical (amputations and spinal cord injuries), psychological (PTSD and TBI's), and social training (communication and behavioral responses).

This project looked at the development of a small home for temporary occupancy of a military person who has PTSD. A mixed-method approach was used to gather and analyze data. An assessment of the literature using reductionist techniques allowed data to inform design thinking measures. This technique identified common triggers, symptoms, and behavioral responses associated with PTSD, which were then used to develop a small temporary home with a high probability of ameliorating symptoms.

The fifth stage of the design thinking method is testing, which involves the testing of the small prototype home for its efficacy. If it has a high degree of effectiveness for people with PTSD, the next phase would be to replicate this process to develop a home for a person affected by a TBI, spinal cord injury, and amputations. The end goal would be a camp where discharged veterans can be helped to address a health condition and reintegrate into civilian life.

Evaluating Vacant Lot Reuse Programs: A before/after Visual Landscape Assessment Approach

Sara Hadavi (Kansas State university); Paul H Gobster (USDA Forest Service); Alessandro Rigolon (University of Illinois Urbana-Champaign); William Stewart (University of Illinois Urbana-Champaign)

Many cities in the US have been facing issues of urban vacancy. While high vacancy in most cases contributes to negative outcomes such as social segregation, lack of visual quality and safety, vacant land greening initiatives are a potential urban asset to alleviate many of these negative outcomes and improve neighborhood conditions. We have examined the impacts of one such initiative, the Large Lot Program in Chicago that transfers vacant lots to neighborhood residents by allowing them to purchase 1-2 city-owned vacant lots on their block for $1 each. The purpose of this study is to examine the impacts of this program through evaluating fine-scale landscape change over time.

We developed a condition-care index through use of aerial and street level imagery to perform visual landscape assessment. This index measures characteristics of land/tree cover and condition/care (Dewar et al., 2013) for lots purchased in two high vacancy areas, one year before and after purchase (N= 424 lots).

Our findings show 8% increase in lots with gardens one year after implementation of the program, and 16% reduction in lots with mature trees, accompanied by a similar increase in the proportion of mature trees rated in good condition. Our work presents an approach that informs planners about the type and degree of changes being made to lots, to evaluate progress for urban greening. This work adds a temporal dimension to the literature on urban vacancy and greening, which has tended to examine social or ecological patterns at a single point in time (Anderson and Minor, 2017). This work extends methods of visual landscape quality assessment by adapting approaches from allied fields of sociology and urban ecology (Li et al., 2015) and applying them to understand how landscape condition and care are visually expressed by the greening of vacant lots.

Nudging Pedestrians to Safer Street-Crossing : An Interdisciplinary Research Initiative for Transport for London

Lily Bernheimer (Space Works Consulting)

In line with Vision Zero, the Mayor of London has set a target of eliminating all deaths and serious injuries from London's transport system by 2041. Although progress is being made, pedestrian casualties in London remain too high. This session will present the results of an interdisciplinary research collaboration between an environmental psychology consultancy, transportation consultancy, and street design specialists to advise Transport for London (TfL, equivalent to Department of Transportation) on interventions to help reduce road casualties by enhancing pedestrian awareness.

This project seeks to fill gaps in understanding around the behavior and risk of people walking, and how this may have changed as a result of new technology such as smart phones. The interdisciplinary team undertook research into pedestrian behavioral factors and light touch street design responses that may help reduce road danger, beginning by identifying trends through an international literature review, including identifying responses trialled across the world, as well as analysis of TFL casualty data, and finally undertook video-based observational surveys at 15 intersections in London.

These findings were used to broaden understanding of how pedestrians respond to their environment, and develop measures TfL may use to help 'nudge' pedestrians into avoiding road danger. This project will complement Transport for London's wider road danger reduction approach, which seeks to reduce the sources of danger to people walking, such as high vehicle speeds and vehicle design.

D25 - Transform Places with People and the Environment: Education

Strategies for Green Building Education in the Green Science Museum: A Multiple Case Study of Eight U.S. Museums

Aysegul Akturk (University of Missouri); Laura Brianna Cole (University of Missouri); Georgia Lindsay (University of Colorado Boulder)

Public green building education (GBE) could enhance public support for green building practices and help building users to engage in sustainable practices within buildings. Few opportunities exist for GBE, however building themselves are increasingly being investigated for their potential teaching tools for environmental sustainability (e.g. USGBC, 2008; Schiller, 2012). While this potential has been explored in buildings like schools and libraries, science museums are also a promising venue.

There is an increasing trend to incorporate LEED standards when science museums renovate or build new buildings. This practice potentially supports GBE by using the built environment to communicate the story of the planet and values of environmental education as part of the institution's mission. Their green facility equips museums with a 3D environment to enhance sustainability education. To date, no literature examines the strategies used by science museums to leverage their facilities for GBE.

This multiple case study research examines how museums with LEED-certified buildings use their buildings for public GBE. Photographs systematically taken by researchers in eight green science museums across the U.S were analyzed for: 1) the interpretation strategies used by museums and 2) green building themes interpreted. Results shows that there was a spectrum of strategies from passive to active in order to communicate green building themes. The passive strategy of educational signage was the most common while several museums provided more active opportunities. The top five interpreted themes were recycling, water, energy, materials and eco-landscaping. After analyzing the data, the team created GBE profiles for each museum to explore higher-level strategies and choices made by museums. These cross-cutting themes were centralized versus decentralized displays, media types, visibility of GBE features, passive to active engagement, and knowledge to behavior learning outcomes. This study concludes by discussing the possible ways that GBE could be enhanced in green museums.

Partnering with Users: How Students Would Modify Campus Outdoor Spaces at the University of California, Irvine

Miryha Gould Runnerstrom (University of California Irvine); Nik Warren (University of California Irvine)

Recent studies show that the usability of outdoor campus spaces has a beneficial impact on the quality of life of college students worldwide (McFarland, Waliczek, & Zajicek, 2008; Mogra & Furlan, 2017). The primary objective of this study is to explore what changes students at the University of California, Irvine (UCI) would make to existing campus outdoor spaces to make these spaces more usable. An anonymous online survey of 143 UCI students ($\bar{x}$ age=21.9 years) was conducted to explore campus green space use and opportunities for psychological restoration. The survey included open-ended questions on what modifications students would like to see in outdoor environments and why they would make such changes. Open coding was conducted on participant responses to these open-ended questions to derive themes from categories and codes.

Barriers to outdoor space use and outdoor space enhancements are the two predominant themes that emerged during analysis. Examples of barriers to outdoor space use include the need for additional seating, shade, tables, power outlets, and WiFi. Outdoor space enhancement examples include the addition of green space, blue space, art, and the creation of poly-functional spaces for studying, sleeping, eating, and socializing. UCI is currently evaluating enhancements to natural spaces on campus with the aim of increasing access and use of these spaces. This study sheds light on the types of modifications student users would like to see in outdoor spaces on campus, including natural spaces. Integrating feedback from students early in the process helps to ensure that proposed enhancements are meaningful for this key user group.

Exploring Sensory Design Approaches to Improve Quality of Life for People with Disabilities

Lean Hatim Fathi (Texas Tech University)

In 2004, the Architectural Barriers Act (ABA) of 1968 was combined with Americans with Disabilities Act Accessibility Guidelines (ADAAG) of 1991 and reformatted into Americans with Disabilities Act (ADA). While ABA focused on accessibility, ADA concentrates on mobility. These design codes address barriers that overlap with standards regarding mobility for people with physical disabilities. Designers endeavor to minimize barriers by following standards such as the width of corridors for wheelchair dimensions; however, ABA and ADA do not focus on wellness. One solution is to augment these codes with a new standard by considering architectural sensory-oriented design strategies. Juhani Pallasmaa claims in The Eyes of the Skin (2005) that the qualities of life can be enhanced through the sense of touch, our primary sense. To test this statement, a research experiment was conducted in two parts. Firstly, an installation Sense-Ability (2018-2019) was constructed to enhance the sense of touch and by eliminating the sense of sight. The installation consisted of five by twelve feet fabric curtains with opposing pairs of sleeves sewn into them. This enabled people to embrace one another blindly. Participants' surveys concluded that by eliminating the sense of sight, the sense of touch was elevated. In Diane Ackerman's A Natural History of the Senses (1995), she claims that the sense of touch has been documented as a wellness factor. The second part addressed the touch sense as architectural design factor of healing. If we are convinced that the quality of life is an important design consideration for people with physical disabilities, then compliance with accessibility and mobility standards in the ABA and ADA codes are not enough. It is our responsibility to think critically about wellness of everyone and how our built environment is experienced with the body.

Interpreting Heatmaps with Coordinate System: A New Method to Analyze Visual Preference

Pai Liu (Clemson University)

Heatmap is becoming a popular method in research relating to visual preference in the field of built environment. However, the traditional analytical methods limit the possibility of interpreting the heatmap data. To create a heatmap question for visual preference survey, the picture is divided into several regions to represent different features. When the participants click on the picture, the points are recorded by the computer using a (X, Y) coordinate. The points were then visualized as a heatmap, typically red color indicating a high influential region while blue color indicating a low influential one. The researchers then interpret the participants' preferences of the pictures forming on the different colors and their radiuses (Duffey, 2014). In some conditions, the heat circles may locate on the boundary of two regions, indicating a vague of data interpretation. This study aims to develop a new method using Rhino and Grasshopper to analyze the (X, Y) values of the points. The researcher's previous study has collected visual preference data of outdoor walking spaces from seniors living in the United States (n=129), which is the database for this study. Before importing data into Rhino, the researcher gave a Z value to each point to demonstrate its region, for example, pathway was 1 and plant was 2. Instead of combining all the points on one picture, the points were layered and illustrated by their Z values. These layered points provided a clearer explanation of people's preference of outdoor spaces. For example, the traditional analysis showed that plants are a factor influencing seniors' preference of walking in the outdoor spaces. The new methods illustrated that the seniors prefer shrubs and small trees instead of lawn and big trees. This method is expected to provide a new perspective of visual preference analysis.

Restorative Places on College Campuses: Results of a Multiple Campus Participatory Mapping Study

Brian Schermer (Workshop Architects)

Restorative Places on College Campuses: Results of a Multiple Campus Participatory Mapping Study
University life is challenging because of the myriad academic and social pressures that students face. To succeed, students need places that they can go to rest, relax, and revitalize (Strange & Banning, 2001). The environment-behavior literature provides strong evidence that natural settings help restore attention and reduce stress (Kaplan & Kaplan, 1995; Herzog et al, 1997). This research seeks to understand how college students utilize physical settings on and near their campuses to help them to relax and restore. Key questions addressed in this research include: Where do students go when they are looking to relax and revitalize? Do they seek natural places? Do they look to places where they can be physically active, such as recreation centers and playing fields? Or, do they seek alone time in their residence hall or home? Moreover, how do different college campuses compare in terms of their capacity to offer restorative places to college students? This research reports on an online participatory mapping study of six different college campuses across the United States. These campuses differ greatly in terms of the amount of nature that is available to students. The results suggest that when natural settings are immediately accessible, students readily identify them as restorative. On such campuses, students can identify about twice as many restorative places as students at campuses which lack access to natural areas. The survey, which was conducted as part of the planning and design process for various student life facilities at different universities, provides evidence that campuses lacking in access to nature are likely to benefit from investment in facilities such as recreation centers and wellness centers. This type of analysis may bolster campus support for building new facilities that enhance restoration.

D26 - Transform Places with People and the Environment: Movement in Designed Environments: Nature & Ecology

Bridging the Environmental Divide: Bringing Environment and Religion Together

Samantha Samples (Arizona State University)

While Lynn White's 1967 article, The Historical Roots of Our Ecologic Crisis, ignited a firestorm of controversy regarding the relationship between religion, particularly Christianity, and the environment, the testing of White's hypothesis that Christians are anthropocentric and anti-environment has produced results that are significantly less clear. Some researchers argue that Christians are less concerned with the environment than non-Christians(Clements, McCright, and Xiao, 2013), while others note that those attending church more regularly are more concerned about climate change (Kilburn, 2014). In addition to the mixed findings, little research has been conducted with experts in the field of Christianity, the clergy, deepening the already significant gaps in the literature. A research study was conducted to help fill that gap by providing the perspective of clergy on the relationship between Christianity and the environment. This qualitative study used in-depth interviews to assess the topics of importance to members of the clergy within the Phoenix, Arizona metropolitan region as well as their professional perspective of the relationship between religion and the environment. The results indicate that community is of great importance to leaders of faith-based organizations, but stewardship is also a primary foundation of the church. While no support was found for White's hypothesis, a willingness to learn and a need for expertise on environmental issues was identified. In this regard, environmental professionals, specifically landscape architects, have been identified as the ideal group to provide the bridge between faith-based organizations and the environment. This presentation will present the background literature, provide a brief overview of the findings, and focus on the actions of one faith-based organization as an example of the potential for change in communities when environmental professionals and faith-based organizations collaborate.

The Role of Design in Psychiatric Care Spaces within the Emergency Department

Hannah Shultz (HKS inc.)

It is estimated that in the United States 1 in 8 visits to the emergency department are for a mental health or substance abuse issue (Weiss, Barrett, Heslin, & Stocks, 2016). The complex history of care for mental and behavioral health (MBH) has systematically influenced the opportunities for crisis care. Emergency departments often act as the front door for many experiencing a MBH crisis. This initial visit has the potential to impact how that individual feels about the healthcare system and their own ability to get better. For this reason, it is a vital step in our nation's mental and behavioral health treatment plan to provide adequate and human-centered care for MBH patients presenting to the emergency department. This study examines the role of emergency psychiatric care spaces (e.g., crisis stabilization units, holding units) and their impact on the health and well-being of individuals presenting in a MBH crisis. Specifically, this study utilizes a review of the literature, industry expert interviews, case studies, and non-participant observations to examine the question: What is the role of design in psychiatric care spaces in the ED on facilitating patient health? The outcome of the study delves into the field of mental and behavioral health in crisis care environments in order to provide insight for future research opportunities, and how we as designers can challenge our own understanding about the environment for someone in a MBH crisis.

Comfortable Parks

Anastasiya Volkova (University of Maryland); Madlen Simon (University of Maryland)

How do parks successfully provide comfort to city dwellers? This research summarizes a literature review conducted on the theory of nature, park typologies, and comfort in public urban spaces (Thompson). Concepts discovered in the literature review are applied to the analysis of three case studies based upon direct observation of three urban parks in different cultural contexts: Yusupov Gardens (St. Petersburg, Russia), Gorky Park (Moscow, Russia), and Bartholdi Park (Washington DC, USA). Field observations are made for each of the parks condensing into a graphic analysis of threshold, circulation, nodes, vegetation, water, characters, and seating in each third place. The participant observation took place over the course of twenty-one visits with a collective study of approximately 3,500 individuals. The differences gathered from the graphic analysis categorizes the three green spaces into four distinct park typologies - Pleasure Ground Park, Reform Park, Spectacle Park, Sustainable Park- that cater to different social goals, activities, and elements (Cranz). Each case study concludes with an overall analysis of the field observations and a critique of successful elements that lead to comfort in that urban park. The paper offers a focused strategy for understanding how a green space can be designed to serve a given area and audience with comfort: diverse seating, opportunity for people watching, and catalysts for social interaction (The Social Life of Small Urban Spaces). A better understanding of the relationship between comfort and designed green space can better inform architects, landscape architects, urban designers, and planners on how to improve the public realm at the scale of a city park.

Low-Tech Tree Protection Strategies in Ladakh, North India

Carey Clouse (University of Massachusetts Amherst); Mark Hamin (University of Massachusetts Amherst)

Throughout time and across cultures, trees have been protected by humans in remarkably clever and cunning ways. From Julia Butterfly Hill's 738-day canopy squat to the Earth First!ers who popularized tree spiking, and from contractors who hang price tags from branches at construction sites to the Buddhist monks who ordain trees in Thailand, individuals have publicly interceded on behalf of arboreal lives.1 These examples of socially embedded transformation reveal not only the passionate intensity with which people commit to protecting trees, but also the creative and wide-ranging approaches that can be availed in the defense of nature.

In Ladakh, north India, trees are vital to the development of agricultural land, to the production of material for vernacular buildings, and to the formation of social and cultural spaces in villages. Yet, in this high-altitude, arctic desert region, trees are incredibly difficult to usher into maturity. Open grazing practices and the challenging environmental conditions of the region limit plant growth and necessitate specific strategies for supporting tree planting efforts.

This paper addresses some of the creative design ideas that have been used in Ladakh, north India, to protect trees, over time. These strategies are presented as a taxonomy of different design approaches, with images to help to illustrate the design thinking that underscores these practices. Trees in Ladakh provide valuable ecosystem services and important environmental management opportunities in the context of climate change. Moreover, as the environmental conditions in Ladakh shift with changing weather patterns, land use, and demographic trends, human-sponsored systems for protection can be leveraged in defense of trees.

Designing in the Amazon: An Immersive Student Experience Understanding and Improving the Field Researcher's Environment As Part of a Scientific Expedition

Laura Claire Kimball (Radford University); Lillie Price (Radford Univeristy)

The Radford Amazonian Research Expedition (RARE) at the Las Piedras Peru Biodiversity Station offers undergraduate students an opportunity to explore primary research in the Amazon rainforest. This immersion experience integrates the physical environment, learning and research practices, and social and cultural interactions. In involving non-science students, how does an interior design undergraduate participate in a scientific research expedition? Researchers primarily focus on the body of their work with little consideration of the function and design of the physical environment in which they are working. The interior design student proposal studied the built environment, researching the researcher's functionality in the rustic biodiversity station, applying a facility improvement project, and facility documentation for future improvement. While researchers are able to perform their tasks within the remote location, limited functionality inhibits research efficiency and social interactions. Improved function and aesthetics of the physical and social workplace environment have a strong impact on worker satisfaction, encourages creativity, and increases productivity for routine tasks (Kopec, 2006). Station improvement has significant impact on the physical and social uses and users. Sustaining Amazonian context and impact were vital considerations. In order to assess the functional needs and use, qualitative data from current and previous participants was gathered through questionnaire and interviews. Data informed the function and purpose of the space from the perspective of faculty, students, and caretakers. Both remote and onsite evaluations were conducted with analysis, site diagrams, and environmental observations. Additional research included information gathering and project development for the hands-on facility improvement project which benefit resident and researchers. The applied improvements included the construction of a clay oven for cooking, collecting specific Amazonian clay and employing appropriate rustic construction methods. The use of the finished project benefited the ease of food preparation, researchers and future researchers with improved safe meal options, and environmental sustainability.

E2 - Transform Processes and Outcomes for Equity: Health

Toward Culturally Enriched Communities - Lessons from Copenhagen

Tasoulla Hadjiyanni (University of Minnesota)

Copenhagen is a model for how healthy cities might be created across the world—it consistently sits at the very top of the United Nation's happiness index and is one of the star performers in the Healthy Cities initiative of the World Health Organization. Yet, Copenhagen also has areas that experience inequality and marginalization—in the diverse poorer district of Nørrebro for example, life expectancy drops by seven years. This complicated picture closely resembles that of Minnesota's Greater Minneapolis/St Paul region—although one of the healthiest states in the country, Minnesota residents of the highest income areas have been found to have an average life expectancy of 82 years, while residents of the lowest income areas have an average life expectancy of 74 years, and these areas typically have high concentrations of people of color (Helmstetter, Brower, & Egbert, 2010).

Through interviews with Danish architects, planners, and design educators, along with field observations in Copenhagen, this study identified best practices and design interventions that can support the creation of Culturally Enriched Communities (CEC), healthy and connected communities in which everyone can thrive. Informing CEC's eight principles are design interventions that emphasize "relationship building" and include: windows that maximize daylighting and views; parks that incorporate activities for all ages and abilities; urban spaces that bring together an addiction clinic and an upscale restaurant or an elementary school and elders' housing; inner courtyards where residents can get to know the neighbors; community gathering spots where anyone can come to share meals and yoga classes; and the humbleness of janteloven. CEC's aim is to expand the tools available to policy makers preparing for a successful and vibrant 2040, when projections show that over 40% of Minnesota's population will be people of color, those experiencing health, income, and educational disparities the most.

Recommendations for Evaluation Metrics for Asthma Home Visiting Programs: Measuring Environmental Management & Health Outcomes

Brendan Brown (Green & Healthy Homes Initiative); Ruth Ann Norton (Green & Healthy Homes Initiative); Kiersten Sweeney (Green & Healthy Homes Initiative); Elsie Andreyev (Green & Healthy Homes Initiative)

Comprehensive Asthma Interventions include environmental health education, assessment, and remediation of asthma triggers in the home which are necessary to remediate environmental factors that contribute to health inequities. Research evidence concludes that home-based, multi-trigger, multicomponent interventions with an environmental component are effective in reducing acute care visits, lowering costs, and improving overall quality of life and productivity in children and adolescents with asthma. Ensuring health equity requires that policymakers and payers "incentivize the reduction of disparities and the promotion of health equity by building health equity measures into new and existing healthcare payment models."

The Environmental Management and Health Outcomes Metrics for Evaluation (EMHOME) Work Group convened a representative group of national stakeholders to offer their input on which measures to include for recommendation. The goal was to produce a standard set of measures that providers, health systems, managed care organizations, and public health departments can utilize to link health outcomes to program intervention measures for the purpose of obtaining sustainable reimbursement for services. This final report of the Work Group presents a set of twenty-eight measures all of which are integral to the evaluation of an asthma home visiting program both in terms of efficacy and cost-effectiveness.

Building Design for Users with Disabilities: Building Performance Evaluation of Ed Roberts Campus

Kyuho Ahn (University of Oregon); Linda Zimmer (University of Oregon); Olivia M. Asuncion (Shah Kawasaki Architects)

All public buildings are required to comply with the American Disability Acts Accessibility Guidelines (US Department of Justice, 2010). This baseline is meant to insure equal accessibility. However, many users with disabilities still feel that public buildings lack support for independent use (Sherman and Sherman, 2012). Ed Roberts Campus (ERC), a civic building that serves individuals with disabilities, is designed to foster independent living by adapting universal design strategies. This project presents the results of a post occupancy evaluation (POE) that examined ERC, how it is perceived relative to public buildings and how people with varied disabilities use ERC features.

An online survey instrument was designed based on Presser's (1983) building performance (BP) criteria: (1) Health, safety and security; (2) Functional efficiency; and (3) Psycho-sociological aspects. Sixty-two responses were identified as valid for analysis. The survey measured perceived qualities of general public BP as a benchmark to compare ERC BP. Survey data were also analyzed based on disability profiles (mobility, vision, hearing, cognitive, and olfactory) for comparison of perceived BP based on the profiles.

The results show enhanced BP of ERC in all three areas ($p<0.001$). Psycho-sociological aspects were regarded most positively by users relative to public buildings. Perceived qualities where ERC was regarded most favorably compared to public buildings include barrier-free, friendly environment, aesthetically pleasing condition, good meeting places, and quick emergency evacuation. Perceived quality of safe place, wayfinding, and efficiency while still positive, showed a smaller deviation between public buildings and the ERC. Survey responses on wayfinding qualities showed a slight positive difference in the ERC compared to general public buildings and the POE shed light on how people with different disability profiles used varied wayfinding cues including signage, sound, floor texture and lighting. Detailed findings for the disability profiles and design implications will be discussed further.

Outdoor Sports Area Design for People Affected By Cerebral Palsy

Gökhan Keskin (Yaşar University)

Cerebral palsy (CP) is a disorder that can affect a person's ability to take part in daily activities. It is argued that the creation of a facility for people with disabilities could help reduce the obstacles that they face and enable them to improve their quality of life. This study aims at developing outdoor sports areas for disabled people affected by cerebral palsy (PCP) to enhance their social life and make them part of the community.

The subject of this study was selected because of the importance of sports for PCP. With the help of the sports and the sense of belonging to a team or a group, the feeling of being different can be overcome (McConkey, et al. 2013). In this study, types of impairments were analyzed according to WHO and IPC to clarify the place of CP in disability. Furthermore, disability in sport was examined to define the benefits of sports on PCP.

Cerebral Palsy Portuguese Association ("APPC," n.d.) believes that a mixed type of community that includes children, teenagers, and adults with and without CP is helpful for PCP to overcome alienation from community. A case study, Villa Urbana Gondomar (VUG), in Portugal, was analyzed. The facility hosts 34 guests affected by CP. It also includes kindergarten, after-school education, atelier and indoor sports area for people with and without CP. The structure of VUG community was an advantage to observe and understand sports area design for all needs. Quality of life surveys were employed to collect data on PCP daily routine and preferences. In addition to surveys, interviews and observations were also used to understand the PCP needs in order to overcome their obstacles, improve their quality of life, and support social integration. The collected data was used to propose a design solution for outdoor sports facilities.

E4 - Transform Processes and Outcomes for Equity / Transform Places with People and the Environment

Enhancing the Engagement of the Community in Contemporary Art through the Development of Public Art Spaces

Silvia Rolla (İzmir University of Economics)

This research aims to investigate the relationship of art spaces as public spaces of different scales and to analyze their impact on the community in terms of engagement and interest in art. Audience engagement means not only increasing the audience size but also improving the quality of the experience and the services offered to the existing audiences (Black, 2005). For this reason, art spaces need to be intellectually and physically accessible to all people (Vergo, 1989) to become a "contact zone" where communities meet and build culture (Clifford, 1999).

The research is developed focusing the attention on contemporary art since a wide appreciation of this type of art is often missing, so it still struggles in reaching a wide audience. By the correct design approach, architecture at different scales might be able to connect the community with contemporary art. For this purpose, supporting the theoretical approach, the study identifies useful tools for the design of such spaces.

The methodology used to conduct this research consists of three parts. Firstly, a case study from Mexico is selected because of its urban context and architectural characteristics, which are designed to establish a strong dialogue with the public. Those peculiarities are analyzed on different architectural scales. Secondly, observations and structured interviews are conducted among visitors or potential visitors of Izmir, Turkey to understand how the audience perceives architectural barriers or stimulus in attending contemporary art events. Individuals' activities and tastes are also taken into account. Finally, because of the similarities with the context of the case study and the potential of Izmir Kulturpark, a proposal for Izmir is offered based on the information acquired.

This research concludes by pointing out insights for the design of contemporary art spaces to enhance community engagement at different architectural scales.

Understanding Rural Tourism Experiences from Coded Social Media Data: Evidences for Three Examples in USA

Bo Zhang (Oklahoma State University); Fanghua Chen (Beijing Jiaotong University); Qiannan Lyu (Beijing Jiaotong University); Zhongzhong Zeng (Beijing Jiaotong University)

Recent decades have produced a boom in social media that enables people to share information and opinions by publicly posting textual (words, tags, and comments) or non-textual (pictures, videos or audio) data online. the environment and society (Hansen et al., 2012). The huge, diverse, and ever-growing data source sheds light on many built environment issues that used to be infeasible for researchers to explore. Currently, there have been a limited number of empirical researches on rural tourism projects (Marcouiller,1997, Forstner, 2004). The research methods that are practiced in the populated urban spaces, such as observation, survey, and interview are not readily applicable in rural environments, which are always remote from city centers and count relatively low visits.

This study seeks to take advantage of the merits of social media data in understanding the usage and experiences of three rural tourism projects. Three projects with distinct characters are selected: Kentucky Horse Park, Woolaroc Museum & Wildlife Preserve, and Shaker Village of Pleasant Hill. 2710, 348, and 952 Googleplace posts and 2563, 563, ,655 Tripadvisor posts are mined for analysis. Stories, facts, opinions, emotions, and moments are communicated and recorded, along with geo-locations, timestamp, and account information to the public. Photos and texts from the posts are coded into three themes respectively, including activities, programs/ objects/scenes, and experiences/facial expressions. The study conducted four separate analyses, such as: (1) popularity of different programs; (2) human activities; (3) human facial expressions in different areas; and (4) human experiences and emotions in different areas. By systematically concealing the dynamics of usage and perception of rural tourism programs, planning strategies and design guidelines are updated to inform future practice.

Empathetic Design: How Elementary School Environments Designed to Reduce Stress Can Foster Inclusion of High Functioning Autistic Children

Robin Z. Puttock (The Catholic University of America)

All people experience stress in their environments. The specific causes of stress vary from person to person as does one's ability to cope with each stressor. These variations are magnified in people who have Autism Spectrum Disorder (ASD) as they exhibit deficits in communication, social skills, flexibility and sensory integration (American Psychiatric Association, 2013). This presentation introduces the author's Theory of Empathetic Design, a design strategy that fosters the inclusion of high functioning autistic children in mainstream educational environments, possible only when empathy for the child with ASD is achieved by the designer. Achievement of empathy is a particular challenge because children with ASD often exhibit communication deficits and cannot convey their environmental stressors which cause behavioral outbursts and eventual exclusion from the mainstream classrooms.

This presentation will demonstrate that achievement of this empathy is possible through the analysis of sketches drawn by children with ASD, gathered via open-ended questionnaires given to their parents. The children with ASD were asked to draw how they perceive their current educational environments during the six parts of a typical school day: approach, arrival, content delivery, transition, recreation and departure, as well as what their ideal educational environments could be. Analysis of these sketches reveals the most common mental, sensory and social stressors for children with ASD in elementary educational environments as well as specific spatial strategies that can prevent, reduce and/or relieve stress by providing individual control, encouraging physical activity and introducing calming surroundings.
This presentation, highlighted by these insightful and unique sketches, and supported by current learning theory and educational inclusion research (Mostafa, 2014), is meant to enlighten a new generation of Empathetic Designers and give children with ASD a voice in the design decision making process of their inclusive educational environments.

Participatory Design Processes in the Education Environments Lab

Nathan Bicak (University of Nebraska Lincoln); Steven Hardy (University of Nebraska Lincoln)

The Education Environments Lab is a fourth-year, collaborative design studio, comprised of multidisciplinary teams of architecture, interior design and landscape architecture student-researchers. The studio is co-taught by professors from different disciplines and focuses on outreach and client engagement. In fall of 2019 the studio worked with the Ruth Staples Child Development Lab to develop programming concepts, design vocabularies, and schematic proposals for a new facility. Throughout the research phase, the studio employed a mixed-methods, participatory approach to understanding client needs and preferences (Walliman, 2016). The student-researchers divided into groups looking at missions and values, quantitative and qualitative benchmarks, and space needs using specific techniques like observation, interviews, blocking and gaming. Student-researchers interviewed and surveyed teachers, parents, and practicum students regarding the perceived successes and challenges of their current facility. They conducted observation and mapping exercises to understand daily routines and designed hands-on gaming exercises from children's drawings to treasure hunts (Derr, 2018) for sets of qualities within the existing facility. These activities engaged teachers and children in envisioning their future facility.

The diverse user group revealed dynamic needs and desires, which were catalogued in a comprehensive programming report including problem statements, program area visualizations, benchmark imagery and ritual mapping. The student-researchers continued working with the clients to develop vision studies and design options for the new facility. The programming document was organized to communicate the needs of the client-community to a variety of audiences, including the Ruth Staples director and faculty, university administration, facilities planning officials, and potential donors. This studio simulated social science research methods to reveal client attitudes related to site and building spaces, organization, materials, and outdoor learning environments. These methods gave stakeholders – young children, educators, and researchers – a strong voice in their design process, ensuring equity throughout the project's information gathering phase.

Design Empathy and Civic Practice of Critical Placemaking: A Case for Women-Only Fitness Centers ('Gendered Design' Thinking & Influences on Community and Public Urban Space) Gender, Culture and Design, Gender Disparities in Use of Space

Taraneh Darvish (Arizona State University); Chingwen Cheng (Arizona State University)

Public space design is not one-size-fits-all. Women and minority groups have been overlooked in current design thinking and practice to reflect diversity and inclusion in public spaces (Chuko and Chen, 2016, Geddes & SOKOLOWSK, 2019). Due to the diversity in cultural differences and religions reflected in regulations, women-only spaces require a novel approach and mindset to be sensitive and tailored to each special group such as Islamic cultures. Participatory design has been identified as an effective strategy to achieve procedural justice through (Hagan, 2019) engaging users in empathy design (Dam & Siang, 2018) to examine the relationship between space and target users. This paper answers the questions 1) how gender plays an edifying role in shaping a designed space? and 2) how does the user's input help shape the design? This paper aims to promote social equity in public spaces by filling the gap of design thinking and practice analyzing case studies of two women-only fitness centers, Jabz Boxing and Curve in Phoenix, Arizona. The proposed framework seeks to theorize the processes by which how emphatic and civic practice design strategy may contribute to transforming community narratives and building more inclusive, participatory, and empowered communities. Qualitative research method including questionnaire surveys, interviews, and field notes were employed. The study observes varied women groups identified by class, age and religion categorized by each group's preferences for locations and activities. The preliminary results demonstrate a range of women's experiences, expectations, memories, and activities varied in particular by age group and culture differences. This research contributes to the entrenchment of social, economic, global health and racial inequality and as well advances using 'design empathy and civic engagement and practice' as tools for critical placemaking studies to promote awareness, empowerment and improvement toward the usage of place or its creation.

E5 - Transform Processes and Outcomes for Equity: Cities & Globalization / Participation

Connect before Collaborate: Learning from Hawaiians for Decolonizing Design

Chingwen Cheng (Arizona State University)

Conventional public participatory process has been using tokenism methods to inform and consult with the public. Such practices are derived from the model of colonization when linear model of decision-making, top-down control, and expert-driven processes are dominant. As a result of lacking meaningful participation in the design process, distrust and disapproval from the community occurs. Decolonizing design provides a critical framework in analyzing the inequity of mainstream participatory practices. This paper draws from lessons learned in the process of redesigning a neighborhood park and reimagining Hawaiian watershed management systems in Honolulu from two landscape architecture design studios. Key concepts of decolonizing design learned from Hawai'i experience are derived from meeting notes and informal conversations with stakeholders and workshop participants from local communities. Major findings include recognition of different process should be anticipated when working with different culture context. Connecting with community before talking about collaboration is a critical first step when working with Hawaiian communities. The principles of decolonizing design learned from the Hawaiian experiences provide a critical framework in working with every other community around the world in order to establish healthy and sound relationship with the community and start to build trust. When deepening the connection with community, designers can generate genuine appreciation of listening and respect to leverage local assets and design with the community, not for the community.

Experiences with (resisting) Green Gentrification: Evidence from Chicago

Tania Schusler (Loyola University Chicago); Amy Krings (Loyola University Chicago); Richard (Max) Melstrom (Loyola University Chicago); Colette Copic (Loyola University Chicago)

Environmental contamination and limited access to green spaces disproportionately burden residents of low-income neighborhoods and communities of color. These environmental injustices can negatively affect residents' health. In response, residents in impacted communities and their allies have come together to improve environmental quality and promote environmental justice. Yet, cleaning up contamination and creating green spaces to improve health equity has, in some cases, been correlated with displacing the very residents intended to benefit, as the neighborhood becomes desirable to more affluent populations. This has been referred to as "environmental," "ecological," or "green gentrification." Gentrification can bring benefits like improved safety but also can push out vulnerable community members, disrupt lives, and degrade a community's cultural fabric. Residents in neighborhoods experiencing environmental injustice face a paradox (Checker, 2011): Must they reject environmental improvements in order to resist the gentrification that may follow? Or do ways exist to support environmental cleanup, green space creation, and the concurrent health benefits without displacing existing residents? Focusing on the city of Chicago, this study employed mixed methods to: (1) examine quantitatively the extent to which green gentrification is occurring, and (2) explore qualitatively how stakeholders perceive the risk of green gentrification as well solutions to advance environmental improvements without displacing existing residents. We present preliminary findings of in-depth, semi-structured interviews conducted with practitioners who possess deep knowledge and experience in one of five areas: environmental justice, public health, housing, urban planning, or community development. We highlight how concerns related to environmental quality, public health, housing affordability, economic opportunity, and culture intersect in land use decisions in Chicago. We also report stakeholders' perceptions of policy, planning, and community organizing interventions that hold the potential for achieving sustainable, equitable urban development.

Designing Democracy: How Taiwan's NGOs Advanced the Forced Relocation Issue with Citizen Participation Process

YuWen Liao (National Sun Yat-sen University)

Since the development of heavy industry in Kaohsiung City in the 1960s, Dalinpu, once a fishing village, has been surrounded by factories. The residents had initiated several protests toward the over-development and the environmental pollution, but all ended in frustration. Although relocation has been discussed for over 25 years, the government had not proposed a concrete plan yet. Because of the depletion of trust and the lack of attention, the movement was stagnant. In 2018, local NGOs launched a citizen involvement project and adopted a deliberative democracy model, the "Scenario Workshop", trying to convergence consensus and propose a citizen version relocation plan. Eventually, in October 2019, the Executive Yuan officially approved the budget for the relocation.

With the framework of deliberative democracy theory, this paper will report an explorative case study of the "Dalinpu Scenario Workshop" to see the impact of the citizen involvement process on the social movement and civil society. This research will adopt qualitative research methods. Through interviews with organizers and practitioners, this research will seek to understand the reasons and challenges of adopting a deliberative democracy model. Also, it will analyze relevant media documents and the meeting minutes to see its influence on residents and the political society. The study is expected to yield the following results. First, as a third-party, the NGOs conducted the participation process impartially, which helped to build the trust in the residents. Second, the participation process established a platform for mutual understanding between stakeholders, including government officials. Third, the citizen participation process empowered the residents. Experts are invited to provide information and technical advice. With the assistance of facilitators, the residents developed a more comprehensive discourse toward relocation. As a result, the citizen involvement process brought attention from civil society and pressured the government to respond.

E6 - Transform Processes and Outcomes for Equity: POE/ Programming

Learning from the Past: Social Justice and Empowerment, the Fun Palace Project

Lubomir Popov (Bowling Green State University); Franklin Goza (University of Wisconsin Whitewater)

Design professionals are fascinated by the possibilities to contribute with architectural means to social justice and equity. Our study is designed to bring the Fun Palace Project to the attention of contemporary architects and design researchers. The project has been in development in the early 1960s in London by architect Cedric Price and theater director Joan Littlewood. There is much to learn from the Fun Palace project in areas such as social innovation, empowerment of underprivileged social groups, and interdisciplinary collaborative practices. Furthermore, recent conversations with influential architects who work on environmental justice issues revealed their unfamiliarity with this landmark project. Filling this knowledge void is another motivation to develop this study. We elaborate the project's progressive way of empowering disadvantaged social groups through both education and professional training. By so doing, we hope to raise current standards of engagement with disadvantaged populations and showcase the interdisciplinary organization of the programming process. This paper is theoretical. Our methodology involves historical analysis and a sociological approach to interpreting the project's programming actions. We also discuss the social foundations of the project, the social factors that directed the process, and the interdisciplinary collaborations of professionals, politicians, scientists and famous intellectuals. We discuss how the inception of a single building by a socially compassionate client led to a much larger social project that focused on social disparities and the ways to overcome them. The presentation will introduce the audience to ways of using the process of facility development for providing disadvantaged populations with resources to change their lives through education and meaningful leisure in an encouraging environment.

Facilities Programming As an Instrument for Environmental Justice

Lubomir Popov (Bowling Green State University); Franklin Goza (University of Wisconsin Whitewater)

The growing disparity in income and power is reflected in the facility delivery process as well as the final product, the building. Social groups with more power, resources, advisors, connections, and knowledge of social processes are more likely to participate in the building delivery process and to sway the design of built environment in their favor. On the other hand, disadvantaged and underprivileged social groups are practically excluded from the building delivery process. This leads to environmental inequality and further marginalization of these people. Our goal is to empower beleaguered social groups to obtain environmental resources and support for their lives and families. Our objective is to provide opportunities for the disadvantaged to provide input into the process of creating built environment and to successfully defend their environmental needs and requirements.

This project is theoretical. Our methodology is based on the sociology of social problems, activity theory, and citizen participation theory. We conceptualize the project delivery process as a system of activities. We analyze this system and search for points where underprivileged social groups can participate, successfully provide input, and defend their environmental requirements. We analyze the project delivery process from inception to building delivery. Our analysis indicates several points where citizen participation is very important for defending the interests of socially disadvantaged user groups and stakeholders. We delimit this study to facility programming as a major stage where user and stakeholder requirements are developed, negotiated, and balanced in a fair and just way. Our presentation will inform both stakeholders and design professionals about the practical potential of programming for motivating disadvantaged social groups to participate in facility development discussions, to explicate their needs and requirements, to successfully negotiate them with other social groups and design professionals, and to obtain a fair share of built environment.

Post-Occupancy Evaluations for Permanent Supportive Housing

Corey Schnobrich (Leddy Maytum Stacy Architects)

Permanent supportive housing, which combines stable shelter with social services, is the most successful model for the health and housing stability of the formerly homeless (Fitzpatrick-Lewis, 2011, p. 10). Designing and programming for this type of housing faces unique challenges. Future residents may be difficult to reliably reach or may not even be homeless during the programming phase. They may also be wary of participating in research; social workers described their clients as an "over-surveyed but under-focused group" (Nolan, 2018, p. 16). In addition, non-resident users, such as building management and social service staff, often offer different opinions on ideal design layout, amenities, and operations.

The presentation will discuss a means of navigating these challenges through post-occupancy evaluations (POEs) at permanent supportive housing facilities in northern California. These POEs used multiple methods for different user groups, including resident focus groups, resident energy use analysis, social service staff interviews, and building management walk-and-talks. The POEs culminated in a report issued by the social service provider addressing four of their buildings: two purpose-built, ground-up projects and two rehabilitations.

This report documents numerous findings for spaces ranging from the main entry to trash rooms and makes recommendations for future building design. The presentation will conclude with a discussion of how these results informed subsequent designs for permanent supportive housing and analyze how research methods could improve the experiences of residents, staff, and management.

E7 - Transform Processes and Outcomes for Equity

City agencies to support creative placemaking in the public realm and lighting specifications that reinforce a healthy and equitable use of that space by pedestrians, cyclists, transit users, and city residents.

Utilizing the Gehl Institute "Inclusive Healthy Places" Framework: A Research Protocol to Investigate "What Physical and Social Elements Are Most Likely to Positively Affect People's Perceptions of the Local Environment?" in Baltimore.

Jennifer Goold (Neighborhood Design Center); Briony Hynson (Neighborhood Design Center)

The Neighborhood Design Center (NDC) is undertaking, or is partnering on, a series of public space interventions in the Station North Arts district in Baltimore, MD. Interventions include lighting and programming funded via the National Endowment of the Arts and transportation improvements that will be undertaken by the Baltimore City Department of Transportation and the Maryland Transit Authority. We will also partner with MICA's Social Design department for pilot projects in the transportation space.

The Neighborhood Design Center will document the impact of these interventions in a pre- and post-assessment of the spaces utilizing the Gehl Institute "Inclusive Healthy Places" Framework and the Gehl Public Life Data Protocol to determine "What Physical and Social Elements Are Most Likely to Positively Affect People's Perceptions of the Local Environment, including safety?" in Baltimore.

This presentation will focus on the development of a customized protocol to answer our research question and the pre-assessment findings. In addition to the framework and the protocol specified above, NDC will utilize "The Built Environment An Assessment Tool and Manual" developed by the Centers for Disease Control and Prevention for evaluation of the transportation spaces. The goal of this work is to influence Baltimore

Engaging Local Residents in Sustainable Placemaking: Findings from Case Studies of Cbpp in the Singapore Public Housing Estate and Ecotourism in the Chinese Historic Village

Jessica Diehl (National University of Singapore); Yuting Su (AECOM); Xuanyu Qian (Ramboll Studio Dreiseitl)

Evidence suggests that community attachment and community-based participatory planning (CBPP) are positively related. Aligned with this, many scholars argue that community-led ecotourism development can install greater place attachment of local residents and also meet the place satisfaction of tourists. This paper compares two research projects that investigated the relationship between local residents' level of participation and place attachment: (1) CBPP in the Singapore public housing estate and (2) ecotourism in the Chinese historic village.

As Singapore shifts from a top-down to community-engaged planning approach, there is the question of whether community attachment can be strengthened. The aim of the first project was to investigate CBPP and community attachment through a comparison of two similar housing estates, one recently involved in a CBPP project and the other with only government-led planning. Semi-structured surveys were conducted with 156 residents. Residents who participated in a CBPP project reported a significantly higher level of community attachment as compared to those involved in government-led planning. This was true even when controlling for residents living in the same estate but who did not participate in the CBPP project.

The second research project investigated historical village tourism in China, considered an effective means to promote the protection of cultural heritage and development of rural communities. Two historical towns near Tai Lake were selected with different approaches: one community-based ecotourism development (CBED) approach, the other government-led. A semi-structured questionnaire measured local residents' place attachment (n=140) and tourists' place satisfaction (n=60). The CBED historical town had significantly greater place attachment of local residents and greater place satisfaction of tourists as compared to the government-led historical town.

Comparing the findings from both studies suggest that improving community participation could have a positive effect on community attachment and ultimately improve the sustainability of planning and development.

Assessing Residents' Satisfaction with Neighborhood Environments in Revitalized Urban Neighborhoods

Eun Young Kim (University of Tennessee Chattanooga)

The city of Chattanooga has been through urban revitalization aimed at re-creating its inner city as a vibrant center for residents. Although the city's initial redevelopment planning was inclusive, the outcomes seem to have a discrepancy in the distribution of renewal benefits among different neighborhoods. Housing and community environments are experienced not only physically, but also socially and psychologically. In particular, residents' perception of their environment influences their quality of life because the neighborhood environment affects residents' perception on overall living conditions. A number of research studies suggest that residents' satisfaction with their neighborhoods has been positively associated with a strong sense of community. Safety and social problems also can significantly affect residents' neighborhood satisfaction and well-being than the physical conditions of housing units. Therefore, consideration of residents' perception of their housing environment is crucial in planning and assessing future redevelopment in Chattanooga.

The goal of this study was to examine the residents' satisfactions with their community environments in order to identify any discrepancies among different neighborhoods and to investigate influential factors that have potentially contributed to residents' perceptions. The residents' satisfaction survey was conducted from spring 2018 to summer 2019. A total of 413 residents of the city of Chattanooga completed the survey. The survey collected data on the degree of satisfaction of community environments by examining perceptions on accessibility to grocery, public transportation, public schools, hospitals, parks, cleanliness, sense of safety, neighbors' friendliness, and police protection. The results suggested that residents were relatively satisfied with their community environments. This study also conducted focus group interview with six residents. The focus group interviews results sued in validating survey results. The presentation will compare satisfaction level among different neighborhoods.

E8 - Transform Processes and Outcomes for Equity: Cultural Aspects

The People's Architects

A. Sameh El Kharbawy (California State University)

Although architecture's social function and power are universally and historically recognized, a social project for architecture remains elusive today, both in the profession and in the academy. Ideological platitudes, elitist banalities, and a thoughtless mob of sloganeers continue to occupy architectural discourse, leaving the mainstream of its community indifferent, uninspired, even apathetic.

Against an architectural canon long untethered from social responsibility, long confined to sanctioned ideas and expressions, stand those who believe that architecture is a human activity, an inherently social praxis, in which a multiplicity of peoples, histories and experiences is (must be) present and manifested. Further complicating this tension is the fact that architecture has expanded so much in recent years, shedding many of its familiar disciplinary skins. With this came a rethinking of architecture's definition and responsibility; a long-overdue attention to the question of whether architects and designers can (and should) act as agents of social change or not, and if so, how? Toward what ends?

Those dilemmas have a long history, and there is a considerable, quite disciplined–perhaps even regulated—academic and professional attention to them today. Out of this discourse this paper was born. As opposed to the ideologies which sanctioned and canonized certain architectures while marginalizing others, in this paper, I shine some light on unique architectural practices (in infrequently visited corners of the world) which managed to evade the discipline' ideological binds, and charted a promising social project for architecture.

In writing about these practices, I maintain a critical consciousness, a vigilance in employing the instruments of historical, humanistic and cultural construction from which architecture's "modern" history has frequently been victim. Focusing on the telling detail rather than detailed telling, my aim is to bring these global architectural practices to bear on our understanding of architecture's social function and responsibility.

A Museum of Anthropology: The Process of Rehabilitation at Contested Land

Jan Haenraets (Boston University); Alyssa Schwann (University of Manitoba)

The Museum of Anthropology (MOA) in Vancouver, BC, Canada is on unceded territory, the ancestral homeland of the Musqueam Nation. Together, the building from 1976 by architect Arthur Erickson and the landscape design by Cornelia Hahn Oberlander, envisioned a 'total' environmental design. Though considered a masterwork of its era, its concept had shortcomings. It spoke of a setting inspired by the Haida Nation who inhabit the north-west coast of Canada; in contrast, MOA resides in the south-west, on Musqueam land. Musqueam relationship to this place has been largely overlooked—in particular, their living culture. A conflict results; reconciliation is needed.

The museum is planning a third expansion that includes landscape remediation. However, typical design processes consistently omit Indigenous voices, and land is often minimized in design discussions, thereby profoundly undermining Indigenous culture. How do we reconcile the past and achieve environmental and social equity in the needed landscape work?

The project aims to transform the status-quo by exploring a research and design methodology that allow for in-depth understanding of traditional uses and values of the place, while ensuring Indigenous voices are part of the design and decision-making process. Literature review, archival research, on-site investigations, together with engaging people are being explored. Ultimately the method must emerge from a partnership with indigenous communities, rooted in an individual commitment to action towards truth and reconciliation. The long relationship that Canada's Indigenous people have had with land since time immemorial holds the depth for understanding the complexity of the natural environment and its systems as cultural landscapes. Lessons are being learned about the challenges in transforming such processes. The vision now strives towards an eco-cultural woodland restoration strategy, to balance the living culture of the Musqueam with the recognition that this is a 'Museum of Anthropology.'

E9 - Transform Processes and Outcomes for Equity

Tracking the Gaze: How the Presence of Natural Green Elements Impacts the Pattern of Spatial Attention

Hossein Saedi (North Carolina State University)

This study aims to explore the potential impact of micro-interactions with natural green elements and how an individual's attention is directed within a space. Access to natural environments provides people with numerous health benefits (Baycan et al., 2009). Research on restorative environments has explored the possible correlation of the existence of natural green elements in an indoor space with a higher sustained attention level (Keniger et al., 2013). However, how the presence of green elements changes the pattern of spatial attention has not been thoroughly explored.

This exploratory study examines how an individual's attention is directed in indoor environments depending upon the presence or absence of natural green elements. This study compares the results of heat-maps generated for 35 participants via VR goggles. These maps evaluate the scene based upon the direction and duration of an individual's gaze. Participants were randomly assigned to experience, for fifty seconds, two versions of a building's threshold modeled in a 3D virtual reality environment. One version included natural green elements and the other did not. The order of experiencing the two versions was also random to compensate for the carry-over effect.

The results indicate that the pattern of attention changes with the presence of natural green elements. With green elements, the participants' gaze was drawn to capture over 50% of the participants' attention. Also, the features that received a higher level of attention in the absence of natural green elements received a considerably lower degree of attention when the natural green elements were introduced. This research suggests that the green elements have the potential to draw more attention, and they do not deviate attention to other elements in space. Moreover, it begins to shed light on how green elements impact an individual's improvement in cognitive functioning and sustained attention.

In Search of an Equitable Community Development System: Integrative Approach and System Elements

Joongsub Kim (Lawrence Technological University)

How can an effective community development system be developed? What constitutes an equitable community system? This paper aims to address both questions by sharing the experience of an initiative based in Detroit and by reporting the results of literature review, site visits to four cities, participatory observations, and interviews. The term "system" is defined tentatively as a multisector partnership-based, support mechanism for community development in Detroit. The author has participated in a series of regularly scheduled "work" meetings for various stakeholder groups. The author has been a participant in these engagements to develop key elements of a community development system for Detroit and a researcher observing the engagement process. Two hundred leaders of various organizations across Detroit have participated in the process over the last two years. The author and leaders representing various sectors participated in site visits and focus groups with key players in Boston, Cleveland, Indianapolis, and Philadelphia. The results of this community-based process suggest that system governance, system capitalization, data and evaluation, city engagement, capacity-building, neighborhood voice, and career pipelines are central to an equitable community development system. Using these seven system elements as a guide, the author conducted an in-depth review of literature. The review aimed to determine whether its results support the outcomes of the above-mentioned community processes. They resulted in seven common themes similar to the seven system elements developed in Detroit. These themes are central to an effective community development system (Kay 2005). Finally, through interviews and additional literature reviews, the author learned that engagements used in Detroit are necessary to develop elements of an equitable community development system, although the community stakeholders in Detroit and scholars do not always use the same terminologies (Phillips & Pittman 2014).

edra15

Tempe, Arizona

Abstracts - Group Presentations

A -Transform Design for Resilience & Sustainability

Arcosanti, Arcology and Green Urbanism: Moving Toward a Sustainable Future (Workshop)

Ruth Rae (New York University); Patrick McWhortor (Cosanti Foundation); Anthony Floyd (City of Scottsdale); Tony Brown (ECOSA Prescott College)

Arcosanti is a prototype city and urban laboratory based on the visionary architect Paolo Soleri's principles of Arcology, which fuses architectural and ecological principles. The design of Arcosanti incorporates biophilic principles that preserve the biodiverse natural landscape, has a compact organic form, and when complete, is intended to function with a circular metabolism that is analogous to nature. Arcology proposes a compact three-dimensional urban form to provide a lean alternative to the unsustainable urban sprawl found in most of America. Through reduction of dependence on the automobile, reliance on pedestrian transportation, proximity to nature and proposal to have agriculture integrated into the city, Arcology illustrates a vision of Green Urbanism.

This workshop will examine and discuss how the concept of Arcology and the development of the Arcosanti prototype encompass principles of Green Urbanism and sustainable development. Speakers will discuss how as an urban laboratory Arcosanti continues to provide provocative and practical opportunities to learn from, the challenges it faces, and the changes it is making today. The discussion will explore how the principles of Arcology are increasingly relevant to today's world.

B -Transform Education for Our Common Future

Body Conscious Design: Transforming Sedentary Behavior, Environments, and Culture through Education (Workshop)

Galen Cranz (University of California Berkeley); Gourab Kar (Cornell University); Chelsea Rushton (University of California Berkeley)

Body Conscious Design (BCD) is a theory, co-design method, and practice that enables designers and users to transform objects, buildings, and places to serve our physical bodies and change our behavior. Contemporary Western culture and design standardize bodies, restrain them in movement and posture, and cause physical pain and deformities; Body Conscious Design considers bodies in all their variations as an important source of design inspiration, and promotes movement, physical ease, and health.

Rethinking sitting, one of BCD's core concepts, is crucial: new studies show that sitting in chairs for more than three hours daily causes premature mortality from heart attack, stroke and cancer—more than being old, fat, or smoking (Zhu and Cooke, 2016). Transforming the practice of sitting with BCD is more effective than only redesigning the chair itself. How would design education change if BCD were a teaching and learning requirement? BCD is a particularly effective pedagogical method for making change because it requires experiential learning. Designers, design educators, and users must all become aware of our own somas to make physiologically supportive designs and to use them properly. Somatic exercises are part of this workshop.

These sessions will bring together designers, educators, and users toward three related goals: to apply BCD in the home, classroom, and office; build pedagogical confidence, and expand collective awareness of BCD.

In Session One, participants will: 1) Learn concepts, principles, guidelines, and terminology for BCD (e.g., embodied cognition, soma, somatics, ergonomics, alignment, comfort, movement, consciousness, sedentary behavior, typology of postures (including sit-stand, perch, lounge), experiential anatomy). 2) Discover the benefits of BCD through experiential exercises (regarding feet, different seat heights, and constructive rest). 3) Appreciate the challenges BCD sets out for designers: to support the body's optimal alignment and make opportunities to change position and move regularly. 4) Learn "guerilla ergonomics" techniques for how to best use standardized furniture in domestic, professional, and public settings.

In Session Two participants will: 1) Envision using BCD in their personal lives, design offices, and classrooms. 2) Brainstorm which experiential (and other) methods for teaching BCD could be most pedagogically effective in design schools. For instance, how can design educators use positive (social conformity, immediate rewards, and progress monitoring) rather than negative motivations to teach BCD? How can we acknowledge cultural variation and issues of resistance? 3) Discuss when to invite mind-body experts to collaborate in teaching, and when to invest in learning new practices (e.g. Alexander Technique, yoga, etc.) for oneself as an educator and user. 4) Become aware of (and join) the Association for Body Conscious Design (bodyconsciousdesign.com)

Environmental designers and researchers can respond in many ways to the call for a more sustainable, equitable and resilient world for all people, regardless of age, gender, race, ethnicity, and abilities. Body Conscious Design is one of them. Our common future will be enhanced if everyone learns the principles of BCD.

Education of Design Activists: Examining Models, Lessons, and Proposals (Workshop)

Jeffrey Hou (University of Washington); Mallika Bose (Penn State University); Chingwen Cheng (Arizona State University); David de la Peña (University of California Davis); Deni Ruggeri (Norwegian University of Life Sciences)

The field of environmental design is experiencing a surge of interest in design for transformative social and environmental outcomes broadly defined as design activism. As evident in recent award-winning student entries as well as common topics among studio projects and study abroad programs, socially engaged design and professional responses to urgent social and environmental challenges has become a growing aspiration among students and emerging practitioners. The aspiration for greater social engagement is evident in a growing number of studio projects and publications (e.g., Angotti, et al., 2011; Bose, et al., 2014; de la Peña, et al., 2017; Abendroth and Bell, 2018). In Landscape Architecture, the Landscape Architecture Foundation's New Landscape Declaration (2017), with a strong focus on social and ecological justice, resilience, and democracy, also represents a response to this growing interest.

Given the growing interest and attention, is the current model of environmental design education meeting the challenges of our time? Are students acquiring the necessary knowledge and perspective to address these urgent issues of equity, justice, and climate adaptation challenges facing vulnerable communities that traditionally lie outside the scope of professional practice? Are we preparing the next generation's leaders who can elevate and position the profession to be at the forefront of the movement? How can we prepare students to become not only competent practitioners but also design activists who are socially and politically engaged to produce transformative outcomes? This workshop will present findings from a yearlong collaborative study that address these questions. Specifically, the study seeks to identify models and lessons from current practices and to generate proposals for expanding and integrating design activism into environmental design education.

By presenting the proposals as a working draft, the workshop is intended to engage the audience in evaluating their efficacy and explore opportunities for implementation. The goal of the workshop is also to further expand the network of scholars and educators who are interested in this topic. The specific agenda of the workshop will include a presentation of the findings from the yearlong study, followed by questions and answers, group discussions, and reporting back. Facilitated by the project collaborators, the group discussion will focus on clarifying and improving the proposals that address curriculum, programs, pedagogy, and engagement. As a working session, the audience will have a chance to share their insights and perspectives and help shape and contribute to the outcome of the study.

C -Transform in Transitions

Adding Value: The Application of LEAN Workshop, Space Syntax, and Systematic Distance Evaluation for Evidence-Based Design Decision Making in a Hospital Emergency Department (Practitioner/ Researcher Collaborations)

Dawn Gum (EwingCole); Zahra Zamani (EwingCole); Kimberly Johnston (EwingCole)

How can architects and interior designers apply LEAN principles in the design development of emergency departments in response to increased client demand to implement strategies for streamlining emergency department flow and efficiencies? LEAN thinking emphasizes reducing waste or non-added value elements from workflow processes to provide greater value for customers and improve staff, patient, and equipment flows.

This study explores an architect-led LEAN event with the client's key stakeholders, creating patient and staff journey mappings. Current state was mapped to identify areas for process improvement. Ideal state mapping formed the basis for floor plan concept development. We then employed space syntax analysis and systematic layout planning distance evaluation to compare two architectural concepts for objective evaluation of cost, patient targeted visibility index (TVI), and accessibility outcomes. The systematic layout planning analysis evaluated nurse, physician, and nurse tech travel distances between nurse stations, triage, trauma, disposition, radiology, CT, and exam rooms.

LEAN workshop findings included co-locating CT with Trauma Rooms to decrease stroke response time, performing triage registration in triage rooms, opportunities to eliminate redundancies, and providing ambulation space for behavioral health patients. Diagramming found additional redundancies in FTE's and activities.

Distance evaluation analysis showed that one of the concepts provided better spatial access for clinicians yielding $57,921 annual operating costs savings. This concept had higher TVI values, indicating better patient head visibilities from nurse stations to exam rooms. Space syntax results showed the same concept provided better visibility in trauma and lobby areas, which are critical in ED environments. This study employed multiple methods to streamline flow and enhance patient outcomes corresponding to LEAN thinking philosophies. The results of this study shed new light into the importance of ED physical environment spatial attributes and ED flow and visibility outcomes as design tools supplementing LEAN design.

Sydney Housing Challenge (Practitioner/Researcher Collaborations)

Nilou Vakil (University of Kansas); Casey Franklin (Cornell University); Gregory Crichlow (University of Kansas)

Sydney Alternative Housing Ideas Challenge
Data is becoming as impactful on the way we configure cities as water and electricity were 100 years ago. (Bennett, 2016) The connectivity made possible by the high-speed low-latency internet of Smart Cities allows our built environment to respond to our needs in exciting ways.

This group presentation illustrates the team's design submittal to the 2019 Sydney Alternative Housing Ideas Challenge and the process utilized to develop the scheme through workshops. Named one of seven finalists, "Smart Home Sydney" introduces affordable and sustainable housing and neighborhood models. The design proposes an investment corridor through Sydney's downtown that utilizes data collection systems to deploy population health initiatives. (Halegoua, 2015) Housing unit designs that incorporate smart home technology as well as proposals for financing and policy to deploy such technology are also discussed.

Walkable mixed-use districts are designed with health and wellness in mind. These lifelong neighborhoods are created to allow one to thrive at all stages of life. (Ball, 2012) They incorporate great parks, great schools, access to multi-modal transit, jobs, cultural districts, and all the amenities needed in later life. Buildings are organized to support vibrant busy streets that hold individual car use as secondary and focus attention on walkable, bikeable neighborhoods. (LaPlante, 2007) Individual dwelling units are designed to collect activity and biometric data and transmit this information to an on-site clinic or Living Lab. These environments are able to collect human vital signs, physical activity, environmental conditions, and pharmaceutical regiments. This data can be collected and analyzed to deploy Population Health strategies to deliver health care more affordably, effectively, and sometimes before we know we need it.

The technological capability provided by the clinic allows the housing operator to leverage institutional resources by leasing space to university centers or medical research entities. The rental rate far exceeds comparable rates for similar commercial facilities thus subsidizing the housing costs, technology upgrades, and even net zero energy construction. In addition to technological connectivity, the scheme's social connectivity supports intergenerational living. The proposal includes a mobile app that connects residents in much the same way as social media. Also, bluetooth beacons incentivize sustainable behavior while facilitating face-to-face interaction. We have also designed a mobile app schematic called CareFree that allows primary caregivers to monitor family members and intervene when necessary.

The presentation will not only describe the housing prototype, but will also outline the innovative competition process. Unlike many solution-based housing competitions, the seven finalist teams were invited to participate in a series of day-long workshops to engage with one another, the Sydney government, and the community in order to develop their ideas into solutions that could be implemented.

The Value of Purposeful Survey Development in Design Research (Symposia)

Organizer: Jeri Brittin (HDR)

Surveys are an efficient and effective way to gather quantitative and qualitative data from large groups of people. Used extensively in design research, surveys are considered a standard method of evaluation within the field. However, there are few survey instruments repeatedly used in design research, leading to a lack of translation and replication across studies. In addition, survey development and application methodology is often not thoroughly reported within the field. Rigorous design research is critical to informing and improving design decisions, and validated instruments that measure occupant perceptions, experience, and outcomes in the built environment need to be explored and developed.
The first presentation will introduce the science of survey research and explain the basic concepts of questionnaire design and methodological techniques that can be applied in survey development. The presentation will highlight tools to improve the quality of data collection and reporting practices and introduce concepts that should be considered when critically reviewing the survey methods and results of other researchers.

The second presentation will illustrate these concepts with a literature review of studies that used surveys to associate the inpatient healthcare environment to nurse outcomes. Attendees will gain a sense of survey development and use in design research, including the level of methodological rigor and testing typically undertaken and the most common survey instruments used with nurse populations.

The third presentation will focus on addressing the gap in current measurement. We will present the conceptual thinking and process in developing a new instrument to measure a construct not represented in currently available instrumentation. We will illustrate the process of rigorous survey scale development including both qualitative and quantitative methods, related to nurse experience in healthcare environments. This example will demonstrate a framework to create valid and reliable survey tools for use in human experience research.

Presentation 1: Reliable and Valid: Appropriate Adaptation, Development, and Testing of Survey Instruments

Abigail Heithoff (HDR)

Surveys are widely used in design research to collect quantitative information from building occupants, typically assessing their level of satisfaction with various aspects of the design or documenting their experience within the built environment. Often, survey scales, which consist of multiple related items, are utilized to capture a latent construct. Both scale items and individual survey questions might originate from a range of source types, from established psychometric instruments to questionnaires written by designers or researchers for a specific study.

When reviewing or applying study results, it is important to understand the methods of data collection, including any survey instruments employed. However, in the current body of design research, there are inconsistent levels of adaptation, development, and testing of survey instruments. In addition, adequate explanations of methods, including surveys, are often not reported in the design research literature.

In this presentation we will examine key considerations in appropriately applying previously developed survey instruments, including assessment of prior psychometric testing and applicability to study populations. We will define and explain the multidimensional concepts of validity and reliability, the standards by which survey-based scales are evaluated, and offer a range of methodologies appropriate for the development and testing of both adapted and new survey instruments (DeVellis, 2012; Fowler, 1995; Krosnick & Fabriagar, 1997). The goal is to provide attendees with tools they can apply to their own survey research to improve the quality of both their data collection and reporting practices and to introduce concepts that should be considered in critical review of survey methods and results of other researchers.

Presentation 2: A Review of Quantitative Measurement Tools for Nurse Experience Related to the Inpatient Care Environment

Francesqca E Jimenez (HDR)

While the experience of nurses is often a focus of healthcare design research and surveys are frequently used to gather quantitative data related to their work within the hospital unit environment, there has not been a focus in the design research literature on appropriate survey questions and scales to best capture this information in a validated way.

A literature review was conducted to assess the range of quantitative survey instruments that have been used to measure nurses' ability to work efficiently and effectively in their environment. The review focused on instruments intended to measure the relationship between nurse outcomes and the built environment in terms of the delivery of patient care, along with constructs that consider physical design or are hypothesized to be related to the environment, such as job satisfaction, teamwork, and organizational factors.

We will summarize the results of the review of studies which employed survey instruments to connect the inpatient environment to nurse outcomes, including the journals where this research is typically found, the environmental dimensions explored in the studies, and the nurse outcomes of interest. Studies will be categorized by whether the instruments had been previously published, adapted, or self-developed by the authors, and the development, testing, and validation methods used will be assessed. More specifically, several nurse survey instruments that have been applied multiple times in design research literature will be explained in detail with a comparison between the original instrument development and its application to healthcare environment research.

Presentation 3: The Development and Rigorous Testing of a Survey Instrument to Measure Nurse Experience Related to the Inpatient Environment

Renae K Rich (HDR)

As an example of appropriate and systematic survey development in practice, we will present the methodology of a phased study to create a new survey instrument to measure nurses' ability to work efficiently and effectively in their environment. The built environment encompasses a largely overlooked set of variables in nursing research, and there is no instrument specifically intended to quantitatively measure nurses' experience specific to the physical environment. As a result, there is inconsistency in measurement across related studies, leading to a lack of replication within the design research field.

As part of this study, partially funded by a grant from the Academy of Architecture for Healthcare Foundation, researchers implemented both quantitative and qualitative testing of the survey instrument, including initial pilot testing, feedback from content experts, and cognitive interviews with a sample from the intended population. Exploratory factor analysis and techniques from item response theory, along with quantitative and qualitative summary of content feedback, were used to refine the survey instrument before final confirmatory pilot testing took place. A rigorous process of this magnitude has not previously been undertaken in the development of a survey instrument specifically intended to study the built environment.

Such an instrument will be valuable in generating quantitative and standardized data, making associations with other factors and comparisons across time and facilities possible. Ultimately, within the healthcare design industry, this capability will lead to greater knowledge about the aspects of a healthcare facility that best support nurses' work and that are associated with their ability to provide quality patient care.

The Decline of Public Space? Exploring a Counter Narrative (Symposia)

Organizer: Karen Franck (New Jersey Institute of Technology)

Since the publication or Davis's City of Quartz(1990) and Sorkin 's Variations on a Theme Park (1992), there has been an ongoing narrative describing the "decline of public space" (see Madden's "Revisiting the End of Public Space: Assembling the Public in an Urban Park," 2011; Mitchell's "People's Park Again: On the End and Ends of Public Space," 2017).What this "decline" means varies. It refers variously to: the loss of actual spaces; more frequent ownership and/or management by the private sector; greater prominence of commercial activities; exclusion of particular activities, most prominently political ones; and the exclusion of certain kinds of people.

A co-existing, counter narrative, which is an alternative perspective but not a replacement, holds that public space is, also, alive and thriving (see for example Franck and Stevens, Loose Space, 2007; Hou, Insurgent Public Space, 2010; and Bishop and Williams, The Temporary City, 2012). This narrative often draws upon cases of public spaces closely observed, often with recognition that particular kinds of activities occur in particular types of spaces and that the uses of a given space may well change over time, particularly in response to changes in design features or management practices.

This symposium will explore this alternative view, focusing on two themes: (1) emergence of new types of public space and (2) changes in the design, management and use of cases of existing types. Following the first theme, Vardi-Topal will report on the design and use of pedestrian plazas in New York City while Yan will describe ecological restoration parks in China. Pursuing the second theme, Huang will present changes in the design use and management of interior privately owned public spaces in New York and Abdelazim will review the history of such changes in Tahrir Square in Cairo.

Presentation 1: Pedestrian Plazas: Uses and Users of a New Type of Public Space

Hanife Vardi Topal (College of Architecture and Design): Karen Franck (New Jersey Institute of Technology)

Pedestrian plazas are public spaces typically created through the transformation of a portion on a roadway through collaboration of city government with a non-government agency. In New York City, since the NYC Plaza Program started with Times Square Plaza in 2009 pedestrian plazas have been created I at seventy-five sites in Manhattan, the Bronx, Queens, and Brooklyn. The program is directed by the Department of Transportation in partnership with local and governmental agencies to provide quality open public space in a ten-minute walking distance of the homes of all New Yorkers.

New York's pedestrian plazas are used in a variety of ways such as gathering with friends and family, people-watching, chatting, eating/drinking, and attending programmed events. Occupied by hundreds of people every day, each plaza has its own characteristic use and users. For instance, some plazas are typically frequented by the same people on a daily basis and used as gathering places whereas others are more likely to be used for shorter period activities such as eating lunch, meeting someone, and brief resting. Drawing on observational data on seven pedestrian plazas in Bronx, Brooklyn, and Queens, this paper explores the varieties of uses while also considering design, management, and user characteristics of these plazas

Presentation 2: Urban Ecological Restoration: How Public Can It be?

Han Yan (New Jersey Institute Of Technology)

Since many environmental issues fail to be resolved solely by engineered solutions, restoring ecological system functions is becoming increasingly popular in urban environments around the globe. Traditionally, ecological restoration has been defined as returning a site to a pre-"disturbance" state. Therefore in most cases public use is not considered a priority for ecological restoration projects. In recent years,however, the growing needs of urban populations and the expansion of urban areas call for innovative approaches to relationships between people and ecological systems.

It is possible for ecological restoration sites to meet some of those needs by providing access to open space. Planners of many ecological restoration projects in China are pursuing that option by choosing particular design features, maintenance routines, public accessibility, and programming of the ecological restoration projects to allow, and encourage, public use.

In this paper, several urban ecological restoration projects in China are described to illustrate this new type of urban public space. Observations of several sites and interviews with staff and visitors are used to identify the design, maintenance, programming and advertising features that promote public use.

Presentation 3: More Welcoming? Three Redesigned Bonus Spaces in New York City

Te-Sheng Huang (Te-Sheng Huang Architect & Associates)

Three interior privately-owned public spaces in New York were redesigned between 2013 and 2017: the space at 180 Maiden Lane, Olympic Tower atrium and Park Avenue Plaza. Given the New York City Planning Commission's commitment to make bonus spaces more accessible and more inviting, one may well ask how the design and use of these of these spaces changed?

Observations of the three spaces before and after the redesign suggest they are indeed more welcoming. As a result of the redesign, the spatial quality each space has improved. The floor plan of each space has been re-configured to provide continuous rather than fragmented spaces. As a result, all spaces look much spacious even though the floor area has not changed.

The incorporation of light and vividly colored furnishings brighten each space. Both the Olympic Tower atrium and Park Avenue Plaza include green walls now. In the Olympic Tower atrium, the water fountain is operating continuously now but the space for art exhibitions remains closed. The space at 180 Maiden Lane has two drinking fountains and art exhibitions are curated three times a year. However, the weekly music performances are suspended. The space at 180 Maiden Lane that was originally open only during the week until five is now open seven days a week and into the evening.

This paper will illustrate the design changes of each space in detail Analysis of observations of use and users conducted in 2013 and again in 2019 will be presented to answer the question of whether the use and users have also changed. The findings will contribute to the question of whether, at least in some ways, public space is alive and well and, in addition, in what ways privately controlled public space can contribute to urban life.

Presentation 4: Closed or Open? The History of a Public Square

Mariam Abdelazim (New Jersey Institute of Technology)

Public space is a gateway to civil society. In the best circumstances, a place where community members can communicate, interact and express themselves and their opinions freely.For this to be possible, at the very least the space, must be physically accessible and open to the public.

The history of Tahrir Square in Cairo presents a revealing story of the changing openness of an urban public space. It was first constructed in 1863 as a traffic circle; then in 1919 it became a platform for political dissent in addition to being a traffic circle. After the 1952 revolution the government redesigned the square to be a spacious roundabout, adding fencing to prevent access by large groups of people. In January 2011, the square changed again to a stage for political protest, becoming an indispensable space for maintaining the momentum of the Revolution as well as for ensuring the safety of protestors who were able, for several weeks, to control access to the square. In 2016 the square was re-designed, fences were reconstructed– closing it from the public. However, in 2019 the square was re-opened and Egyptians now use it as a recreational space for picnics and social activities including playing and socializing. The purpose of this paper is to examine this history, exploring how the design of the square, its manifest function, and its management– both by government agencies and members of the public– changed from 1863 until today. It will make clear how dramatic changes in openness supported the emergence and then the extinction of a political movement and now, again, supports recreation.

Presentation 5: Active photography for environmental design: Methods for engaged practice (Workshop)

Emma Froh (Associate Ten x Ten); Anne C Godfrey (State University of New York)

Photography is the most important

representational medium used within the environmental design disciplines. Yet it is often treated as a copy machine instead of a tool for knowledge creation. We passively think photography shows us what we see instead of embracing the fact that photography is a process of making, just like any other representational medium commonly used to create environments. Currently, the default approach to photographing for site investigation and design is simply showing up, making "a lot" of photographs and leaving (point, shoot, leave). Thus, critical methods easily applied to daily practices are very much needed.

This workshop presents a series of practicable and accessible photographic methodologies that develop active engagement for critical study and design of places. The methods are organized around three primary concepts: making (not taking); building in small amounts of time to consider the "why" behind each photograph before releasing the shutter; and limiting the number of photographs made during any site visit. A set of very specific, yet easily practicable, methods will be presented for each of these concepts. These methods, developed by the organizer, include: the five second rule, asking two questions, scanning, changing pov, and circumnavigation. The workshop will additionally present conceptional origins and rational behind the methods and allow participants to test the methods during the workshop. Once the workshop starts and primary concepts and ideas are covered, the workshop will become a mobile classroom. Specific methods will continue to be presented as the workshop photographically explores the environs of the conference area. Participates will receive handouts describing the methods and will get to try each of them in the field. Workshop facilitators will both present information to the group and individually check in with participants as they are photographing. All methods and exercises can be done with one's personal smart phone camera.

These methods address faulty ideas about the photograph as a literal window onto the world and the camera doing all of the work. The relationship between personal values, biases and past experiences and how one photographs will also be examined. The methods presented shift cognitive interaction with photography into thoughtful and active engagement with time, place and personal motivations. These methods encourage those who participate in the design of environments to embrace the fact that we are all photographers, and thus can learn from and apply core knowledge from the discipline of photography into our daily practices. These methods also cultivate a critical understanding about how our daily photograph practices are highly influential to our decisions for place making.

D -Transform Places with People and the Environment

Preserving the Cultural Landscape of Taliesin West (Practitioner/Researcher Collaborations)

Elena Rocchi (Arizona State University); Fred Prozzillo (Frank Lloyd Wright Foundation)

In recent decades, educators and practitioners have filled the gap between practice and research by connecting the theoretical with the empirical using reflective inquiry. This session reports on the results from a practice-based research and interpretive-historical research seminar in 2015, using both on-site experiences and an archival fond. The researcher and the practitioner built a partnership to enhance students' knowledge and research skills using a professional situation as a unique client situation. On the one hand, the researcher produced a cultural landscape's knowledge using research; on the other hand, the practitioner used the researcher consultation to develop understanding in practice. The practitioner set the research agenda, and the research was adapted to practice requisites and active listening. The practice-based research utilized students' group reflective journals, research, and final projects. Results show that the partnership developed out of shared interests enhanced through positive relationships based on appreciation.

The proposed session aims to show how successful research is a hypothetical situation in which benefits and expectations concerning the type of understanding are shared and inform the practice research. By reflecting on a broader question, "How much adaptable are the assumptions we make as researchers as well as practitioners?" the main objective is to reflect on ways to nurture imagination, creativity, and non-linear modes of thinking using collaboration to work together.

The research developed knowledge in research and practice using a reflective inquiry method. Methods for reflective practice used were: 1) a group reflective journal: the journal was created a) to reflect through documenting ideas, feelings, observations, and visions to explore and map historical artifacts of the cultural landscape listed by the practitioner; b) to reflect upon processes behind them, and c) to develop a conversation with others; 2) a co-operative inquiry: the group of students met on a regular basis with the practitioner to reflect together, to try out a form of collective reflective practice, and to support the reflective practice of participatory research; 3) a co-operative inquiry: research into the human condition. The group worked through the semester following a cycle of action and reflection to develop personal understanding.

Outcomes of the collaboration: 1) Bringing together academics and professionals helped to rethink the meaning and practice of cultural heritage conservation, encouraging cooperation. 2) The practice-based research challenged the traditional notion of historic preservation, revealing a new approach to conservation. 3) The research identified the recognition of both cultural and natural values as a successful approach in the preservation of the past.

Impacts of Decentralized Nursing Unit Typology: A Comparative Case Study of Two ICU Units in an Academic Medical Center (Practitioner/ Researcher Collaborations)

Hui Cai (University of Kansas); Intisar Ameen Tyne (University of Kansas); Kent Spreckelmeyer (University of Kansas); Frank Zilm (University of Kansas)

Previous studies tend to compare the impacts of centralized vs decentralized nursing unit design on nurses' work efficiency. However, a limited amount of studies has identified the impacts of different typologies of decentralized nursing unit design on nurses' work efficiency.

This research aims to empirically evaluate the impacts of nursing unit typology through a comparative study in a newly designed ICU (unit A) with another existing ICU (unit B) of the same academic medical center. The research intends use Post-Occupancy Evaluation (POE) to compare two ICUs with different typologies of decentralized nursing units but similar staffing model and patient acuity level to evaluate the impacts of the physical environment on nurses' efficiency, teamwork, and patient satisfaction.

Both the units have hybrid layout with four sub-nurse stations and nursing alcoves in-between each pair of patient rooms. Unit A has 28 patient rooms, linear rectangle in shape with a bend in the middle. The two sets of nursing units are located at the two ends of the central support area. Additionally, in unit A, two team rooms were designed in the middle of each set of sub-nurse stations. Team room is mainly used by therapists, nurses, dietitians, patient counselors and other visiting health professionals to consult on the patient treatment procedures. Whereas, Unit B has 14 patient rooms, compact and square shaped. The four nurse substations are located at the four corners of the central core.

The research is based on a mixed method quasi-experimental comparative study. Research methods include visibility analyses, noise and lighting measurements, pedometer readings, behavior mapping and behavior tracking, nurse interviews and patient satisfaction and outcome analyses.

Based on the Center for Healthcare Design standardized patient room POE toolkit, the average rating of the patient rooms in unit A were 24% higher than unit B (4.41 vs 3.33 respectively). The noise levels at unit A was significantly lower than the noise level of unit B. Nurses of unit A had more frequent work-related interactions and less social interaction compared to the nurses of unit B. Unit A had reduced CALBSI and C. Diff total incidents and monthly rates after moving to the new design, while the control unit (Unit B) had higher total incidents and average monthly rates after the study period than before. The patient satisfaction rating improved in unit A when compared to the patient satisfaction rating of unit B. However, the overall visibility of unit B is higher than the overall visibility of unit A (7.119 vs 5 respectively). Nurses' of unit B walk less than their peers at unit A (4030 steps vs 5168 steps respectively).

The findings demonstrates that Unit A had improved lighting quality, reduced noise, refined teamwork environment, and enhanced patient satisfaction level. However, the shape and configuration of unit A caused difficulty in visibility, accessibility and increased nurses walking distances.

Detention Facilities: Research, Ethics & Impacts on Community (Symposia)

Organizer: Melissa M. Farling (Gould Evans)

Incarceration environments are "total environments" where those detained spend long periods and are completely dependent on the institution for their needs. These facilities have profound impacts on those who experience them, including staff and visitors. In the US, approximately 2.3 million people, including juveniles, are held in federal and state prisons and local jails (Prison Policy Initiative). According to the National Alliance on Mental Health, more than 2 million arrests involving people with serious mental illnesses are estimated. Although no official count exists, a 2015 report from the Vera Institute for Justice estimates between 80,000-100,000 inmates are housed in solitary confinement at any given time in the US, and there is abundant evidence prolonged isolation can be mentally and psychologically traumatic.

The AIA Academy of Architecture for Justice, along with collaborators, has been exploring a variety of conditions in correctional and detention facilities as well as the ethics of design and research. The guidelines and limits for doing IRB-approved research with prisoners are strict, and exist for very good reasons, in part because of the difficulty in assuring true voluntary consent by people who have been involuntarily detained. These restrictions, of course, also add to the difficulty of conducting the kind of true experimental research on the nature of solitary confinement that would add to our confidence about its impacts. We will discuss these limitations and how they have affected the nature of our understanding to date. Incarceration impacts the health and well-being of those inside and the surrounding community in which they are embedded. Panel moderation will include discussion on research as well as the ethics of research and design. The issues associated with humane detention, solitary confinement and impacts on mental health will be examined. Connections between research and policy will be considered while offering a possible future of corrections.

Presentation 1: What Is Humane Detention? Processes, Principles and Alternatives

Richard Elliot Wener (New York University)

The International Committee of the Red Cross' (ICRC) "Persons Deprived of Liberty" unit commissioned the volume Towards Humane Prisons: A Principled and Participatory Approach to Prison Planning and Design to support better planning processes that could lead to more humane detention facilities. The volume attempts to summarize critical issues in the design and operation of detention facilities that affect humane conditions. It provides descriptions of planning processes that encourage jurisdictions to consider alternatives to detention, since alternatives are typically less traumatic and more economical than incarceration. Where such alternatives are not possible, the book provides specific recommendations for ways to provide decent, respectful and safe environments that help support reintegration to society. This presentation will review the ICRC process and summarize key principles presented in the book.

Presentation 2: Adverse Effects of Lighting in Typical Correctional Settings – What Can We Do about It?

Jay Farbstein (Jay Farbstein & Associates)

There is a wide variety of conditions in typical correctional facilities that affect the quality and quantity of lighting available both to residents and staff. These include generally limited access to adequate natural daylighting, the limited (and often incorrect) spectrum artificial lighting that is provided, and the fact that at least some lighting is kept on at all hours of the night (in cells and dormitories). It was hypothesized, based on research in other settings, that the typical correctional lighting provisions result in negative impacts on both inmates and staff (especially those staff who work night shift or who rotate shifts on a regular basis). The expected impacts include disruption of sleep patterns and circadian rhythms with attendant psychological and health outcomes, which can be very powerful. The potential of improved design layouts together with new kinds of artificial lighting (programmable LEDs) to remedy the situation are worthy of discussion (and study), given the potential beneficial impacts, which could include improved physical and mental health and reduced utilization of health care and mental health services, improved attentiveness when attending classes and other programs, reduced stress and related behavioral manifestations including conflict, fights and assaults, and many others.

Presentation 3: The Mental Health Crisis – Advancing Treatment in Correction

Erin Persky (Erin Persky, Assoc. AIA)

According to the Bureau of Justice Statistics (BJS), almost one-third of those in jails and 15% of those in state prisons have a diagnosable serious mental illness. However, the number of incarcerated individuals who have a history of trauma and/or other mental health issue is much higher; for example, BJS reports that two-thirds of incarcerated women have a diagnosed history of mental illness. Architects and Planners have been tasked with collaborating with custody staff, health care staff, community organizations, and others to create built environments that support optimal mental health and medical treatment of confined individuals. While clients expect architecture professionals to recommend and apply up-to-date evidence-based design to their justice projects, there is an unfortunate dearth of applied research in these environments due to complications with obtaining informed consent from those who are incarcerated.

The mental health crisis impacts jail and prison operations in different ways. Jails, for example, are the largest mental health provider in the United States and jail detainees have a substantially shorter average length-of-stay than those incarcerated in prisons, necessitating distinct approaches to mental health treatment. This presentation will consist of a discussion of the mental health crisis in federal, state, and local correctional/detention environments, associated design impacts, examples of applied design best practices, and will explore existing research and opportunities for further investigation.

Presentation 4: Solitary Confinement: Research and Impacts

Melissa M. Farling (Gould Evans)

No official count exists, however the Vera Institute for Justice estimates between 80,000-100,000 inmates are housed in solitary confinement (also known as isolation or restricted housing) at any given time in the United States (Shane, Wilcox, & Subramanian, 2015). There is abundant evidence prolonged solitary confinement can be mentally, psychologically, and physiologically traumatic. Such impacts have been documented by psychiatrists and neuroscientists (Grassian, 2012; Stanford University Human Rights in Mental Health Lab, 2017; Akil, 2018). The United Nations Nelson Mandela Rules define prolonged solitary confinement as a period in solitary confinement in excess of 15 consecutive days. While some jurisdictions are changing policy, a majority of jurisdictions in the country have individuals in prolonged solitary confinement – at least 22 hours per day in excess of 15 consecutive days (The Marshall Project, 2016). This discussion will focus on the issues and theories associated with the physical design of solitary confinement and its psychological and behavioral impacts as well as current neuroscience research. It will also tie into the work presented by other panelists regarding humane detention and mental health, as well as the factors that may impact an individual's successful reentry back into the community after experiencing solitary confinement and possibly subsequent recidivism.

Building Performance Evaluation: Transform Processes for Environmental, Economic, and Social Settings around the World (Symposia)

Organizers: Ulrich Schramm (Bielefeld University of Applied Sciences); Akikazu Kato (Mie University)

Based on its evolution from Post-Occupancy Evaluation (POE) over the past 50 years, the Building Performance Evaluation (BPE) model was developed to provide a systematic process to evaluate buildings. It provides feedback not only after a building is occupied, but at every stage of building delivery and lifecycle, from conception to demolition.

Formed over 25 years ago, the International Building Performance Evaluation (IBPE) consortium is a group of academics, practitioners, and consultants from around the world who undertake performance-based building evaluation research. The IBPE consortium has published Assessing Building Performance (Preiser and Vischer, 2005), Enhancing Building Performance (Mallory-Hill, Preiser, and Watson, 2012), and most recently Building Performance Evaluation (Preiser, Hardy, and Schramm, 2018).

The IBPE Symposium will provide attendees with an excellent opportunity for learning about strengths and weaknesses, chances, and challenges in performance-based planning, design, and occupancy. The session contains presentations from different perspectives such as practice, research, and teaching concerned with both procedural issues of BPE and its application toward more environmental, economic, and social settings around the world. Following the paper presentations, the audience is invited to participate in an interactive discussion related to the conference theme of Transform: Socially Embedded Collaboration to consider what role BPE plays in the development and understanding of current strategies intended to ensure the quality of built environment and to optimize user satisfaction during occupancy.

Presentation 1: Teaching Building Performance Evaluation for Sustainable Buildings: Chances and Challenges

Ulrich Schramm (Bielefeld University of Applied Sciences)

With its consideration of the entire building life cycle, the Building Performance Evaluation (BPE) process model constituted a challenge for planners and other professionals in the building sector when it was introduced in the late 1990s: Traditionally, the building delivery process has been understood as a linear course of activities that ends with the opening of the building. At that time, with the new understanding of the building life as a cyclic sequence of phases – including occupancy – and internal review loops, the establishment of performance criteria and their evaluation became crucial. In the meantime, this basic idea of the BPE process model has been verified with the emergence of similar concepts in the building industries, such as certification systems for sustainable buildings (like LEED or Cradle-to-Cradle), circular economy, urban mining, or BIM. Thus, universities connected to the building sector in Germany have to take these changes and developments into consideration and adapt their existing degree programs – or even develop new ones! But what is the main content and who are the addressees for BPE?

Some of the unresolved challenges are: Which are the pivotal processes, phases, and review loops to focus on in the context of BPE? What are the relevant topics, methodologies, and instruments to be taught? Does BPE fit into programs for architects, construction project managers, or facility managers? Should BPE become part of other transdisciplinary programs for new specialists to appear yet in the building sector? How can the collaboration with building users be strengthened throughout the BPE process in order to meet environmental, economic, and social challenges in general and to optimize user satisfaction in particular? The presentation will discuss these issues based on the author's professional background as an architect, his teaching experience, and his research activities in the area of BPE.

Presentation 2: Building Performance Evaluation and Indoor Environmental Quality Research: Evidence-Based Guidelines for Sustainable Buildings Design

Ihab Elzeyadi (University of Oregon)

How can we design a better indoor environmental quality (IEQ) in sustainable buildings? The current approach for achieving IEQ in high performance buildings relies on a segregated metric that predicts user independent comfort attributes--such as visual, thermal, and acoustical comfort--with failed accuracy. In real settings, especially those designed with sustainable objectives, the occupant interacts with the overall ambiance of the environment combining space performance, indoor comfort metrics, user experience, as well as spatial functionality and aesthetics. This presentation reports on a series of comparative Building Performance Evaluation (BPE) projects of LEED™ rated buildings that provide insights into the complex nature of IEQ findings and how BPE protocols can help clarify the interrelationships between multi-comfort phenomena and building performance from the user's perspective. For these studies, we employed a systemic BPE protocol for assessing IEQ parameters, building performance, and occupants' multi-comfort and satisfaction of ten LEED rated buildings. A comparative analysis of the design strategies and spatial performance of the buildings and patterns of indoor experiences are outlined and compared. In addition, the presentation will focus on six different sampled spaces --of different spatial configurations and IEQ strategies—to develop a framework for the relationship between the various IEQ parameters from the occupant's perspective. The session will present the patterns and LEED credits that contribute to a better IEQ as perceived by the occupants and measured through validated metrics. Methodological innovations for conducting comprehensive BPE studies as a research paradigm will be summarized. The hope is to provide a decision support process and methods for environmental researchers, practitioners, occupants, and owners that would help them design and evaluate green buildings and IEQ in a comprehensive way.

Presentation 3: How to Utilize Building Performance Evaluation (BPE) in the Building Life Cycle of Hospital Inpatient Quarters: Bpe from the Perspective of Practice in Case Studies of Japanese Healthcare Facilities

Akikazu Kato (Mie University)

One of the major research topic for Japanese team of IBPE has been the BPE of hospital inpatient quarters. In the recent research, the movement survey of nurses has been conducted in conjunctions to surveys on user satisfactions of nurses and patients as Post-Occupancy Evaluation (POE). In the movement survey, sample nurses are selected to represent the nursing team and all movements of these nurses and their causations are recorded by the examiners. The results of analysis have been utilized in the building life cycle of hospital inpatient quarters.

In this presentation, a ten-year long case study of a prefectural hospital in a mountain area of Gifu Prefecture is shown, which was to rebuild the entire hospital. The presenter served on the jury to select the architect in 2009/2010, advised the design and construction processes during 2010 to 2014, and carried out a POE process in 2014 to 2019. The design concept of inpatient quarters of this hospital is the usage of all single rooms and the unit-grouping of those rooms, which is still quite unique in Japan.

The presentation shows the outline of evaluation activities for each stage of BPE process model. It focuses on the recent POE where the results of findings were used to fine-tune the newly adopted Partnership Nursing System and to develop management rules for daily facility usage. Furthermore, it refers to the database used in the future development of similar facilities. The simulation research to predict nurses' movements shall enhance the productivity and creativity in the future projects especially by incorporating the process in a Building Information Modeling (BIM) system. Finally, it shows the significance of the process in the realization of Evidence Based Design (EBD).

Shape the future: where engagement, youth partnership, and planning intersect (Symposia)

Organizer: Alice Huang (City of Boulder)
Presentations:

Engagement and Partnerships with Minority Residents, Sarah Huntley (City of Boulder); Engagement and Planning, Alice Huang (City of Boulder)

In 2019, the City of Boulder, Colorado began its first subcommunity plan in more than 20 years, promising to collaborate with the public. This entailed working together to formulate solutions and incorporating community advice and recommendations into decisions to the maximum extent possible. A subcommunity plan is a tool for residents, land owners, business owners, city officials and city staff that communicates expectations about the future of a subcommunity and guides decision-making about subcommunity resilience and evolution into the future. The East Boulder Subcommunity plan is serving as a prototype for future plans and collaborative community projects. Our presentation brings together three viewpoints: engagement expertise, partnership with minority residents, and community planning.

We looked to community members to identify sites, spaces, and issues that are either valued or need improvement. We have accomplished this by engaging people where they meet, eat, recreate, and work; convening a working group of stakeholders, who represent the unique blend of interests; and piloting the Community Connector Program. Community Connectors partner with the city, connect with residents from underrepresented communities, and share lived experiences of their neighbors.

When planning the future of East Boulder, the community needs to guide how and which Boulder Valley Comprehensive Plan goals will be met. A guiding document is only as successful as the community support and excitement it generates. The value of community-driven planning reflects the rich legacy of planning in Boulder, which has resulted in a national model of open space and connections to nature. Urban form and green spaces still shape and are shaped by our community today.

Transforming Experience: Coming to your senses (Workshop)

Susan Drucker (Hofstra University); Richard Elliot Wener (New York University); Gary Gumpert (Urban Communication Foundation); Sheila Gobes Ryan (University of Souther Florida); Sue Weidemann (University at Buffalo/BOSTI Associates)

From Aristotle, and Plato to McLuhan and Palasmaa, the senses have been seen as created the interaction between body and the external world. Whether it is sight, hearing, smell, touch, and sound the idealized structure nurtures and addresses each of them. The senses work together closely to enable the mind to better understand its surroundings, and in so doing construct, rather than reflect, reality. It has been argued that Berger and Luckmann's Social Construction of Reality should be considered when dealing with sensory perception (Friedman, 2015). Focusing on the senses can enhance our understanding of the process of the social construction of reality. What is the relationship of authenticity and inauthenticity with our sensory experiences in communities?

In evaluating places, what role do the senses play in the lived experience. Has our sensory approach to the city changed with the mediatization of the city? How are the senses engaged to facilitate the desired behavior or reaction to a place? Can an authentic community be built in an "inauthentic" place. This interdisciplinary panel will consider how the senses may influence the behavior and choices of designers as well as the experience of users of settings, particularly those which recreate or represent other places.

Let's do it together! - exploring the potential of games in democratic landscape planning and design (Workshop)

Anna Szilagyi-Nagy (Eberhard Karls Universität Tübingen); Saeed Ahmadi Oloonabadi (North Carolina State University); Gabriel Kaprielian (Temple University)

There is an increasing trend to use games in planning and design to encourage a wide range of stakeholders to participate in making decisions about the future of their environment. Such games might be digital or analog and could be played by a diverse or homogeneous group of stakeholders. Moreover, games can address various environmental issues, from/ small-scale neighborhoods improvements to advocating for climate movements. Rules and mechanics of the game can facilitate dialogue between stakeholders and may lead to a fruitful discussion or consensus. Players can also work on developing proposals and solutions for specific environmental issues.

The workshop aims to bring the EDRA community together and encourage international and multidisciplinary participants to discuss various meanings of the game, design, landscape, participation, and democracy, and how these concepts may be interrelated. The workshop will start with five mini-talks about the implementation of games in participation and democratic design. Case studies will explore analog and digital games in the context of European and American participatory planning and design. During the talks, workshop participants will take part in interactive and playful learning experiences that are specifically designed for the workshop and create collaborative concept maps. Concept maps will lead to group discussions, where participants share their experiences and incorporate their ideas and thoughts related to design, landscape, democracy, engagement, games, and other presented concepts. Discussion topics may include: 1) How do participants relate to landscape democracy and games? How are these concepts interrelated? 2) Why and how are games implemented in participatory planning and design? What could be the benefits of such games on personal, community and environmental levels? 3) What could be the challenges of using games in participatory planning and design?

E -Transform Processes and Outcomes for Equity

What Is Your Favorite Research Method? Why? (Workshop)

Karen Franck (New Jersey Institute of Technology); Sharmin Kader (TreanorHL); Linda Nubani (Michigan State University); Claudia Bernasconi (University of Detroit Mercy)

How can architects and interior designers apply LEAN principles in the design development of emergency departments in response to increased client demand to implement strategies for streamlining emergency department flow and efficiencies? LEAN thinking emphasizes reducing waste or non-added value elements from workflow processes to provide greater value for customers and improve staff, patient, and equipment flows.

This study explores an architect-led LEAN event with the client's key stakeholders, creating patient and staff journey mappings. Current state was mapped to identify areas for process improvement. Ideal state mapping formed the basis for floor plan concept development.

We then employed space syntax analysis and systematic layout planning distance evaluation to compare two architectural concepts for objective evaluation of cost, patient targeted visibility index (TVI), and accessibility outcomes. The systematic layout planning analysis evaluated nurse, physician, and nurse tech travel distances between nurse stations, triage, trauma, disposition, radiology, CT, and exam rooms.

LEAN workshop findings included co-locating CT with Trauma Rooms to decrease stroke response time, performing triage registration in triage rooms, opportunities to eliminate redundancies, and providing ambulation space for behavioral health patients. Diagramming found additional redundancies in FTE's and activities. Distance evaluation analysis showed that one of the concepts provided better spatial access for clinicians yielding $57,921 annual operating costs savings. This concept had higher TVI values, indicating better patient head visibilities from nurse stations to examples.

edra51

Tempe, Arizona

Abstracts - Pre-Conference Intensives

A -Transform Design for Resilience & Sustainability

Dimensions of Residential Environments in Transforming Communities Toward Resiliency & Sustainability

Organizers: Ali Momen Heravi (University of Illinois Urbana-Champaign); Keith A Miller (University of Illinois Urbana-Champaign)

Building resilient communities requires more than reasonable bureaucratic planning policies; it must have input and feedback from local stakeholders in order to adapt to ever-changing contingencies, facilitating greater well-being for individuals and their communities. Well-being is intimately entangled with quality of life, giving the world's residential settings a central role in the achievement of well-being. Acknowledged as the primary setting for human habitation, residential environments should enable us to relax, be ourselves, and have the highest degree of control on our interactions with the outside world. Residential environments should also offer optimal support for well-being's many facets. In the context of creating resilient communities, specific physical requirements for residential environments change as social, political and economic forces instigate cultural transformations.

Six independent presentations spanning a range of housing types, user groups, and geographic locations are slated for this session, highlighting core research concerns and methods at the intersection of sustainable resiliency and well-being in urban residential environments. Ranging in scope from room design to community planning, this intensive explores questions of indoor living enhancements, lifespan housing alternatives/ageing-in-place, strategies to improve social capital in residential environments of vulnerable populations, and policy considerations to encourage residential environments that are more supportive of human well-being broadly. Following individual presentations, we will conclude with a round-table discussion of specific topics presented, aimed at subsequent collaboration for future research projects.

Social Immediacy in Planned Community Design: The Problem of Macro-Scale Architectural Programming and Lost Opportunities for Social Capital

Keith A Miller (University of Illinois at Urbana-Champaign)

As New Urbanism gains popularity within the field of real estate development, a greater number of urban designers seek facility with this stylistic vocabulary. A common tendency for practitioners is to quickly comprehend the traits of the language without first understanding the purpose behind those traits. Specifically, New Urbanism design standards require a porch or balcony for every dwelling, situated within close proximity to a common walkway along streets; however, the fundamental purpose for such amenities can easily be overlooked.

Adhering to Edward T. Hall's well-established dimensions of the "social-space" zone (8-12 feet), these elements are expressly intended to foster casual interactions within a compressed behavior setting. The assumption is that within social-space proximity, pedestrians would readily make visual and verbal contact with those inhabiting these ubiquitous porches. While this assumption is valid in most cases, there are a number of common features that in reality can easily thwart this goal (such as vegetative growth and vertical distance not comprehended in a common site plan), thus inhibiting neighborhood-wide Social Capital.

This exploratory study critically examines the assumptions that underpin many New Urbanist design guidelines, taking examples from a current case study in Los Angeles to illustrate potential problem areas that result from over-generalizing programmatic goals across a large-scale development. The situated nature of "social space" will be closely examined, guided by Foucauldian descriptions of bio-power as influenced by characteristics of the physical environment.

Confluence of Design and Technology to Support Healthy Aging and Longevity in Older Adults

Altaf Engineer (University of Arizona)

Global aging statistics indicate that by 2050, 65 nations will have 30 percent of their populations aged over 60 years. Using a multidisciplinary point of view from the fields of architecture, interior design, health and wellness, and engineering, this paper will comprise a systematic review of relevant literature along with the author's own expert opinion to better understand opportunities and challenges associated with designing homes and living communities for older adults. It will discuss how wearable devices and environmental sensors that collect real-time individual physiological and psychological data inform smart and adaptive environmental control systems that help promote independent living and healthy aging in existing and new homes, as well as senior living communities. Potential for new areas of research in the field will also be discussed.

Utilizing Digital Visualization Technology to Support Aging in Place in the Interest of Sustainable Communities

Denise McAllister Wilder (NCIDQ)

Empowering aging individuals to safely remain in their homes and continue to be active in their neighborhoods contributes to the sustainability of America's communities. Strengthening housing opportunities is listed on the Urban Land Institute's Center for Sustainability and Economic Performance as one of the ten principals for building resilient communities (Urban Land Institute, 2017). The process to support aging in place begins with an assessment of a person's individual needs followed by designing appropriate modifications to the existing home (Cumming, 1999). Those modification must then be effectively communicated to the aging individual and their friends and family members. Without faith in the effectiveness of the proposed modifications, people are often moved unnecessarily from their homes where they are vital members of the community into big box retirement communities which often has a negative impact on the related economic environment (Chappel, 2004). The presented study looked at the effectiveness of using architectural visualization technology to better communicate recommendations. Existing communication methods were considered when gathering data regarding the confidence level in various common home modifications. Information regarding confidence in the same modifications was then gathered to ascertain the effectiveness of the digital walk through used in the study. Views of construction, health care and design professionals were considered as well as those of aging individuals and their friends and family members. Thoughts regarding the safety, attractiveness, constructability and ease of use of eight different modifications often used to support aging in place will be discussed.

C -Transform in Transitions

Transforming Inclusive Communities through Authentic Thresholds and Nature Based Solutions

Organizers: Leila Tolderlund (University of Colorado Denver); Robert Flanagan (University of Colorado Denver)

In this 1/2 day intensive, presenters will continue an on-going discussion about ways to expand the framework for defining and understanding authenticity as it relates to design and transformation of inclusive places and communities in today's world. Thresholds are gateways where transformation (can) occur. In these transition zones, time moves at a different pace unlike the world above, below and around. These portals of nature based solutions hold a latent potential of discovery, absorption, filtration, release, decay, and renewal. As we transition between time and place; inside and out; perception and truth; nature and city; landscape and architecture; and memories and dreams, we aim for authentic ways to root, re-root, and restore a sense of place respecting the spirit of place and the spirituality of its indigenous people.

In this session we will reveal alternative ways to thicken, grow, use, and occupy transitions zones. We will explore: 1) transitions and thresholds in place making as they relate to finding honesty and openings into new models for nature based solutions; 2) strategies for using local governments to incentivize public participation from communities and citizens supporting democratic futures; 3) participation and design of ever-changing cultural transformations free of secondary process; 4) challenges related to 'authentic place-making' vs. place simply being co-authored by human and other-than-human transformational processes over time; 5) how thresholds and connections to 'new' places can be made after dis-placement; 6) circular understanding of place and time supporting transformative threshold; 7) the idea of inclusive 'of-the-place' communities taking into account the spirit of place and the spirituality of its indigenous and native people; and 8) place-based design and transit thresholds to evolve and transform local communities to more authentic ways of being, seeing, hearing, smelling, touching and feeling in the 21st century.

We Are Such Stuff As Dreams Are Made on

Joseph Juhasz (University of Colorado Denver)

It would appear that we human beings are at the threshold of transforming this planet into hell on earth. Fortunately, we have not yet crossed the threshold. Multiple thresholds are open to us, and as environmental designers one of our tasks is to steer constructed environments in non-suicidal directions. One of these paths is the one I choose to explore: how rooting and grounding places into the specificities of their ecological, natural, genetic, historical-cultural, setting can avoid the inauthenticity and lack of community that produces suicidal tendencies like global warming.

In effect this requires being in harmony with the spirit and the spirits of the specific locality. This is the way to overcome the deracination of contemporary communities—their selfishness and short sightedness. The Faustian bargain not been sealed yet. We have to have hope and we have to act with faith and charity hope. We can transform ourselves by transiting available thresholds.

The avenue that I propose and that I pursue is one that involves all of the senses in unison and thus is not "purely rational" in a verbal-visual-logical path. A blend of reason, emotion, faith, dreams, visions, and hard-nosed realism. In other words creativity, not discovery, nor invention, and much less, innovation. In doing this, we cannot borrow spirits, faiths, visions, and vibrations from other places, other times, or from people who are in one sense or another not rooted to and in a particular place. Certainly one does not have to be "religious" in order to found, maintain, transform, and evolve authentic human communities. That truth that the spirits of a place are real is not based in religion. It is based on universal experience. We are on the THE threshold.

Authenticity in the Cloud

Robert Flanagan (University of Colorado Denver)

Successful implementation of a collaborative and diverse professional/academic strategy in environmental design is a three-part consideration: first, identify diversity goals and objectives; second, define collaborative work practices, means and methods; and third, empower a curriculum simulating collaborative work practices in the cloud. This discussion addresses the latter, where work practices challenge and sometimes mute collaborative voices of engagement in the profession and the academy. Diversity, Technology, and Authenticity are an incongruous mix. In "Diversity in the Profession of Architecture," Elizabeth Chu Richter spoke to the AIA's collective mission, "To grow a robust and valued profession prepared to serve the needs of people young and old, rich and poor—all hungry for better communities, better infrastructure, and better lives—our profession requires talents as diverse as life itself." She identifies an enduring conundrum, "In a world where technology seems to be the driving force in how we act and react, maintaining the human touch has never been more important." (Richter, 2015).

In a vexing complication, the driving forces of technology often negate voices of participation. BIM focuses on tangible metrics of productivity – but often at the expense of authenticity, the domain of conscious and unconscious thought. Process devolves into past practices of the sequencing of paper drawings, a disjointed approach favoring a pastiche interpretation - simulacrum derived of common practice.

Authenticity, Pedagogy, and Technology

"[In] a 2015 AIA survey ... half reported that finding candidates with either the required technical skills or project management skills was a major problem..." (AIA Architect, Baker, 2018). BIM technology (the vehicle) is a superb modeler/consolidator, but it is not facile in accommodating multifaceted collaborative input (content). This paper seeks discussion on contemporary pedagogy and collaboration in the cloud, its impact on diversity of design voices, and innovations to empower next-generation designers.

Against Spectacle and Placemaking: Authentic, Eidetic and Aesthetic

Joern Langhorst (University of Colorado, Denver)

The term "placemaking" has a long and unfortunate tradition in spatial design and planning disciplines. This paper presents a critical position and offers an alternative by exploring "authentic" as value and relational phenomenon that underpins meaningful human-environment relations. Debord's concept of "spectacle", and Baudrillard's notion of "hyperrality" provide critical perspectives on many design projects, in particular those that occur in urban conditions and are set within various diverse communities and neighborhoods.

Corollary to such projects' insidious instrumentality within the neoliberal social engineering inherent in most contemporary urban renewal projects, they represent the "inauthentic" -what Dovey calls the "phenomenon of fakery" which is the "replication of environmental meaning through manipulation of appearances....which breeds doubt and deception in person-environment interactions." (Dovey 1985, 33) and ultimately resulting in "placelessness" (Relph 1976).

The paper argues the necessity of authenticity, "a direct and genuine experience of the entire complex of the identity of places – not mediated and distorted through a series of (quite arbitrary) social and intellectual fashions about how that experience should be, not following stereotyped conventions." (Relph, 64) is a central component of the emergence of place, and that authentic places cannot be made. Authenticity here is "not of environmental form, but of process and relationship... and as a process is characterized by appropriation and an indigenous quality" (Dovey, 33), in fact "authenticity is the very source from which form gains meaning (ibid. 34).

Authentic then is relational in many ways, not just between subject and object, and "eidetic" and "aesthetic" are two ways to frame the processes of experience and interaction by which "meaning in place," a sense of place and self, and ultimately place itself are formed. This approach firmly establishes place as co-authored by human and other-than-human processes and subjects over time, and as something that cannot be "made".

Indigenous Places and Place-Making and Thresholds of Time and Place

Lynn Paxson (Iowa State University)

Place-making and places that celebrate Indigenous core values and respect the cultures and environments they are part of can be supported using thresholds of transformation. Indigenous peoples see themselves as part of places, just one part of places. Although many of our Indigenous peoples were moved off their traditional lands, they made place in the places they were forced into. Also, as land-based practitioners they recognize the spirit and power of places that is part of spaces themselves. They see themselves as just one of the communities or 'peoples' that populate the earth.

T his means that there is a recognition of all of the peoples of nature – the 4 legged ones, the winged ones, the plants etc. Understanding place and time as circular rather than linear supports many transformative threshold options. Using community based design processes, and explorations of thresholds can Native Nations create places that not only support their spirituality but also the spirit of places themselves? Can this also support the design of ever-changing cultures while resisting colonization, assimilation, adoption, or adaptation? These topics will be explored using various contemporary Indigenous projects and examples.

Speculative Authenticity in Design through Nature Based Solutions

Leila Tolderlund (University of Colorado Denver)

Beyond the need for defining landscape (and place) democracy, inclusion, and participation (Knudtzon 2018), Haraway (2016) suggests a necessity to find alternative 'ways of living for which there are no particular models'. She goes on to describe alternative combinations of living beings in 'an environment constructed with human and nonhuman companions'. This paper proposes an expanded framework for authentic, democratic, living systems design, inspired (defined as 'inhaling'/taking in a new breath) by the flow, patters, and guidance of the force of nature, 'Veriditas', and inclusion of all species and beings in a given place and space.

In a study of 12th century visionary Hildegard, historian of medicine, physician and author of 'Slow Medicine", Dr. Sweet, describes Hildegard's use the word 'Veriditas' as both: 1) the power of plants to put forth leaves and fruit for human survival, and 2) the sense of a related intrinsic universal power of human beings to grow and to heal, symbolizing the spiritual and physical connection to place we depend on with nature (Sweet 2018).

This paper points to alternative design methods using 'Veriditas', nature based solutions, intuition, and spirituality (defined here as not 'fashionable' spirituality, but a spirituality grounded, rooted, and grown out of and into place), paying close attention to local thresholds and transitional democratic (in the sense of 'for all species') relationships with nature and the landscape around and within us.

Case studies from five advanced living systems studios, guided by the idea of 'Veriditas' and biomimicry studying new potentials for inside/outside/, city/nature, and landscape/dwelling transitions zones, will be shared to initiate a discussion about alternative approaches for future human habitat design, using holistic, integrated, and immersive methods to optimize and honor democratic inclusion and authenticity of place and space.

edra 51

Tempe, Arizona

Abstracts - Video Shorts

D -Transform Places with People and the Environment Transform Design for Resilience & Sustainability

BLUE SPACE As a Therapeutic Concept "the Applicability in Healthcare Environments".

Fares Alsaygh (Texas Tech University)

Healthcare waiting areas are one of the most stressful places for the patient families which provide care and support to the patients. The literature on evidence-based design aims to provide an efficient solution to reduce stress and anxiety in such public spaces. However, the impact of bluespace (water feature) in healthcare waiting areas is yet to be explored. The water feature provides multiple sensory benefits; such as sound, sight, etc.

The method of the study depended on practice-based research and evidence-based design. A local existing space (courtyard) has been chosen in a medical building to be improved by implementing the research design recommendations.

The empirical evidence clearly indicates the therapeutic effect of green (vegetation) and blue space (water features) in the field of landscaping. A systematic review of the literature was conducted to scrutinize the impact of water features and water spaces (blue spaces) as natural design elements on occupant's health and well-being. Considering the complexity and strict regulation of infection control requirements, this study aims to present set of design recommendations for including blue space in a healthcare environments.

Findings indicate positive psychological response such as alleviating stress and anxiety. These recommendations may lead to a compelling physical design which leads to the well-being of the family and support for the patients. The aim of this study was to answer the following research questions:

Question 1. How can blue space, as a design concept, be applied in a healthcare facility? Question 2. Based on the relationship between green & blue spaces, what are the restrictions that might prevent blue space concept from being applied in certain indoor environments?

edra15
Tempe, Arizona

Abstracts-Posters

Role of Physical Environment of Long-Term Care Facilities on Dementia Care in Sweden and Canada: Focus Group Interviews with Staff

Sook Young Lee (Yonsei University); Lillian Hung (University of British Columbia); Habib Chaudhury (Simon Fraser University)

This study explores staff perceptions of the physical environment in Swedish and Canadian dementia care facilities. We highlight complexities of how care was organized and influenced by the physical environment of the setting. An exploratory, descriptive qualitative research design based on focus groups was used. A total of 24 staff members from four facilities, two in Sweden and two in Canada, participated. Data were analysed by thematic analysis. Staff in both countries reported similar physical environmental characteristics that enabled and hindered them to deliver good care. The deficits in the physical environmental characteristics prevented staff from providing effective person-centred care. Our findings identified substantial differences between the facilities of the two countries, although it is possible that greater differences might exist between the range of facilities in each country with regard to the quality of environment and care. The quality of environment contributed to a high job satisfaction reported by staff in Sweden. The unsupportive and problematic features of physical environment for persons with dementia seemed to be the primary factor that triggers agitation among the residents with dementia in Canada. The variations in the physical environmental characteristics and quality of care suggest the value of comparative research in identifying and exploring the possible causes and consequences. Future development in nursing home requires a better understanding of staff experiences and staff involvement in design.

Destruction and Recovery: Flight 93 National Memorial and the Transformation of the Rural Landscape

Diane Nicodemus (University of Pittsburgh Johnstown)

On September 11, 2001, United Flight 93 was hijacked by Islamic extremists and crashed into an active coal mining site in Shanskville, Pennsylvania (Emile, Pizarchik and Mahan, 2016). This study examines how the designers of Flight 93 National Memorial materially and symbolically developed this highly-polluted rural landscape to support its interpretive theme of heroism and its mission of collective psychological healing. Further, it considers how the memorial shapes visitors' beliefs and understanding of the historic ecologically and socially destructive mining-industry practices of rural places

This analysis is based on a semiotic reading of Flight 93 National Memorial and its immediate surroundings (Barthes, 1968, 1977; Ledin and Machin, 2018). It describes the way in which this specific site constructs and transmits its own values and system of meaning, as well as the way visitors are instructed, emotionally involved and guided through the space on their visit.

It concludes that Flight 93 National Memorial is a performative space where visitors are immersed in a world of horror from which they emerge transformed through a process of sanctification (Violi and McEwen, 2017). The trauma is reenacted through a series of devices (i.e. reconstruction of settings) that create distinct physical and psychological experiences that may transform visitors from empathetic visitors to vicarious witnesses or even secondary victims of the trauma.
Thereafter, visitors begin their transcendence from grief to renewed hope of a better future as the degraded landscape becomes a place of environmental and symbolic healing (i.e., wetlands and reforestation). Restoration of this natural landscape reminds visitors that places are not static. This is a landscape open to change and to new ways of understanding the relationship between ourselves and the natural environment. As fossil fuels become obsolete, we must re-evaluate old attitudes and develop new attitudes about a place and an event.

Bridging the Gap between Idealized and Attainable Infrastructure Sustainability and Resilience

Yeinn Oh (Georgia Institute of Technology); Jun Wang (Georgia Institute of Technology); Nevedita Sankararaman (Georgia Institute of Technology); Osvaldo A. Broesicke (Georgia Institute of Technology); Yilun Zha (Georgia Institute of Technology); John Crittenden (Georgia Institute of Technology); Ellen Dunham-Jones (Georgia Institute of Technology)

Many suburban projects within the United States that have been redeveloped to improve sustainability do not measure the actual performance. The study measures the performance of various suburban retrofits and quantifies their before and after performance across energy, water, transportation, and social infrastructures. We analyze a database of 1,500+ case studies by two different methods. In the first method, we count the frequency of the appearance of relevant keywords in the database. In the second method, we perform a deep-dive analysis into eight case studies investigating the actual performance in each topic. Our keyword analysis reveals that changes in transportation infrastructure are the most common improvement, followed by improvements to social, water, and energy. Our deep-dive analysis reveals that: 1) the energy use intensity decreases by up to 40%; 2) permeable area in most cases increases by 10%, surface water runoff decreases by 0.02 gallons per acre; 3) intersection density increases by 54%, block size decreases by 65%; and, 4) average home prices increase by 60%. In most cases, the selected deep-dives improve upon two or more infrastructures. The result implies that the sustainability retrofit in suburban projects improves among energy, water, transportation, and social infrastructures.

Thermal Delight, Impact of Landscape Design on Building Energy Performance Using Computer Simulation Tools

Hossein Entezari (Iowa State University); Victoria Goetz (Iowa State University); Ulrike Passe (Iowa State University); Mira Engler (Iowa State University)

The likelihood of extreme heat events is predicted to exacerbate the heat storage capacity of the densely built environment of urban areas. Meanwhile, many residences in low-income neighborhoods may lack central air-conditioning systems (e.g. up to 50% of low-income homes in Polk County, Iowa) to face this harsh climate. This study investigates the influence of landscape design strategies on building energy consumption in various climates and urban typologies. The goal is to evaluate microclimate effects exerted by planting design in/on/around buildings to improve landscape design prediction for energy saving. Although efforts to integrate these effects in combined building-microclimate energy models have been attempted (Taleghani et al., 2016), there are no convenient tools which integrate landscape consideration with a construction energy performance (Hashemi et al, 2018). This research collected data assessment from DIVA (Design-Iterate-Validate-Adapt), radiation plug-in program for Rhinoceros. By utilizing EPW (Energy-Plus-Weather) files, assigning materiality and energy simulation nodes, and producing the false-color radiation maps, we were able to determine the radiation impact on building surfaces in four different urban density scenarios (Passe & Battaglia, 2015) in three cities–Des Moines, IA; Phoenix, AZ; Miami, Fl–and in three seasons. We then developed an applicable design matrix and added 3D forms of plant systems to the models. The scenarios illustrate a considerable reduction in radiation per area of exterior building surface and heat gain using the green wall strategy, thus reducing building energy consumption in summer. The research further developed landscape guidelines for the aforementioned spatial conditions. Finally, the urban modeling interface (UMI) demonstrates a method to predict correlation between a landscape strategy and the anticipated reduction in building energy consumption for a chosen urban density. The findings enhance our understanding of the relationships between natural infrastructures, building energy efficiency, and urban microclimate.

Using Biomimicry to Respond to the Effects of Sea-Level Rise in Coastal Cities

Cynthia Fishman (Biomimicry Design Allaince)

As climate change progresses, coastal regions and cities are feeling the effects of sea-level rise caused by rapidly-melting glaciers and ice caps. These regions are experiencing "sunny day flooding", where the sea-level rise regularly pushes high-tide levels past the point that coastal infrastructure was designed to accommodate, even without a storm event. Roads, sidewalks, and green areas are inundated by sea water coming over sea barriers, up through sewers, and even up through the ground water. This situation will only worsen - it is estimated that by 2100, sea levels could rise by a foot (Meyer 2019). With over 126 million people, or 40 percent of the nation's total population living in coastal cities, sea-level rise will have a major impact (National Oceanic and Atmospheric Administration [NOAA] 2013). Using biomimicry, (the process by which humans look to nature to solve design challenges), B.L.O.T.S. is a prototype technology designed to mitigate the effects of "sunny day flooding".

Learning from numerous organisms, the design manages water by relying on structure, material, and earth's operating conditions, in order to provide an adaptable and resilient solution. From the small intestine, the team learned how to absorb fluid; from lymph vessels the team learned how to passively direct fluid; from mangrove roots the team learned how to filter particulates from fluid; and from the combination of the hottentot bread, resurrection fern, and lungfish, the team learned how to store fluid. B.L.O.T.S. has numerous economic and social sustainability wins including reducing the need for rapid infrastructure improvement, reducing the risk of lack of access to basic necessities, and helping to protect potable water sources and other essential utilities while empowering residents to take control of an emergency situation.

Pedestrian Thermal Exposure in Urban Parks in Tempe, AZ

Ananth Udupa (Arizona State University); Florian Arwed Schneider (Arizona State University); Jennifer Vanos (Arizona State University); Paul Coseo (Arizona State University); David Hondula (Arizona State University); Braden Kay (City of Tempe, Arizona); Ariane Middel (Arizona State University)

Extreme heat poses many risks to human health and is a major public health concern. In hot desert cities, pedestrian thermal exposure is mainly driven by mean radiant temperature (MRT), a parameter that integrates the longwave and shortwave radiation a body is exposed to into a temperature value. This study investigates shade coverage and resulting outdoor human thermal comfort at playgrounds and busy pedestrian routes that lead to those playgrounds in four public parks in Tempe, Arizona. Hourly microclimate transects were conducted using a mobile biometeorological weather station during four hot, sunny days in July and August 2019. Additionally, during the transects, hourly surface temperatures of shaded and sun-exposed playground equipment were sampled using an infrared gun. Thermal conditions were evaluated for different times of day and various shade situations. At Kiwanis Park, one of the four parks of interest, MRT averaged 57.1°C along the noon transect with a maximum of 74.4°C and a minimum of 35.9°C, which equals air temperature in the shade. A newly designed multi-use path provided the least amount of shade with an average MRT of 70.2°C, while a tree-lined path in the park exhibited an average MRT of 47.7°C. In contrast, mid-day playground equipment surface temperatures peaked at 44.4°C under a shade sail. These results were then compared to survey results from local community members who participated in a community outreach program with ASU to further justify the findings. For instance, the extremely hot conditions at the newly designed multi-use path and unshaded parts of the playground demonstrates the necessity of shade and other heat-protective measures to reduce thermal stress on park users in hot desert environments during the summer months. The findings from this research will inform 'cool' space design guidelines for playgrounds and pedestrian walkways in the City of Tempe.

Bridgin' the Burg: Revitalization of Downtown Williamsburg

Jeongok Park (University of Kentucky); Melissa Habermehl (University of Kentucky); Jayoung Koo (University of Kentucky)

Williamsburg, Kentucky, is the county seat for Whitley County which is located in the southeastern Appalachian region of Kentucky. Although a small rural campus town, Williamsburg lacked connections between the downtown and the University of the Cumberlands, a central area of town with a high potential for economic activity. The surrounding areas in the county include popular tourist destinations such as the original Kentucky Fried Chicken (KFC) cafe and the Lake Cumberland State Resort Park. Therefore, the focus of "Bridgin' The Burg" was to facilitate economic development opportunities by enhancing downtown Williamsburg and the connections to the Cumberland River riverfront. The proposed design interlinks the two areas to improve connections and walkability in downtown Williamsburg, and ultimately promotes a healthier lifestyle through increased physical activity. The design processes included inventory and analysis, community engagement, and case studies focused on strategies such as streetscape enhancements, and riverfront development and restoration.

"Bridgin' The Burg" emphasized connecting two essential elements that could improve economic activity in central Williamsburg: a walkable downtown and connections to the Cumberland River trails. Objectives of the proposal focused on strategies that would positively impact the presence of people in the downtown. Information gathered during the community workshop and online surveys identified key connection points throughout the town. Design strategies included 1) changing traffic patterns to increase the number of people flowing directly into the central area of the downtown; 2) enhancing the streetscape and amenities for a safer and comfortable pedestrian environment; 3) connecting the downtown to the trails along the Cumberland River which could become a stronger attraction for tourism and education; and 4) providing multiple access points to the Cumberland River. Overall, the design proposal aimed to improve social and physical connections between downtown Williamsburg, the campus, residential neighborhoods, and the Cumberland River.

Place-Making in New Geographies of Displacement: Serial Relocation after Wildfires in Southern California

Mariana Junqueira (University of California Irvine)

The Woolsey Fire (2018) burned 96,949 acres in Los Angeles and Ventura counties, prompted the evacuation of 295,000 people, and destroyed 1,643 structures, among which, at least 400 were homes. Fire, ecology, and life in California are intimately connected (Halsey, 2008), akin to other disasters that are typical in other states. However, amid extended housing shortage (California was ranked 49th among the 50 United States in housing units per resident in 2018), the abrupt loss of homes to disasters worsens the existing imbalance between supply and demand.

Proposing a four-stage process of post-disaster housing that is adequate to this context, Quarantelli (1995) differentiated sheltering from temporary housing by describing the latter as an essential step to recovery, as a household regains some sense of normalcy. Studies show that people navigate sudden changes in home environments with difficulty, but the commitment to remake places remains strong (Fullilove, 2013). This research, however, initially aimed at investigating place-making in temporary housing after the Woolsey Fire, found that "burned-out" residents have often been displaced for months, and potentially years, rarely settling into long-term temporary housing. Further, their relocation efforts have overlapped to those of people displaced in other fires (Thomas, 2017, Hill, 2018, and Getty, 2019), intensifying the issue.

Data from qualitative interviews inform these preliminary findings suggesting that barriers to making suitable living arrangements amid scarce housing options (1) led to serial relocation and (2) prejudiced peoples' experiences in remaking places. Concluding, the desire to remain local exacerbated constraints and revealed the geographies of displacement through systematic mapping of participants' serial relocation. Also, participants' experiences revealed how Southern Californians view and cope with disasters, see themselves and their environments, and remake places. Lessons inform the creation of more suitable temporary housing processes and solutions, providing displaced populations with better opportunities to recover appropriately.

Ready! SET! Activate! Using Green Infrastructure to Promote Health Equity in South Phoenix.

Amanda Trakas (Arizona State University); Elizabeth Ferguson (Arizona State University); Kevin Scholfield (Arizona State University); Chingwen Cheng (Arizona State University)

Ready! Set! Activate! Is a research-based schoolyard redesign proposal as part of an ongoing collaboration between Arizona State University and Paideia Academy, a K-8 public charter school located in south Phoenix, Arizona. The five-acre school property experiences persistent environmental justice issues endemic to the greater South Phoenix area, including flooding, excessive urban heat, and poor air quality in the context of a historically racially segregated area of the city with limited access to public green space1. Because of the school's location in an alluvial floodplain of South Mountain, green infrastructure interventions, including the creation of bioswales and bioretention for stormwater, are critical to mitigate seasonal flooding and improve environmental site conditions while making the site a regional model for effective implementation of these low impact development techniques.

Beginning with a series of community engagement workshops, this project has evolved to envision the use of green infrastructure at Paideia to create resilient natural learning2 and play landscapes that promote health equity3 by improving the quality and access of green space for the student community. Using Padeia's whole-person education paradigm to guide the goals of the project, the aim is to begin the conversation on how school landscapes can address the immediate health challenges of the site while creating a welcoming and healthful green space amenity for the community.

Promote Green Transportation through Evidence-Based Design of the Signage in Railway Stations – Case Studies in China

Hongbo Wu (Henan University); Zhe Wang (Henan University)

The Signages of traffic stations are the bridge between network navigation and realistic wayfinding. Because the links between various transportations are not close enough, the exit halls and transfer spaces of many rail transit stations are considered as the end point of a trip, thereby reducing the design of the Signages. Intercity trains and high-speed railways are large-volume, convenient and fast public transportation. Taking high-speed rail and intercity itself is already a green trip. However, their coverage is limited, and many travelers still have to change to other modes. Establishing a guidance system based on green traffic signs at interchange stations can effectively increase the utilization rate of low-polluting vehicles and mass transportation modes. Study aims at the problem of the low efficiency of the Signages of the transfer space at rail transit stations, and uses evidence-based design research methods to use the orientation signs of travelers in the interchange spaces of multiple hub sites in China. Collect empirical data and evaluate the current status of the transfer space infrastructure.

The observation method is used to record the interaction behavior of the outbound pedestrians with the Signages during the transfer process, and the residence time. Then, the questionnaire survey method is used to collect data on the travel methods of the Signages users through visual tracking and score. The degree of influence of the Signages in the interchange space on the passenger's transfer to green vehicles is obtained. Finally, the analysis is based on the relationship between the same kind of infrastructure scores and observations and questionnaire data between the transfer spaces at different stations. The research aims to improve the utilization rate of green transportation in the transfer process through the Signages design, to provide ideas for similar interchange station guidance sign design.

Research on Traffic Conflicts on Hospital Sites and the Environmental-Design Intervention - Case Studies in Central China

Zhe Wang (Henan University); Erping Hu (Henan University)

In urban areas, the rapid increase of populations and varied vehicles leads to frequent conflicts between active (e.g. driving) and passive traffics (e.g. walking). This phenomenon is common in small cities in China, especially on the sites of local hospitals, due to limited healthcare resources. Hospitals are challenged by heavy and complicated traffic flows associated with patient visits and services.

The aim was to identify active-passive traffic conflicts in and around the hospitals and investigate possible solutions from the perspective of environmental design. Focusing on site-level environmental design factors and traffics, case studies were conducted in three hospitals in Kaifeng and Sanmenxia. In three workdays, field observations were conducted to collect the data of environmental design (e.g., site layouts, buildings, roads, and parking lots) and on-site traffic (e.g., motor vehicle/ non-motor vehicle/ pedestrian flows and conflicts) in two-to-three observation zones, with an emphasis on the areas near to the main entrances/exits. These were conducted for 90 minutes in each zone during three time periods (morning, mid-day, and afternoon). Factors of site planning and design, traffic pattern, active-passive conflict, and parking-lot accessibility and usability were analyzed.

The hospital entrances and surrounding areas suffered serious traffic congestion and active-passive conflicts. The average number of parking spots was 330 in each of the hospitals and the occupancy rates were averaged to 89%. Traffic congestion and conflicts were associated with the inappropriate design of motor parking entries and the limited accessibility of spots for non-motor parking. To promote sustainable urban development and reduce traffic conflicts, the site planning of a hospital should pay more attentions on the entrances/ exits and the need of non-motor drivers for parking in the context of local cultures. From the perspective of environmental design, design and renovation guidelines for creating efficient traffics in hospitals are discussed.

Investigate Energy-Saving Standard-Floor Plans for an Inpatient Ward Building in Zhengzhou

Zhe Wang (Henan University); Fuhua Yang (Henan University)

Environmental problems such as heat island effects and smog in urban areas have plagued people, and the attention on green buildings has gradually increased. Can the standard-floor plan of a ward building contribute to energy efficiency? Which kind of floor plan consumes the least energy? For building energy efficiency, scholars have studied and discussed the energy efficiency of residential buildings and office buildings, but the research on hospital ward buildings is very limited.

Based on the theory of Evidence-based Design, this paper used combined methods of data collection, including on-site observation, interviews, and software simulation. Firstly, on-site observations of three representative ward buildings in Zhengzhou city were conducted and the empirical data of 11 features/indexes identified for software simulation were collected. Based on the observations, eight standard-floor plans were created for the comparison of energy consumption and relevant 3D models of buildings applied these plans were developed for software simulation.

The software 'Design Builder' was used to simulate the energy consumption of each model. By comparison, the three most energy-efficient floor plans were identified as follows: Double Corridor, Radiation, and Single Corridor. It was found that the double- corridor plans is more suitable for the ward building in the Zhengzhou urban area of this study. The double-corridor plane has better accessibility to the nurses' station than the single corridor plane, the work efficiency of medical staff is higher.

From the perspective of environmental design, this research suggests to promote natural lighting, spatial applicability, proximity between medical and nursing streamlines, and visibility of nurse stations for energy saving in ward buildings. The findings can help medical designers and managers to consider the impact of building layout on energy saving in hospitals. More is needed to investigate energy-saving design solutions for energy consuming buildings.

Climate Change Adaptation in Ladakh

Carey Clouse (University of Massachusetts Amherst)

Drawing upon a collection of traditional and contemporary design interventions, the poster highlights the variety of examples of climate adaptive responses in Ladakh. Ladakh, a high-altitude region of India, has been on the front lines of climate change due for over a decade (Grossman, 2015; Mingle, 2015). In this challenging physical context, cultural values of self-sufficiency and community care stand out as central features of the long-term functioning of village groups (Crook & Osmaston, 1994; Gutschow, 1993; Norberg-Hodge, 1991).

The eight strategies explored in this poster include: Artificial Glaciers: ice stockpiles at 14,000';Ice Stupas: ice towers created to retain water (as ice) for use in the spring; Snow Barrier Bands: stone walls used to catch and divert snowfall at 17,000'; Solar Prospects: passive solar, photovoltaic electricity, and solar hot water; Greenhouses: year-round agriculture; Zings, Reservoirs and Canals: water management and dispersal; Food Aid: the current state of the Indian ration program and other forms of assistance; and Tree Planting and Armor: husbandry, cultivation and creative armoring devices. The content for the poster builds upon the observation field research conducted in Ladakh by the author over a five-year period.

Study on the Mechanism of Sustainable Community Cultural Construction Based on Cultural Heritage

Feifan Jiang (Zhengzhou University)

Cultural heritage is the precious wealth left by history to mankind. In order to fully respect its historical value, the construction intensity around it is often strictly controlled, so it is often regarded as the resistance of urban development. In the new development period, cultural heritage, as an intuitive material carrier of urban culture, has played an important role in building the spirit of the city and establishing the unique cultural image of the city, and has become an important driving force for the sustainable development of the city.

With community residents as the main body and culture as the path, the cultural heritage will be integrated into the "people-oriented" sustainable community renewal construction, so that the citizens can fully understand and cherish the precious cultural heritage. Based on the experience of Singapore, London and other places, combined with the practice of cultural heritage protection planning in Zhengzhou, a sustainable community cultural construction policy and planning with a long-term cultural vision is formed to attract the active participation of community residents, so that community residents can become participants and creators of culture.

Take cultural heritage as the link of community cohesion and enhance the pride of community residents. Actively improve the traffic to promote the systematic display of cultural heritage in the community; hold high-quality cultural heritage community discussion activities to form a multi-cultural communication path; build community cultural facilities to guide urban space construction, promote community renewal, and strengthen the popularization of cultural heritage; etc. With these methods, we can provide more vitality planning and design for cultural heritage and more quality living space for community residents.

Transcendence- Safe and Sound for All

Phu Trinh (University of Texas at San Antonio); Neda Norouzi (University of Texas at San Antonio)

Client integration is important in creating a holistic design that responds to user's needs and affords architect's creativity in discovering the best solution. This house was designed for a young adult who described themselves as reserved, trustworthy, adoptable, talented, competitive, direct, focused, open-minded, creative and analytical. The client expressed the desire of aging-in-place in an environmentally sustainable home. An intelligent, universally designed, passivehaus was proposed to respond to the client's personality and living requests. A randomly selected group of 64 young (ages 20-32) and old (ages 65+) adults were interviewed to better understand how technology can serve the needs of people of different ages. The results of the interviews influenced the use of technology in this house. Building a house with universal design standards far exceeds the aging-in-place specifications as it makes the house usable by all people, to the greatest extent possible, without the need for adaptation or specialized design. The spatial design, product selection, and color choices were intended to accommodate young children, older adults, and people with variations in physical and cognitive abilities. Some of these features are non-slip flooring, adjustable counter tops, drawer storage, pull down shelves, pull out, and open shelves. In respond to the client's request for a sustainable building, the designer applied the passivehaus standards that increases user's comfort all while lowering energy use and environmental impact. The construction of the house is though high performing, airtight walls and windows to control the thermal heat gain with an energy recover ventilation unit that supplies the house with constant fresh air. Another main component of the house is incorporating artificial intelligence (AI) in order to connect the link between universal design and sustainability. AI offers the necessary tools for building systems that identify behavioral patterns, monitor and surveillance the residents, and maintains energy sustainability.

Healing Village: Improving Sustainability of a Hospital and Well-Being of Users in Terms of Site-Plan Design

Hamed Yekita (Texas Tech University); Rana Hassan Bazaid (Texas Tech University); Amani Khan (Texas Tech University); Debajyoti Pati (Texas Tech University)

Sustainability is considered a critical issue in a healthcare facility in order to have the maximum outcomes. Healing Village student project was designed not only for patients inside the building but also for all people living in the surrounding community and are benefited from the site advantages. The goals of this project are spreading long-term health and well-being among neighbor, implementing surrounding natural energy resources, and encouraging people to develop their social activities.

In this redesign process, a new center was presented which is completely adaptive with its surrounding environment. This center will produce, consume, recycle, and influence based on its resources and needs. Analysis of the climate and the surrounding neighborhoods as well as challenging issues (such as air pollution) made this new center a good example of well-being for a city with high air pollution.

In this project, planting areas were considered to develop natural planting and encourage people to have healthy foods which can also be used for educational purposes for agricultural students. Additionally, a community center was considered for people are either inside or outside of the center that could improve the overall health of people through social interactions.

In addition, other features such as water and landscape were considered based on the history of the city, in order to reinforce the sense of belonging for users. While people can walk or seat on the shaded surfaces in warm seasons, the sunlight is being converted to clean energy in the same time by implementing new technology. As a result, the final design does show the meaning of the Healing Village which was inspired by the natural environmental resources around the project. The main significance of this project is to cause a huge impact on the conception, planning, and execution of the project community center activities.

B -Transform Education for Our Common Future

Teaching Landscape Graphics in Higher Vocational Education: The Benefits of Using Picture Books

Po-Sheng Hsieh (National Chung Hsing University); Hung-Ming Tu (National Chung Hsing University)

Landscape graphics is an important basic course to establish graphic techniques for landscape in higher vocational education. The first course of landscape graphics affects students' learning motivation toward landscape. In practice, traditional imitating drawing continue to be the major teaching methods in landscape graphics. However, traditional educational methods can lead to functional fixedness, obstruct creative thinking, and reduce learning motivation. The use of picture books has been an effective tool to promote cognition and knowledge, and induce learning interest and creative thinking in child education. Therefore, the purpose of this study is to explore the benefits of picture books in teaching landscape graphics in higher vocational education.

The study focused on the landscape graphics course (35 students) in the Taiwan's senior vocational high school. The 18-week course with picture books included line, circle, texture, gradient, font design, triangle, point–line–plane, and rapid design. The researchers used various picture books as teaching material based on the different course topics. The qualitative (observation and interview) and quantitative (formative and summative) evaluations were used to assess the benefit and change of learning motivation, creative thinking, problem-solving capability, and improvement of educational method.

The results showed that using picture books induced learning motivation and creative performance, and resulted in thinking about how to vary ideas in classwork and homework. With regard to improvement in educational method, using leading questions promoted the understanding and acceptance of the picture book quickly, after the initial first to third lessons; several students found it difficult to understand the meaning of the picture book in the first lesson. Although most students were not used to creative thinking, using picture books with several interactive and encouraging strategies can increase student confidence, promote hand-painted technology, and achieve the goal of the course in terms of graphic creativity and performance.

Exploring Building Information Modeling (BIM) in Interior Design Studio

Faten Yanksari (University of Minnesota); Abimbola Asojo (University of Minnesota)

The purpose is to explore the influence of using Building Information Modeling (BIM) in the schematic design (SD) and the design development (DD) phases in the interior design studio class. The data was collected from five senior interior design students. Few BIM studies have been done in the interior design discipline. Thus, there is a gap in the interior design body of knowledge. This study is a qualitative case study and the data collected are direct observations, documentation of the design process, and interviews.

The findings reveal a significant influence of using Revit on the design process. In the SD phase, Revit provides different representations and accurate measurements in the existing structure, which helps to explore and develop the chosen concept, spatial forms, spatial relationships, and conflicts identification. Different design methods could be used during the design exploration. In addition, Revit helps the forms of the design become more legible, which helps in human needs accommodation and design communication. In the DD phase, Revit provides a visual illustration of 3D modeling and simulation of virtual reality that helps to explore and develop design details, including the RCP, materials, and lighting. The trial and error approach could be used during detail exploration. In addition, Revit helps to conduct a series of renderings to reach the final design execution. This study suggests a new conceptual framework of Revit in the interior design process to facilitate the intellectual design work in studios.

The interpretation of this study is that students become more thoughtful in the design decision. The suggested new conceptual framework of Revit in the interior design process could provide a strong guideline and help to construct prompts. Using these prompts in design education can raise students' awareness about the design process.

Protocol Analysis of the Design Process Based on Extrinsic and Inherent Values

Mushfika Tasnim Mica (Oklahoma State University); Molly Elizabeth Jackson (Oklahoma State University)

The study provides the results of an experiment that analyzes the creativity design students would bring into their problem-solving process, based on their inherent and extrinsic values. The study's results show a greater effort on the researchers' part to identify how design students create, regardless of their learned or naturally creative abilities. For the purpose of this research, it will be assumed that anyone in a design field is innately creative. Freshman and senior level design students from two different programs were asked to solve a simple ill-structured problem. The researchers then analyzed the data provided from the studies, given from verbal protocol analysis and fNIR (Functional Near-Infrared Spectroscopy) data, to better understand how inherent abilities and learned abilities affect creativity. The information derived from the experiments was then coded to compare and contrast the abilities of the freshman who only have inherent knowledge versus the senior level students who have both inherent and extrinsic knowledge. The information provided from the study is vital to design faculty who must teach the design process to novice design students.

Ideation Generates Steam for Middle School in Columbia, MO.

Lyria Bartlett (University of Missouri)

Opportunities for students to work with communities during times of cultural shift are rare. Curating the delivery of a design studio to be mutually beneficial for students and clients is difficult. Architectural Studies students at the University of Missouri partnered with the Columbia STEM Alliance and Columbia Public Schools to explore the transition of Jefferson Middle School (JMS) to a STEAM-based magnet school. We were able to collaborate with JMS during 'year zero', the inaugural year of weaving a STEAM curriculum into existing pedagogy. Our design program was uniquely situated to facilitate the discussion between all parties and to implement Universal Design for Learning strategies (UDL). An excellent example of project-based and active learning, this partnership provided student engagement with clients as well as real-world troubleshooting for existing conditions.

Students were able to translate their knowledge of effective learning environments to aid in the transformation of a neglected school with a new agenda. Our team developed a comprehensive scope of this transition: including identification of key elements in a STEAM magnet school, curation of curriculum that will support this mission and the effect of existing and new infrastructure on these goals. Building Information Modeling and collaborative learning were critical components of our process. Balancing tangible and visionary elements of the program with desired technical and creative components, we were able to prioritize essential issues and propose meaningful interventions that will positively influence the conditions at JMS.

In its 120 year history, various non-comprehensive additions and renovations to JMS have resulted in a maze of inefficient spaces, minimal accessibility, confusing wayfinding, limited security, and challenging travel distances for occupants. Our partnership allowed us to generate data and prototypes that enable the school district to capitalize on their opportunity for growth, intelligently support effective learning environments, and drastically impact their community.

C -Transform in Transitions

Smart Home Design: Industry Engagement Project

Kyoungmee Kate Byun (Northern Arizona University)

The demand of high-tech devices facilitated within homes is increasing. While commercial design projects such as offices, hospitals, and schools have been developed as innovative high-tech facilities, smart home designs have been left behind. Additionally, commercial design projects tend to be planned from early stages to completion, and as a result, there are consistent design concepts for these innovative high-tech spaces. On the other hand, it is challenging to build a new house as a smart home from the beginning because typically, homes that are already constructed are purchased. Therefore, it is rare to see not only residential projects designed as smart homes but also its process as a collaborative team. The purpose of this project is to address how a designer can perform a smart home renovation with a collaborative team. A 4235sq.ft traditional house (basement: 1910sq. ft and main floor: 2325sq. ft) located in a mid-western urban city is provided to renovate into a smart home. An industry, which is well-known for advertising smart technology and devices, is engaged with this project to co-work with one main designer and four funded undergraduate students in interior design program. The design process including the programming phase, schematic or concept development phase, and design development phase is implemented. When design problems arise at each phase, the discussion to resolve them is recorded chronologically to describe the process through which a design team makes a decision. Upon completion, four funded students have semi-constructed interviews to gather what they learned from this project. As a result, this project describes the process of a smart home design and demonstrates the learning values for students.

Continuity, Change and Adaptation: Understanding the Transformation of Ahmedabad Fort Wall, India

Sweta Kandari (SPA Bhopal)

In 2017, UNESCO declared the 600-year-old walled city of Ahmedabad as India's first Heritage City for its outstanding universal values such as integrity, authenticity and the multi-cultural character of the urban fabric. The fort wall encapsulating Ahmedabad has been an integral part of the heritage city and has constantly evolved with it.

Patrick Geddes's 'Notes on Ahmedabad', an extremely descriptive report for the Ahmedabad municipality about the fort wall written in 1915 is taken as the starting point for the study. After a century, this paper revisits and explores the various influences of his document that may have led to the existing footprint of the fort wall and the conditions around it. Through the primary study of the fort wall and its adjacent area, the paper explores and understands the multiple ways in which people have appropriated and accommodated themselves along the fort wall. The study focuses on understanding the evolution and identifying the neglect, decay, repair and demolition of the fort wall surrounding the city of Ahmedabad.

Through the study of archival data, documentation and mapping, this study also analyses and brings forth the existing fragmented (disregarded, demolished, repaired and appropriated) remains of the fort wall that acts as an urban artefact. These adaptations represent the socio-economic conditions and depict various historical narratives. The research concludes with an understanding of the changing values and functions of the wall that includes the gates, the bastions and the settlement around it as a part of the built environment.

Symbolic Meaning of House Styles in the 21st Century, across Geographies and Disciplines

Kathryn Terzano (Arizona State University)

Built environment researchers have long understood that architects, planners, and other built environment professionals evaluate housing styles differently than lay people (Nasar, 1989; Stamps, 1999; Gifford, et al., 2002; Montañana, Llinares, & Navarro, 2013). The classic study of the symbolic meaning of housing styles (Nasar, 1989) used black and white artistic renderings of six housing styles (Farmhouse, Tudor, Mediterranean, Saltbox, Colonial, and Contemporary), surveying built environment professionals and lay people in two cities (Los Angeles and Columbus, Ohio) about three areas of symbolic meaning (evaluation, friendliness, and status). Using the same methods, this study builds on the classic study and expands it two important ways: first, this study included respondents in both New England and Arizona, which is important considering two of the housing styles predominant in those areas (Saltbox and Mediterranean, respectively) and these housing styles tend to be among the least favored in other locations. Second, this study surveyed built environment students as well as students majoring in leadership, the latter of which is a field that would be classified as "built environment lay people" but is important because of the symbolic meaning of status, where both the classic study and this present study ask the respondents to choose which housing style best represents where a community's leader would live. (Results are pending, as the survey is still open.)

Reduce Interruptions and Support Collaboration: Can the Workstation Design Support Physician Handovers in the Emergency Department?

Rutali Joshi (Clemson University); Anjali Joseph (Clemson University); Michelle Ossmann (Herman Miller); Larissa Coldebella (University of South Carolina)

End-of-shift handovers between physicians in the emergency department (ED) are crucial episodes of patient-related information exchange achieved through collaboration and communication. Communication breakdowns are known to occur in fast paced and dynamic ED setting due to frequent interruptions (Blocker et al., 2017) that lead to medical errors eventually. Approximately 11% of these medical errors are attributed to handovers. While some interruptions may be important to the functioning of the ED, others may be unnecessary, more so during handovers. Inspite of the impact interruption can have on work processes, no study has examined them in the context of cognitively demanding and crucial tasks like handovers. To date, healthcare research in the ED has focused primarily on the content and process of handovers, and little on the impact larger system comprised of the organizational policies as well as built environment, tools and technology on interruptions (Werner & Holden, 2015).

The objective of this study is to understand the relationship between ED built environment and interruptions faced by physicians during handovers. 3 ED pods with varying physician workspace configurations – (a) open plan layout, (b) enclosed physician workspace, and (c) semi-enclosed physician workspace were compared. Unnecessary and critical interruptions were captured in real time during 60 handovers using an observation tool. Physicians perception of interruptions, collaboration and impact of built environment on these behaviors were garnered using a survey. In addition, space syntax was used as an analytical tool to measure mutual visibility to and from the handover location and visibility to artefacts central to handovers like status boards.

This correlational study reports differences in interruptions during handovers with variation in syntactic characteristics offered by the ED design. This poster will present findings to support design decisions that may help mitigate unnecessary and support only critical interruptions within the ED during physician handovers.

Gardening Spaces in a Residential Drug and Mental Health Treatment Center

Caila Dawn Sveinson (University of Manitoba)

The poster documents a restorative landscape design for a transitional residential drug and mental health treatment center. The design is informed by case studies, site analysis, and interviews with the treatment center's staff and residents as well as literature on Attention Restoration Theory (Kaplan,1995), Prospect Refuge Theory(Appleton, 1975), and the structure of therapeutic gardens(Stigsdotter and Grahn, 2002). The intention is to develop the landscape so that work routines can include horticulture tasks, create restorative outdoor spaces for staff and residents, and grow food supportive of healthier lifestyles. This complements the center's goals to transition residents into society by addressing mental health needs, teaching social skills to build healthy relationships, providing a daily work routine and implementation of healthy diets.

The interviews identified positive and negative aspects of the landscape perceived by staff and residents as well as interest for expanding the work routines to include gardening activities. The responses indicated a high interest in growing food and traditional medicines, but that a simple and straightforward design was needed due to high turnover of residents and the staff's already heavy workloads.In the site analysis wind, sun and shadow determined placement for the greenhouse area. Historical characteristics in the landscape such as the treed allées were identified and are preserved and emphasized in the design. Site drainage is poor due to the large dike built for flood protection, therefore a prairie swale which encourages wildlife into the site is proposed. Background literature's influence in the design shows in the site's overall plan in the outdoor rooms' prospects and in the refuge spaces in the allée edges. The provision of sensory experiences such as of sounds, smells and color in the detailed areas and thresholds that make users feel as though they are entering another world were informed by attention restoration theory (Kaplan, 1995).

Brick: Re-Envisioning the Module

Lisa Patusky (Drexel University)

Brick is a tried and true building material; simple yet refined, from its low-tech formation to the way it perfectly sits in a bricklayer's hand (Sovinski, 1999, p.11). This poster presents an exploration of brick that re-envisions the module and how it can be configured in nuanced ways. It features a unique interlocking form that has four corresponding channels and crenulations that can, in a system, turn a corner, extend into a cantilever, and be arranged in a multitude of expressive patterns.

The design was inspired by the seam pattern between a cast concrete relief and its wooden formwork before it was released. The contours of the re-imagined brick required the plasticity of concrete cast in a rubber mold. Since each brick is produced by hand, color can be customized to enhance the rhythmic profile, making the brick highly configurable and continually visually interesting. Currently being explored are methods for increasing production, as well as potential applications including exterior facing, interior partitions, furniture, and custom elements.

Developing a brick unit-form has been an opportunity to analyze time-honored craft traditions and test the boundaries of its shape, configuration, and application. Though architectural pedagogy is increasingly reliant on digital technology, designing and making bricks by hand is instructive for understanding how to utilize materials in creative and sustainable ways. It is also a lesson on the craft of architecture and design, which encompasses both the micro and macro aspects of construction from the minutiae of the work of a maker's hand to the compositional whole of a building (MacKay-Lyons, 2015). The persistent honing of craft creates a material consciousness combined with integrity that is fundamental in shaping a more humane world (Sennett, 2009, p. 9). Brick is a timeless building material and a continual generator of innovation.

D -Transform Places with People and the Environment

The Potential of Learning Landscapes at Public Schools: Applying the Findings of a Generation of Studies on the Benefits of Contact with Nature to Facilitate Academic Enhancement, Fitness and Well-Being

Cynthia Hron (Penn State University)

This essay presents an overarching philosophy, gleaned from previous research, that can shape learning landscape design strategies for public education environments. The advantages of this extensive and growing body of knowledge illuminates ways in which nature can be beneficial for students attending public schools. The challenge for landscape architecture is to synthesize the information to inform a committed learning landscape design philosophy with studies indicating correlation between the campus spatial design and student outcomes, while at the same time honoring the individualistic school district identity. Landscape architecture intervention can be a catalyst to activate public school campus sites to address academic enhancement, fitness and student well-being through greater contact with nature.

Design Guidelines for Sensory Environments: Inclusive Children's Treatment Centres Supporting Children and Youth with Special Needs

Pui Yee Nikkie To (OCAD University); Peter Coppin (OCAD University); Maya Desai (OCAD University)

In Canada, over 540,000 youth have one or more disabilities with developmental and learning disabilities accounting for 60% of disabled youth (Statistics Canada, 2019). Special needs individuals face diverse challenges in their physical, intellectual, emotional, social, verbal, and/or behavioural development. Moreover, many special needs individuals experience sensory processing disorders that impact their sense of sight, sound, smell, taste, touch, and/or proprioception (Grandin, 2013). Children's treatment centres are integral in supporting special needs children and youth aged 0 to 21 by providing specialized services, programs, and treatments. The design of such spaces can facilitate or impede the experiences and well-being of the children that they serve (Devlin, 2018). While treatment centres seek to support special needs children and youth to live at their full potential, there currently does not exist a design guideline that addresses the diversity of the children's needs through the design of inclusive sensory environments. Further, the diversity of special needs poses a challenge to designing spaces that benefit all equally while fostering diversity and inclusion. This ongoing research aims to address these challenges and propose guidelines for designing inclusive sensory environments in children's treatment centres with the goal of improving the experiences and well-being of special needs children and youth. Specifically, it will address auditory, visual, tactile, olfactory, and proprioceptive design attributes within four main categories of spaces: public gathering spaces and amenities, recreational spaces, treatment spaces, and transitional spaces. The use of participatory research methods including interviews and co-design sessions have been chosen to ensure that the voices and perspectives of individuals with lived experiences are represented in the research and development of the design guidelines. Among the preliminary findings, the combination of murals, text, and tactile elements in hallways are shown to support transitions and help facilitate therapy outside of treatment spaces.

Can Biophilic Design Components Impact the Mood and Behavior of Dementia Patients?

Susan G. Booher (Ohio State University)

My research study is investigating the potential application of biophilic design components to improve the moods and activities of daily living (ADL) of older adults with dementia. Biophilic design brings elements of nature into the built environment where they are sensorially experienced by those within the space. There were positive results in individuals with dementia achieving independence in ADLs with the use of multi-sensory behavior therapy of Snoezelen and Sonas programme in Staal et al.'s study (2005). This study explores the potential impact of biophilic sensorial stimuli.

The study's approach is interventional and inclusive of dementia patients and their caregivers. The interventions will include scenes and sounds of nature, scents and tastes of food, and kinesthetic activities. The methods used to measure the effects of the interventions will include observations by the caregivers as well as interviews and discussion with the dementia patients.

Observation guides will be used by the caregivers to record the behaviors of the dementia patients both before and after the interventions. During the study's first session, patients will be interviewed about indoor and outdoor activities and be asked to rate biophilic sensorial images using emotion stickers (e.g., smile, frown, and indifference). The sticker activity will generate discussions and the results of this activity will inform the stimuli to be used in the interventions.

One intervention per day will be received by each group over the period of seven days. The control situation is not sensorial, i.e., the patients will agree to participate in conversations of whatever topic the group may choose to discuss. The pre- and post-intervention data will be analyzed to learn whether any of the biophilic components affected the mood and/or behavior of the elderly patients with dementia. The study will be conducted before February 2020.

Adolescent Care Center: Serving Staff and Patients Simultaneously

Zayra Rico (University of Texas San Antonio); Neda Norouzi (University of Texas San Antonio); Antonio Martinez (University of Texas San Antonio)

Environmental design is an important factor in psychiatric healthcare and has a direct impact on health, safety, attitude, and behavior of end-users. Psychiatric patients spend most of their time in a patient unit under strict staff supervision. The most recent direction of psychiatric facilities is towards patient-centered care, encouraging patients' involvement in their own recovery. These models which focus on autonomy and collaboration can only be successful if the environment in which the programs take place are also supportive of this notion. Therefore, there are many scholarly articles on the influence of the physical environment towards the wellbeing and recovery of patients. However, there seems to be a gap on architectural design of psychiatric facilities that support health, safety, attitude, and behavior of staff and patients.

This study is in response to a psychiatric hospital's request for information on architectural design of a person-centered care facility. By reviewing and analyzing the most relevant scholarly articles, our team discovered the importance of biophilic design in relationship to person-centered care facilities. The principle investigator also interviewed 30 members of a psychiatric hospital to learn about the specific needs of the end-users.

The results revealed that both patients and staff members need and appreciate direct and indirect access to nature (color, sound, smell) and controllable natural light. This was aligned with biophilic principles such as direct and indirect sensory experiences, natural daylighting, and contact with natural materials proven to speed up the mental recovery process.

The outcome of this study presented a design solution focused on creating an adolescent psychiatric center utilizing the research that was collected. The facility takes heavy inspiration from biophilic design principles that maximizes connections to nature through courtyards, as well as meditation and recreational activities that supports autonomy and agency for all occupants.

Preserving Heritage

Ifan Payne (New Mexico Tech)

This presentation focuses on the importance of preserving heritage through the integration of existing population considerations into the design and development of environments characterized by their historical, cultural and natural assets. The poster displays studies of the effect of the loss of heritage on emotion and behavior.

By heritage, we mean the physical manifestation of environmental history, urban and rural, natural and built, large and small scale. Using case studies, we will show that the eradication of heritage severs the past from the present and from the future. Obliterating historical environments results in the loss of cultural memory and is a method for controlling the future; a small minority, which can in many cases be equated with the 1% comprised of the wealthy – funders, developers, commerce – is able to control the population through personal and group alienation from the newly formed environments.

The integration of representatives of existing populations – ranging from native Americans identifying sacred sites to members of preservation societies identifying historic building and sites to neighborhood residents identifying territorial space – is essential to counteract the social dysfunction, individual disorientation, breakdown of social structures and support systems and a sense of deep-rooted sadness for loss of environmental memory. Alienated from its heritage, severed from its environmental history, a society can be molded and reformed into a malleable and controllable populace which has lost its sense of place and the stability of its roots in the past.

The ABC House Renovation: Trauma-Sensitive Design to Benefit a Child-Abuse Intervention Center.

Davis Harte (Boston Architectural College)

This practice and research project is an adaptive reuse project designed for a child abuse-intervention center. The center purchased a large (8,000 sq. ft.) building built in 1980, originally designed for financial services. The abuse-intervention center led a multi-year design process to remodel and expand the building, to facilitate an increase in quality and quantity of services. The project showcased used both a rapid ethnography process, and trauma-sensitive design knowledge, which may be viewed as an exemplar of balance between practicality of practice and epistemological standards.

Although there is some evidence to inform the physical design for domestic abuse residential shelters and for substance-abuse centers, there is a notable lack of evidence to inform child abuse-intervention center design. Similarly, although there have been some advances recently, there is still sparse trauma-sensitive design evidence for any setting. The current design process for child abuse intervention centers often seems to be determined by center staff preferences and the architectural teams' understanding of precedence. This project seeks to address this gap in the trauma-sensitive evidence-based design (EBD) knowledge.

Upon engagement as the environmental designer for the project, the researcher applied both trauma-sensitive design evidence and a rapid ethnographic process. Expectations for an on-going design iteration process amongst the transdisciplinary team, strengthened the design outcomes. Photographic documentation of the existing center, the new building before renovation, and the 18-month long design journey as it led to the final adaptive reuse project are highlighted in the poster. Examples of the evidence translation, via images of the building before and after, as well as the final designs, are provided. An understanding of the trauma-sensitive design process in a child-abuse intervention setting will arm both scholars and practitioners with creative, timely and pragmatic tools for complex environments.

Design Thinking as a Form of Participatory Action Research

Amin Mojtahedi (HGA Architects & Engineers); Susan Foong (HGA Architects & Engineers)

Rooted in the participatory design tradition evolved since the early 1970s (Bjögvinsson et al., 2012; Gobble, 2014) and the ongoing design discourse about the nature of design since the 1960s (Rhinow & Meinel, 2014; Johansson-Sköldberg et al., 2013), Design thinking has been largely framed and popularized by IDEO in partnership with several other progressive design companies in the 1990s and Stanford University d.school during the mid-2000s as a way of enabling innovation and dealing with complex problems by making designers' way of thinking accessible to non-designers (Camacho, 2016; Luchs, 2016; Katoppo & Sudradjat, 2014; Johansson-Sköldberg et al., 2013; Bjögvinsson et al., 2012; Melles et al., 2012).

Several studies have recognized the many parallels between Design Thinking and Action Research. For example, Katoppo and Sudradjat (2015) adopted design thinking as a tool to complement participatory action research in architectural research method. Trullen and Bartunek (2007) also recognized the similarities and wrote that design approaches "follow steps established in action research interventions – data collection, diagnosis, planning, taking action, and evaluating needs which may lead to another cycle of action" (p. 33). Tonkinwise (2010) went one step further by framing design thinking as a form of action research.

Toma-the Digital and Spatial Installation Curating a Dialogue between Northwest Arkansas and the Marshall Islands.

Torrey Tracy (University of Arkansas); Jacob Costello (University of Arkansas)

Northwest Arkansas is home to the second largest population of Marshallese people in the world. The U.S.' relationship with the Marshall Islands first began when the countries made an agreement allowing the U.S. to test nuclear bombs on a portion of the country's land in 1946. In the decades since, another agreement, The Compact of Free Association, has allowed thousands of Marshallese to leave their home country and settle in the United States. After Tyson Foods, the largest poultry meat distributor in the world, began employing a great deal of people in the Marshall Islands, many Marshallese would find an opportunity to relocate to Springdale, Arkansas, the world headquarters for Tyson. Today, the Marshallese community of over 12,000 represents a vibrant portion of the Northwest Arkansas society (Brown, A. L.). Most of these people have close friends or family still residing in the Marshall Islands. It is more important than ever for the Northwest Arkansas community to have an awareness that their neighbors may be climate change refugees. TOMA-a schematically designed engaging and technologically forward installation will invite visitors to the Crystal Bridges Art Museum in Bentonville, AR to stand in unity with the Marshallese people. The digital spatial experience will invoke a sense of interconnectedness between visitors in Northwest Arkansas and the lives of those affected by climate change in the Marshall Islands. The installation is not meant to instill a sense of fear or arouse emotions of guilt surrounding the concepts of human consumption or climate change. We feel that environmental or activist art often leads to people shutting off their emotions to cope with the frightening realities of climate change. Alternatively, this piece is a sobering, yet beautiful experience, inviting people to stay and, alongside others, empathetically contemplate their place in the world.

The Impact of Religious Iconography in Inpatient Hospital Rooms

Ann Sloan Devlin (Connecticut College); Cammi Marcin (Connecticut College); Lilly Noble (Connecticut College); Gabrielle Peterson (Connecticut College); Mo Zhou (Connecticut College)

Representational art in hospital rooms has positive effects (e.g., Hathorn, 1993; Ulrich & Gilpin, 2003), yet no research to our knowledge examines the impact of religious iconography. At least 15% of the hospitals in the US are religiously affiliated. In 10 states, this number exceeds 30%. Patients may not have religious preferences or their preferences may not match the religious tradition of symbols displayed, with implications for supportive design. In a between-subjects design, Amazon MTurk participants (N = 343) read a description of hospital room amenities and viewed a photograph of that room with either a crucifix & framed art (crucifix condition, n = 178) or two pieces of framed art (art condition, n = 165). They gave qualitative comments for 3 aspects of the room, indicating whether the element was positive or negative. Results showed significant differences by condition for Aspect 1 (p = .001): 73.9% positive comments in the art vs. 45.3% positive in the crucifix condition. In the crucifix condition, 37 comments were specifically about the crucifix, 86.5% negative. For Aspect 2 (p = .025), percentages for the room were 75.4% positive (art) and 65.2% positive (crucifix), with 14 comments about the crucifix (64.3% negative). There was no significant difference for Aspect 3 (p = .478), with 75.0% and 71.6%, positive comments, respectively. When the presence of the cross was assessed negatively, people often connected it to religion: "I don't like the cross on the wall. I'm not religious. I worry someone may say something to make me uncomfortable. I don't want to discuss religion;" "the cross on the wall was what I saw immediately, it made me very uncomfortable as I am not religious and I wouldn't want to stare at that thing every day." Religious iconography may negatively impact patients' perceptions of inpatient rooms.

The Impact of Campus Outdoor Spaces on Student Happiness: A Case Study of the ASU Tempe Campus

Kara Davis (Arizona State University)

College and university campuses can play an important role in a student's life, and campus outdoor spaces have the ability to positively impact various aspects of student health and well-being. It has long been understood that natural environments can promote health and well-being, and in recent years research has begun to examine the impact of parks and landscapes in urban settings on subjective well-being (SWB). Subjective well-being (aka "happiness") refers to one's self-reported measure of well-being and is thought of as having a high level of positive affect, low level of negative affect, and high degree of life satisfaction (Diener, 1984).

This study was conducted to assess the interrelationships between affective experiences, SWB, and usage of campus outdoor spaces in order to learn how outdoor spaces on the Arizona State University (ASU) Tempe campus can be enhanced to increase SWB and usage. In total, 832 students completed a survey questionnaire 1,140 times for six campus outdoor spaces. The results showed that students experience the greatest amount of happiness in the Secret Garden and James Turrell ASU Skyspace, relaxation/restoration is the affective experience most strongly related to SWB, and SWB is negatively correlated with frequency of visits but positively linked with duration of visits. To improve student happiness and usage of outdoor spaces on campuses, planners and designers should work on increasing the relaxing/restorative qualities of existing locations, creating new spaces for relaxation/restoration around campus, reducing the perception of crowding and noise in large spaces, increasing fun/excitement by adding stimuli and/or opportunities for activity and entertainment, and adding equipment necessary for students to perform the activities they want. In addition to the ASU Tempe campus, the methodology and findings of this research could be used to improve outdoor spaces on other college and university campuses and other types of outdoor environments.

Students' Integrated Sensory Analysis Results: A Case Study of the Main Lobby Area in College Buildings

Yongyeon Cho (Iowa State University)

The main lobby area in college buildings, including, an atrium, and lounge areas can be a socially embedded collaborative learning area, a transition space from outside to inside, a casual meeting space, an individual study area, and a rest zone. A well-designed space influence on users' overall positive sensory experience (Winer & Keim, 2018, August 7). Human sensory inputs have a stronger influence on both cognitive and emotional responses (Augustin et al., 2009) which closely related to design a learning environment. Although students' sensory experiences are associated with their learning outcome, little research has been investigated student's perception of sensory design elements in campus main lobby areas. Sophomores participated in a survey as a part of his/her course assignment. Participants visited a total of three different college main lobby areas and asked their own seven sensory experiences. During the survey, participants used a modified version of Malnar & Vodvarka's sensory slider published in 2004, to evaluate which level of sensory stimuli optimal and the participants experienced, and which design elements impact on your decisions. Demographic questions, visiting pattern questions were followed. As a part of qualitative research, the contents of the survey data were analyzed to identify what environmental considerations impact the users each sensory experience. As a part of quantitative research, the data used for statistical analysis to discover the relationship between the visiting patterns and the user's sensory experiences. The survey result indicates a shape of space impacts students' visual sense and basic orienting sense, sound comes from people impacts students' auditory sense, presence of café impacts students' taste-small sense, furniture impact students' haptic-touch and kinesthesia, and weather impact students' haptic-temperature and humidity. The results can be utilized to determine which design features need to be carefully considered for students' positive sensory experiences in future learning environments.

Reciprocity: Architecture and Landscape

Erika Zekos (University of Massachusetts Amherst)

This poster will present a series of introductory exercises developed to help students in an early core architecture design studio develop a critical appreciation for a reciprocal relationship between building and landscape [1]. Throughout the semester-long studio, students explore traditional architectural design concepts of composition, form, space, light, and program, while engaging with a landscape for the first time.

Preliminary studies invite them to be inspired and informed by both abstract and diagnostic understandings of an existing site. In both cases, they consider the influence of environmental (topography, vegetation, habitat, solar orientation, hydrology) and human (history, infrastructure, perception, experience, culture) forces.

Design methods begin with a study that encourages students to employ the imagination to move beyond the hierarchy of the visual [2] through the use of digital and physical collage to translate and interpret their site observations and research. Through the process of layering, removing, and remaking using inputs such as photographs, drawings, maps, and text, they discover new, invisible relationships that inform their conceptual thinking.

Abstract methods are paired with traditional analysis of site conditions to develop an understanding of concrete constraints and potentials that impact siting and architectural form making. Students learn techniques for diagramming and documenting existing conditions to communicate their findings. These early studies provide a foundation for an architectural design outcome that is fully integrated with its site; incorporating ideas far beyond where the building meets the ground. The spatial language developed by the students allows for the communication of abstract and experiential ideas while also managing the demands of the physical context. Links between imagination, site, and program are developed and strengthened.

Evaluating the Usability of a Participatory Mobile Augmented Reality App: Engaging Residents Digitally in Assessing Neighborhood Walkability

Saeed Ahmadi Oloonabadi (North Carolina State University); Perver K Baran (North Carolina State University)

This poster will present the process and preliminary findings of an ongoing study focusing on developing and evaluating an Augmented Reality Participatory Platform. This novel digital engagement platform aims at assessing streetscape features, such as sidewalks or bike lanes, for their suitability in improving walking in neighborhoods. In this study, we evaluate the usability of an Augmented Reality (AR) mobile app prototyped as the main component of the platform, with Human-Computer Interaction (HCI) experts and volunteer community members in Charlotte, NC. CommunitAR enables users to virtually place streetscape features or the future street designs while in the field, take images of the scene, and share these images as well as associated feedback through social media platforms and multimedia messaging systems. Like other systems, our mobile app will not be successful unless users accept it, and the quality of this acceptance is primarily determined by evaluating its usability. While recent technological advances in AR are promising, the lack of studies focusing on capabilities of AR - and mobile AR - in enhancing participation requires a considerable attention. Therefore, this study contributes to the very limited evidences on using mobile AR as an interactive, portable, and in-situ visualization apparatus in digital engagement. In this study, we will evaluate the AR app's capabilities by testing how well the participants can use it for undertaking tasks such as feature selection, sharing images of the scenes, and rating alternative designs in the field. We will explore whether the app provides effective, accurate, and satisfactory user experience to visualize, interact with, and comment on walking-related streetscape features and future street design alternatives. Overall, the study will create new knowledge about innovative approaches in digital engagement methods that will inform urban designers and planners in public and private sectors, and therefore, will benefit the design community.

Image of the City Represented in Individual Cognitive Maps - a Study in Atlanta

Yeinn Oh (Georgia Institute of Technology)

This study is an attempt to compare people's cognitive maps to see how they perceive the city differently by having different experiences in the city. Various researchers used cognitive map (or mental map) method to look at how people structure the city in their minds. People make a mental representation of spaces which are systematically simplified or distorted (Tversky, 2003). When it comes to the city and people are trying to remember a larger space, people make their own version of maps that has errors. An image of a city inside one person is made by the unique memories and meanings connected to the city and the person (Lynch, 1960). However, not many of them tried to see how the individual experience related to cities' physical forms takes part in the image of the city. This study focuses on bringing the experience of living in the city and tie them into the topic of the image of the city. People living in Atlanta were asked to draw a map of the places that they have visited, and mark the places on an actual map. From this study, we found that different exposure that the city gives to the resident influences the image of the city. Especially, the longer people live inside the city and they grow older, they tend to explore more parts of the city away from where they live. Transportation mode also gives a significant difference in understanding the city. People tend to remember the streets and roads that they frequently use, and more importantly, use them as a strategy to remember the city. Image of Atlanta has been found in people's maps. Clusters of neighborhoods were aligned by major highways - and different purposes of places appear street by streets.

Active and Frequent Use of Campus Green Space By Students at the University of California, Irvine

Nik Warren (University of California Irvine); Brandon Yoon (University of California Irvine); Miryha Gould Runnerstrom (University of California Irvine)

Studies have shown that green spaces not only contribute positively to student experiences, but that frequent and active use of campus green space is associated with lower stress, more positive health, and higher quality of life (Holt, Lombard, Best, Smiley-Smith, & Quinn, 2019; Speake, Edmondson, & Nawaz, 2013). The goal of this study is to understand which existing outdoor campus spaces are used most frequently and how they are used by students at the University of California, Irvine (UCI).

An anonymous online survey of 211 UCI students ($\bar{x}$ age=21.4) was conducted to explore student use and preferences of campus green space for psychological restoration. Included in the survey were open-ended questions about favorite on-campus spaces, favorite outdoor on-campus spaces for particular activities, and other activities performed outside on campus. Open coding was used to derive categories and themes from participant responses to open-ended questions.

Socialization and active use of natural spaces are two emergent categories/themes that arose during data analysis. Students actively use campus green spaces for exercising and organized sports. They also show a preference for socializing with friends in natural spaces such as Aldrich park. Interestingly, 24.86% of respondents report taking daily or weekly naps in the park. Other favorite outdoor passive pastimes mentioned were eating, studying, and using social media.

UCI is in the process of evaluating enhancements to natural spaces on campus. This study highlights how students are currently using outdoor spaces on campus as well as with what frequency. By integrating student feedback into the possible enhancements process, students will have a voice, ensuring that a key demographic is heard and able to contribute to future changes in the campus climate.

Urban Regeneration through University Campus Reorganization

Kengo Mitsui (Tokyo Institute of Technology); Naoko Saio (Tokyo Institute of Technology)

In this globalized century, transformation of urban space to university campus promoted by university reorganization has been progressing in cities around the world. The Belval site is located in the second largest city in Luxembourg. The steel industry wasteland area was transformed into a city of science, with the University of Luxembourg positioned as an important resource of this conversion project.

Urban regeneration promoted by universities can create an environment that enhances the attractiveness of the city. In this research, the authors focus on the trends of reorganization of university campuses. There are only a few studies that provide an overview of such trends. This research uses an international sample of more than 20 campuses.

University official documents - such as campus master-plans and framework plans, including neighborhood plans, which define a vision for the future of the campus and its surroundings - were reviewed, and the campus reorganization case studies were assessed based on the following points: 1) Planning Purpose, 2) Scale of Reorganization and 3) Land Use and Buildings before Conversion. Furthermore, these case studies were arranged based on 1) Campus Form on the Growth Process, and 2) Campus Formation. Based on data obtained, the study tried to categorize campuses according to location and scale and find out what kind of spatial features the different types have and how they are associated with the regeneration.

The results of this research can be summarized thus: Some campuses were found to have expanded throughout cities through renovation of idle assets such as industrial plants. Urban sustainability can be enhanced by flexibly integrating campuses into urban contexts.
Transforming areas of urban decline into campuses generates not only educational sites but also living areas that create new activities in the city and can change the structure of the city.

Wellness Design Assessment Criteria: Content Analysis of Workplace Design Guidelines

Yongyeon Cho (Iowa State University); Huiwon Lim (Penn State University); Hye Jeong Park (Iowa State University)

Recent studies have addressed that design elements can enhance users' health and well-being in their workplaces. In creating a mindful workplace, designers commonly refer to evidence-based guidelines, evaluation tools, and rating systems to supports each user's physical, emotional, and social wellness. Although many rating systems are useful to experts, the systems are challenging to laypersons to use because of text-based content with jargon and complexity. Moreover, many such systems don't tend to support non-expert users' awareness of designing a workplace because all design elements are equally weighted even though some design elements are more essential than others. For this limitation, we focus on a design rating system for workplaces. This study is designed with three parts: The purpose of the first study is to develop inclusive wellness design criteria for office environments; the second study is to understand current office user's perception of wellness regarding design features; the third study is to build a new framework to create a real-time based user-friendly mobile application for evaluating office users' wellness about design experiences. This study is the first part of a total of three parts. To filter the criteria related to workplace design settings and users' perceptions of wellness experiences this study reviewed existing design evaluation guidelines and design-rating systems that related to users' workplace wellness experiences including WEEL V1, Active design guidelines, and the Centers for Disease Control and Prevention (CDC) worksite health scorecard through qualitative content analysis. The results identified 20 design features which are classified with technical, functional, and aesthetic categories. The top three wellness design features are accessibility, indoor activity, and health-related information. Our findings could be valuable designers, project managers, and clients as well as non-expert users to support their decision-making in the design process of the new workplace environments.

What Are Historical Buildings in Japan?

Chikage Sunaga (Tokyo Denki University); Hiroshi Tsuchida (Tokyo Denki University); Yuka Honda (Tokyo Denki University)

There are many historical buildings in Japan. Moreover most of its historical buildings made of wood. Therefore it's impossible to leave original form. A style of building is also changing by the present life. Historical buildings have been used, changed and maintained. However, the present streets and old historical streets differ in architectural style and lifestyle. On the other hand, in Japan, if the standard of a building is met, it will become historical streets scape.

The target place is made Kauemon-cho, Tochigi-shi which prospers as Machiba (Town) from the old days. In the Edo period, it flourished in commerce. However, it's maintained as a residential area at present. The research method examines the standard of the historical streets and the consistency of the present streets.

The structure of the historical buildings in Kauemon-cho is made of wood. In addition a wide opening is the character. The land allotment which continues from the Edo Period is also the feature. It's enacted the height, the slope, the color and the location of the building by the landscaping standard. In Kauemon-cho, the living space as a residence is given priority. Therefore a wide opening isn't used. What is actually taken into consideration is only color and height. When its standard is maintained, it can be preserved as historical streets scape. However, Kauemon-cho has 123 rebuilt buildings. Is it possible to form a historic streets scape only by design control?

The Influential Factors and the Role of Physical Design on Improving Patient Experience in Healthcare Environments.

Fares Alsaygh (Texas Tech University)

Healthcare as a built environment has shown impacts on its users and affects their care and services. The quality of care and services are influenced by several indicators, the most common two are patient satisfaction (PS) & patient experience (PX). Recently, patient experience became a more important and critical clinical aspect that helps in improving the quality of care and services of healthcare facilities. The literature review explored many factors and aspects influencing patient experience such as physical designs and stuff communication skills.

The main three objectives of the study: 1) To review and understand every aspect of patient experience such as when, where and how it started and defined, and what could influence it. 2) To explore and collect every factor that influences patient experience which was mentioned in existing articles. 3) To filter and categorize the factors based on the findings from prior studies, based on their definitions or their impact on patient experience. What are the most influential factors on patient experience and how can physical designs improve it?

The study began with an extensive semi-structural and systematic review that was done through several steps: 1) Identify keywords 2) Select the core databases 3) Filter the limitations 4) Collect and review the most relevant articles 5) Report all findings from articles.

The finding showed many influential factors that have been categorized into nine groups. The physical and environmental category has the greatest number of these factors.

The results showed that to improve the patient experience in healthcare environments, there are several categories of factors that have to be considered earlier. The physical and environmental category showed the greatest number of factors that can be used for future improvements. We need future research to identify the design recommendations that might enhance the patient experience in healthcare facilities.

Transformation of Japanese Regional Cities: Focus on the Land Use and Municipal Merger

Hitomi Shimada (Tokyo Denki University); Hiroshi Tsuchida (Tokyo Denki University); Yuka Honda (Tokyo Denki University)

A city in Japan has grown and expanded attendant on the postwar population growth and industrial development. However, the population has been decreasing recently. A city is going to the reduction by the population decline.

A city is administrative unit which is constitute of a core town and agricultural community areas and established since Meiji Period (1868-1912). The castle towns which daimyo (territorial loads) governed and the villages (for example agricultural community) spread out the core town constituted the Japanese urban structure of Edo period (1603-1867).

The shogunate system ended and Japan entered Meiji Period. Since the Meiji Restoration (1868), the city have been formed by planning and developing to aim modernization. Most pf Japanese cities were established at Haihan-chiken (abolishment of feudal domains and establishment of prefectures) in 1871. The city repeated the Great municipality mergers, namely Meiji (1889), Showa (1953-1961) and Heisei (1999-2010), for searching the proper scale of the control area as a municipality. As a result, a city is expanding by a merger repeatedly.

However, the sprawl phenomenon in city has been regarded as a problem in recent years. In the future, cities need to consolidate their spread communities. The city planning requires planning methods to shrinking cities.

A city is a combination of regions with different characteristics. An old castle town has been a core town. There are urban functions in agricultural community area which has formed a community. The present city has more than one urban area. The Japanese cities had controlled a city area and agricultural community area in city planning and an agricultural promotion plan, respectively. But a city has both areas at present. I propose to connect a core town and urban areas and aim for a network-type urban structure.

Learnscape: Proposal for an Innovative Approach to Design a Sustainable School in La Cincelada, Colombia

Gisou Salkhi Khasraghi (Texas Tech University); Atina Kia (Texas Tech University); Ivan dario Iturregui (Texas Tech University)

Teaching Sustainability to children can be achieved by respecting nature and environment. This project is the redesign of a school in La Cincelada (Colombia) to promote ecological and social sustainability.

The redesign starts from a theoretical stance in which a paradigm shift in the implementation of design strategies is necessary. In this process We found that even using biomimicry it is necessary to consider the strategies in a hierarchical order, the simpler the strategy, the implementation seems to be more passive. A comprehensive site, environmental and climatically analysis was conducted. A Biomimicry approach was selected to reduce the building footprint and to find new solutions contextual problems. Accordingly, a project ´s SWOC analysis was developed and the core concept of "Hiding" is generated. In order to achieve sustainability in three major aspects: economic, environmental and societal, Hiding is translated into accurate design strategies. The main prospective toward the project redesign was to reduce anthropic intervention and apply passive design. As a result, it is expected to reach ecological connectivity, a more sustainable community, better user interaction with nature and less disturbance to the natural context.

Revitalization of Public Housing and Around Areas

Kosuke Kuwabara (Tokyo Denki Univercity); Chiaki Konishi (Tokyo Denki University)

Japan has a problem with the declining birth rate and aging population. According to research of the government, many people who live in the public housing are the elderly living alone. We think demands of the public housing will increase because public housing it have been supported the elderly in term of rent. It is difficult to make a sustainable community only the elderly.

It is important to live with various people different ages, occupations, and income levels. It is necessary to update the functions of public housing, improving the environment where various generations live together. However, the floor area ratio is set lower than the surround area, there is a situation which reconstruction is not progressing for economic reasons. Public housing has abundant plants and a large number of households, so redevelopment of public housing may contribute to revitalization of the surrounding area.

The purpose of this study is to extract the effects of the construction of public housing on the surrounding area, prove that public housing is a local resource, and suggest the possibility of rebuilding. The research method analyzes the situation before public housing was built, the situation immediately after it was built, and the current situation nearly 60 years after construction.
To looking at the map, we focused on roads and open spaces and summarized surrounding area changes. As a result, they were classified into three types of processes. The case shown on the poster is a case where public housing was built utilizing the historical topography of the area. In addition, it was identified that a special form of open space was formed because a single housing complex was formed by the accumulation of planned public housing..

Active Design of Parks: A Case Study in Oklahoma

Rylie Davidson (Oklahoma State University); Aditya Jayadas (Oklahoma State University)

The World Health Organization (WHO) estimates that 3.2 million individuals die each year globally resulting from physical inactivity (WHO, 2010). The cost associated with physical inactivity was estimated at $53.8 billion globally with the public sector bearing 60.3% of this cost. Physical activity interventions are necessary to reduce both incidence of deaths from physical inactivity and the cost associated with physical inactivity. One approach is to integrate active design principles at the town and city levels in any country. One example on the implementation of active design is the renovation of the Blue Hole Regional Park in Wimberly, Texas to promote physical activity by promoting exercise in children through the design of nature-based play features from repurposed materials (Center for Active Design, n.d.). The goal of this research study was to explore opportunities for Active Design integration into parks in order to improve health and wellbeing of individuals who reside in Stillwater, Oklahoma. The methodology involved the use of observation and site visits to document park accessibility and park facilities, along with photo documentation. In terms of the results, specific details on the 16 parks in the city are summarized as it relates to good attributes that promote physical activity, and not so good attributes that could be enhanced through the integration of active design principles. The use of better signage, improved lighting, adding amenities and holding community events are all being recommended to improve active design of the parks. In terms of future studies, the next phase of this study will include surveys on walkability, park utility, and perceived safety along with focus groups with open ended questions centered on increasing park utility.

Interaction and Participation in Community Cinema As Places for Community

Mai Horikoshi (Waseda university); Masayuki Sato (Waseda University)

We clarify the interaction participation in community cinema as places for community.
In Japan, large-scale multiplex movie theaters with more than 6 screens are common with, at the same time small-scale movie theaters are being forced to close one after another. However, in Japan, new type movie theaters called for "community cinema" performing public screenings has been increasing. We examine three cases with different forms such as scale and functions among these theaters built nationwide since the 2000s. For the theater managers, we conducted interview surveys, and in the case A with the largest scale among the three cases, we conducted a free-form questionnaire survey, interview surveys, and behavior observation surveys for 20 visitors.

In addition, we classify trial (programs, workshops, events, and so on) in each theater as "cycle" and "duration" to measure the participation in planning of visitors (Table 1). As a result, all movie theaters had a donation event. A, C have many regularly and singly events, and also have discussion events talking about movies, which promote the interactional communication between visitors. Furthermore, the around B whole area have many permanent programs and some events. The check of such a classification is able to create transformation.

The results of the survey are summarized as "Ladder of Participation", and it turned out that it is important to have connections with visitors. As the stage of participation goes higher, the visitor's connection is not only the conversation with the operator or staff but also the conversation among the visitors. In addition, in order to have a sense of ownership for visitors, a long-running event is required. Furthermore, visitors get the chance for participation to events of various cycles and periods and participate having a sense of ownership to them.

Who Benefits from Walkable Downtown Sidewalks?

Barbara Brown (University of Utah); Wyatt Jensen (University of Utah); Carol Werner (University of Utah)

Walkability-design features intended to encourage walking--provides a number of community benefits, such as less reliance on polluting cars, activation of streets and their destinations, and physically healthy walking opportunities (Werner et al., 2018). Designers have added walkability to hundreds of new "complete streets," which focus on better designs for active travelers. However, few walkability evaluations actually count pedestrians. We examined four street corridors in Salt Lake City, Utah, over three different years. These four segments included a less walkable street, a less urban complete street, a more urban complete street, and a high-walkable street. Complete streets received their street improvements by year two. Walkability was assessed with the Irvine-Minneapolis Inventory (Day, Boarnet, Alfonzo, & Forsyth, 2006). We tested whether pedestrian counts reflected walkability and whether both genders benefit from walkability. Mixed models tested 94 street segments measured across weekdays (4 time blocks) and weekends (3 time blocks) in fall. Our results showed that streets varied in walkability, with the most walkable street showing the highest levels of environmental indicators of population density, diverse destinations, accessibility, attractiveness, traffic safety, and crime safety (Jensen, Stump, Brown, Werner, & Smith, 2017). In terms of total pedestrian usage, the low-walkable street had the fewest pedestrians, the high-walkable street and the urban complete-street had the most pedestrians. Complete street improvements were followed by more pedestrian use, especially for the complete less-urban street. More males than females used all four streets. The gap was narrowest for the most walkable street, where females were 40% of weekday and 42% of weekend users. The gap was widest for the low-walkable street, where females were 19% of weekday and 21% of weekday users. These results show that complete street improvements can support more walking, but that design professionals need to address gender disparities in street use.

Color Renovation of Urban Street Buildings' Facades — Case Study of Daliang Road in Kaifeng

Qian Qiao (Zhengzhou, China); Hongyi Lv (Zhengzhou, China)

Streets are part of urban public spaces. Colors of street buildings represent a city's culture, especially in historic cities. However, many street buildings have chaotic colors and in need for visual beauty and cultural characteristics. This research focused on color renovation of building facades along the streets in historic cities. Based on a literature review and empirical data collection and analysis, it investigates the color rules of visual esthetic sense and the color elements forming visual attraction on building facades. On-site study of the buildings along Kaifeng Daliang Road collected the value of building-facade colors and the data of local residents, pedestrians, and passengers' evaluation of building-facade colors. The methods of color cards comparison, questionnaire, and photo interview were applied. Findings from this research identified the color rule of street building facades, color preference, and significant color elements for the renovation of urban street buildings. It can be used as a guideline for designing and renovating of street building s' facades in historic cities in China to promote city cultures.

Small Spaces, Meaningful Transformations: Transformable Treatment Therapy Room

Cristina Michelle Garza (University of Texas San Antonio); Neda Norouzi (University of Texas San Antonio)

The differences in disabilities and treatment methods that exist in pediatric rehabilitation today vary from caregivers and providers to the facilities where treatment takes place in. These differences may not only affect treatment outcomes, but also affect the way clinicians apply therapeutic methods and intervention.
This study analyzes the interior architecture of a therapeutic room and how to better design these spaces in order to best serve both patients and therapists.

The principle investigators, a healthcare architectural designer and a licensed therapist, wrote and conducted surveys to collect data on the use of therapeutic spaces. The therapeutic professionals provided information based on direct personal work experience used as a basis for reasoning, discussion, and research on how to improve health and rehabilitation spaces.

The results of the study present the needs of patients and therapists which led the presented design solution for a transformable therapy space. These spaces would allow the therapists to accommodate the specific needs of each patients by easily shifting or moving objects and materials needed for each session. This design will allow for essential objects and items to be readily accessible, and quickly stored away or hidden which will be a tremendous help for the therapist who must change and prep the room in a very short time period. This design treatment model can be used across a variety of settings including rehab facilities, hospitals, and schools.

Case Study of the Landscape Color in West Lake Park in Kaifeng

Xiongfeng Xiao (Henan University); Zhe Wang (Henan University)

Color landscape design to use in the construction of city park can largely increase the charm of wetland park to form unique and rich visual aesthetic feeling and the study of urban color landscape park and utilization is still in a primary stage, the application of urban wetland park color landscape lack theoretical guidance, among the factors resulted in the urban wetland park landscape colour collocation chaos, poor visual effect, failed to reflect the park features such as a series of problems to be solved in Kaifeng west lake park, for example, this paper adopts the method of questionnaire investigation, analysis of different gender to colorific be fond of all ages Based on the moncel color system, the HSB model is used to analyze landscape chromatography, summarize the application of color, and then put forward the key points of color design suitable for urban parks, hoping to play a positive reference role in the future urban park color landscape design.

Case Study of the Slow Traffic Systems in West Lake Park in Kaifeng

Wei Tang (Henan University); Zhe Wang (Henan University)

In recent years, people began to pursue a healthy and green way of life. As a place for leisure and entertainment, urban park studies the slow traffic system in the park. And then better meet the needs of people's use. Taking Kaifeng West Lake slow traffic system as an example, the practicability of the traffic system, the comfort and beauty of the surrounding environment of the road, and the setting of supporting facilities are analyzed comprehensively,

On the basis of drawings, through the combination of observation method and questionnaire survey method, with people's use experience and relevant specifications of road design as the evaluation criteria, the collected data are summarized and analyzed. Finally, combined with the slow traffic system score, the main factors influencing the slow traffic system design are analyzed. Based on the comprehensive study of Kaifeng West Lake Park, the design principles of slow traffic system are summarized. 1. The design of slow traffic system needs to be analyzed from two aspects of substance and spirit, and adopt appropriate walking space design strategy. 2. The entrance and exit of the park should be closely connected with the urban road. According to the flow direction of people, set the location and number of entrances and exits to meet people's needs. 3. The road setting in the park shall be connected with the landscape node, and the connection and transformation between roads shall be considered. 4. The main road in the park shall fully consider the aesthetic degree of road pavement, the quality of road materials, the convenience of supporting facilities, the degree of environmental beautification on both sides of the road and other aspects.

Develop Efficient Emergency Departments of General Hospitals in Large Cities in China

Zhe Wang (Henan University); Li Yue (Henan University)

Large-size general hospitals and their emergency departments have been fast developed in China. Based on a literature review, this research focused on the architectural and environmental design of emergency departments in these large hospitals, including the factors of overall planning, functional plan, and spatial details. To promote work efficiency in emergency departments, the overall hospital campus planning should be convenient; the traffic streamlines should be functional; the internal and external ambulance traffic should help to reduce inflection; and the distances between places that patients need to visit should be short to reduce waiting; the streamlines of patients, medical services including the dirt and clean streamlines should be separated and/or with suitable lengths and crosses. If the waiting areas and transitional environments for emergency care are healing, the stress on patients should be reduced, leading to calm and peace in mind and the effect of secondary cure. Sound and illumination design should be coefficient in accordance with the specific needs of emergency services, including appropriate indoor lighting and noise controls.

Field investigation and analysis were conducted to investigate the environments in nine emergency departments of large-size general hospitals in seven cities in China. By conducting field investigations, the overall hospital layouts, functional spaces for emergency services, internal circulations in the departments were summarized. Observations focusing on the waiting behavior of patients and accompanies were conducted and the data was statistically analyzed. The demands of patients and accompanies for healing environments were deeply analyzed and summarized. Guidelines for designing healing and efficient emergency departments were developed. Methods used in this research can be refined for future studies on emergency care.

Designing for All Senses: Empowering Brand Identity through Multi-Sensory Design

Ye Jin Chang (Iowa State University); Victoria Faccin (Iowa State University); Eliza Malloy (Iowa State University); Daejin Kim (Iowa State University)

It is evident that multi-sensory design has a significant effect on consumer's perception, judgement, and behavior (Krishna, 2012). Multi-sensory designs have numerous positive effects; they celebrate qualities of place (Lupton & Lipps, 2018) and enhance brand interaction (Wheeler, 2018). However, few research has investigated the method of documenting and analyzing users' sensory experience and incorporating them into design projects. The premise of this research is to redesign and empower brand identity of an cafe in the Midwestern region with the objective of maximizing customers' sensory experience.

The presentation categorizes of two parts; the analysis of sensory information in the café through observational research and proposing a new multi-sensory design strategy as an interdisciplinary design team. Using sensory sliders (Malnar & Vodvarka, 2004), the authors evaluated the physical environment; they visited 2-3 times during the weekend, both day and night time for an hour.

Based on the use of the sensory sliders, these senses were underrepresented: olfaction, haptic, and auditory. In terms of the olfactory experience, the space lacked immediate smell; coffee or baked scents during the analysis. Due to an unpurposed of materials and textures, this space failed to provide memorable haptical experience. Noise was one of the most negative experiences for the authors as the material choices acted as poor dampeners. Thus, creating an undesired environment for unwinding.

From the analysis, the designers' intention is to enhance the sensory experience by proposing a modernized design while maintaining the comfort and warmth of the existing characteristics. In this poster, the author(s) elicits an open discussion about impacts of multi-sensory design, sensory analysis processes and share important findings to create brand identity through multi-sensory approaches. As a research-informed design project, the poster will provide a redesigned prototype for an existing café.

Sensory Path; An Immersive Natural Refuge

Fahimeh Farhadi (Penn State University)

As the modern life gets more hectic, the need to connect with nature becomes more crucial as nature is a resource for peace and serenity. Aesthetics in Landscape is usually affiliated with visual comprehension and overall scenic values of the space; however, landscape goes far beyond vision. Humans use five senses; sight, scent, sound, touch and taste to experience an environment and get a deeper understanding of it.

Sensory gardens have established as therapeutic designs for nursing homes and for children with disabilities, however, this concept can be applied to urban areas to bring people closer to nature. Using these five senses in landscape design can create a temporary natural refuge from the city and impact spatial perception of humans and their interactions with nature. Every sense can be presented in a zone with related characteristics. Sight fosters a heightened awareness of spatial relationship and provides for emotional connectivity. Landscape inherently involves the visual beauty due to plants and pollinators. Fragrant plants help recall memories or an association to new experience using scent and create feeling of comfort and relaxation. Urban agriculture cultivates taste and reinforces the character development of a place. Sound of birds and rattling of leaves, whether it's loud and energetic or still and calming defines the surrounding landscape. Unique textures in plants and hardscape create sensation of warmness, coolness and raised, help create an innate richness with the environment. Using available literature and design method, "Sensory Path" considers all elements of sensory spectrum to create an impactful and immersive natural space in an urban area in Baltimore, for people. The design is composed of five zones, each contributing to deeper connection of people to nature using each sense. It demonstrates how literature and research of five senses can be accommodated in design.

Where Can I Smoke in Tokyo 2020?

Kohei Sasaki (Tokyo Denki University); Hiroshi Tsuchida (Tokyo Denki University); Chiaki Konishi (Tokyo Denki University)

In the 23 cities of Tokyo, outdoor smoking is prohibited by city rules. It is possible to smoke in only permitted smoking spaces. In addition, indoor smoking is also prohibited by Metropolitan rule. In Tokyo, smoking is banned both indoors and outdoors. So, it is important for co-existance of smokers and non-smokers to make comfortable smoking space.

In this study, the location and purpose of outdoor smoking cessation in Tokyo's 23 cities were organized. We also organized the types of outdoor smoking areas. The method of smoking cessation outdoors differs in each of the 23 cities of Tokyo.

The area of non-smoking outside varies depending on the act and the range. The range was divided into 4 types: "Whole city", "Part of the city", "Station entrance / park".The purpose of the outdoor smoking cessation is "improvement of environmental beautification" is mentioned in all cities and "prevention of burn damage" in 14 cities.On the other hand, "Prevention of passive smoking" is mentioned in Tokyo. Organized focusing on outdoor smoking areas in non-smoking area. There were three trends: smoking area completely surrounded by glass, smoking area surrounded by trees, and smoking area not enclosed.

Environmental beautification can be achieved with an ashtray holder. However, achieving other objectives in the current smoking area is considered difficult. In the future, indoor smoking cessation will progress and smoking outside will increase. I think the design of the smoking space will be important to achieve the goal.

Sooner Spout: Propose a Design Solution to Integrate Stormwater Percolation with Community Engagement on University Campus

Matthew Kesseler (University of Oklahoma); Anu Konda (University of Oklahoma); Kviza Samani (University of Oklahoma); Yashaswini Dasarapalli (University of Oklahoma); Yung-Chieh Liu (University of Oklahoma); Alice Calmon (University of Oklahoma); Mingzheng Chen (University of Oklahoma); Pai Liu (Clemson University)

The University of Oklahoma, Norman Campus, has many flooding issues during rainfall and poses a problem to the users. Along Asp Avenue, the 6"high flood water and run off decrease the performance capacity of many mechanical appliances which has its machinery exposed to the runoff and disturbs plant life due to the increased amount of water (EPA, 2016). In addition, the large number of impermeable surfaces on the campus obstructs water from being absorbed into the ground. In addition, there are very limited social infrastructures along the Asp Avenue, which limits people's use of this area as a social space. This lack of green and social infrastructures led us to think of a new solution (Walk, 2003).

To address these issues, this study proposes a design solution along the Asp Avenue in Norman, OK to combat stormwater run-off issues by proposing sustainable design solutions that benefit the environment, along with promoting educational and social values among the user group. To identify the low points on campus for green infrastructures, the team integrated the GIS high flood risk map with observation of low points to locate the rain gardens, bioswales, green and blue roof, bio retention cells, and green parking (Culligan, 2019). In addition, mixed methods research was conducted to analyze the environmental, social, and economic needs on OU campus through interviews with stakeholders such as students, faculty and facility management specialists, online survey, observation, and archival. The study proposes a series of design solutions to mange stormwater percolation while developing the green infrastructures as social spaces to improve education, recreation, and social integration on University Campus. The results indicate a high preference of combining urban agriculture with green infrastructure design. It also provides an alternative of creating a green social campus to promote stormwater management and community engagement simultaneously.

Study on Meeting Facilities and Social Relationships of the Elderly in Collective Relocation Sites for Disaster Prevention of Ofunato City, Japan

Mitoko Nakashima (Kagawa University)

Most of the stricken areas of the 2011 Tōhoku Earthquake entered the post- temporary housing stage. However, actual conditions of the post- temporary housing stage are regionally diverse in relocation processes and community conditions. Author elucidated how continuity of original community influenced on the elderly's social relationships in collective relocation sites (Nakashima 2018). With focusing on the relocation processes from temporary housings to collective relocation sites of Ofunato City, this paper attempts to clarify social relationships of the elderly and roles of meeting facilities in the community of relocation sites.

On the basis of data obtained from questionnaire survey and in-depth interviews conducted at two collective relocation sites (S site with a meeting facility and R site without it), this paper examines: 1) How the relocation from temporary housings to the collective relocation sites influenced their local community and social relationships of the elderly. 2) Significance of existence/nonexistence of the meeting facility in collective relocation sites. And 3) Residents' needs for gathering place and expectations for the community.

As a result, we found that 1) Different relocation processes of two relocation sites influences on contrasting conditions of local community of the relocation sites although few differences in social relationships of the elderly. 2) A meeting facility of S site doubles as a community center and played important roles in activating the community while R site has no meeting facility and inactive community conditions. And 3) most of the residents of both S and R sites require gathering place while more expectations of community support for the elderly can be seen in R site.

Examining the Role of Non-Traditional Learning Environments: Focusing on Three Lobby Design on a University Campus

Qingrou Lin (Iowa State University); Jae Hwa Lee (Iowa State University)

Now, designing places outside the classroom is more important than ever in higher education. A key role of educational space design has been suggested to provide a community center that tightens the relationship among separated facilities across campus and supports students' daily events (Uduku, 2015). Moreover, good quality social spaces with flexible and interesting furniture stimulate students' daily activities (Edgerton, McKechnie, & McEwen, 2011). As casual learning spaces, this research sought to explore student reactions to newly transformed public learning environments intended to better support behaviors of the current generation.

The main purpose of this study is, therefore, to investigate students' use of the lobby spaces in three different buildings on one university campus in the Midwest area of the US. Through an intensive behavior mapping technique, this research conducted three times of ten-minute observations in each lobby and recorded the frequencies of activities and gender of those associated users at multiple different time frames.

The most frequent student activities were using electronic devices, talking to others, and having food & drinks in all three buildings. Through the observations, the study also found that all activities were highly associated with the furniture setting in the environments. The most frequent furniture that was in use during the observations included: benches for three, wide armless chair with a small table, and a round table for six. Based on the frequencies of the furniture uses, the authors concluded that the considerable factors that affect students' use in these spaces are the functionality and comfort of provided furniture, the distance between seats and outlets, and the amount of natural light. This poster presentation would present these observational data with visual images and charts as well as discuss design implications to contribute to a transformed campus for the current students' needs and beyond.

From Functionalist to Experientialist: A Comparative Image of Seattle's Chinatown-International District through Social Media

Bo Peng (University of Washington)

The research uses the Chinatown-International District in Seattle as an example to demonstrate the shifting image of the urban environment in the era of social media. The main methodology is inspired by Jack Nassar's research on the evaluative image of the city, which establishes evaluative maps of people's preference of space according to Kevin Lynch's theory on "the Image of the City". The methodology of this research is a comparative case study of city images using Lynch's theory and a more contemporary evaluative approach through social media. The purpose of this presentation is to discuss the rationality of Lynch's theory on "the Image of the City" through the platform of social media, which offers a more contemporary perception of urban dynamics and people's behaviors. It argues that, concurrent with the shifting focus to a consumer economy and society, people's perception of a place also transfers from physical attributes to a more experiential aspect; therefore, the fundamental and traditional approach to recognize city images needs to be updated by adding social and cultural aspects of the city including city culture, civic life, and natural landscape.

Planning and Promoting Opportunities for Lifelong Communities

Marilyn Bruin (University of Minnesota)

We had one hundred and forty-five baby boomers complete an online housing and lifestyle survey out of which 94% were homeowners, 65% were married, 18% were living with a child in their home, and 22% lived alone. We found that 72% of the survey takers worked with 54% having a household income of over $90,000 while 32% earned over $120,000 a year. Among the working population, 68% planned to retire by the age of 68; 63% percent of these individuals have a pension/ retirement account, 62% of them have IRAs, 47% had real-estate developments, and 54% had other retirement savings.

A large majority (92%) were satisfied with their housing with 88% wanting to remain in their neighborhood.13% of the responders reported chronic conditions that might affect housing and 55% said their home needed modification to facilitate aging. Suggested modifications included ramps, grab bars, and main floor bedroom, bath, and laundry. Sixty-five percent 65% said less home maintenance was motivation to move. Future senior housing options varied between condominiums (25%), single-family houses (24%), townhouses (20%), independent senior community (16%), affordable apartments (4%), duplex/four-plex (4%), luxury rental (3%), and assisted living (1%), and nobody chose nursing homes as an option.

While this group of urban baby boomers realized that home modifications were necessary to accommodate aging they had some common ideas about what these were and this could be attributed to their unfamiliarity with technological advances that promote safety and independence. We found that there was diversity in their expectations for future housing options, and concluded that there is a need for educational outreach related to home modifications as well as to develop more condominiums, townhouses, apartments including independent and assisted living communities.

Common Accessibility Issues in Design and Construction

Karen Kim (United Spinal Association)

Accessibility laws and regulation are transforming the U.S. design and construction industry. In particular, housing and public facilities are increasingly being demanded to show evidence of accessibility compliance for civil rights and funding purposes. The potential for lawsuits alleging noncompliance is therefore great. This makes architects, developers and building owners nervous about a lawsuit being placed against them for inaccessible conditions or, worse, injuries resulting from the use of noncompliant spaces. Such concerns bring accessibility services at the forefront of design and construction developments to create more inclusive environments for underserved (disability) communities. Organizations specializing in accessibility compliance serve building professionals by helping them understand their legal obligations for accessibility. It is important to be aware of accessibility programs and resources that are available to facilitate compliance. While accessibility laws and regulations provide technical understanding of how to make building facilities accessible, the interpretation and enforcement of the standards is challenging. In this presentation of "Common Accessibility Issues in Design and Construction," I discuss the top mistakes found in design and construction through case studies and the corrective measures that were applied to achieve compliance. This visual poster presentation provides photo examples and short story discussions on the issues in relationship to current industry standards.

E - Transform Processes and Outcomes for Equity

Traffic Space of Bicycle Made from Leftovers

Kai Ito (Tokyo Denki University)

Currently, bicycle road planning is being promoted in Tokyo, as bicycles raise awareness of the environment and health and are a convenient means of transportation in tourist areas. In Japan, rapid road development was once promoted, and urban infrastructure was greatly improved. Currently, bicycles generally run on the left side of the road due to the road structure centered on the road. However, it is hard to say that it is a bicycle-friendly road space. Therefore, the aim is to gain knowledge about the bicycle space issues that arise when developing bicycles from existing roads and how they will be developed in the future.

Bicycle road planning can be broadly divided into three categories, based on the necessity of separating road, sidewalk, bicycle path, traffic volume and regulated speed of roadway, and advance to the stage where maintenance is possible now--A Bicycle path, bicycle lane on the road, and instruction lane.

We surveyed three planned roads in Tokyo. Stopped cars hindered bicycles, and it was difficult to notice how to guide them to the bicycle path. There was a possibility that a cyclist could not cross the intersection or that a pedestrian might be blocked in the middle of a bicycle path. Both the regulation and the design of the bicycle traffic seemed to prioritize pedestrian- and car. And the complexity of the road structure, there were sections where it was difficult to grasp where to pass. In addition, when making long-term plans, there were sections where existing road structures that relied on cars could be harmful. In each case, a glimpse of the complexity of the traffic space of Japanese bicycles was seen.

Welcoming Church Spaces: Queer Inclusion in the Christian Church Building

Taylor Mortorff (University at Buffalo)

The current socio-political climate in the U.S. has caused tension between Christian and queer communities. Social politics and the fight for gay rights has exacerbated the apparent conflicts (Kreitzer, 2014). Church architecture is so far unresponsive to this socio-political shift, and queer, religious people are challenged with creating queer-friendly spiritual space in search of religious and sexual identity integration (Browne, 2010; Rodriquez, 2009).

What makes a spiritual space feel welcoming for queer people? How can church design better foster a queer inclusive, community-based worship experience? Welcoming design strategies, traditional Christian church design tenants, and queer space precedents are used to guide discussions on queer-welcoming church environments in focus groups with two Christian churches in Buffalo, New York. Using photographs, focus group participants are asked to rate church facades according to how welcoming they appear. Additionally, participants are asked to identify the factors that play a role in their decisions, and the environmental factors that create a queer-inclusive space at the churches in which they attend. Focus group responses are used to discuss the perceived effectiveness of possible queer-inclusive Christian church design strategies deduced from the literature.

Home, Community, and Well-Being: A Qualitative Study Among African American Elders

Marilyn Bruin (University of Minnesota)

Research suggests aging in place is desired by the majority of elders (AARP, 2010). Aging in place means growing old in a familiar environment (Rowles, 2008). For the majority of Americans, this means staying in the home where they lived the majority of adulthood rather than moving to a supportive institution-like environment, such as assisted living or a nursing home. Home symbolizes independence, home is unique to each individual, and the environment where residents have a sense of control, mastery, and familiarity (Lawton, 1990; Sixsmith, 1990). Recognizing the importance of home, this study describes the residential satisfaction of 17 elder African Americans in three housing situations: 202 low-income rental, market-rate rental, and homeownership.

The interview protocol was designed to encourage participants to describe their housing and its meaning. Participants were shown a ladder with four rungs and asked to place their health status on the ladder with the top rung being the best possible health for them and the bottom rung their worst possible health. Individual structured interviews were recorded and transcribed verbatim. Participants were asked to review and revise a summary. The revised transcripts were reviewed inductively. Overall themes and themes correlated to housing type emerged.

Feeling valued and being rooted in a community rich with African-American history contributed to the HUD 202 participants' well-being. Being able to shut out the rest of the world and retreat to their own property provided homeowners great satisfaction. The relationship between place and well-being is correlated with physical health, however the findings emphasized the psychosocial impacts of place on well-being. The participants' voices demonstrated how sense of place deeply impacted identifies, self of self, and residential satisfaction.

Ed Roberts Collaborative: Creating Better Independent Living Scenarios for All Levels of Ability

Jack Ross Wakeley (University of Oregon); Kyuho Ahn (University of Oregon)

This study was conducted as a research analysis of universal design principles based on a case study of the Ed Roberts Campus. Findings would be used as a catalyst for designing better independent living scenarios for a local independent living advocate agency. The case study was conducted through the lens of fifteen able-bodied undergraduate and graduate Interior Architecture students at a Pacific Northwest university toward the goal of empathetic design that involved community members with varying states of ability and service agencies. The findings would be used to create programs and building designs for the advancement of these communities in an adaptive-reuse project.

Research remained primarily qualitative with focus on experiential findings for users divided into the disability profiles cognitive, auditory, visual and mobile. Disability as defined through ADA standards is "A physical or mental impairment that substantially limits one or more of the major life activities of such individual." Independent living as defined through research is "having the right to make the wrong decision". The ability to have assurance on decisions regarding one's own well-being is what individuals within these profiles would seek. Research was collected through compiled interview dialogue, field survey, literature review, and observation as part of a post-occupancy evaluation.

The patterns recorded indicated that the varying disability profiles rely on devices that can hinder the spatial use and intuition for other user groups such as, the use of water features as wayfinding devices can aid those who are visual impaired, but disorient ones who are cognitively impaired.

The resulting program created was a cooperative living environment with the primary goal of societal integration for those pursuing higher education. Program elements created would be life-skills classrooms, drop-in clinics, makers spaces, short/long-term living quarters, and mock domestic rooms.

A Study to Evaluate the Simulation Performance of Two Lateral Boundary Configurations Using Envi-Met

Abdullah Aldakheelallah (Jones Studio Architects)

The use of microscale atmospheric models by architects, landscape architects, urban planners, and designers in general, have increased in the last two decades to simulate the performance of urban heat mitigation strategies. Crank et al., (2018) found that 30% of the literature modeling the atmospheric microclimate in the built environment were using ENVI-met since 2006. However, the performance of mitigation strategies can be inconsistent depending on the accuracy of the validation process to the simulated results of the model compared to field observations (Chow, Brennan, & Brazel, 2011; Emmanuel & Fernando, 2007; Middel, Häb, Brazel, Martin, & Guhathakurta, 2014). Additionally, different studies have indicated the shortcomings of earlier versions of ENVI-met concerning the initial atmospheric settings, specifically fixed wind speed and direction. Yet, ENVI-met version 4.4.3 accounts for sub-hourly variability of both wind speed and direction using the Forcing Manager application. The aim of this study is to validate the simulation results of an ENVI-met model under two forcing configurations of lateral boundary conditions (simple and full-forced) using field measurements from Phoenix, Arizona. The validation results of simple-forcing configurations were found adequate and comparable to other studies validating ENVI-met for Phoenix, Arizona. Although the simple forced model showed higher agreement between simulated and observed measurements, the full forced model simulated a more realistic thermal behavior across the studied domain.

edra 15

Tempe, Arizona

Reviewers

Marwa Abdelmonem
Rana Abudayyeh
Ahmed Agiel
Kyuho Ahn
Saud AlKhaled
Arash Alborzi
Andrew Alden
Fawaz Alshatti
Elke Altenburger
Cherif Amor
Frederic Bally
Jonathan Bean
Claudia Bernasconi
Sahera Bleibleh
David Boeck
Djamel Boussaa
Aaron Brakke
Richard le Brasseur
Heather Burpee
Peter Butler
Kyoungmee Byun
Chingwen Cheng
David de la Peña
Ozlem Demir
Jessica Diehl
Suining Ding
Gerardo Escalona
Alshimaa Farag
Melissa Farling
Lean Fathi
Michelle Fehler
Beth Ferguson
Polly Fong
Kristi Gaines
Amanda Gale
Morten Gjerde
Kutay Guler
Sara Hadavi
Erin Hamilton
Li Han
Davis Harte
Hui He
Lorraine Hiatt
Ana Hidalgo
Cynthia Hron
Hao Huang
Yiwei Huang
Zahid Islam
Roxana Jafarifiroozabadi
Aditya Jayadas
Rutali Joshi
Mariana Junqueira
Sweta Kandari
Sara Khorshidifard
Daejin Kim
Eun Young Kim
Joongsub Kim
Jingyi LI
Jae Hwa Lee
Sung-Jin Lee
Mark Lindquist
Silvina Lopez Barrera
Lesa Lorusso
Yanhua Lu
Qing Luo
Debarati Majumdar Narayan
Renae Mantooth
G. Mauricio Mejia
Luis Mejia Puig
Sahar Mihandoust
Neda Norouzi
Linda Nubani
Crispino Ochieng
Indrajit Pal
Giyoung Park
Jin Gyu Park
Ifan Payne
Gabriela Pereira
Lubomir Popov
Robin Puttock
Maryam Rezaei Ghaleh
Marzieh Rezaei Ghaleh
Renae Rich
Hossein Saedi
Emily Schlickman
Corey Schnobrich
Ulrich Schramm
Stacey Schulte
Kapila Silva
Dietmar Straub
Mohammad Suleiman
Chiara Tagliaro
Ashley Tannebaum
Isabella Trindade
Tzuhui Tseng
Elif Tural
Eli Typhina
Elizabeth Umbanhowar
Uchita Vaid
Olivier Vallerand
Zhe Wang
Andrea Wheeler
Tera Williams
Jane Winslow
Bo Yang
Terri Zborowsky
Bo Zhang
LEI Zhang
Parisa Ziaesaeidi
Milagros Zingoni

edraw

Tempe, Arizona

Author Index

Author Index

A

B

C

D

E

F

G

H

I

J

K

L

M

N

O

P

Q

R

S

T

U

V

W

X

Y

Z

The Environmental Research Design Association (EDRA)
2010 Pear St.
#1060 Pinole, CA
94564-9998
www.edra.org

ISBN 978-1-7138-1577-8

9 781713 815778

CURRAN ASSOCIATES INC.
proceedings.com